IFAC

International Federation of Automatic Control

# MICROCOMPUTER APPLICATION IN PROCESS CONTROL

IFAC Proceedings Series, 1987, Number 7

# IFAC PROCEEDINGS SERIES

*Editor-in-Chief*

JANOS GERTLER, Department of Computer and Electrical Engineering, George Mason University, Fairfax, Virginia, USA

GERTLER & KEVICZKY (*General Editors*): A Bridge Between Control Science and Technology (*Ninth Triennial World Congress, in 6 volumes*)

Analysis and Synthesis of Control Systems (*1985, No. 1*)
Identification, Adaptive and Stochastic Control (*1985, No. 2*)
Large-scale Systems, Decision-making, Mathematics of Control (*1985, No. 3*)
Process Industries, Power Systems (*1985, No. 4*)
Manufacturing, Man–Machine Systems, Computers, Components, Traffic Control, Space Applications (*1985, No. 5*)
Biomedical Applications, Water Resources, Environment, Energy Systems, Development, Social Effects, SWIIS, Education (*1985, No. 6*)

BARKER & YOUNG: Identification and System Parameter Estimation (1985) (*1985, No. 7*)

NORRIE & TURNER: Automation for Mineral Resource Development (*1986, No. 1*)
CHRETIEN: Automatic Control in Space (*1986, No. 2*)
DA CUNHA: Planning and Operation of Electric Energy Systems (*1986, No. 3*)
VALADARES TAVARES & EVARISTO DA SILVA: Systems Analysis Applied to Water and Related Land Resources (*1986, No. 4*)
LARSEN & HANSEN: Computer Aided Design in Control and Engineering Systems (*1986, No. 5*)
PAUL: Digital Computer Applications to Process Control (*1986, No. 6*)
YANG JIACHI: Control Science and Technology for Development (*1986, No. 7*)
MANCINI, JOHANNSEN & MARTENSSON: Analysis, Design and Evaluation of Man–Machine Systems (*1986, No. 8*)
BASANEZ, FERRATE & SARIDIS: Robot Control "Syroco '85" (*1986, No. 9*)
JOHNSON: Modelling and Control of Biotechnological Processes (*1986, No. 10*)

TAL': Information Control Problems in Manufacturing Technology (*1987, No. 1*)
SINHA & TELKSNYS: Stochastic Control (*1987, No. 2*)
RAUCH: Control of Distributed Parameter Systems (*1987, No. 3*)
FLORIAN & HAASE: Software for Computer Control (*1987, No. 4*)
MARTOS, PAU & ZIERMANN: Modelling and Control of National Economies (*1987, No. 5*)
GENSER, ETSCHMAIER, HASEGAWA & STROBEL: Control in Transportation Systems (*1987, No. 6*)
ADALI & TUNALI: Microcomputer Application in Process Control (*1987, No. 7*)
WANG PINGYANG: Power Systems and Power Plant Control (*1987, No. 8*)
BALCHEN: Automation and Data Processing in Aquaculture (*1987, No. 9*)
YOSHITANI: Automation in Mining, Mineral and Metal Processing (*1987, No. 10*)
GEERING & MANSOUR: Large Scale Systems; Theory and Applications (*1987, No. 11*)
ROOS: Economics and Artificial Intelligence (*1987, No. 12*)
TROCH, KOPACEK & BREITENECKER: Simulation of Control Systems (*1987, No. 13*)
KAYA & WILLIAMS: Instrumentation and Automation in the Paper, Rubber, Plastic and Polymerization Industries (*1987, No. 14*)





*IFAC Related Titles*

BROADBENT & MASUBUCHI: Multilingual Glossary of Automatic Control Technology
EYKHOFF: Trends and Progress in System Identification
ISERMANN: System Identification Tutorials (*Automatica Special Issue*)

# MICROCOMPUTER APPLICATION IN PROCESS CONTROL

*Selected Papers from the IFAC Symposium*
*Istanbul, Turkey, 22–25 July 1986*

Edited by

E. ADALI and F. TUNALI

*Technical University of Istanbul, Ayazaga, Istanbul, Turkey*

Published for the

INTERNATIONAL FEDERATION OF AUTOMATIC CONTROL

by

PERGAMON PRESS

OXFORD · NEW YORK · BEIJING · FRANKFURT
SÃO PAULO · SYDNEY · TOKYO · TORONTO

| | |
|---|---|
| U.K. | Pergamon Press, Headington Hill Hall, Oxford OX3 0BW, England |
| U.S.A. | Pergamon Press, Maxwell House, Fairview Park, Elmsford, New York 10523, U.S.A. |
| PEOPLE'S REPUBLIC OF CHINA | Pergamon Press, Room 4037, Qianmen Hotel, Beijing, People's Republic of China |
| FEDERAL REPUBLIC OF GERMANY | Pergamon Press, Hammerweg 6, D-6242 Kronberg, Federal Republic of Germany |
| BRAZIL | Pergamon Editora, Rua Eça de Queiros, 346, CEP 04011, Paraiso, São Paulo, Brazil |
| AUSTRALIA | Pergamon Press Australia, P.O. Box 544, Potts Point, N.S.W. 2011, Australia |
| JAPAN | Pergamon Press, 8th Floor, Matsuoka Central Building, 1-7-1 Nishishinjuku, Shinjuku-ku, Tokyo 160, Japan |
| CANADA | Pergamon Press Canada, Suite No. 271, 253 College Street, Toronto, Ontario, Canada M5T 1R5 |



First edition 1987

**British Library Cataloguing in Publication Data**
Microcomputer application in process control:
selected papers from the IFAC symposium,
Istanbul, Turkey, 22–25 July 1986.—(IFAC
proceedings series; 1987, no. 7).
1. Process control—Data processing
2. Microcomputers
I. Adali, E. II. Tunali, F.
III. International Federation of Automatic
Control IV. Series
670.42'7 TS156.8
ISBN 0–08–034340–6

*These proceedings were reproduced by means of the photo-offset process using the manuscripts supplied by the authors of the different papers. The manuscripts have been typed using different typewriters and typefaces. The lay-out, figures and tables of some papers did not agree completely with the standard requirements: consequently the reproduction does not display complete uniformity. To ensure rapid publication this discrepancy could not be changed: nor could the English be checked completely. Therefore, the readers are asked to excuse any deficiencies of this publication which may be due to the above mentioned reasons.*

*The Editors*



*Printed in Great Britain by A. Wheaton & Co. Ltd., Exeter*

# IFAC SYMPOSIUM ON MICROCOMPUTER APPLICATION IN PROCESS CONTROL

*Sponsored by*
IFAC Committee on Developing Countries

*Co-sponsored by*
IFAC Committee on Computers
IFAC Committee on Applications
IFAC Committee on Education
IFAC Committee on System

*National Sponsors*
ITU – Technical University of Istanbul
BU – Bogazici University
YU – Yildiz University

*Supported by*
OKV – Automatic Control Foundation
UNESCO
TUBITAK

# PREFACE

The object of this symposium was to bring together researchers from different disciplines working on process control to exchange ideas and discuss problems encountered in applications. Papers related to theoretical results, new design methodologies, new technologies, and new fields of application were presented.

A total of 110 abstracts were submitted and 80 were accepted for presentation. In addition, 7 survey papers were presented. A total of 25 nationalities participated in this symposium.

The symposium was organized into four sections: (i) Surveys and Tutorials, (ii) Theoretical Sessions, (iii) Application Sessions, and (iv) Case Studies. In addition, three panel discussions were arranged.

We gratefully acknowledge the financial help we received from several institutions. Specifically, we thank UNESCO, the Automatic Control Foundation, and the Turkish National Research Council. We also would like to thank the Faculty of Electrical and Electronics of Istanbul Technical University for the production of the preprints.

Finally, the help we received from the students of the Technical University of Istanbul, Bogazici University, and Yildiz University is gratefully acknowledged.

A. F. KONAR
Chairman of IPC

E. ADALI
Chairman of NOC

# CONTENTS

SURVEYS/TUTORIALS

Trends in Instrumentation and Control 1
M. THOMA

Fast Algorithms and Architectures for Signal/Image Processing and Understanding in Intelligent Automation Systems 5
O. ERSOY

Failure Detection and Isolation in Complex Process Plants 13
J. GERTLER

Education for Microprocessor Application in Control 25
W. SCHAUFELBERGER, H. GOOD, A. ITTEN

Methodology for Large-scale Process Control System Design 33
A.F. KONAR

The Three Waves of Automatic Control 47
M. CUENOD

Energy Management for Process Industry: How much control is needed? 53
A. KAYA, M.A. KEYES

ADAPTIVE SYSTEMS

A Microprocessor-based Model Reference Adaptive Control System for Hydroelectric Generating Set 59
W. BAILIN

Adaptive Robust Speed Regulator for Hydroelectric Generating Set 63
W. HONG

New Control Algorithms Based on the Sliding Mode Control Methodology 67
S.Z. SARPTURK, O. KAYNAK

IDENTIFICATION AND MODELLING

A Generative Model for Asynchronous Finite-valued Time Signals as Applied to Computer-based Process Control 73
S.Y. KWANKAM, M.E. KALISKI, A.T. MILLER

Assembler Generated Binary Test Signals for Process Identification 77
I.A. HENDERSON, A.A. IBRAHIM, J. MCGHEE, D. SANKOWSKI

HARDWARE, INTERFACE AND COMPUTER ARCHITECTURE

A Multiprocessor Architecture and its Application to Adaptive Control 83
F. CESARINI, G. SODA, G. ZAPPA

OPTIMIZATION, EXPERT SYSTEMS AND LEARNING SYSTEMS

Microprocessor Implemented Sub-optimum Smith Controllers for Temperature Control 89
C. BATUR, J. BRAUN

Using Expert Systems to Troubleshoot Microprocessor-based Control Systems: An expectation-based approach 95
P. SCHAEFER, H.A. GUVENIR

An Associative Memory Based Learning Control Scheme with PI-controller for SISO-nonlinear Processes 99
E. ERSU, S. WIENAND

CEMENT INDUSTRY

Application of Modern Numerical Control to an Industrial Rotary Cement Kiln 107
J.-D. MARCUARD, A. MEISTER

Software Development and On Line Tests of a Microprocessor Based Automatic Start Up System for a Fluidized Bed Lignite Combustor 113
H. MIDOGLU, E. TULUNAY

CHEMICAL, PETRO-CHEMICAL INDUSTRY

Costoptimized Energy Consumption in Sewage Treatment Plants On-line Controlled by Distributed Microprocessor Systems 117
G. FLOTHO

Distributed Microprocessor-based Control Systems for Flue Gas Cleaning 123
W. LATZEL, H. KAHLE

Fault-tolerant Microcomputer Control of a Heat Exchanger Network 131
R.A. MARTINI, A. CINAR

ENERGY AND POWER SYSTEMS

Microcomputer Application in Large Power Plants 137
B. BACHMANN, H. CVETKO, M. SCHURTENBERGER

Microcomputer and Computer Applications in the Hydro-electric Power Plants Karakaya and Ataturk 143
M. CUENOD, H. REGLI, M. SCHURTENBERGER

The Application of Intel 8080 Microcomputer System to Measurement and Control of Energy and Power of Steam 151
R. ZARZYCKI

Speed Control of Kaplan Turbine by Using Microprocessor 157
B.T. AKGUN, A.E. HARMANCI, M.K. SARIOGLU, M. GULEC

Real Time Control of the Power Provided by Two Aerogenerators from a Single Specialized Processor 163
C. KRAIF, S. BILLO, P. ROUAN, L. PROTIN

A Centralized Agricultural Irrigation System 167
A.F. CALIKOGLU, G. TANYERI, O. ARKUT

ROBOTICS

Multiprocessor Controlled Robots 173
D. NAUNIN, K. CEVIK, H.-C. REUSS

Multi-microcomputer Robot Control System Using a Special Robot-arithmetic-processor for Advanced Applications of Robot Manipulators 179
K. RATHGEBER, M. SCHNELL

External Control and Graphics Simulation of a Tracking Robot 185
M.B. ZAREMBA, W.J. BOCK

Optical Proximity Sensor Systems for Intelligent Robot Hands 191
E. ERSU, G. KEGEL

AUTOMATION AND MANUFACTURING

Development of a Control System for a Film Processing Device and Technology Transfer to an Industrial Company 197
M. ARDAC, B. SCHURMANN, H. RAKE

Substantial Improvements of the Performance of Axial Winders by Adaptive State Space Control 203
W. HOGER, D. SCHRODER

Fault-tolerant Microprocessor-based Control System for Multiple Lift Installations 209
J. DUATO

Distributed Control System for Machines in Manufacturing 215
G. HAIDEGGER, G. CSURGAI, L. FRITTMANN, L. REVESZ

Self-tuning Cross Profile Control for a Paper Machine 219
A. GRASER, W. NEDDERMEYER

MACHINE TOOLS

The Experimental Classifications of the Linear Models of the Lathe Cutting Mechanisms 225
N. OLGAC, G. ZHAO, J. GUTTERMUTH

Adaptive Tracking Algorithms and their Application to a Machine-tool 231
P. HULLIGER, CH. KOHLER

Modelling and Analysis of a Stepper Motor Driven Feed-Drive System of a CNC Lathe 237
B.E. PLATIN, B. CELIK

Fast Position Control with Microprocessor for Machine Tools 243
U. KUNZ, W. OBERSCHELP

Pantograph with Microcomputer 249
E. ADALI

A Language for CNC 253
E. ADALI, F. TUNALI

Author Index 261

Subject Index 263

# TRENDS IN INSTRUMENTATION AND CONTROL

**M. Thoma**

*President, International Federation of Automatic Control*

It is amazing how Automatic Control in its widest sense did develop in the last 25 years. It has an increasins impact not only on technology but also on other fields like society, economy, developing countries and others. The present development of micro electronics is causing far- reaching changes which have in the past been characterized by the expression INDUSTRIAL REVOLUTIONS. The start of industrialization saw engines replacing man's muscle power and incorporated the first revolution. Feedback control indeed had already beed used at that time - let me just mention the steam engines- but it took quite some time until Automatic Control was accepted as an independent special discipline. It was a laborious path before Automatic Control gained in the 40th year of this century recognition as a basic science and was granted appropriate consideration in the curricula of all technical universities. The 2nd industrial revolution whereby work was made easier and more efficient by use of automation was characterized particularly by developments in control technology. Nobody at that time could ever predict the collosal progress which was achieved up to now where man's intelligence is being supported by microelectronics and he is being released from tedious routine brain work. There is no need to point out that modern Automatic Control technology and theory is playing a major role in this 3rd industrial revolution, as it is sometimes called. We just need to think of modelling, of the use of modern algorithmic control methods, of output distributed computer control, and of automation of production processes with INTELLIGENT ROBOTS.

The classical control of single - input single - output servo- mechanism or fixed point control was for two reasons very effective. On one hand the classical methods like frequency response and root locus methods give for single loop control a very good insight in the real (physical) behaviour of the systems. In other words, the applied methods are very closely related to the systems dynamics. On the other hand the dynamic behaviour of single loop control systems is rather robust concerning parameter changes and/ or small changes of the systems structure.

However, if one has to deal with complex systems like multivariable control systems or one likes to implement higher control strategies like optimal or adaptive control the classical methods are no more sufficient. New methods for the analysis or synthesis of such systems were created which are usually characterized by the expression state space methods. However, they imply much more advanced mathematical methods whereby the intuition was more or less lost.

For economic reasons, however, one was forced to automize e.g. complete energy supply networks or production processes. But the use of single control strategy did for complex systems lead to very obscure and inflexible structures so that the safety was reduced to a great extent. The results were large control rooms, where all control measurement and protection equipments were centralized. This trend was supported by the development of reliable and fast process computers.

Until few years ago the process computer was a rather expensive equipment. Its use was therefore limited to processes with a very high turnover like hot strip rolling mills, large chemical plants, extensive electric energy supply systems and so on. For economic reasons one was forced to utilize fully the expensive process computer which did lead to a centralized concept.

But the explosion in micro computer technology did lead to a price decay. Nowadays the expenses for process computers are almost negligible. Sophisticated logic circuitry and minicomputers will soon be a fact in every day life. With the help of cheap computers more and more decentralized control concepts will be used in the future. Besides a better understanding of an entire plant high control strategies for a more economic and safe operation can easily be implemented.

Three examples are used to demonstrate how modern control strategiess can be implemented with the help of process computers.

The first example is an energy and time optimal on-line control of a railway system: the train No.1 is running with constant speed at the inner loop (long distance traffic). At any desired time the train No.2 can be started and it should in the optimal way complete two full rounds (local traffic) in the opposite direction of train 1. Of course, the two trains should never meet during the three conflicting sections. With the help of an infra-red transmission a realtime feed forward control with constraints is realized. By using of spline functions the optimal trajectory of the 2nd train is computed in 0.5 seconds. A feedback control algorithm is implemented in the process computer which forces the train No.2 to follow the optimal trajectory. See Figures 1 and 2.

The second example is the control of a continuous stirred tank reactor. As an example the two component chemical reaction in liquid phase of

ethyl acetate+sodium hydroxide -
sodium acetate + ethanol

is used. Thereby the flow rate of ethyl acetate is disturbance, the flow rate of sodium hydroxide is the controller output, and the concentration of ethanol is the controlled variable. The necessary on-line measurement of the concentration is, however, rather difficult to obtain or at least very expensive. Instead o non-linear observer is used to observe the concentration by measurements of the pH value and the conductivity of the liquid in the reactor. See Figures 3 and 4.

The third example is the observer based control of a double pendulum, hinged to a cart on the rail. The aim is to stabilize the system in the neighbourhood of an equilibrium point. Only one angle of the double pendulum and the position of the cart are measured. The crucial second angle is not measured, it is reconstructed with the help of an observer. Besides by means of a model reference control the position dependent friction on the rail can be neglected; the cart behaviour is forced to match the output. See Figure 5.

These examples demonstrate very clearly the power of modern control theory. However, without the use of computers these problems could not be solved economically.

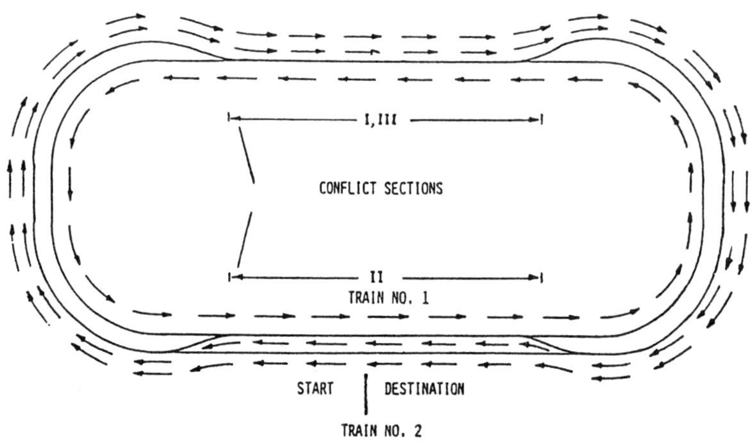


Fig. 1 Considered Railway System

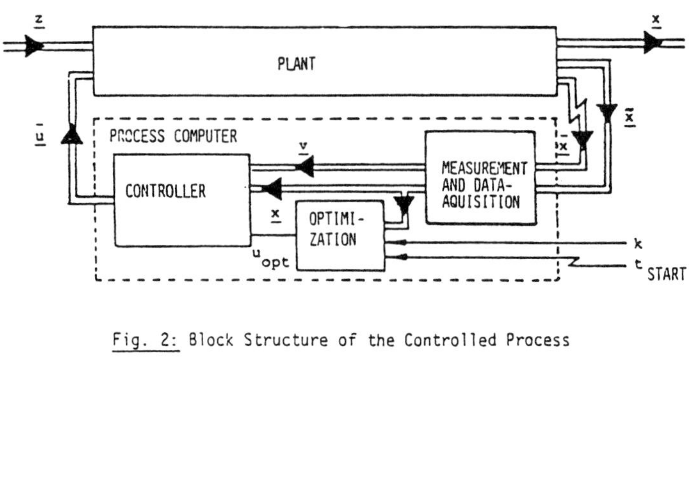


Fig. 2: Block Structure of the Controlled Process

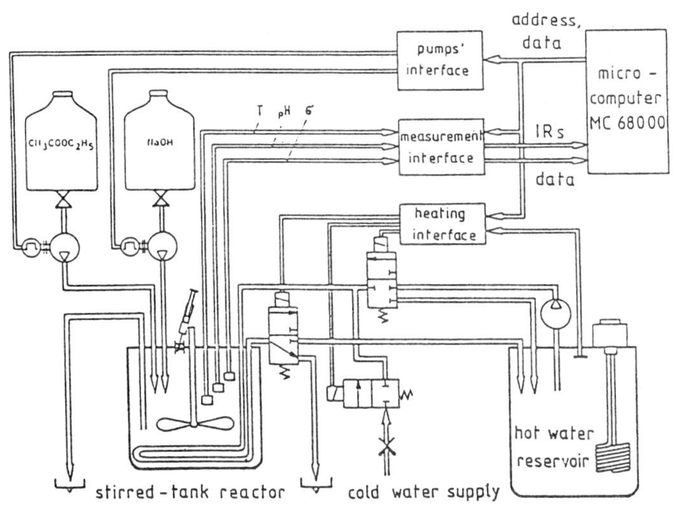


Fig. 3.: Schematic diagram of the plant and control equipment

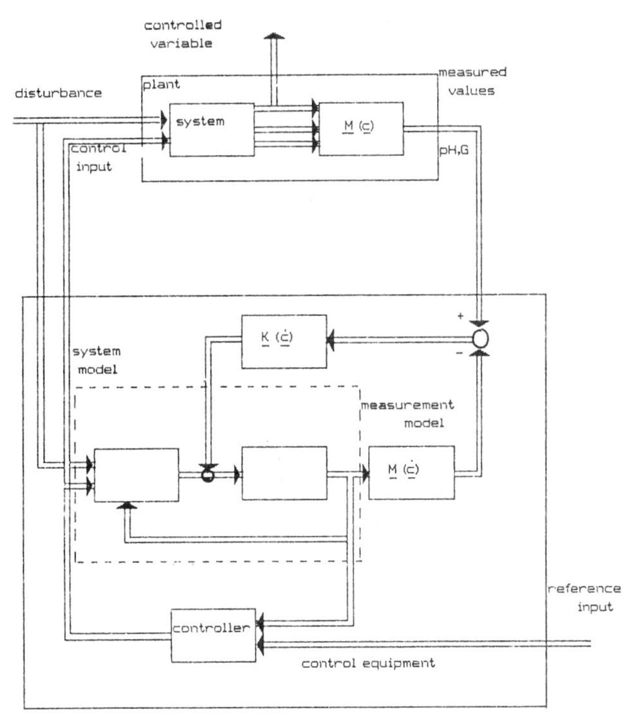


Fig. 4. Block diagram of the controlled process

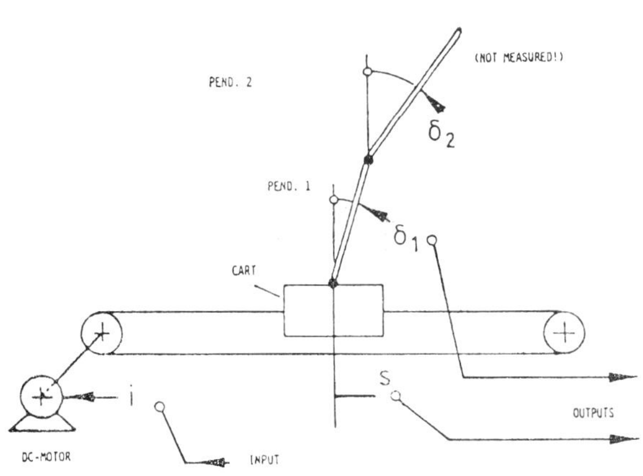


Fig. 5: Double Pendulum

# FAST ALGORITHMS AND ARCHITECTURES FOR SIGNAL/IMAGE PROCESSING AND UNDERSTANDING IN INTELLIGENT AUTOMATION SYSTEMS

**O. Ersoy**

*Purdue University, School of Electrical Engineering, West Lafayette, IN 47907, USA*

*Abstract.* Fast algorithms and architectures covering new and old classes of Fourier-like transforms, convolutions, Toeplitz-like matrices, their eigenvalues, eigenvectors and their inversions as well as their applications in real-time signal/image processing and understanding are discussed. The decomposition of transforms into preprocessing and postprocessing leads to new concepts of signal/image processing and understanding. This approach also suits well with coarse and fine sensing of signals and their processing with microcomputers and VLSI chips built into the sensor itself.

*Keywords.* Fast Fourier transforms; Fourier analysis; discrete time systems; artificial intelligence; signal processing.

## 1. INTRODUCTION

The purpose of this article is to come up with some answers to the need of fast processing of signals and images obtained from all sorts of sensors as well as fast processing of control functions so that the real time operation of the various components of an automated system is achieved. Fast processing is possible in a number of ways. In a sequential machine, it is important to come up with algorithms which have the least number of operations. Parallel processing promises highest speed increases, but algorithms have to be considered which best satisfy the constraints of parallel processing. Cost should also be an important criterion since one can always have a very large number of processors to do a job within the required time, but the cost may be too high for a particular system.

Which technology to use is another important consideration. VLSI and multiprocessor systems are definitely the most promising. However, the optical technology also has the promise of satisfying a number of requirements such as very fast operation, low cost and flexible arrangements in a number of applications.

One of the greatest needs in robots and automation is improved and more versatile sensors of every kind, and fast processing of signals obtained from the sensors for real-time operation. In general, sensing may be divided into coarse and fine sensing. The required algorithms for processing should also be divided into two such classes. There is a great need for integration of sensors and controls into much more complex systems. This process will require microcomputers and VLSI chips to be built into the sensor itself and it will be common to do preprocessing of information at the sensor level before transmitting information to the next higher level.

Present computational methods are not capable of providing the necessary real-time rate for many advanced operations. For example, the real-time solution of the dynamic equations which control the motion of a manipulator, adaptive operation of robots where the robot modifies its operation in response to information obtained from its environment, expert system modules to identify an initial state, set up a goal and proceed to the goal by appropriate means, self-programming and learning abilities for such systems, image processing, and the automation of robot vision are among the problems which require extremely high processing rates.

In our work for fast algorithms and architectures, we have so far concentrated upon those methods and algorithms which occur most often in different sections of an automated system. These can be considered to be linear transforms such as Fourier, Laplace and Karhunen-Loeve transforms, convolutions, correlations, special types of matrices such as Toeplitz matrices, symmetric banded or sparse matrices, covariance matrices, their eigenvalues and eigenvectors, and their inversions. Some of our work to achieve very fast processing leads to simple algorithms and new transforms which perform amazingly well in terms of accuracy as well. Experiments explaining some of these results are discussed in section 9.

## 2. REAL DISCRETE FOURIER TRANSFORM

Almost all methods for the fast computation of the Fourier transform are based upon the discrete Fourier transform (DFT). We have instead developed methods starting with the real discrete Fourier transform (RDFT) (Ersoy, 1985c). RDFT corresponds to Fourier series for sampled periodic frequency responses just as DFT corresponds to complex Fourier series for the same type of signals. It can be written as

$$y(n) = \sum_{k=1}^{N-1} x(k)\cos(2\pi nk/N + \theta(n)) \tag{1}$$

where

$$\theta(n) = 0 \quad 0 \le n \le N_1 \tag{2}$$
$$= \pi/2 \quad N_1 < n < N$$

$N_1$ being N/2 when N is even and (N-1)/2 when N is odd. The inverse transform is

$$x(n) = (2/N)\sum_{k=0}^{N-1} y(k)v(k)\cos(2\pi nk/N + \theta(k)) \tag{3}$$

where

$$v(k) = 1/2 \quad k = 0, N/2$$
$$= 1 \quad \text{otherwise} \tag{4}$$

RDFT has better performance than DFT in data compression and filtering for all signals because it achieves a higher degree of data compression of signals than the DFT. It has also better performance than DFT in the computation of real convolutions because of its use of real arithmetic instead of complex arithmetic, the reduced number of operations, and the fact that forward and inverse transforms can be implemented with the same signal flowgraph, thereby facilitating hardware and software design.

Some of the properties of RDFT listed above have increasing importance as the dimension of the signals increases. For example, in transforming a multi-dimensional signal, there is no need to worry about the development of complex signals when the input signals are real. This also means that the number of operations are minimized from the start, and there is no need to incorporate special algorithms for real signals.

As an example of the efficiency of RDFT, the table below shows the comparison of the number of operations required by RDFT and by DFT methods to compute real circular convolutions, assuming the most efficient algorithms are used for the transforms (Sorensen and others), (Ersoy, Hu):

| | Real Additions Per Point | | Real Multiplications Per Point | |
|---|---|---|---|---|
| N | by DFT | by RDFT | by DFT | by RDFT |
| 4 | 30 | 15 | 3 | 3 |
| 8 | 95 | 49 | 15 | 13 |
| 16 | 261 | 141 | 53 | 41 |
| 32 | 665 | 373 | 153 | 113 |
| 64 | 1,617 | 933 | 401 | 289 |
| 128 | 3,809 | 2,245 | 993 | 705 |
| 256 | 8,769 | 5,253 | 2,369 | 1,665 |
| 512 | 19,841 | 12,037 | 5,505 | 3,841 |
| 1024 | 44,209 | 27,141 | 12,545 | 8,705 |
| 2048 | 97,793 | 60,421 | 28,161 | 19,457 |
| 4096 | 214,017 | 133,125 | 62,465 | 43,009 |

We have recently discovered that other discrete trigonometric transforms, such as discrete cosine (DCT) and sine (DST) transforms, can best be computed using RDFT. The operation counts using RDFT and Chen's method (Chen, 1977) to compute DCT and DST are compared in the table below:

| | Number of Adds in DCT | | Number of Multiplies in DCT | |
|---|---|---|---|---|
| N | Chen's Method | By RDFT | Chen's Method | By RDFT |
| 64 | 482 | 513 | 292 | 192 |
| 128 | 1154 | 1217 | 708 | 448 |
| 256 | 2690 | 2817 | 1668 | 1024 |
| 512 | 6146 | 6401 | 3844 | 2304 |
| 1024 | 13826 | 14337 | 8708 | 5120 |
| 2048 | 30722 | 31745 | 19460 | 11264 |
| 4096 | 67586 | 69633 | 43012 | 24576 |

For real signals, RDFT is better than DFT in terms of number of operations. For complex signals, DFT itself can best be computed using 2 RDFT's (Ersoy, Hu, 1986).

These conclusions really mean that the philosophy behind hardware and software development in technology for applications related to fast transforms may be erroneous. This is because all work has evolved around the fast computation of DFT. Perhaps a major reason for this is that all literature on Fourier techniques stress the complex formulation because of ease of notation, ease of operation with the complex exponential and the power of complex variables. For reasons discussed above, we think that it should have been RDFT rather than DFT chosen as the basic building block of computation.

Our further investigations of RDFT have led us to new families of fast transforms. One family can be considered to be the transforms which arise from the factorization of the RDFT matrix into two other matrices. The first matrix is a simple transform, and thereby very fast. The second transform is circular and connects the Fourier transform coefficients to simple transform coefficients. The second family of transforms, which we call the generalized real discrete Fourier transform (GRDFT) (Ersoy, 1986), is obtained by developing the most general formulation of the argument of the trigonometric functions.

## 3. REPRESENTATION OF RDFT IN TERMS OF A NEW SET OF BASIS FUNCTIONS

The development of the first family of transforms starts with the representation of RDFT in terms of a new set of basis functions in place of the usual sines and cosines (Ersoy, 1985d). This was achieved by using Möbius inversion of certain types of series in number theory. Further generalization of this approach resulted in the new family of fast transforms which are actually components of the discrete trigonometric transforms (Ersoy, 1985b).

We will first summarize the highlights of the method using Möbius inversion. The end result is the factorization of the RDFT matrix into two other matrices. The first matrix consists of the samples of the bipolar rectangular wave function r(x) with values of $\pm 1$, and 0, defined by

$$\begin{aligned} r(x) &= 1 \quad 0 < x \bmod 1 < .5 \\ &= -1 \quad .5 < x \bmod 1 < 1. \\ &= 0 \quad x = 0 \text{ or } .5 \end{aligned} \tag{5}$$

The first matrix is obtained by substituting $r(n/N+1/4)$ in place of $\cos 2\pi n/N$ and $r(n/N)$ in place of $\sin 2\pi n/N$ in the RDFT equations. Operation with this matrix can be considered to be preprocessing of the input data vector $x(\cdot)$. This preprocessing can be done quickly since there are only additions -- no multiplications -- in this operation and there exist fast structures for its implementation (Ersoy, 1983).

The preprocessing of the input is followed by operations with the new basis functions which are linear combinations of sines and cosines. The new basis functions $b(m_1,N)$ are obtained by solving the following simultaneous system of equations:

$$\sin\frac{2\pi n}{N} = \sum_{m_1=1}^{M} b(m_1,N) r(m_1 n/N) \tag{6}$$

$$\cos\frac{2\pi n}{N} = \sum_{m_1=1}^{M} b(m_1,N) r(m_1 n/N + 1/4) \tag{7}$$

where $m_1=1,5,9,\ldots$ and M is a whole number depending on N, such as N-3 when N is a power of 2.

The operations with the new basis functions are in the form of direct sum of blocks of circular correlations, the largest circular correlation block being of size N/4. Since many fast and parallel algorithms such as systolic arrays have been developed in recent years for the fast computation of linear and circular convolutions (Ersoy, 1985a), these circular correlation computations can be performed very quickly.

Both preprocessing and circular convolutions are totally different data structures compared to the butterfly structure of the Cooley-Tukey type FFT algorithms.

## 4. PROPERTIES OF THE NEW BASIS FUNCTIONS

A very interesting property of the new basis functions which are used in the computation of circular convolutions in our fast algorithms is the following: they are number theoretic series; in addition, most of them approach zero as N grows. The reason for this property is that $b(m_1,N)$ asymptotically approach the following value for large N:

$$b(m_1,N) = \frac{\pi}{4}\left(\frac{\mu(m_1)}{m_1} - \frac{\mu(N-m_1)}{N-m_1}\right) \tag{8}$$

where $\mu(n)$ is the Möbius number defined by

$$\mu(n) = 0 \qquad n = mp^k, \quad p \equiv \text{prime} \quad k > 1 \tag{9}$$

and, if $n = p_1 p_2 \cdots p_k$, $p_i$'s being prime,

$$\begin{aligned} \mu(n) &= 1 \qquad k \text{ even} \\ \mu(n) &= -1 \qquad k \text{ odd} \end{aligned} \tag{10}$$

One of the terms in (8) will be negligible if $m_1$ or $N-m_1$ is large. If both $m_1$ and $N-m_1$ are large, $b(m_1,N)$ will be close to zero. In other words, as N grows, there are only a few $b(m_1,N)$ which are significantly different from 0. When N is a power of 2, the ones that are significantly different from 0 can be considered to be b(1,N), b(N-3),N), b(5,N) and b(N-7,N), approaching the values $\pi/4$ times 1, 1/3, -1/5 and 1/7 as N grows. Factoring out $\pi/4$, each number requires only 3 bits to be represented. In many applications, speed is of prime importance, and approximate computations are permissible. Taking most of the new basis functions as zero would increase the speed of computation considerably. Especially when only a small percentage of the Fourier coefficients are used in the end such as in data reduction, feature extraction, followed by detection and classification algorithms, which is commonly the case in problems like robot vision, this approach would allow few simple direct multiplications instead of complete computations of blocks of circular convolutions. Recall also that the preprocessing involves only additions. As a result, this approach would allow much faster transform computation. Some recent experiments supporting these arguments are discussed in section 9.

## 5. THE NEW TRANSFORMS

The matrices of factorization described above are not orthonormal. In order to achieve matrices of factorization which are orthonormal, the bipolar rectangular wave function r(x) had to be replaced by another function, say, z(x), which was designed from scratch. z(x) was chosen to have the same basic properties of r(x). As such, z(n/N + 1/4) is isomorphic to $\cos(2\pi n/N)$, and z(n/N) is isomorphic to $\sin(2\pi n/N)$. Thus, the preprocessing matrix is composed of samples of $z(\cdot)$ and it has the same structure as the RDFT matrix. There are 3 possible cases which determine the values of $z(\cdot)$. When N is a power of 2, cases I and II are as follows:

$N \leq 8$ (both cases)

$$z(0) = z(\frac{1}{2}) = 0$$

$$z(\frac{1}{4}) = 1$$

$$z(\frac{1}{8}) = z(\frac{3}{8}) = \frac{1}{\sqrt{2}}$$

$n \geq 16$

*Case I*

| k | 0 | N/16 | N/8 | 3N/16 | Otherwise |
|---|---|---|---|---|---|
| $z(a^k \bmod N/N)$ | $F_1$ | $-F_1$ | $-F_1$ | $F_1$ | 0 |

where a equals -3 or N-3 and $F_1$ equals $\sqrt{N/32}$.

*Case II*

| k | 0 | N/8 | Otherwise |
|---|---|---|---|
| $z(a^k \bmod N/N)$ | $F_2$ | $-F_2$ | 0 |

where $F_2$ equals $\sqrt{N}/4$.

Case III is the combination of the cases I and II such that only powers of 2 are allowed as the nonzero values of $z(\cdot)$. Which case to choose is a matter of design considerations. It is seen that many of the values of $z(\cdot)$ are zero. Those which are not zero are $\pm 1$ and powers of $\sqrt{2}$ or powers of 2. RDFT can be computed by skew-circular correlations (Ersoy, 1983). Preprocessing can also be computed fast by the same technique. When $z(\cdot)$ replaces $\mu(\cdot)$, the following happens: the skew-circular correlations further reduce to the direct sum of single multiplications if powers of $\sqrt{2}$ are used, or the direct sum of skew-circular correlations of size 2 when powers of 2 only are used, involving shift and add operations without multiplications. Thus, the function $z(\cdot)$ eases the computation of preprocessing considerably.

In addition, preprocessing also corresponds to a very fast simple transform since the matrix is orthonormal. If one replaces $\sin(2\pi n/N)$ by z(n/N) and $\cos(2\pi n/N)$ by z(n/N + 1/4) in any discrete trigonometric transform, one achieves a simple fast transform. For the lack of a better name so far, we refer to these transforms as T2, T3, etc. Thus, the following correspondence can be defined:

T2 $\Longleftrightarrow$ real discrete Fourier transform
T3 $\Longleftrightarrow$ discrete symmetric cosine transform
T4 $\Longleftrightarrow$ discrete sine transform
T5 $\Longleftrightarrow$ discrete Fourier transform
T6 $\Longleftrightarrow$ discrete Hartley transform

and so on.

For example, T2 and its inverse are given by (1) and (3) if $\cos 2\pi nk/N$ and $\sin 2\pi nk/N$ are replaced by z(nk/N+1/4) and z(nk/N), respectively.

The new simple transforms have the same data structures as the corresponding discrete trigonometric transforms. Therefore the fast algorithms developed previously can be directly utilized for their fast computation. The fact that they involve simple values will be utilized to achieve efficient implementations as discussed in section 8.

Our original philosophy has been that a number of coarse detection, classification and filtering algorithms can be achieved with these simple transforms. Processing of finer resolution can be done only if needed, for example, when a signal of interest is detected. If so, the Fourier spectral components can be obtained by multiplying the results of preprocessing by the second matrix of factorization consisting of blocks of circular correlations as before where the new basis functions are used instead of sines and cosines. In addition, each circular correlation matrix is orthonormal, thereby defining a transform, which is referred to as T1. Since T1 is circular, it is very suitable for fast computation and parallel processing.

Very recent experiments we carried out showed that preprocessing only actually gives better results than the total processing in some of the cases investigated. These startling results are discussed in section 9.

## 6. GENERALIZED RDFT: SECOND FAMILY OF TRANSFORMS

RDFT is first modified by including a phase angle $\Theta(\cdot)$ in the kernel functions (Ersoy, 1986). Symmetry conditions on the total phase leads to the most general definitions of discrete cosine and sine transforms and show that they can be classified into 4 classes. $GRDFT(M_1,M_2,\Theta)$ is built further from a combination of discrete cosine and sine transforms belonging to one of the four classes. Each combination of $M_1$,$M_2$ and $\Theta(\cdot)$ as well as the number of data points N being even or odd determine a particular transform. $M_1$ and $M_2$ are nonnegative integers less than N, and define the types of symmetry properties in the frequency and signal domains, respectively.

One importance of GRDFT stems from the following fact:

In the continuous domain, Fourier, cosine and sine transforms are the 3 major transforms. In the discrete domain, there has been developed many more numbers of discrete trigonometric transforms in recent years. The work on GRDFT has shown that all the different types of discrete sine, cosine and Fourier related transforms actually stem from the discretization of the three continuous transforms, but branch further into four classes as a function of the properties of whole numbers, and consequent symmetry properties. GRDFT reflects the most general way of discretizing real Fourier series, incorporates all the circular symmetry properties possible in the signal and frequency domains, and achieves a high degree of flexibility by the inclusion of a phase vector $\Theta(\cdot)$, which can be chosen at will.

As a specific application of GRDFT, we have shown that GRDFT (odd $M_1$,0,0) is the transform for the fast computation of skew-circular convolution, using real arithmetic, just as RDFT is the transform for the fast computation of circular convolution with real arithmetic.

A number of algorithms developed recently for the fast computation of one-dimensional and multidimensional convolutions reduce the problem into a number of smaller size circular and skew-circular convolutions (Blahut, 1985). As a consequence, GRDFT (odd $M_1$,0,0) is expected to be very useful in conjunction with these algorithms.

## 7. FAST TRANSFORMS FOR COMPUTING EIGENVALUES, EIGENVECTORS AND APPROXIMATING KLT

Optimal signal representation is very important in a lot of applications. The optimal transform which satisfies a number of criteria is the Karhunen-Loeve transform (KLT) (Fukunaga, 1972).

The main disadvantage of KLT is the fact that it is signal dependent. As a consequence, it is slow and difficult to use in real-time applications and applications in which the data length is long.

Lack of efficient algorithms to implement KLT has promoted the use of suboptimal transforms such as DFT and DCT (Elliott and others, 1982).

The fast representation of KLT can be considered to be equivalent to exact or approximate diagonalization of the covariance matrix C such that

$$C = U^* \Lambda U \qquad (11)$$

where U is the fast transform matrix which is exactly or approximately the same as the KLT matrix. $U^*$ is the complex conjugate matrix of U, or when U is real, it is the transpose of U. $\Lambda$ is the diagonal matrix whose diagonal elements $\lambda_0,\lambda_1 \cdots \lambda_{N-1}$ are the exact or the approximate eigenvalues of C.

It is interesting to observe that the representation of C as in (11) is important in all problems involving eigenvalue, eigenvector computations and matrix inversions. For example, in filtering, smoothing and prediction problems, a number of spectral estimation algorithms as well as problems involving partial differential equations (Priestley, 1981), (Vemuri, Karplus, 1981), it is necessary to find $C^{-1}$. $C^{-1}$ and/or C operate on a vector, say, x. $C^{-1}$ can be written as

$$C^{-1} = U^* \Lambda^{-1} U \qquad (12)$$

where $\Lambda^{-1}$ is diagonal with diagonal elements $\lambda_0^{-1},\lambda_1^{-1} \cdots \lambda_{N-1}^{-1}$.

Both (11) and (12) mean that multiplying x by C or $C^{-1}$ can be carried out fast according to the following procedure:

A. Compute $X = [X_0,X_1,...X_{N-1}]^T = Ux$

B. Do the pointwise multiplications $X_0\lambda_0,X_1\lambda_1 \cdots X_{N-1}\lambda_{N-1}$ or $X_0\lambda_0^{-1},X_1\lambda_1^{-1} \cdots X_{N-1}\lambda_{N-1}^{-1}$. Let the resulting vector be z.

C. Compute $y = U^*z$. This procedure is analogous to the computation of circular convolution by DFT or RDFT (Oppenheim, Schafer, 1975), (Ersoy, 1985c). In what follows, all problems as discussed above will be referred to as approximating KLT.

A number of fast trigonometric transforms have been used as approximation to KLT. The results are often tested with Markov-I signals whose covariance matrix has elements $c_{m,n}$ given by

$$c_{m,n} = \rho^{|n-m|} \qquad (13)$$

where $0<\rho<1$.

Markov-I model is quite often used in a number of applications such as image processing, in which $\rho$ is found to be close to .9 in most cases (Jain, 1977). For such large values of $\rho$, DCT is a good approximation to KLT.

Because of their importance in approximating KLT as well as a number of other applications, the development of new discrete sine and cosine transforms have been an active area of research. However, the relationship between these different transforms has not been previously appreciated. Instead one has investigated which one performs best for certain types of signals.

To this end we are working with approximating KLT for a particular signal or class of signals with an average covariance matrix according to a particular choice of $GRDFT(M_1,M_2,\Theta)$. In order to show why this is a promising approach, we will discuss 2 cases in which GRDFT is the exact KLT next.

A covariance matrix which is circular has elements C(i,j) given by

$$C(i,j) = c((i+j) \bmod N) \qquad (14)$$

where c(n), n = 0,1,...N-1 makes up the first row of c. In this case, it can be shown that $GRDFT(0,0,\theta)$ is the exact KLT if $\Theta(\cdot)$ is chosen according to

$$\cot 2\Theta(n) = \frac{C_1(n)}{C_0(n)} \qquad (15)$$

where $C_1(n)$ and $C_0(n)$ are the RDFT coefficients of $c(\cdot)$ (Ersoy, 1986).

A covariance matrix which is skew-circular has elements C(i,j) given by

$$C(i,j) = sgn(N-1-i-j)c((n+k) \bmod N) \qquad (16)$$

where sgn(m) is the function

$$sgn(m) = 1 \quad m \geq 0$$
$$= -1 \quad m < 0 \qquad (17)$$

In this case it can be shown that GRDFT(odd $M_1,0,\Theta$) is the exact KLT if $\Theta(\cdot)$ is chosen according to (13).

Our original work on approximating KLT by a fast transform was before the discovery of GRDFT (Ersoy, 1984). Here DCT was found to factorize into a permutation matrix, followed by RDFT and then a matrix of Jacobi rotations which correspond to choosing $\Theta(n)$ in RDFT(0,0, $\Theta$) as

$$\Theta(n) = \pi n/N \qquad (18)$$

Let the transform corresponding to the initial permutations and RDFT be denoted by $R'$. Consider the transformation of the covariance matrix C according to

$$C' = R'CR'^T \qquad (19)$$

We replaced the last step of Jacobi rotations where $\Theta(\cdot)$ is defined by (19) by another set of plane rotations in which the largest off-diagonal elements of $C'$ were successively reduced to 0. This was very efficient since DCT missed the largest off-diagonal elements altogether. One plane rotation gave better results than 8 plane rotations of DCT when N was 16.

These results are shown in figures 1 and 2. Q is the normalized correlation measure which shows the amount of energy left in the off-diagonal elements (Pearl, 1973). We did not want to be restricted to a single signal with a specific $\rho$ according to the Markov-I model. Instead a range of $\rho$'s was chosen, denoted by $(\rho_1,\rho_2)$ in figures 1 and 2, and an average covariance matrix was generated. After designing of the transform, the result was applied to a signal with $\rho_1 < \rho < \rho_2$. If the covariance matrix was generated for the particular signal in mind, the results would even be more in favor of the new technique.

These results imply that it is possible to approximate KLT by fast transforms very closely in a wide variety of cases. We believe that $GRDFT(M_1,M_2,\Theta)$ is one promising approach to the problem.

## 8. COMPUTATIONAL ALGORITHMS, VLSI, PARALLEL PROCESSING AND RELATED COMPUTER ARCHITECTURES

We have developed new algorithms for the fast computation of our transforms with due regard to the developments in technology in the areas of VLSI, parallel processing and related computer architectures.

These algorithms can best be described in conjunction with FFT algorithms grouped in two major classes. The first class consists of algorithms based on the butterfly technology, and they are variants of Cooley-Tukey algorithm. The second class of algorithms have been developed in more recent years and are based upon circular correlations (CC's). Among these, prime-factor algorithm and Winograd's FFT are the most well-known. The first class often uses powers of 2 as the number of data points and this results in regular data structures. The complexity of operations (additions and multiplica-

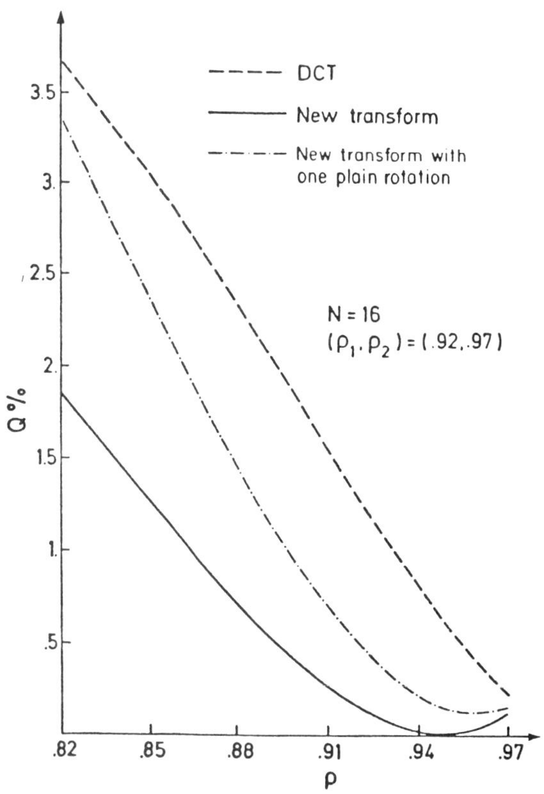


Fig. 1. Comparison of DCT and New Transforms for .92 < ρ < .97

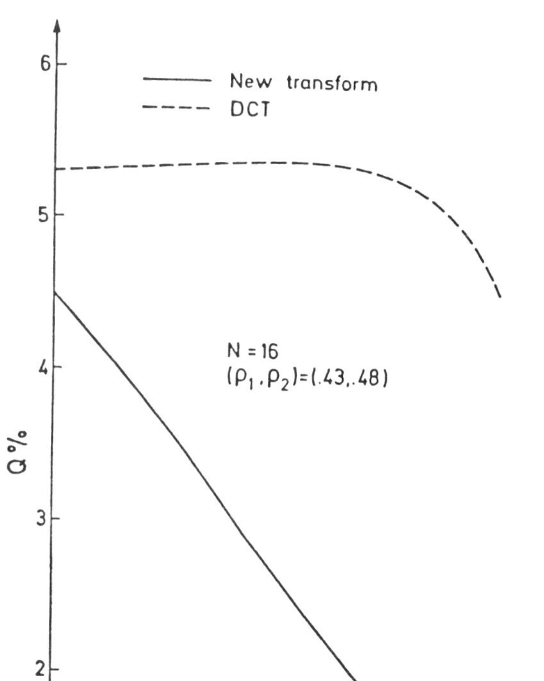


Fig. 2. Comparison of DCT and New Transforms for .43 < ρ < .48

tions) is of the order of N $\log_2 N$. We have recently shown that the minimum number of operations as well as regular data structures are obtained if RDFT rather than DFT is computed in this class of algorithms (Ersoy, Hu, 1986). In the second class, N is chosen as a whole number whose factors are integers prime to each other. This causes data structures to be complex and nonregular even though the number of multiplications can be reduced to the order of N.

In our algorithms, we were able to choose N to be powers of 2, as well as replace butterflies by circular convolutions. The circular correlations reduce to skew-circular correlations (SCC) if RDFT is considered as a whole. The largest SCC is of size N/8. SCC's are distributed throughout the ($\log_2 N$–2) innermost stages of the fast data structure, which is regular regardless of N.

When RDFT is factorized into a simple transform and diagonal blocks of circular transforms, the simple transform (T2) has the same data structure as RDFT. However, SCC's reduce to size 2 or 1 because most values of the function z(•) are zero. T2 is followed by T1, involving circular correlations. The largest CC size is N/4.

The data structures obtained are very suitable for latest technological developments. Simple transforms like T2 are very attractive for VLSI or special-purpose processor implementation since they do not involve ordinary multiplications. The circular correlations are also candidates for parallel implementation, especially with algorithms such as systolic arrays. We have developed semisystolic algorithms for the fast computation of circular and linear convolutions (correlations) (Ersoy, 1985a). The circular correlations can also be implemented using number-theoretic transforms (NTT's) (Ersoy, 1983). In this case, the total number of multiplications are N instead of $3N\log_2 N$ since NTT's can be chosen without ordinary multiplications. There is a lot of activity to integrate NTT's in VLSI. Such chips would simplify the implementation of postprocessing, and what remains would be to integrate the simple transforms like T2 in VLSI. The same concepts can be discussed in terms of parallel multiprocessor systems as well as their optical counterparts.

Perhaps the most novel concept here is the division of processing into preprocessing and postprocessing. The reason for this is twofold. The first one is that one obtains totally new data structures as outlined above; these structures make it possible to design novel architectures both for sequential and parallel processing. The second one is that the division of processing into pre and postprocessing significantly simplifies many tasks such as coarse detection, filtering, image understanding, etc. Postprocessing is applied only when it is needed. This reduces overall processing time by eliminating the unnecessary steps that would otherwise have to be executed. All these reasons are sufficient to take the concept of the division of processing seriously. However, it turned out to be even better than that as discussed in the next section.

## 9. SOME RECENT EXPERIMENTS

In order to show the feasibility of operating with the preprocessing only, we have recently carried out a number of experiments using the preprocessing matrix obtained when the Möbius inversion technique is used.

The first problem involved the Fourier descriptor method of shape recognition. In this method, one takes the DFT of the contour of a closed boundary which is modelled in terms of a chain-code (Grogan, Mitchell, 1983). The DFT equation is

$$X(n) = \sum_{k=0}^{N-1} x(k) \left[\cos 2\pi nk/N - j\sin 2\pi nk/N\right] \quad (20)$$

The preprocessing amounts to modifying (20) to

$$X(n) = \sum_{k=0}^{N-1} x(k) \left[r\left[mn/N + 1/4\right] - jr\left[mn/n\right]\right] \quad (21)$$

where r(•) is the bipolar rectangular wave function as

given in (5). Except for this modification, the analysis procedure is preserved exactly the same as before. We will refer to the resulting method as the DRWT method.

The first experiment was to classify the measured contour as one of 6 different aircraft. The total number of features available was 512. The 32 largest features were chosen in the transform space and the Euclidean distances were calculated according to both methods.

We investigated the performance of the two methods in the presence of noise. The x and y coordinates of the contour of each aircraft were corrupted independently by Gaussian noise. An example of what happens to the shape in such noise is shown in Fig. 3 with the contour of the B57 aircraft. The figure at the top left is the no noise contour. The others are contours corresponding to 30 dB, 20 dB, 10 dB and 3 dB SNR, respectively.

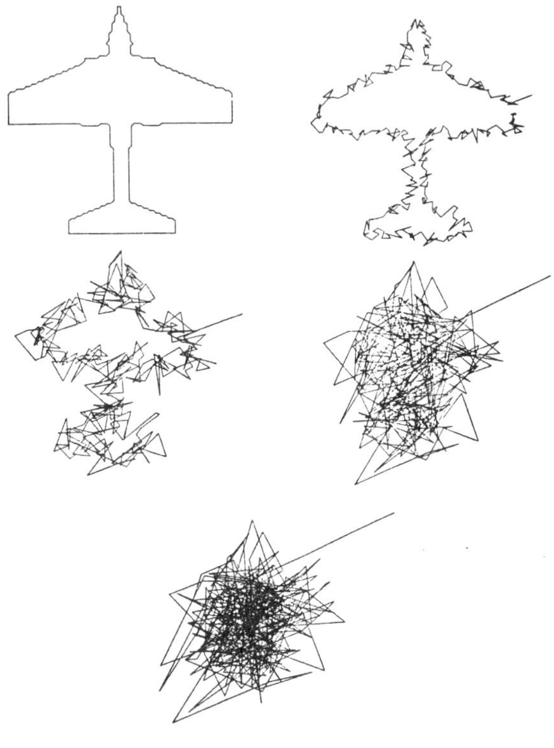

Fig. 3. Object Contour in White Gaussian Noise.

The table below shows the percentage of correct recognition with the two methods.

| SNR | Fourier Descriptors | DRWT Method |
|---|---|---|
| 3 dB | 50.0% | 83.3% |
| 10 dB | 66.7% | 100.0% |
| 20 dB | 66.7% | 100.0% |
| 30 dB | 83.37% | 100.0% |
| ∞ | 100.0% | 100.0% |

It is clear from the table above that the DRWT method achieves considerably better results than the Fourier descriptor method in the presence of noise. We will give a brief explanation of these starting results. In both methods, 32 largest coefficients are chosen as the features. In the method of the Fourier descriptors, these are the low frequency components, meaning that high frequency information is discarded. In the DRWT method, most of the largest coefficients are also the low order coefficients in the transform domain, but these are also combinations of Fourier spectral components such that they contain both low and high frequency Fourier spectral information. As such, they represent the true shape more faithfully. Especially the information related to sharp edges and corners, which are the most different from contour to contour, is more preserved in the DRWT method.

One strength of the Fourier descriptor method is the fact that it can be made insensitive to object rotation and/or the choice of the starting point on the contour by normalizing the transform coefficients. We developed a similar normalization procedure for the DRWT method. Then, the two methods were compared when the test objects were rotated by 150 degrees and the starting point was shifted by 200 points. The results are shown in the following table:

| SNR | Fourier Descriptors | DRWT Method |
|---|---|---|
| 3 dB | 50.0% | 83.3% |
| 10 dB | 66.7% | 100.0% |
| 20 dB | 66.7% | 100.0% |
| 30 dB | 83.37% | 100.0% |
| ∞ | 100.0% | 100.0% |

It is seen that the normalization procedure is very effective in both methods since the percent accuracies are basically the same as the ones in the previous table. It is interesting to observe that, at 3 dB SNR, the accuracy of the Fourier descriptors has gone down to 33.3% whereas the accuracy of the DRWT method remains the same at 83.3%. Thus, the DRWT method seems to be more robust in the presence of object rotation and different starting point.

The next experiment is related to matched filtering. In section 4, we have discussed that most basis function values are close to 0 in the Möbius scheme. In the following experiment, the small values were set to zero.

We first computed matched filtering for a square wave in Gaussian white noise with SNR of 10 dB using DFT. The computations were then repeated using the Möbius basis functions, with many of them set to zero. The results are shown in Figs. 4 and 5.

The information in these two figures is in the peak of zero lag. Thus, the two results are basically the same. Since we are simulating these results using floating-point arithmetic, the quantization errors are negligible. Matched filtering is often used in applications where speed is of crucial importance such that short wordlengths such as 8 bits are commonly used. In such a case, quantization noise in the computation of DFT is considerable since sines and cosines have irrational values. The approximate computations as outlined above are expected to yield better results because of few multiplications in such cases. This conclusion is further supported by the fact that the values of the basis functions are simple numbers once $\pi/4$ is factored out, as discussed in section 4. Keeping only b(1,N) and setting b(m,N) equal to zero for $m \neq 1$ should give sufficient accuracy in many applications. Then postprocessing corresponds to a diagonal matrix and can be combined with the values of the transfer function in the transform domain. As a consequence, processing will be done only with the preprocessing matrix, resulting in drastic simplication and speed increases.

## 10. FAST ALGORITHMS AND ARTIFICIAL INTELLIGENCE (AI)

So far algorithms which are numerical in nature have been discussed. AI techniques to be used in the automated systems of the future will often be symbolic and search oriented. However, we expect a lot of interaction between these algorithms and AI methods and tasks, as well as integration of both types of knowledge in terms of expert systems.

For example, it should be clear from the discussion in the previous sections that the search for the best fast algorithm for a particular task can be quite considerable since there are many possibilities, and rules have to be defined in order to go through the search path and find the best alternative. This problem can best be handled in terms of an expert system.

Such an expert system will involve more than just searching the right algorithm. Given a certain task with well defined goals, the expert system may be required to synthesize the best fast algorithm out of its knowledge base with existing parts and rules. The expert system

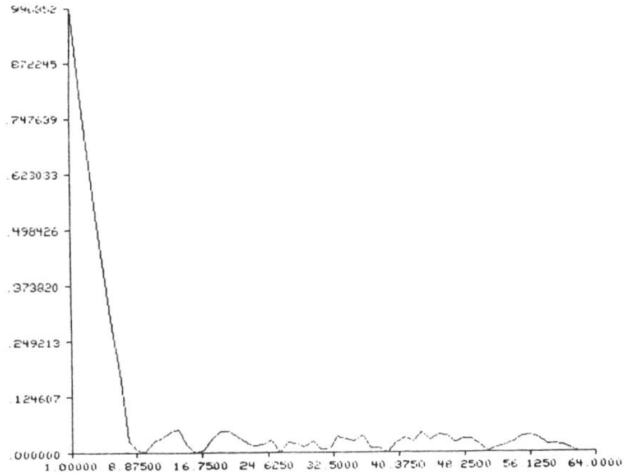

Fig. 4. DFT Matched Filtering

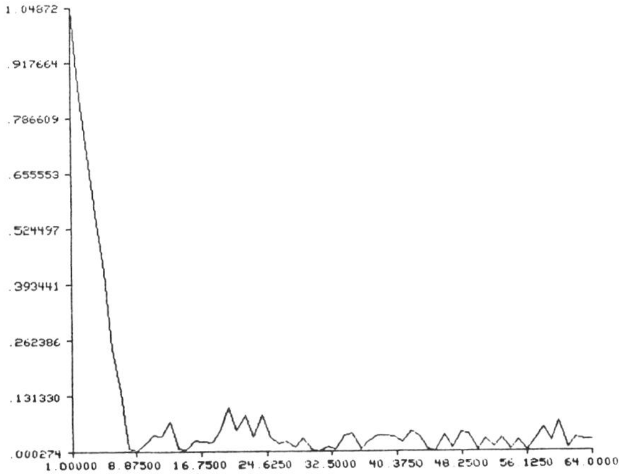

Fig. 5. Matched Filtering When Small Basis Function Values are Set to Zero in the Möbius Scheme.

can also be equipped with learning capabilities such that its knowledge base is routinely updated. For example, certain simple transforms can be parts to be used to build up a fast algorithm. It may be necessitated that such transforms have to be followed by a number of Jacobi rotations whose variables have to be chosen by the expert system, which uses the existing or measured statistical knowledge in order to determine these values. Once a fast algorithm is designed, it is tested, and if satisfactory, it can be used until the dynamic conditions require a new design.

One aim of this work has been building up an understanding of various basic parts of what makes up a fast algorithm suitable for a particular application. A challenge now is to build expert systems which can put the parts together and use the results for various applications involving all sorts of disciplines such as pattern recognition, image processing, speech understanding, robot vision, partial differential equations, and so on. The basic parts are actually quite advanced numerical algorithms. By putting such parts together, the expert system would indeed be very intelligent, ready to solve problems of very high degree of complexity.

The tools necessary to build such an expert system are becoming available nowadays. For example, there are Lisp modules with a robust foreign function interface, allowing for the dynamic use of compiled Fortran, C and Pascal modules. We think that it is best to code numerical algorithms in such high-level languages. Thus, they would be accessed using the interface to Lisp.

## 11. CONCLUSIONS

The major aim of our research is to come up with some cost effective answers to the need of fast processing of signals and images obtained from all sorts of sensors as well as fast processing of control functions so that the real time operation of the various components of an automated system is possible. Our answers are closely related to some primitives of fast algorithms and transforms which we have been carefully investigating and which can be put together according to some specific rules and measures in order to achieve a given task very fast and cost effectively. We have addressed both software and hardware issues.

## REFERENCES


Blahut, R. E. (1985). *Fast Algorithms for Digital Signal Processing,* Addison-Wesley, 1985.

Chen, W., H. Smith, S. C. Fralick (1977). A Fast Computational Algorithm for the Discrete Cosine Transform. *IEEE Trans. Communication,* COM-25, p. 1004.

Elliott, D. F., K. R. Rao (1982). *Fast Transforms.* Academic Press.

Ersoy, O. (1983). A Real Formalism of Discrete Fourier Transform in Terms of Skew-Circular Correlations and Its Computation by Fast Correlation Techniques. *SPIE Real Time Signal Processing VI Proceedings,* p. 239.

Ersoy, O. (1985a). Semi-Systolic Array Implementation of Circular, Skew-Circular and Linear Convolutions, *IEEE Trans. Computers,* p. 190.

Ersoy, O. (1985b). A New Family of Discrete Unitary Transforms and Their Potential Applications. *Tenth Symposium on Signal Processing and Applications,* Nice, France.

Ersoy, O. (1985c). Real Discrete Fourier Transform. *IEEE Trans. Acoustics, Speech, Signal Processing,* ASSP-33, No. 4, p. 880.

Ersoy, O. (1985d). Representation of Real Discrete Fourier Transform in Terms of a New Set of Functions Based upon Möbius Inversion. *Bit,* V. 25, p. 541.

Ersoy, O. (1986). Generalized Real Discrete Fourier Transform: A Family of Transforms. *The Twentieth Annual Conference on Information Sciences and Systems,* Princeton University, and submitted to *IEEE Trans. Acoustics, Speech, Signal Processing.*

Ersoy, O., N. C. Hu (1986). A Unified Approach to the Fast Computation of All Discrete Trigonometric Transforms. In preparation.

Fukunaga, K. (1972). *Introduction to Statistical Pattern Recognition,* Academic Press, 1972.

Grogan, T. A., O. R. Mitchell (1983). Shape Recognition and Description: A Comparative Study. *Tech. Report No.* TR-EE 83-22, Purdue University.

Jain, A. K. (1977). Partial Differential Equations and Finite Difference Methods in Image Processing. *J. Opt. Theory and Applications.* v. 23, No. 1, p. 65.

Oppenheim, A. V., R. W. Schafer (1979). *Digital Signal Processing,* Prentice Hall.

Pearl, J. (1973). On Coding and Filtering Stationary Signals by Discrete Fourier Transform. *IEEE Trans. Inform. Theory* IT-19, p. 298.

Priestley, M. B. (1981). *Spectral Analysis and Time Series.* Academic Press.

Rao, K. R. (1985). *Discrete Transforms and Their Applications,* Van Nostrand.

Sorensen, H. V., M. T. Heideman, C. S. Burrus (1986). On Computing the Split-Radix FFT. *IEEE Trans. Acous., Speech, Signal Processing,* ASSP-34, No. 1, 152-156.

Vemuri, V., D. J. Karplus (1981). *Digital Computer Treatment of Partial Differential Equations,* Prentice Hall.

# FAILURE DETECTION AND ISOLATION IN COMPLEX PROCESS PLANTS A SURVEY

**J. Gertler**

*School of Information Technology, George Mason University, Fairfax, VA 22030, USA*

Abstract. Techniques to detect and isolate failures in complex technological systems, such as sensor biases, leaks and equipment deterioration, are discussed. The surveyed methods are based on analytical redundancy afforded by a mathematical model of the system. The main components of such techniques are residual generation using the model, signature generation via statistical testing and signature analysis. Model-structural conditions for failure isolability are introduced together with transformation methods to attain them. Sensitivity and robustness considerations are presented and an artificial intelligence approach proposed based on model redundancy.

Keywords: failure/fault detection; failure isolation; process diagnostics; filtering; statistical testing; robustness.

## INTRODUCTION

The detection and diagnosis of faults in complex process plants and in their control instrumentation is one of the most important tasks assigned to the computers controlling/supervising such plants. By early indication of onsetting failures, major plant breakdowns and catastrophies can be avoided, ones that could result in substantial material damage and even claim human lives. By assisting the human operator in assessing the nature and extent of the fault, automatic diagnostic systems may contribute significantly to the fast and proper reaction to failure situations, such reactions ranging from immediate emergency actions to modifications of the maintenance schedule.

From the very beginning of computer process control, most practical systems have contained some form of failure detection and diagnosis. In the majority of these systems, the detection and diagnostic function is rather simple and is based on straight limit-checking. The well-known development of computational equipment and techniques has set the scene for the general application of more sophisticated and powerful methods.

The literature of process fault diagnosis is not very extensive, especially if compared to some other areas of control engineering/theory. The first major survey was written by Willsky (1976). A couple of significant books were published on the subject (Himmelblau, 1978; Pau, 1981). Recently, Isermann (1984) surveyed the field.

### The nature of the task

According to the generally accepted terminology failure detection and diagnostics consist of the following tasks:

1. Failure detection, that is, the indication that something is going wrong in the system.

2. Failure isolation, that is, the determination of where exactly the failure is.

3. Failure identification, that is, the determination of the size of the failure.

While any relative weights attached to the above three components are obviously subjective, one may venture to say that detection is an absolute must in any practical system and isolation is almost equally important. Failure identification, on the other hand, while undoubtedly helpful, may not be worth the extra effort it requires. (As we will show later, much of the detection and isolation problem can be handled in a Boolean-logical framework. Failure identification, however, seeks numerical estimates and therefore does not render itself to such treatment).

Note that failure detection and isolation are intimately related to the filtering of plant measurements: in certain schemes filtering is used to facilitate isolation. At the same time, filtering performed to reduce noise effects and isolation are, in a sense, opposite activities: while isolation aimes at pinpointing irregularities, filtering tends to hide (smear) them. In process engineering practice, it is customary to filter the measurements under constraints so that they satisfy energy and material balance equations or other plant models; this approach is often called data reconciliation (see e.g. Gertler and Almasy, 1973; Mah, 1976; Gertler, 1979; Stanley and Mah, 1981). Reconciliation and isolation may use the same or very similar techniques. However, isolation should precede reconciliation and the latter should only be applied after any isolated failure has been removed (at least computationally) from the system.

### The nature of failures

We will use the words failure and fault as synonyms (though strictly speaking, the term failure suggests complete breakdown while a fault may sound as something tolerable).

Almost any classification of failures (faults) reflects, explicitly or implicitly, the framework the detection/isolation problem is posed in. Our classification corresponds to a model-based framework and is general enough to serve as a basis for further discussion.

We will consider three classes of failures:

1. Additive measurement faults. These are discrepancies between the measured and true values of plant output or input variables. Such faults best describe sensor biases.

2. Additive process faults. These are disturbances (unmeasured inputs) acting on the plant, that are normally zero and that cause a shift in the plant outputs independent of the measured inputs. Such faults best describe plant leaks, loads, etc.

3. Multiplicative process faults. These are changes (abrupt or gradual) of the plant parameters. Such faults best describe deterioration of plant equipment, such as partial or total loss of power, surface contamination, etc.

Additive faults (whether measurement or process) are distinguished from noise. Noise is considered random with zero mean. Additive faults are considered deterministic (like a constant bias or drift) or semi-deterministic (jumps occuring at random intervals with random amplitudes).

Note that the above fault categories, though convenient from an analytical point of view, may not describe some practical failure situations in a natural way. A complete sensor failure (zero output) would have to be described either as a variable bias (equal to the true value) or as a multiplicative fault (some parameter changing to zero). Many disturbances are not completely additive: the intensity of a leak, for example, may depend on a pressure, a process variable. An accurate description of this situation would require non-linear modelling. Further, an actuator failure can only be handled as an additive or multiplicative process fault, depending on the nature of the failure. However, the actuator can be easily separated from the plant if its output is measured.

### Approaches to failure detection and isolation (FDI)

The approaches to the problem of FDI fall into two major groups:
-- methods that do not make use of a plant model;
-- methods that do make use of a plant model.

The rest of this paper will be devoted to techniques that rely on a mathematical model of the plant. The model-free methods are only briefly reviewed.

1. Limit checking. Plant measurements are compared to pre-set limits; exceeding a limit indicates a failure-situation. In many systems there are two levels of limits: the first level serves for pre-warning only while the second level triggers emergency action.

2. Installation of special sensors. These may be limit-sensors basically performing limit-checking in hardware (like limit temperature or pressure) or ones measuring some special variables (like sound, vibration, elongation).

3. Installation of multiple sensors (physical redundancy). This measure is especially aimed at detecting and isolating sensor failures. Measurements of the same variable from different sensors are compared. Any serious discrepancy is an indication of the failure of at least one sensor. The measurement that is likely to be correct may be selected in a voting system.

4. Frequency analysis of plant measurements. Some plant measurements have a typical frequency spectrum under normal operating conditions; any deviation from this is an indication of abnormalcy. Certain kinds of failure may even have a characteristic signature in the spectrum that can be used for failure isolation.

5. Expert system approach. The expert system approach is orthogonal to the above methods in that it is aimed at evaluating the symptoms obtained by the detection hardware or software. The system usually consists of a combination of logical rules of the [IF symptom AND symptom THEN conclusion] type, where each conclusion can in turn serve as a symptom in the next rule until the real conclusion (the specific failure) is reached. The expert system may work on the information presented to it by the detection hardware/software, or may interact with a human operator, inquiring from him about particular symptoms and guiding him through the entire logical process. (Hakami and Newborn, 1983; Kumamoto et al, 1984.)

## MODEL-BASED METHODS

A broad class of FDI methods make explicit use of a mathematical model of the plant. In the following, the general structure of such methods and of the models they use will be briefly introduced, and the problems of isolability, sensitivity and robustness mentioned. A more detailed treatment of these subjects will be given in the succeeding sections.

### The general structure of model-based methods.

Most model-based FDI methods rely on the idea of analytical redundancy (Chow and Willsky, 1984b). In contrast to physical redundancy, when measurements from different sensors are compared, now sensory measurements are compared to analytically obtained values of the respective variable. Such computations use present and/or previous measurements of other variables and the mathematical model describing their relationship. The idea can be extended to the comparison of analytically generated quantities only, each one obtained through a different computation. In either case, the resulting differences are called residuals.

While residuals are zero in ideal situations, in practice this is seldom so. Their deviation from zero is the combined result of noise and faults. If the noise is neglible, residuals can be analyzed directly. With any significant noise present, statistical analysis (statistical testing) is necessary. In either case, a logical pattern is generated, showing which residuals can be considered normal and which ones indicate fault. Such a pattern is called the signature of the failure. It should be noted that most FDI methods do not utilize the information embodied in the size of the residuals, beyond their relation to test thresholds.

The final step of the procedure is the analysis of the logical patterns obtained from the residuals, with the aim of isolating the failure or failures that cause them. Such analysis may be performed by comparison to a set of patterns (signatures) known to belong to simple failures or by the use of some more complex logical procedure.

The three stages of the FDI procedure, namely residual generation, statistical testing and logical analysis, are depicted in Fig. 1 (Chow and Willsky, 1984b.).

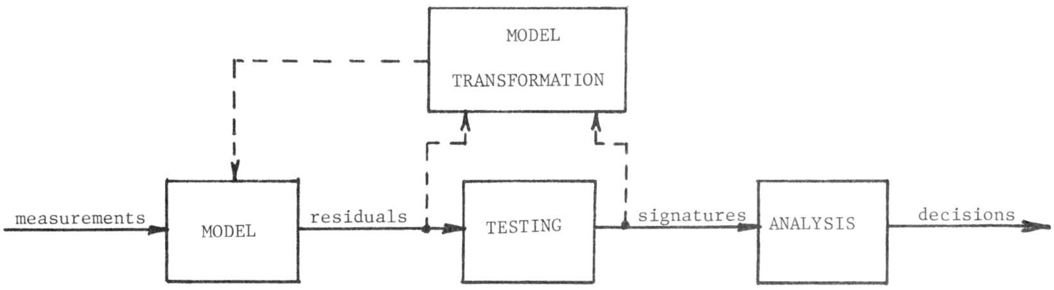


Figure 1

Note that FDI procedures relying on systems identification techniques can be considered as a special class of model-based methods. Instead of plant variables, the residuals now refer to plant parameters.

## Plant and failure models

Most model-based FDI methods rely on linear discrete-time models. This implies that any non-linearity is linearized around some operating point. Also, continuous-time plants are represented by their discretized model.

In the well-known state-space model

$$\underline{x}(t+1) = \underline{\underline{A}}\underline{x}(t) + \underline{\underline{B}}\,\underline{u}(t) \qquad (1)$$

$$\underline{y}(t) = \underline{\underline{C}}\underline{x}(t) \qquad (2)$$

$\underline{x}(t) = [x_1(t),\ldots,x_n(t)]^T$ is the state-vector, $\underline{u}(t) = [u_1(t),\ldots,u_k(t)]^T$ is the input vector, $\underline{y}(t) = [y_1(t),\ldots,y_m(t)]^T$ is the output vector, $\underline{t}$ is the (discrete) time variable, $\underline{\underline{A}}$, $\underline{\underline{B}}$ and $\underline{\underline{C}}$ are the system matrices of appropriate dimension. The system matricies are assumed to be known.

An equivalent input-output model can be presented in the form of

$$\underline{\underline{H}}(z)\underline{y}(t) = \underline{\underline{G}}(z)\,\underline{u}(t) \qquad (3)$$

where $\underline{\underline{H}}(z)$ and $\underline{\underline{G}}(z)$ are matrix-polynomials in the backward shift operator $z^{-1}$ and $\underline{\underline{H}}(z)$ is diagonal. These are related to the matrices in the state-model as

$$\begin{aligned}\underline{\underline{G}}'(z) &= \underline{\underline{C}}\,\mathrm{Adj}(\underline{\underline{I}}z - \underline{\underline{A}})\underline{\underline{B}}\\ \underline{\underline{H}}'(z) &= |\underline{\underline{I}}z - \underline{\underline{A}}|\underline{\underline{I}}\end{aligned} \qquad (4)$$

with $\underline{\underline{G}}(z)$ and $\underline{\underline{H}}(z)$ obtained from $\underline{\underline{G}}'(z)$ and $H'(z)$ by eliminating, row-by-row, any common factor (that is, by simplifying each scalar equation in (3)).

A special case of Eqn. (3) is the material balance model, widely used in chemical engineering. The input and output variables are restricted to extensive quantities (material and energy flow, etc.). The relationship is static and $\underline{\underline{H}}$ is not necessarily diagonal.

Introducing $\underline{v}(t)$ for additive process faults and $\underline{w}(t)$ for (additive) process noise, the state-equation (1) becomes

$$\underline{x}(t+1) = \underline{\underline{A}}\underline{x}(t) + \underline{\underline{B}}\underline{u}(t) + \underline{\underline{P}}\underline{v}(t) + \underline{\underline{Q}}\underline{w}(t) \qquad (5)$$

where $\underline{\underline{P}}$ and $\underline{\underline{Q}}$ are the respective system matrices. The input-output equation (3) becomes

$$\underline{\underline{H}}(z)\underline{y}(t) = \underline{\underline{G}}(z)\underline{u}(t) + \underline{\underline{L}}(z)\underline{v}(t) + \underline{\underline{M}}(z)\underline{w}(t) \qquad (6)$$

Here the matrices $\underline{\underline{L}}$ and $\underline{\underline{M}}$ are obtained in accordance with Eqn. (4), with $\underline{\underline{B}}$ replaced by $\underline{\underline{P}}$ and $\underline{\underline{Q}}$, respectively. Note that the presence of the new terms in Equation (5) may influence $\underline{\underline{H}}(z)$ and $\underline{\underline{G}}(z)$ since $\underline{\underline{L}}(z)$ and $\underline{\underline{M}}(z)$ interfere with the simplification of the equations.

Introduce now $\Delta\underline{u}(t)$ and $\Delta\underline{y}(t)$ for the additive measurement fault (bias) on the input $\underline{u}(t)$ and output $\underline{y}(t)$ and $\delta\underline{u}(t)$ and $\delta\underline{y}(t)$ for the measurement noise on the same. With these the measured input $\tilde{\underline{u}}(t)$ and output $\tilde{\underline{y}}(t)$ are

$$\begin{aligned}\tilde{\underline{u}}(t) &= \underline{u}(t) + \Delta\underline{u}(t) + \delta\underline{u}(t)\\ \tilde{\underline{y}}(t) &= \underline{y}(t) + \Delta\underline{y}(t) + \delta\underline{y}(t)\end{aligned} \qquad (7)$$

Writing the state-equations (1) and (2) for the measured input and output yields

$$\underline{x}(t+1) = \underline{\underline{A}}\,\underline{x}(t) + \underline{\underline{B}}\,\tilde{\underline{u}}(t) - \underline{\underline{B}}\,\Delta\underline{u}(t) - \underline{\underline{B}}\,\delta\underline{u}(t) \qquad (8)$$

$$\tilde{\underline{y}}(t) = \underline{\underline{C}}\,\underline{x}(t) + \Delta\underline{y}(t) + \delta\underline{y}(t) \qquad (9)$$

Similarly, the input-output equation (3) becomes

$$\underline{\underline{H}}(z)\tilde{\underline{y}}(t) = \underline{\underline{G}}(z)[\tilde{\underline{u}}(t) - \Delta\underline{u}(t) - \delta\underline{u}(t)] - \underline{\underline{H}}(z)[\Delta\underline{y}(t) + \delta\underline{y}(t)] \qquad (10)$$

Finally, introduce $\Delta\underline{\underline{A}}$, $\Delta\underline{\underline{B}}$ and $\Delta\underline{\underline{C}}$ for the discrepancies between the model matrices $\hat{\underline{\underline{A}}}$, $\hat{\underline{\underline{B}}}$ and $\hat{\underline{\underline{C}}}$ and the true system matrices $\underline{\underline{A}}$, $\underline{\underline{B}}$ and $\underline{\underline{C}}$:

$$\hat{\underline{\underline{A}}} = \underline{\underline{A}} + \Delta\underline{\underline{A}}; \qquad \hat{\underline{\underline{B}}} = \underline{\underline{B}} + \Delta\underline{\underline{B}}; \qquad \hat{\underline{\underline{C}}} = \underline{\underline{C}} + \Delta\underline{\underline{C}} \qquad (11)$$

Such discrepancies may account for multiplicative process faults (e.g., parameters in $\underline{\underline{A}}$ changing from $\underline{a}$ to $\hat{\underline{a}}$). The state-equations written with the model matrices (while, infact, the plant operates with the true matrices) are

$$\underline{x}(t+1) = (\hat{\underline{\underline{A}}} - \Delta\underline{\underline{A}})\underline{x}(t) + (\hat{\underline{\underline{B}}} - \Delta\underline{\underline{B}})\underline{u}(t) \qquad (12)$$

$$\underline{y}(t) = (\hat{\underline{\underline{C}}} - \Delta\underline{\underline{C}})\underline{x}(t) \qquad (13)$$

Similarly, the input-output equation is

$$[\hat{\underline{\underline{H}}}(z) - \Delta\underline{\underline{H}}(z)]\underline{y}(t) = [\hat{\underline{\underline{G}}}(z) - \Delta\underline{\underline{G}}(z)]\underline{u}(t) \qquad (14)$$

In general, the different failure and noise forms may act together, therefore their effect should be combined into a single (state-space or input-output) equation. This we defer to the next section of this paper.

Isolability, sensitivity, robustness

Isolability, sensitivity and robustness are quality properties of any failure isolation procedure that strongly influence the usefulness of such procedures.

Isolability is the ability of a procedure to distinguish (isolate) certain specific faults, provided their size is large enough. Faults are considered non-isolable if they cannot be distinguished however large they are. Clearly, this ability is linked to the structure of the system matrices.

Sensitivity is a more qualitative measure, characterizing the size of faults that can be isolated under certain conditions. Sensitivity is related to the size of the respective elements in the system matrices and to the properties of the noise.

Robustness is the ability of a procedure to isolate faults in the presence of modeling errors. This is a serious problem since such errors are practically inevitable and may seriously interfere with fault isolation. In particular, they appear the same as multiplicative process faults and can only be distinguished from those based on the time history of the discrepancy.

Isolability, sensitivity and robustness can, in most cases, be influenced by the careful selection and/or transformation of the plant model. Such model design would usually happen in the off-line design phase of the procedure. It may, however, be done on-line, based on the performance of the procedure under certain conditions. In this sense, an on-line model re-design level can be added to the structure of the failure isolation procedure (Fig. 1.).

## GENERATING THE RESIDUALS

Residuals may be generated in a number of different ways. We will discuss below the straightforward use of the input-output and the state-equation and the application of observer, Kalman-filtering and identification concepts. We will also indicate how faults and noise appear in the residuals obtained by some of the listed techniques.

Straight input-output residuals

After a slight re-arrangement, Eqn. (3) is suitable for generating residuals. Introduce the combined input-output vector $\underline{q}(t)$ and the combined system matrix $\underline{\underline{F}}(z)$:

$$\underline{q}(t) = \begin{bmatrix} \underline{u}(t) \\ \underline{y}(t) \end{bmatrix} \qquad \underline{\underline{F}}(z) = \left[\underline{\underline{G}}(z) \quad -\underline{\underline{H}}(z)\right] \tag{15}$$

Now the ideal input-output equation can be written as

$$\underline{\underline{F}}(z)\ \underline{q}(t) = \underline{0} \tag{16}$$

Applying the equation to the measurement $\tilde{\underline{q}}(t)$ with the model matrix $\hat{\underline{\underline{F}}}(z)$, with possible plant faults and noise, the equation becomes

$$\hat{\underline{\underline{F}}}(z)\ \tilde{\underline{q}}(t) = \underline{e}(t) \tag{17}$$

where $\underline{e}(t)$ is the vector of residuals.

A simplified version of this approach, known as static balance equations, has been popular in the chemical engineering literature. Originally, such equations contain a static system matrix $\underline{\underline{F}}$ and the variables in $\underline{q}$ are restricted to material and energy flows (Vaclavek, 1974; Almasy and Sztano, 1975; Stanley and Mah, 1977; Romagnoli and Stephanopoulos, 1981). A first-order dynamic extension is proposed by Rooney et. al., (1978). The full-blown dynamic case is discussed in (Gertler and Singer, 1985; Gertler, Singer and Sundar, 1985). A detailed analysis of the properties of residuals obtained from dynamic input-output equations is given also by Chow and Willsky (1984b).

The residuals $\underline{e}(t)$ are the result of the combined noises, additive faults and model discrepancies (multiplicative faults/modeling errors). According to Eqns. (6), (10) and (14), this relationship can be expressed as

$$\begin{aligned} \underline{e}(t) = {} & \hat{\underline{\underline{F}}}(z)\ \delta\underline{q}(t) - \underline{\underline{M}}(z)\ \underline{w}(t) \\ & + \hat{\underline{\underline{F}}}(z)\ \Delta\underline{q}(t) - \underline{\underline{L}}(z)\ \underline{v}(t) \\ & + \Delta\underline{\underline{F}}(z)\ \underline{q}(t) \end{aligned} \tag{18}$$

Here the first line contains the noise-terms, the second line the additive fault terms and the third line the model discrepancy term. This situation is depicted in Fig. 2.

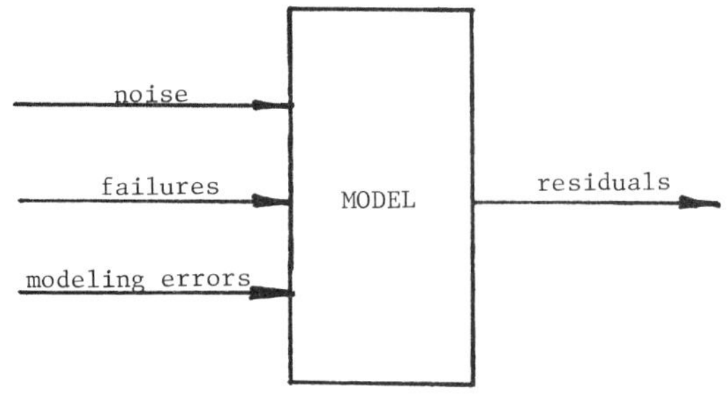


Figure 2

State-related residuals

If the state-variables are directly measurable or if they are computable from the output measurements (matrix $\underline{\underline{C}}$ is invertible) then state-variable residuals can be defined as

$$\underline{e}(t) = \tilde{\underline{x}}(t) - \bar{\underline{x}}(t) \tag{19}$$

Here $\tilde{\underline{x}}(t)$ is the measured state, obtained as

$$\tilde{\underline{x}}(t) = \underline{\underline{C}}^{-1}\tilde{\underline{y}}(t) \tag{20}$$

and $\bar{\underline{x}}(t)$ is the nominal state, generated as

$$\bar{\underline{x}}(t) = \underline{\underline{A}}\ \bar{\underline{x}}(t-1) + \underline{\underline{B}}\ \underline{u}(t-1) \tag{21}$$

or as

$$\bar{\underline{x}}(t) = \underline{\underline{A}}\ \tilde{\underline{x}}(t-1) + \underline{\underline{B}}\ \underline{u}(t-1) \tag{22}$$

(see e.g. Ben-Haim, 1983). The nominal state $\bar{\underline{x}}(t)$ can also be computed by Kalman filtering or by an observer.

Matrix $\underline{\underline{C}}$ is usually not invertible and thus direct comparison of the state is not possible. A workable approach in such cases is to define the residuals in terms of the output. If

$$\underline{e}(t) = \tilde{\underline{y}}(t) - \underline{\underline{C}}\ \hat{\underline{x}}(t|t-1) \tag{23}$$

where $\hat{\mathbf{x}}(t|t-1)$ is an estimate of the state obtained by Kalman filtering, the ensuing residual is called the innovation. This method has been very popular (Willsky, 1976). Under certain assumptions for the noise, the innovation sequence of a fault-free system is white with zero mean. More complex Kalman filtering schemes are reported e.g., in Ben-Haim (1983).

Another popular approach has been the use of parallel observers. If the state-vector (or part of it) is estimated by two observers based on different (though possibly overlaping) sets of outputs, a residual can be defined as

$$\mathbf{e}(t) = \hat{\mathbf{x}}_1(t) - \hat{\mathbf{x}}_2(t) \tag{24}$$

where $\hat{\mathbf{x}}_1$ and $\hat{\mathbf{x}}_2$ are the two estimates. Since the observer errors are $\mathbf{e}_i(t) = \hat{\mathbf{x}}_i(t) - \mathbf{x}(t)$, i=1,2, the residuals are, in fact,

$$\mathbf{e}(t) = \mathbf{e}_1(t) - \mathbf{e}_2(t) \tag{25}$$

The observers can be constructed so that their errors $\mathbf{e}_1$ and $\mathbf{e}_2$ depend on some noise and modeling errors in identical or at least similar ways, thus leading to a robust fault isolation scheme (Patton and Willcox, 1985; Wunnenberg et. al., 1985). Similar schemes can be constructed using parallel Kalman filters (Tylee, 1983). The application of parallel observers to physically redundant sensors has been reported by Stuckenberg (1985). The combination of observers with identification techniques was proposed by Watanabe and Himmelblau (1983).

Identification-based methods

In identification-based methods, a residual-like quantity is defined in relation to the plant parameters. The plant is identified in a fault-free reference situation, then repeatedly on-line. The results of the latter are compared to the reference values and a parameter-error (residual) is formed.

In the works of Isermann and coworkers (Isermann, 1984 and 1985; Geiger, 1984), the underlying physical parameters of the plant are computed from the identification results and the comparison is performed on these. Rault et. al., (1984) compute the covariance matrix of the identified parameters recursively and compare it to reference convariances. Kumamaru et. al., (1985) work with parameter estimates obtained from repeated identifiation runs and compare the consecutive results to each other, without resorting to a reference value.

Identification appears as an auxiliary method in a number of approaches. In Watanabe and Himmelblau (1983), a least-square identifier is used to estimate the multiplicative process faults, based on variable-estimates obtained by an observer. In Halme and Selkainaho (1984), plant parameters are estimated for an adaptive Kalman filter that, in turn, generates innovation-type residuals.

## STATISTICAL TESTING

Whichever of the approaches discussed in the previous section is used, the resulting residual vector is the combined effect of faults and noise (and of modeling errors but these will be neglected at this point). If there is no noise, any nonzero residual is an indication of a fault-condition so the logical analysis of the fault-situation can be performed directly on the residuals. The assumption of zero noise is, however, not a realistic one in most cases; even the round-up errors in the residual computation may cause non-zero residuals. So, in general, the effect of faults in the residuals has to be separated from that of noise. This is done by statistical testing, making use of the fact (or assumption) that the noise is random with zero mean while failures are deterministic or semi-deterministic.

Direct parallel testing

The most straightforward approach is the direct parallel testing of the residuals. Following each computation of the residuals, a separate test is applied to each element of the residual vector. Based on the outcomes of the individual tests, a Boolean signature vector $\boldsymbol{\varepsilon}(t)$ is formed so that $\varepsilon_i(t)=1$ if $e_i(t)$ fired the test and $\varepsilon_i(t) = 0$ if it did not.

Such a test is especially easy to apply if the residuals are generated from an input-output model. This is because, in this case, the residuals of the fault-free system are outputs of a discrete moving average (non-recursive) system driven by the noise. This implies that the fault-free distribution of the residuals can be derived from the noise distribution. In particular, if the noise is normally distributed, it is enough to obtain the noise-free variances of the residuals. From Eqn. (18), the fault-free residuals are

$$\mathbf{e}(t) = \hat{\mathbf{F}}(z)\ \delta\mathbf{q}(t) - \mathbf{M}(z)\ \mathbf{w}(t) \tag{26}$$

Assuming that $\delta\mathbf{q}(t)$ and $\mathbf{w}(t)$ are independent sequences, the covariance matrix $\mathbf{S}_e$ of the residuals is obtained as

$$\begin{aligned}\mathbf{S}_e &= E\{\mathbf{e}(t)\mathbf{e}^T(t)\} \\ &= \sum_{i=0}^{n}\sum_{j=0}^{n}\left[\hat{\mathbf{F}}_i\mathbf{S}_q(i-j)\hat{\mathbf{F}}_j^T + \mathbf{M}_i\mathbf{S}_w(i-j)\mathbf{M}_j^T\right]\end{aligned} \tag{27}$$

where

$$\mathbf{S}_q(i-j) = E\{\delta\mathbf{q}(t-i)\delta\mathbf{q}^T(t-j)\} \tag{28}$$

$$\mathbf{S}_w(i-j) = E\{\mathbf{w}(t-i)\mathbf{w}^T(t-j)\} \tag{29}$$

are the (i-j) shift noise covaviance matrices and $\hat{\mathbf{F}}_i$ and $\mathbf{M}_i$ (i=0,1,...,n) are coefficients of the matrix polynomials $\hat{\mathbf{F}}(z)$ and $\mathbf{M}(z)$. If the noises are white, Equation (27) simplifies to

$$\mathbf{S}_e = \sum_{i=0}^{n}\left[\hat{\mathbf{F}}_i\mathbf{S}_q\hat{\mathbf{F}}_i + \mathbf{M}_i\mathbf{S}_w\mathbf{M}_i^T\right] \tag{30}$$

where $\mathbf{S}_q = \mathbf{S}_q(0)$ and $\mathbf{S}_w = \mathbf{S}_w(0)$.

The diagonal elements of $\mathbf{S}_e$ are the residual variances. Based on these and on a selected level of confidence, the triggering limits can be obtained, using normal distribution. To avoid frequent false alarms, usually a high level of confidence (98% or 99%) is selected.

If the noise-covariances are not available, the residual-variances may be estimated from observations of fault-free periods. In this case, t-distribution is to be used (Vaclavek, 1974), though with a large sample size this approaches normal distribution.

The parallel test is relatively easy to administer and yields a distinctive binary signature that may serve as a basis for failure isolation. A disadvantage of this approach is that it does not utilize the additional information represented by the off-diagonal elements of the covariance matrix $\mathbf{S}_e$.

Multi-dimensional testing

Obviously, the elements of the fault-free residual vector are not independent of each other, even if the noise is white and uncorrelated. To utilize this additional information, Newman and Perkins (1979) suggested a multidimensional statistical test. In the ρ-dimensional space of the residuals $\underline{\underline{e}}(t) = [e_1(t),\ldots,e_\rho(t)]^T$ any constant probability density $p(e_1,\ldots,e_\rho)$ = const. describes a closed hyper-surface. Selecting a level of confidence implies choosing one such surface. If the point defined by $\underline{\underline{e}}(t)$ is outside the limit surface, the system is declared faulty. Beyond the obvious difficulty of administering this test to higher dimensional residual vectors, the main disadvantage of this approach is that it provides a single binary inference that does not facilitate the isolation of the failure.

Compound scalar testing

Another approach to testing the residuals is to introduce a single scalar statistic, e.g., as

$$\lambda(t) = \underline{\underline{e}}^T(t)\underline{\underline{S}}_e^{-1}\underline{\underline{e}}(t) \tag{31}$$

(Romagnoli and Stephanopoulos, 1981). If the noises are normally distributed and there are no faults, then the statistic λ follows the chi-square distribution with ρ degrees if freedom (ρ is the number of residuals).

This test is easy to administer but, just like the multi-dimensional test, provides a single binary inference. Romagnoli and Stephanopoulos proposed a sequential procedure, whereas equations are systematically eliminated until the test indicates a fault-free system. An important component of their method is an algorithm to obtain the new λ following each elimination step recursively, without complete re-computation.

Sequential likelihood ratio test

This test has been applied exclusively to the Kalman filter innovations (Chien and Adams, 1976; Willsky and Jones, 1976; Bonivento and Tonielli, 1984; Uosaki, 1985), though it could be used in other contexts as well. The purpose is to test the hypothesis $H_1$ of nonzero innovations mean against the null-hypothesis $H_o$ of zero mean. The decision is based on the likelihood ratio

$$\lambda_i(t) = \ln \frac{p\{e_i(0),\ldots,\ e_i(t)|H_1\}}{p\{e_i(0),\ldots,\ e_i(t)|H_o\}} \tag{32}$$

where $e_i$ is the i-th element of the innovation vector (see Eqn. (23)). If the innovations are normally distributed, the likelihood ratio is relatively easy to compute (it becomes the difference of two sums).

In most methods, the likelihood ratio is compared to two limits (normal/faulty), with a gray range between them. When it lies in the gray range, there is no decision and further innovation values are taken. In this sense, the test is sequential.

One difficulty in applying this test is the need for a numerical value of the innovation mean in $H_1$. This is either a pre-set nominal value (Chien and Adams, 1976) or is obtained by maximum likelihood estimation from the innovations (Willsky and Jones, 1976).

Another difficulty is the accumlation of normal-state information in the likelihood ratio before the actual situation changes from $H_o$ to $H_1$, resulting in a delayed detection of the change. Chien and Adams propose a scheme in which such accumulations are automatically reset to zero. Uosaki's solution is a backward evaluation of the likelihood ratio, startig with $e_i(t)$ and continuing backward until a decision is reached. Willsky and Johnes include the time of the change as an explicit but unknown variable, making the problem formulation more precise but the algorithm much more complex.

The test limits are derived either from specified probabilities for false/missed alarms (Uosaki) or from the expected time between two false alarms and the expected time of detecting a failure.

Failure isolation is possible by running a set of parallel tests, with different alternative hypotheses $H_1$, $H_2$, etc., based on the same innovations sequence (Bonivento and Tonielli).

Bayesian approach

Some failure detection and isolation schemes utilize an apriori probability distribution of the occurence of a set of failures. Such apriori distribution may be obtained from the observation of an extended history of the plant or may be assumed as design parameters.

Apriori distributions may be used to supplement likelihood data in a non-linear filtering scheme (Loparo et al.; Eckert et al., 1986). A bank of Kalman-filters, each designed for a specified failure mode, generate state estimates in a parallel fashion. Based on these and the output observations, likelihood functions $\ell_j$ are computed, as solutions of a stochastic differential equation, for the j=1,2,... possible failure modes. Then the conditional probabilities of the failures are obtained as

$$p_j = \frac{\ell_j \pi_j}{\sum_i \ell_i \pi_i} \tag{33}$$

where $\pi_j$ (j = 1,2,...) are the apriori probabilities. This algorithm is computationally rather complex but exhibits good robustness properties.

Based on apriori probabilities, a Bayesian framework for the design and evaluation of FDI algorithms may also be developed (Chow and Willsky, 1984a). An overall cost function is constructed, that takes into account the cost of false alarms, of complete misses or delayed detections and of mis-isolation of different failures. In principle, FDI algorithms could be sought by minimizing this cost-function. While the complexity of this approach is prohibitive of direct application, it may serve as a conceptual framework and approximate solutions may be possible.

## ISOLABILITY CONDITIONS

One of the main questions to be asked in connection with any FDI method is whether it is capable of isolating specific failure types from each other. The answer lies primarily in the structure of the model used in residual generation and, to a lesser degree, in the statistical test applied. In the following section, the concepts of deterministic and statistical isolability will be introduced, together with a transformation technique to bring the model to the desired structure. To simplify the discussion, only bias-faults in an input-output model will be considered. The concepts can be readily extended to process faults in the same model structure. They apply to the state-variable formulations as well but the details depend on the method of residual generation. It will be assumed that no modeling errors are present.

### Incidence matrices

The structure of a model matrix $\underline{\underline{F}}$ is characterized by its incidence matrix

$$\underline{\underline{\Phi}} = \text{Inc}\ (\underline{\underline{F}}) \tag{34}$$

The incidence matrix is a Boolean matrix with its elements defined as

$$\begin{aligned} \phi_{ij} &= 1 \text{ if } f_{ij} \neq 0 \\ \phi_{ij} &= 0 \text{ if } f_{ij} = 0 \end{aligned} \tag{35}$$

If $\underline{\underline{F}}$ is the system matrix of an input-output model as defined in Equation (3) and (15), then $\phi_{ij} = 1$ means that a bias on the j-th variable influences the i-th residual and $\phi_{ij} = 0$ means that it does not.

### Deterministic isolability

Consider Eqn. (18), describing the residuals in an input-output setting. Assume that the measurement biases $\underline{\underline{\Delta q}}(t)$ are the only possible failures and there is no noise. Thus

$$\underline{\underline{e}}(t) = \hat{\underline{\underline{F}}}(z)\ \underline{\underline{\Delta q}}(t) \tag{36}$$

In such deterministic (noise-free) situation, the Boolean signature vector $\underline{\underline{\varepsilon}}(t)$ is obtained simply as

$$\begin{aligned} \varepsilon_i(t) &= 1 \text{ if } e_i(t) \neq 0 \\ \varepsilon_i(t) &= 0 \text{ if } e_i(t) = 0 \end{aligned} \tag{37}$$

Obviously, the signature vector in response to a single fault $\Delta q_j \neq 0$ is

$$\underline{\underline{\varepsilon}}(t|\Delta q_j) = \underline{\underline{\phi}}_{.j} \tag{38}$$

where $\underline{\underline{\phi}}_{.j}$ is the j-th column of the incidence matrix $\underline{\underline{\Phi}}$. This implies that for the detection of any single bias fault $\Delta q_j$, its column $\underline{\underline{\phi}}_{.j}$ must be nonzero, and for the isolation of the same, $\underline{\underline{\phi}}_{.j}$ must be different from all other columns of $\underline{\underline{\Phi}}$.

The above condition, that we call <u>deterministic isolability</u>, is rather obvious and has been recognized by several authors (e.g., Ben-Haim, 1980; Gertler and Singer, 1985).

### Statistical isolability

Consider now Eqn. (35) with measurement noise $\underline{\underline{\delta q}}(t)$:

$$\underline{\underline{e}}(t) = \hat{\underline{\underline{F}}}(z)\ \underline{\underline{\Delta q}}(t) + \hat{\underline{\underline{F}}}(z)\ \underline{\underline{\delta q}}(t) \tag{39}$$

Assume that a direct parallel test is applied to the residuals. The signature vector $\underline{\underline{\varepsilon}}(t)$ is now obtained as

$$\begin{aligned} \varepsilon_i(t) &= 1 \text{ if } e_i(t) > e_{ith} \\ \varepsilon_i(t) &= 0 \text{ if } e_i(t) \leq e_{ith} \end{aligned} \tag{40}$$

where $e_{ith}$ is the threshold for the i-th residual. If a bias $\Delta q_j$ is large enough, it will trigger the test on all the residuals it influences, that is, Eqn. (38) holds. This situation will be referred to as <u>full firing</u>.

In general, there is more penalty on a false alarm than on missing a small fault. Therefore the thresholds will be set relatively high. On the other hand, the thresholds are determined by the fault-free distribution of the residuals, therefore the limit value of a fault $\Delta q_j$ that triggers the test will, in general, be different for each residual. Thus a fault of intermediate size may not trigger the test on all the residuals it influences. This phenomenon, called <u>partial firing</u>, results in a degraded signature $\underline{\underline{\varepsilon}}(t|\Delta q_j)$, in that some of the 1-s in $\underline{\underline{\phi}}_{.j}$ will be replaced by 0-s. Now if such a degraded signature is identical with another column $\underline{\underline{\phi}}_{.\ell}$, then the fault $\Delta q_j$ is mis-isolated as $\Delta q_\ell$.

A model structure that guarantees that partial firing does not lead to misisolation is called <u>statiscally isolable</u> (Gertler adn Singer, 1985; Gertler, Singer and Sundar, 1985). A sufficient condition for statistical isolability of single bias failures is that $\underline{\underline{\Phi}}$ has a <u>column-canonical</u> structure (that is, each column has the same number of 0-s, each in a different configuration).

Barring other disturbing effects (such as modeling errors or extreme noise-values), a fault in a statistically isolable structure

-- is either detected and properly isolated (if large enough);

-- or it leads to a signature that does not belong to any known fault (if intermediate size);

-- or it is completely missed (if too small).

### Model transformation

System models in their original (primary) form seldom satisfy the conditions of deterministic isolability, let alone statistical isolability. To attain the desired structure, a model transformation may be performed. A secondary model so obtained may have the same number of equations (residuals) as the primary model, or fewer or more. In the latter case, some of the equations are linearly dependent on the others but their incidence-structure is different. (This implies that such augmented sets can be used for fault isolation but not for fault identification.)

Model transformation may be looked upon as a re-shuffling of zeroes in the residual equations. The effect of some variables (faults) is eliminated from certain equations, at the expense of introducing other variables. In case of structure (3), this implies eliminating input variables and introducing more than one output per equation.

Technically, the transformation requires the solution of the matrix equation

$$\underline{\underline{F}}^x(z) = \underline{\underline{T}}(z)\ \underline{\underline{F}}(z) \tag{41}$$

for the transforming matrix $\underline{\underline{T}}(z)$, with the desired model $\underline{\underline{F}}^x(z)$ specified in terms of its incidence matrix $\underline{\underline{\Phi}}^x$. Each zero in $\underline{\underline{\Phi}}^x$ determines a linear algebraic equation; the solution of these yields the elements of $\underline{\underline{T}}(z)$ (the solution is separable according to the rows of $\underline{\underline{T}}(z)$).

As it is easy to see, this model transformation is governed by combinatorial rules. For anything but the most trivial systems, an extremely large number (hundreds or thousands) of secondary models may be derived, each satisfying the isolability conditions.

Zeroes and/or linear dependence (even if it concerns parts of rows/columns) in the primary model matrix reduce the number of feasible secondary models. In particular, full column dependence indicates that isolability cannot be attained by analytical means; additional sensors are necessary.

The above transformation technique was introduced in (Gertler and Singer, 1985; Gertler, Singer and Sundar, 1985). Some aspects were first developed in (Shutty, 1985).

A somewhat different approach to model generation/transformation was reported by Chow and Willsky (1984b). In their paper, complete linearly independent sets of input-output equations are derived, based on the underlying state-space model and observability considerations. Residual equations are then generated as linear combinations of elements from a basic set. Ben-Haim (1983) applied a different generalization (transformation) to the state vector, to taylor the signature structure of state-variable-based residuals to certain types of failure.

## SENSITIVITY AND ROBUSTNESS

While an isolable model structure guarantees isolation if a failure is large enough, it is also of interest what failure size is really necessary to trigger the tests. This question may be answered by sensitivity analysis. Sensitivity aspects may also play a role in the selection of the desired model and filtering may be applied to improve sensitivity within a certain model framework.

Closely related to sensitivity is the aspect of robustness of the FDI algorithm in face of modeling errors. Such errors are present almost inevitably and may seriously interfere with the isolation of failures.

Modeling errors appear as discrepancies between the true parameters of the plant and the parameters of the model (they may reflect real parameter discrepancies or an improper model structure). Such modeling errors are indistinguishable from certain multiplicative process faults, especially on the basis of momentary signatures. The temporal behaviour of the signatures may provide some clue since changes may most likely be caused by failures, though a change of the operating point may also result in a modeling error in the linear approximation of a non-linear system.

Modeling errors may also interfere with the isolation of additive failures since they contribute to the residuals and may falsify the failure signatures. With the extension of the isoability concept to certain modeling errors, the FDI algorithm can be desensitized with respect to those errors. Robustness considerations may also be taken into account in the model selection/transformation procedure.

### A measure of sensitivity

In a statistical testing framework, the marginal value of a fault that triggers the test under zero momentary noise may serve as a measure of sensitivity. (Strictly speaking, this is an inverse sensitivity measure.) If the residuals are tested in parallel, this marginal triggering value of a fault is different, in general, for each residual:

$$\tau_{ij} = \frac{e_{ith}}{f_{ij}} \qquad (42)$$

Here $\tau_{ij}$ is the triggering value of the j-th fault relative to the i-th residual and $e_{ith}$ is the threshold for the i-th residual. This latter is related to the fault-free variance of the residual that depends on the noise and on the system matrices; see e.g., Eqn. (30). The parameter $f_{ij}$ characterizes the influence of the j-th fault on the i-th residual; for an additive fault it is constant. In a static system, $f_{ij}$ is the respective parameter of the system matrix $\underline{\underline{F}}$. In a dynamic input-output model (Eqn. (17) or (18)), this may be the absolute (steady-state) value of the respective element of the matrix polynomial $\underline{\underline{F}}(z)$:

$$f_{ij} = [f_{ij}(z)\approx_{z=1} \qquad (43)$$

It is not certain, however, that the maximum fault effect on the residual will occur in steady state: $f_{ij}$ may, for example, be of differentiating nature (Sundar, 1985).

Even if we disregard of the above uncertainties in the definition, the triggering values are difficult to work with since they constitute a matrix characteristic of failure sensitivity. The selection of any scalar measure of sensitivity implies (arbitrary) preferences. These sensitivity considerations do not lead to a straightforward framework for FDI algorithm design.

### Improved sensitivity via filtering

Failure sensitivity can be improved by filtering the residuals (Gertler and Singer, 1985). The simple first-order filter

$$e_{Fi}(t) = a\, e_{Fi}(t-1) + (1-a)e_i(t) \qquad (44)$$

can be designed so that the fault-free variance of $e_{Fi}(t)$ is much smaller than that of $e_i(t)$ while their mean is the same. In fact, the variance can be reduced to any desired value, at the expense of a delay in the detection/isolation.

This kind of filter works well if the residual equations are static or, in case of dynamic equations, if the residuals depend on the faults in a proportional fashion. If the relationship is of differentiating nature, filtering may reduce failure sensitivity. Therefore, there may be a need for a number of parallel filters with different coefficients (Sundar, 1985). On-line adaptation of the filter, based on the momentary signature pattern, may be an even more promising approach. This may be achieved in a rule- and knowledge-based artificial intelligence framework.

### Enhanced robustness via algebraic cancellation

Robustness of the FDI algorithm can be improved by de-sensitizing residuals with respect to certain modeling errors. This may be achieved by explicit algebraic cancellation of some terms in the residual equations. This is basically the extension to modeling errors of the concept of isolability and model transformation.

If the residuals are obtained from input-output relationships, uncertainties (errors) of the model matrix $\underline{\underline{F}}(z)$ cannot be eliminated by cancellation, only those of some underlying parameters (Gertler, Singer and Sundar, 1985). Such underlying parameters may be those of a state-space model or of a physical system model. Since the "coefficients" of the uncertainties to be eliminated are process variables (measurements), the goodness of the cancellation depends on the model dynamics. While perfect cancellation can be achieved with a static model and, in certain configurations, with a dynamic model, in some other configurations the cancellation is "good" only if the process variables vary slowly.

The cancellation concept has been applied to observer based residuals as well. If the residuals are obtained as the difference between two observer outputs, the observers can be so designed that some uncertainties of the state model matrices are ineffective (Patton and Willcox, 1985). This method, too, is subject to limitations depending on the frequency (speed of change) of plant variables. Observers can be designed to exhibit exact zero sensitivity with respect to modeling errors in the state-equation, provided that all plant inputs and some outputs are guaranteed to be fault free (Watanabe and Himmelblau, 1982).

Model redundancy

As it was pointed out before, usually a large number of different models can be generated by transformation, that all describe the same physical system and all satisfy the structural isolability conditions. These models have different structure and parameters and, of course, different sensitivity and robustness properties. This great redundancy of models makes it possible, at least in principle, to select the one that is the best from the point of view of sensitivity and robustness.

There are two major reasons why this selection is extremely difficult to realize. First, it is generally impossible to formulate a single measure describing the required sensitivity and robustness properties, especially since these requirements usually vary with time (with the failure/noise/modeling error situation). Second, even if the requirements are clearly formulated, the combinatorial multitude of possible models rules out an analytical solution.

Chow and Willsky (1984b) formalized the robustness requirement as

$$\min_{\underline{\Theta}_i} \max_{\underline{\xi} \in \underline{\Xi}} E\{e_i(t,\underline{\xi})\} \tag{45}$$

where $\underline{\Theta}_i$ is the parameter vector of the i-th residual equation and $\underline{\xi}$ is the vector of uncertain model parameters. This formulation is conceptually clear but only partial yet too complex for practical design. Approximate solutions in simple cases are, however, possible.

A more promising approach is to place the model selection/transformation procedure in an artificial intelligence framework. While most simple failures might be isolated using a single model, some others may require the use of several, sequentially and/or in parallel. A catalog of basic models would be generated off-line. Also, logical rules would be provided concerning the selection of models depending on the failure/noise/modeling error situation at hand, as indicated by the signatures/residuals. This procedure may lead outside the cataloged basic models and imply generating additional ones; this, in turn, calls for logical rules on model transformation. The knowledge base of the system would be continually expanded: experience gained in each search would be utilized in future FDI attempts. The artificial intelligence framework may be extended to initiating new system identification, if modeling errors are suspected.

## CONCLUSION

The main features of model-based failure detection and isolation methods have been surveyed in this paper. Several techniques to generate residuals from plant measurements and to obtain failure signatures via the statistical testing of the residuals have been discussed. It has been pointed out that the major quality issues of FDI algorithms are isolability, sensitivity and robustness. Isolability is related primarily to the structure of the residual generating model and can be attained by appropriate model transformation. Sensitivity and robustness requirements may vary with the failure/noise/modeling error situation and are best taken into consideration in a rule- and knowledge-based framework, making use of the model redundancy afforded by transformation.

## REFERENCES


Almasy, G. A. and T. Sztano (1975). Checking and Correction of Measurements on the Basis of Linear System Model. Problems of Control and Information Theory, Vol. 4, pp. 57-69.

Ben-Haim, Y (1980). An Algorithm for Failure Location in a Complex Network. Nuclear Science and Engineering, Vol. 75, pp. 191-199.

Ben-Haim, Y. (1983). Malfunction Isolation in Linear Stochastic Systems: Application to Nuclear Power Plants. Nuclear Science and Engineering, Vol. 85, pp. 156-166.

Billmann, L. and R. Isermann (1984). Leak Detection Methods for Pipelines. Proceedings of the 9th World Congress of IFAC, Budapest, Hungary, pp. 1813-1818.

Bonivento, C. and A. Tonielli (1984). A Detection-Estimation Multifilter Approach with Nuclear Application. Proceedings of the 9th World Congress of IFAC, Budapest, Hungary, pp. 1771-1776.

Chien, T. T. and M. B. Adams (1976). A Sequential Failure Detection Technique and Its Application. IEEE Transactions on Automatic Control, Vol. AC-21, pp. 750-757.

Chow, E. Y. and A. S. Willsky (1984a). Bayesian Design of Decision Rules for Failure Detection. IEEE Transactions on Aerospace and Electronic Systems, Vol. AES-20, pp. 761-773.

Chow, E. Y. and A. S. Willsky (1984b). Analytical Redundancy and the Design of Robust Failure Detection Systems. IEEE Transactions on Automatic Control, Vol. AC-29, pp. 603-614.

Clark, R. N. and B. Campbell (1982). Instrument Fault Detection in a Pressurized Water Reactor Pressurizer. Nuclear Technology, Vol. 56, pp. 23-32.

Eckert, S., K. A. Loparo, and F. Roth (1986). An Application of Non-Linear Filtering to Instrument Failure Detection in a Pressurized Reactor. Nuclear Technology, to appear.

Fortin, D. A., T. B. Rooney, and E. H. Bristol (1983). Of Christmas Trees and Sweaty Palms. Proceedings of the 9th Annual Advanced Control Conference, West Lafayette, Indiana, pp. 49-54.

Fox, M. S., S. Lowenfeld, and P. Kleinosky (1983). Techniques for Sensor-Based Diagnosis. Proceedings of the International Joint Conference on Artificial Intelligence, Karlsruhe, West Germany.

Geiger, G. (1984). Fault Identification of a Motor-Pump System Using Parameter Estimation and Pattern Classification. Proceedings of the 9th World Congress of IFAC, Budapest, Hungary, pp. 1783-1788.

Gertler, J. and G. Almasy (1971 and 1973). Balance Calculations through Dynamic System Modelling. IFAC Symposium on Digital Simulation of Continuous Processes, Gyor, Hungary. Also Automatica, Vol. 9, pp. 79-85.

Gertler, J. (1979). A Constrained Minimum Variance Input-Output Estimator for Linear Dynamic Systems. Automatica, Vol. 15, pp. 353-358.

Gertler, J. and D. Singer (1985). Augmented Models for Statistical Fault Isolation in Complex Dynamic Systems. Proceedings of the 1985 American Control Conference, Boston, Mass., pp. 317-322.

Gertler, J., D. Singer, and A. Sundar (1985). A Robustified Linear Fault Isolation Technique for Complex Dynamic Systems. Preprints of the 7th IFAC Symposium on Digital Computer Applications to Process Control, Vienna, Austria, pp. 493-498.

Gertler, J. and D. Singer (1986?). An Improved Model Error Approach to Fault Detection and Isolation in Complex Dynamic Systems. Automatica, submitted.

Hakami, B. and J. Newborn (1983). Expert Systems in Heavy Industry: An Application of ICLX in a British Steel Corporation Works. ICL Technical Journal, Nov. 1983, pp. 347-359.

Halme, A. and J. Selkainaho (1984). Instrument Fault Detection Using an Adaptive Filtering Method. Proceedings of the 9th World Congress of IFAC, Budapest, Hungary, pp. 1765-1770.

Himmelblau, D. M. (1978). Fault Detection and Diagnosis in Chemical and Petrochemical Processes. Chemical Engineering Monograph 8, Elsevier Scientific Pub. Co.

Isermann, R. (1984). Process Fault Detection Based on Modelling and Estimation Methods. Automatica, Vol. 20, pp. 387-404.

Isermann, R. (1985). Process Failure Diagnosis Based on Modelling and Identificaiton Methods. IFAC-IFIP-IMACS 7th Conference on Digital Computer Applications to Process Control, Vienna, Austria. Preprints, pp. 49-58.

Ishida, Y. et al. (1985). A Topological Approach to Failure Diagnosis of Large-Scale Systems. IEEE Transactions on Systems, Man and Cybernetics, Vol. SMC-15, pp. 327-333.

Janssen, K. and P. M. Frank (1984). Component Failure Detection via State Estimation. IFAC 9th World Congress, Budapest, Hungary. Proceedings, pp. 2213-2268.

Kumamaru, H. et al. (1982). A Hierarchical Diagnosis for Failure Detection of Dynamical Systems. 6th IFAC Symposium on Identification and Process Parameter Estimation, Washington, D.C. Preprints, pp. 515-520.

Kumamaru, H. et al. (1985). Fault Detection of Dynamical Systems Based on a Recognition Approach to Model Discrimination. Proceedings of the 7th IFAC/IFORS Symposium on Identification and Process Parameter Estimation, York, UK, pp. 1625-1630.

Kumamoto, H. et al. (1984). Application of Expert System Techniques to Fault Diagnosis. The Chemical Engineering Journal, Vol. 29, pp. 1-9.

Loparo, K. A., F. Roth, and S. Eckert (to appear). Non-linear Filtering for Systems with Random Structure; Explicit Solution of the DMZ Equation and its Application to Nonlinear Filtering. IEEE Transactions on Automatic Control.

Mah, R. S. et al. (1976). Reconcilation and Rectification of Process Flow and Inventory Data. Ind. Eng. Chem., Process Des., Vol. 15, pp. 175-183.

Mah, R. S. and A. C. Tamhane (1982). Detection of Gross Errors in Process Data. AIChE Journal, Vol. 28, pp. 828-830.

Newman, R. S. and J. D. Perkins (1979). Some Tests of a Method for Fault Detection on Chemical Plants. Conference on the Profitable Use of Statistics in Chemical Engineering, Manchester, UK.

Patton, R. J. and S. W. Willcox (1985). Comparison of Two Techniques of I.F.D. Based on a Non-Linear Stochastic Model of an Aircraft. IFAC Symposium on Identification and System Parameter Estimation, York, United Kingdom. Proceedings, 711-717.

Pau, L. (1981). Failure Diagnosis and Performance Monitoring. Marcel Dekker Inc.

Rault, A. et al. (1984). Industrial Processes Fault Detection and Localization. Proceedings of the 9th World Congress of IFAC, Budapest, Hungary, pp. 1789-1794.

Romagnoli, J. A. and G. Stephanopoulos (1981). Rectification of Process Measurement Data in the Presence of Gross Errors. Chem. Eng. Sci., Vol. 36, pp. 1849-1863.

Rooney, T. B. et al. (1978). Estimating Plant Flows and Inventories Using Material Balance Relations. IFAC World Congress, Helsinki, Finland.

Shutty, J. (1985). A Multilevel Approach to Fault Detection. Master Thesis. Case Western Reserve University, Cleveland, Ohio.

Stanley, G. M. and R. S. Mah (1977). Estimation of Flows and Temperatures in Process Networks. AIChE Journal, Vol. 23, pp. 642-650.

Stanley, G. M. and R. S. Mah (1981). Observability and Redundancy in Process Data Estimation. Chem. Eng. Sci., Vol. 36, pp. 259-272.

Stuckenberg, N. (1985). Sensor Failure Detection in Flight Control Systems Using Deterministic Observers. IFAC Symposium on Identification and System Parameter Estimation, York, United Kingdom. Proceedings, pp. 705-710.

Sundar, A. (1985). Process Fault Detection Using the Augmented System Model Approach. Master Thesis. Case Western Reserve University, Cleveland, Ohio.

Tylee, J. L. (1983). On-line Failure Detection in Nuclear Power Plant Instrumentation. IEEE Trans. on Automatic Control, Vol. AC-28, pp. 406-415.

Uosaki, K. (1985). Failure Detection Using Backward SPRT. IFAC Symposium on Identification and System Parameter Estimation, York, United Kingdom. Proceedings, pp. 1619-1624.

Vaclavek, V. (1974). Gross Systematic Errors or Biases in the Balance Calculations. Papers of the Prague Inst. of Techn., Prague, Czechoslovakia.

Watanabe, U. and D. M. Himmelblau (1982). Instrument Fault Detection in Systems with Uncertainties. International Journal of System Science, Vol. 13, pp. 137-158.

Watanabe, U. and D. M. Himmelblau (1983). Fault Diagnosis in Non-Linear Chemical Processes. AIChE Journal, Vol. 29, pp. 243-249.

Willsky, A. S. (1976). A Survey of Design Methods for Failure Detection in Dynamic Systems. Automatica, Vol. 12, pp. 601-611.

Willsky, A. S. and H. L. Jones (1976). A Generalized Likelihood Ratio Approach to the Detection and Estimation of Jumps in Linear Systems. IEEE Transactions on Automatic Control, Vol. AC-21, pp. 108-112.

Wunnenberg, J., R. N. Clark, and P. M. Frank (1985). An Application of Instrument Fault Detection. IFAC Symposium on Identification and System Parameter Estimation, York, United Kingdom. Proceedings, pp. 699-704.

Yamada, N. and H. Motoda (1983). A Diagnosis Method of Dynamic Systems Using the Knowledge on System Description. Proceedings of the Eighth International Joint Conference on Artificial Intelligence, Karlsruhe, Germany, pp. 225-229.

# EDUCATION FOR MICROPROCESSOR APPLICATION IN CONTROL

**W. Schaufelberger, H. Good and A. Itten**

*Department of Automatic Control and Industrial Electronics,
Swiss Federal Institute of Technology (ETH), ETH-Zentrum,
CH-8092 Zürich, Switzerland*

Abstract. Computers are extremely useful for a thorough control education. This is shown by providing some general information about their use and by describing three students' assignments in detail.

Keywords. Control education, computers in control education, laboratory education.

## INTRODUCTION

The development of computer science has a strong and increasing influence on the development of control theory and practice. Computer-based production units are the core of many modern industrial enterprises. Based on this, control engineers are required to have a thorough knowledge in the following fields:

measurement and control / artificial intelligence, expert systems / computer aided control system design, simulation / realizations: programming of microprocessors, local area networks, realtime languages and operating systems.

At all levels, the teaching process must take this into consideration. The computer as an easily available and versatile tool, ideally suited for teaching control system theory, will be shown in the sequel.

## USE OF WORKSTATIONS IN CONTROL TEACHING

Workstations or personal computers can be used in a variety of ways to improve teaching:

Commercially available packages may be used in areas which are already standardized, such as: writing and spelling, document preparation (including graphics), table calculations (spread sheets), data-base management, simulation, computer aided control system design.

Students may be asked to **write their own programs** using sequential programming languages (BASIC, FORTRAN, FORTH, PASCAL), real-time languages (Modula-2, Ada), or the languages of artificial intelligence (LISP, PROLOG).

Exercise and problem sessions may be organized using general software packages especially developed for teaching.

Personal computers are also very well suited for laboratory control exercises. Interfaces may be bought or developed at the institute (Allenspach & Schaufelberger, 1985).

The actual problems depend on the time that the student has available for exercises. The following examples are taken from the teaching experience of the control group of the Swiss Federal Institute of Technology in Zürich. Only two hours of one afternoon's assignment are presented. These may be seen in the larger context as described in Mansour et al. (1984), and Mansour & Schaufelberger (1981) where the general attitude of the group towards the use of computers in teaching is described.

Using either the available VAX or Macintosh computers, every student has access to the following languages and packages to do his control exercises:

PASCAL
Modula-2
MATLAB
CTRL/C
PROLOG
LISP
ACSL

MATLAB and CTRL/C are primarily used in the design phase for feedback control systems. LISP and PROLOG are used to design and simulate sequential control systems, PASCAL and Modula-2 are used for implementations and for the design of stand-alone exercises, and ACSL is used for simulations. A set of some 20 small models (traffic corner, elevator, drives, model train, level control systems, etc.) are available to carry out "real world experiments".

## EXAMPLE 1

### A Courseware Program for the Development of a State Feedback Controller

In many cases, it is possible to obtain a better transient behavior of a system by using a state feedback controller. In reality, this type of controller is rarely used, perhaps due to the difficulties of developing such a controller. The problem is complex, and it is not always possible to guarantee satisfactory results which justify the expense. Some of the difficulties lie in the analysis and quantification, and some lie in the quite complex theories of linear algebra and of the different transformations that have to be used.

With the aid of computers it is possible to separate the different types of difficulties, and to analyze each one in sequence.

For this courseware program, we chose the following problem area: The program should help students in understanding the linear algebra calculations used in the design and their relation to the final result, as given by time-responses.

During the lectures, students learn the theory of control system design. In particular, they learn how to derive the control parameters from the desired root locations.

The students then are required to do exercises which in part take place in a "laboratory" on Macintosh personal computers. They can enter the calculated parameters, and then simulate the system. Thus it is possible to apply the theory and get instant feedback by observing the simulated results. Students begin to understand the relation between the root locations and the time behavior. Further, they recognize that it is extremely difficult to tune the controller parameters without calculations.

## Goals

In contrast to many commercial software products dealing with control system design, one of the main goals during the development was to restrict the scope of the program and to make the program easy to use. Every result is presented as clearly as possible to enhance the learning effect. The subject of the controller design exercise is a simple unstable linear system consisting of two cascading integrators.

Due to the limited complexity, it is possible to use very simple algorithms for all numerical calculations.

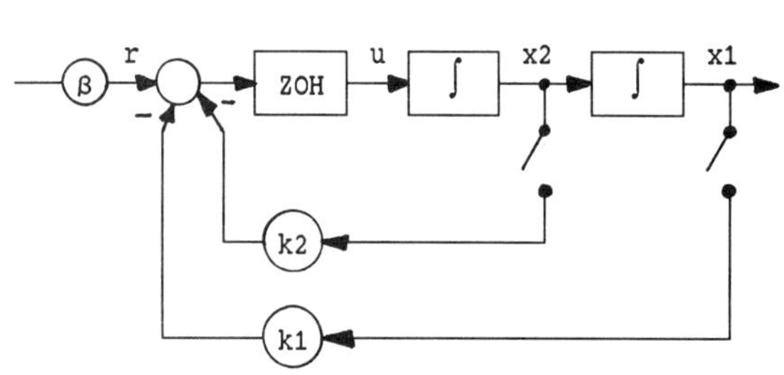


Fig. 1. Block diagram of control system.

With the same program it is also possible to develop an observer. This part is similar; thus, it is not described in detail.

A sample exercise session is described to provide insight into the students' behavior during a problem session:

## A Sample Exercise Session

One week prior to the practical course, the students are given a paper describing two different parts of their task. Part 1 they have to solve at home:

The homework:

Homework Question 1:

What are the discrete system equations for the given system as a function of the sampling interval:

$$\underline{x}(k+1) = \tilde{A}\underline{x}(k) + \tilde{\underline{b}}u(k)$$

$$y(k) = \underline{c}'\underline{x}(k)$$

Solution 1:

Starting from the equations of the continuous time description

$$x_2(t) = x_2(0) + \int u\,dt = x_2(0) + ut$$

(u = constant between two sampling events)

$$x_1(t) = x_2(0) + \int x_2(t)\,dt$$

$$= x_1(0) + x_2(0)t + t^2/2u$$

obtain the discrete time description with the sampling interval T:

$$\underline{x}(k+1) = \begin{bmatrix} 1 & T \\ 0 & 1 \end{bmatrix} \underline{x}(k) + \begin{bmatrix} T^2/2 \\ T \end{bmatrix} \underline{u}(k)$$

Homework Question 2:

Determine for the state feedback controller the general relation

$$k_1 = f_1(p_1,p_2,T)$$

$$k_2 = f_2(p_1,p_2,T)$$

$$\beta = f_3(p_1,p_2,T)$$

for given closed loop root locations $p_1$ and $p_2$, and a static closed loop gain $v = 1$.

Solution 2:

Comparison of coefficients of the characteristic polynomial of the desired system

$$\lambda^2 + \lambda\left[k_1 \frac{T^2}{2} + k_2T - 2\right] + 1 - k_2T + k_1 \frac{T^2}{2} = 0 =$$

$$= (\lambda - p_1)(\lambda - p_2)$$

leads to the following equations:

$$k_1 = (1-p_1-p_2+p_1p_2)/T^2$$

$$k_2 = (3-p_1-p_2-p_1p_2)/2T$$

$$\beta = k_1$$

Similar equations for the observer are derived before the main exercise begins.

## The "Laboratory" Work

During the exercises, two students work together on one Macintosh computer. They familiarize themselves with the program by entering input and running simulations. Although the Macintosh is very easy to use, students need some basic instructions on how to use the mouse.

The second task is to choose sampling time and root locations, and to enter the calculated feedback parameter values. The entering of the parameter values is accomplished with the four levers in the lower left corner on the screen (Fig. 2). Pulling the white box over its possible range will change the value of the parameter. It is also possible to point with the cursor on one of the two arrows, which will change the value by a small amount.

The output of each simulation run is shown in four rectangular windows: In the upper left window is the input signal (nominal value). It can be changed by the menu command "Eingang" (= "input"). Below that screen, the control value u is displayed, and in the two screens on the right side, the time behaviour of the two states are shown. In the lower right corner of the laboratory place, the z-plane is displayed with the unit circle, and the location of the roots are marked with two symbols "x".

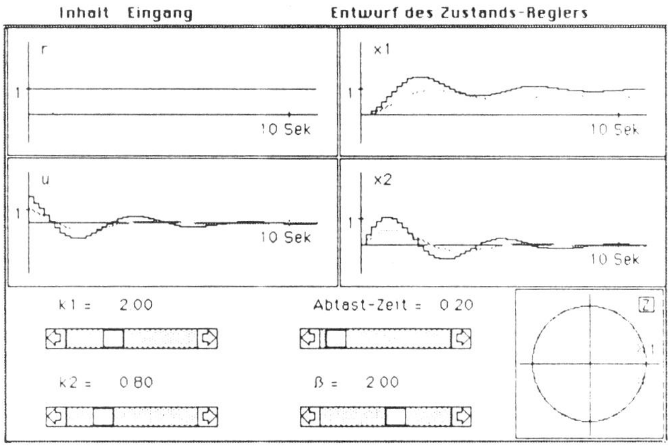


Fig. 2. Screen layout

The students continue the exercise in a laboratory section, and answer the following questions:

Lab Question 1:

Starting from the general equations derived in homework question 3, calculate for a given sampling time the interval T = 0.5 sec and the following root locations of the corresponding state feedback parameter values. Complete the following table:

| T | $p_1$ | $p_2$ | $k_1$ | $k_2$ | ß |
|---|---|---|---|---|---|
| 0.5 | 1.0 | 1.0 | | | |
| 0.5 | 0.0 | 0.0 | | | |
| 0.5 | 0.5 | 0.5 | | | |
| 0.5 | 0.5+j0.5 | 0.5-j0.5 | | | |
| 0.5 | -0.5 | -0.5 | | | |
| | | | | | |
| | | | | | |
| | | | | | |
| | | | | | |

Improve your calculation by simulating the controlled system.

Lab Question 2:

Try to describe the results for root loctions in the following five given areas. Pay special attention to:

- settling time
- oscillation
- energy required

Where would the root locations be placed? Why?

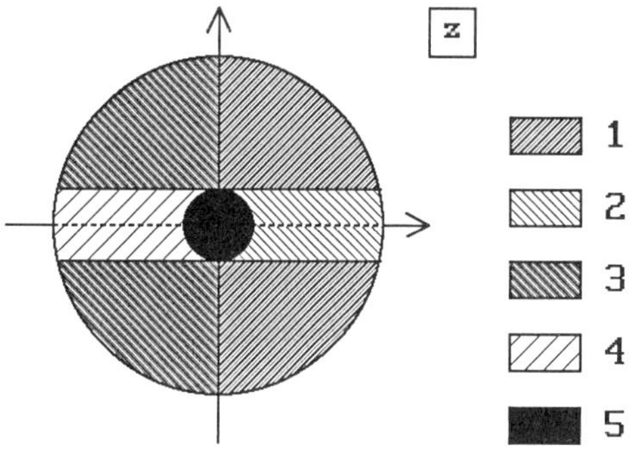


Fig. 3: Unit circle in the z-plane and five given areas.

<u>Comments</u>

The students find the computer a useful tool for learning. They are quite satisfied with the ease of using the Macintosh, which spares them the reading of extensive manuals to solve an exercise.

The expense to produce such a program is quite high. It is very important that the author first defines exactly the learning goals; they should be adjusted to the time available.

The programmer should avoid everything which does not directly address the goals, even if it would be very easy to supplement the program with a few statements. This makes the use of the program simpler and enables the student to focus on the main subject.

EXAMPLE 2

A Brief View on PC-MATLAB

Good interactive computer programs offer an excellent support in solving many problems. The arising problems when designing control systems are of algebraic nature, and the help of a computer is needed to solve systems of large order. PC-MATLAB (Little & Mooler, 1985) is a highly optimized interactive computer program for the IBM personal computer (or equivalent) that can help in scientific and engineering numeric calculations. The command format is natural and corresponds to the "syntax" of algebraic expressions so that no arbitrary command syntax has to be learned. PC-MATLAB works with (real or complex) matrices and vectors, and offers many features for design and analysis. Consider, for example, the transfer function

$$G(s) = \frac{K}{(s+3)(s^2+2s+3)} = \frac{K}{s^3+5s^2+8s+6} \quad ; \; K = [20..50]$$

The following is the corresponding input to the prompt (% introduce comments) to produce the Nyquist diagram in Fig. 4.

```
<> num = [0 0 0 1]; den = [1 5 8 6]
   % num = numerator, den = denominator

<> w = 0:0.1:10;
   % create frequency vector from 0 to 10 rad/sec
<> [re1,im1] = nyquist(20*num,den,w);
   % re = real part, im = imaginary part
<> [re2,im2] = nyquist(34*num,den,w);
<> [re3,im3] = nyquist(50*num,den,w);
<> plot(re1,im1,re2,im2,re3,im3),title('Nyquist
   diagram');
```

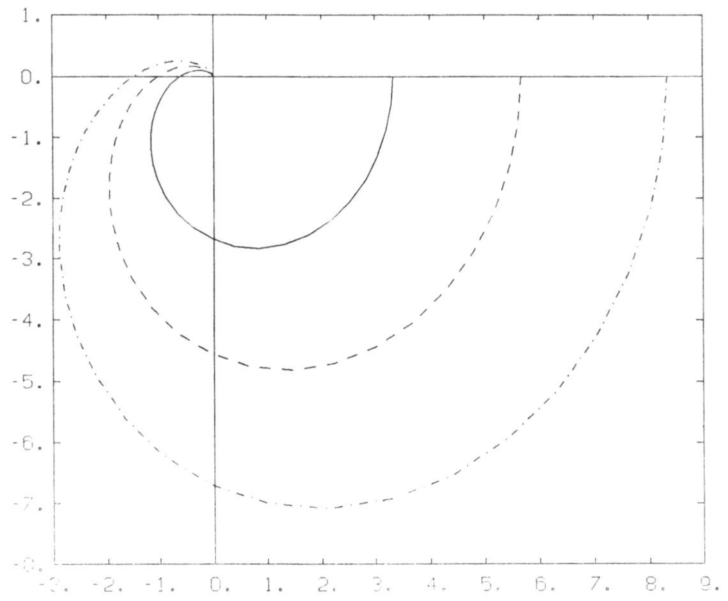


Fig. 4. Nyquist diagram.

For the design of state feedback controllers, the state space description of the system is needed:

```
<> [a,b,c,d] = tf2ss(num,den)
   % tf2ss = transfer function to state space

   A =
        0      1      0
        0      0      1
       -6.000 -8.000 -5.000

   B =
        0
        0
        1

   C =
        1    0    0

   D =
        0

<> eig(a)   % compute eigenvalues of system matrix

   ANS =
       -1.000 + 1.000i
       -1.000 - 1.000i
       -3.000 + 0.000i
```

The intrinsic function 'lqr' solves the linear quadratic regulator design problem which minimizes the cost function

$$J = \int_0^{\infty} (x'Qx + u'Ru)dt \qquad u = -Kx$$

subject to the constraint equation

$$\dot{x} = Ax + Bu$$

```
<> q = diag([100 0 0]);r=1;
   % choose weighting matrices
<> k = lqr(a,b,q,r)
   % solve Riccati equation

   K =
       5.6217    3.6812    0.6897
```

A macro facility can be invoked by writing 'function files'. This provides a powerful aid and is useful for extending PC-MATLAB, i.e., creating new PC-MATLAB functions using the language itself. Variables may be passed, and variables defined and manipulated inside the file are local and do not operate globally on the workspace.

All functions and matrix operations are independent of the system order and thus provide a reliable foundation for control systems engineering.

EXAMPLE 3

Design Station

Introduction. It is well known that better models of industrial processes (i.e. description of linear systems by state variables) are needed for the application of state feedback controllers. In practice, however, there are only restricted domains where state feedback controllers are in use. One reason might be that the design of such controllers implies a sound theoretical knowledge combined with practical experience. Control theory deals with small and well defined problems based on a purely mathematical description. Therefore, the education of control engineers is often confronted with a lack of physical interpretation. The step from a pure mathematical equation describing the control algorithm to the final realization is hardly addressed.

In practice, classical controllers (PID controllers) are very often adjusted using hand-tuning or rules of thumb. In industrial processes, these methods are employed due to their simplicity. State feedback controllers, however, are based on a mathematical model of the process; many coupled parameters cannot be adjusted by hand. Thus, practical engineers consider the design procedure to be too complicated and too time-consuming. Theoretical knowledge alone is not enough to create state feedback controllers; it should be complemented by a general procedure. For example, if the continuous time description in state space of a process is known, the design procedure can roughly be formulated in the following way:

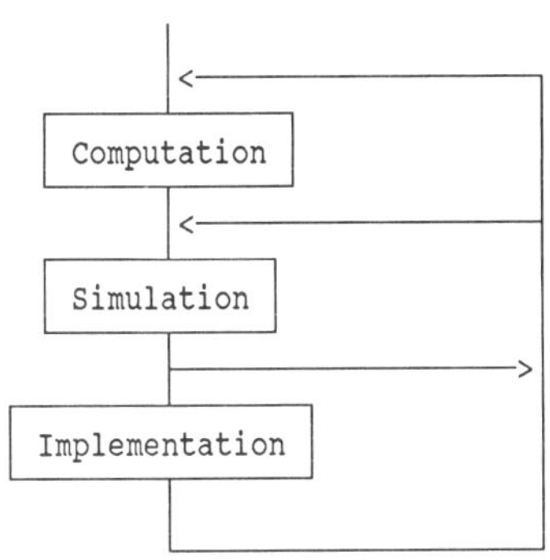


Fig. 5. Design cycle for a state variable feedback controller

In each part of this diagram, the digital computer plays an important role, i.e., it is used in the computation and simulation as well as in the implementation phase. The problem of designing a state feedback controller should be solved with a minimum of 'work' such as programming or building hardware. This implies good and efficient software tools for all three stages. The power and reliability of modern personal computers makes it possible to realize such a design station for portable use.

In the following, the possibilities of a LQG design station are presented, as realized on a HP-1000 process computer and used for tutorial courses in automatic control. Further, it offers a concept for the design of complex modern algorithms for industrial use. The three stages of computation, simulation and implementation; additional possibilities to invoke the compiler, linker and operating system commands are all driven by a menu program with a user friendly interface. Hence, it is possible to create modern controllers on the same station with connection to the process.

Computation. It was shown in example 2 that computation of state-feedback controllers can be done very efficiently with MATLAB. A control system toolbox (pole placement, optimal control, Kalman filter design, etc.) offers different ways to do all the necessary computations. The results from a MATLAB session are saved on a disk file for further use.

Simulation. Controllers can hardly be designed without simulation. The simulation of a state feedback and an observer-based state-feedback controller are therefore tremendously helpful parts before implementation with the real process:

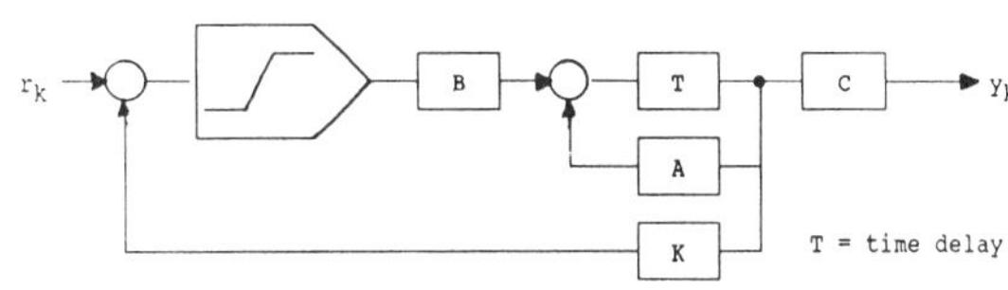


Fig. 6. Block diagram.

In any case, the discrete time description of the system (A,B,C) and the controller (K) are available for implementation on a microprocessor; a high speed recursive simulation shows the effect of the state-feedback (K). All parameters are stored on a file from a MATLAB session. A similar simulation program is also available for the simulation of an observer-based state-feedback to illustrate the effect of an observer.

Implementation. The creation of a digital controller essentially consists of numerical data and an algorithm. More and more, the algorithm is programmed in a high-level programming language like PASCAL or Modula-2. PASCAL is common in education but lacks the real-time properties available in Modula-2. (Cf. Maier & Schaufelberger (1985), and Maier (1984) for a professional implementation of complex controllers in Modula-2 with an exception handling mechanism.)

PASCAL allows a clear style of programming. For the implementation of controllers, some general procedures and functions imply a large reduction of programming work:

- parameter setting (numerical data) from a MATLAB session
- time scheduling of the program from an external hardware clock or from a real-time operating system
- process peripheral communication routines like AD-converters
- user communication routines

Design example. To illustrate the design procedure and some practical effects, a little two-degree of freedom, mechanical helicopter model is provided as an example of process. The linearized continuous time description is:

$$A_c = \begin{bmatrix} .00 & 1.00 & .00 & -.00 & .00 & .00 \\ 0.00 & -2.06 & -.00 & .00 & .03 & .01 \\ 0.00 & .00 & .00 & 1.00 & -.00 & .00 \\ 0.00 & .00 & 10.03 & -.50 & .14 & .02 \\ 0.00 & 2.06 & .00 & -.00 & -1.45 & .01 \\ 0.00 & .00 & -10.03 & .50 & -.14 & -7.62 \end{bmatrix}$$

$$B_c = \begin{bmatrix} 0.00 & 0.00 \\ -1.80 & 0.00 \\ 0.00 & 0.00 \\ 0.00 & -1.41 \\ 63.13 & 0.00 \\ 0.00 & 600.30 \end{bmatrix}$$

$$C_c = \begin{bmatrix} 1. & 0. & 0. & 0. & 0. & 0. \\ 0. & 0. & 1. & 0. & 0. & 0. \end{bmatrix}$$

The unstable helicopter model has two inputs and would be a difficult problem to solve with classical methods. With respect to the knowledge of the continuous time description, the following steps are advisable:

Step 1: Compute the discrete time description of the process in MATLAB.

```
<> t = 0.05;
   % t = sampling interval in seconds
<> [a,b] = c2d(ac,bc,t);
```

where $A = \exp(A_c t)$

and $B = \int_0^T \exp(A_c v) B dv$

Step 2: Compute a state-feedback matrix K for the system

$$x_{k+1} = Ax_k + Bu_k \quad , \quad u_k = -Kx_k$$

by solving an associated discrete matrix Riccati equation:

```
<> qk = diag([2000;1;1000;1;0;0])
<> rk = [100 0;0 100]
<> k = dlqr(a,b,qk,rk);
```

$$K = \begin{bmatrix} 3.41 & 1.61 & 14.91 & 4.31 & .15 & .81 \\ 1.97 & .85 & -1.77 & -.44 & .01 & .00 \end{bmatrix}$$

```
<> save heli.dat
          % stores results on file heli.dat
```

Step 3: A simulation of a zero input response according to an initial state x(0) may result in the following plot:

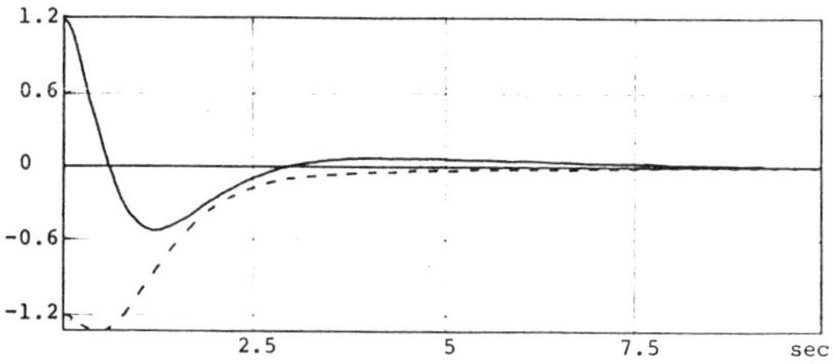


Fig. 7. Time response of state variable feedback control system

It is possible that steps 2 and 3 have to be repeated until the result meets all requirements.

Step 4: Omitting the assumption of full state measurement, an observer is introduced to control the process. Introducing process and measurement noise:

$$x_{k+1} = Ax_k + Bu_k + Gs_k \quad , \quad y_k = Cx_k + w_k$$

$$Q_h = \mathrm{Var}[s_k] \quad , \quad R_h = \mathrm{Var}[w_k]$$

the design of a Kalman filter in MATLAB is done as follows:

```
<> g = eye(A);
<> qh = diag([0 1 0 1 1000 1000]);
<> rh = [100 0;0 100];
```

The call

```
<> h = dlqe(a,g,c,qh,rh);
```

returns the gain matrix H such that the discrete stationary Kalman filter

$$z_{k+1} = Az_k + Bu_k + H(y_k - Cz_k)$$

$$H = \begin{bmatrix} .85 & .03 \\ 4.69 & .43 \\ .03 & 1.07 \\ .53 & 8.25 \\ 27.45 & 99.80 \\ .43 & -.62 \end{bmatrix}$$

produces an LQG optimal estimate of x.

Step 5: The same situation as in step 3 is simulated with the introduction of an observer. Of course, the initial state of the simulated process differs from the initial state of the observer. See the following plot for x(0) = [1.2 0 -1.2 0 0 0]' and z(0) = 0:

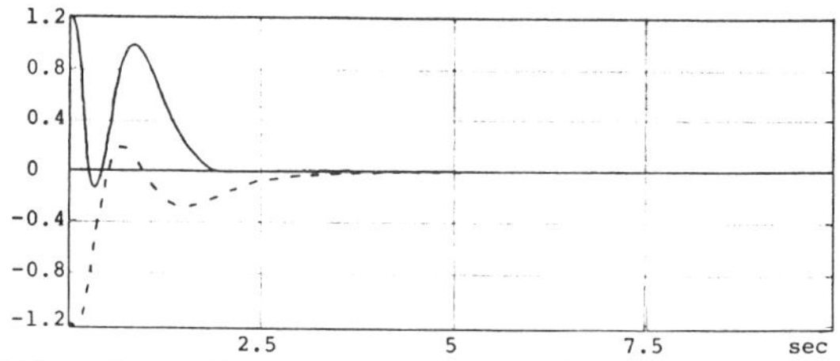


Fig. 8. Time response of observer-based control System

If necessary, repeat steps 4 and 5 to possibly meet the requirements.

Step 6: The simulated process is now replaced by the real process, and the observer (A,B,C,H) including the controller (K) is programmed in PASCAL. With the use of the routines mentioned in the part 'Implementation' of this example, a main PASCAL program could look like this:

```
BEGIN (* Main *)
  Init_Variables;             {fetch A,B,C,K,H and T}
  Set_clock(T);               {set sampling interval}
  WHILE NOT (Stop) DO
    BEGIN
      Read_AD;                {read input signal}
      Write_DA;               {output control signal}
      Measurement_update;
      Observer_update_cycle;
      Compute_control_signal;
      Wait_for_interrupt;
    END;                      {While}
  Exit;
END. (* Main *)
```

After compilation and linking, the controller is tested under real conditions. The following plot shows the zero input response due to the initial condition x(0) = [1.2 0 -1.1 0 0 0]'

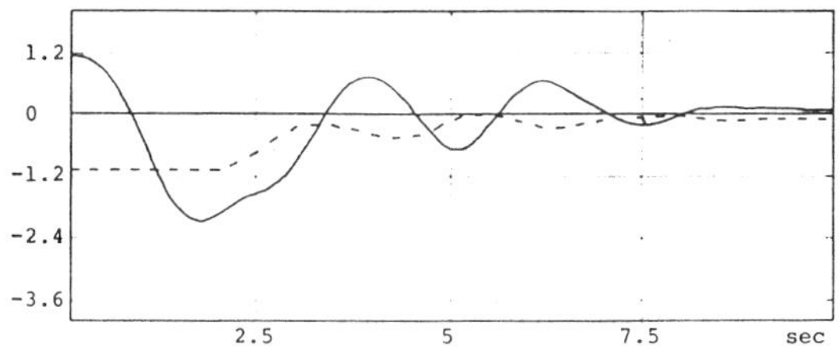


Fig. 9. Measurements from the real process

Obviously, there is a great difference between the simulation and implementation. Model uncertainties, nonlinearities and disturbances lead to worse operation, and may even tend to instability. This fact forces the designer to repeat the calculations by observing the controller under real operating conditions. The algorithm remains the same, and a new set of parameters may be tested quickly by starting the controller program. The simultaneous use of the computer as a computation and implementation tool should foster the application of modern control algorithms.

Comments

The LQG design station presented was found to be very suitable for the design of a first controller and for instructional use. It leads to a more systematic procedure and makes it possible to test the controller under real operating conditions. The availability of modern powerful software renders a user-friendly interface.

## CONCLUSIONS

It is shown in the paper that personal or microcomputers can be used to improve the teaching of control systems. This is done by providing a general overview and by describing three students' assignments in detail.

## REFERENCES

Allenspach, H., Schaufelberger, W. (1985). The Use of a Home Computer for Computer Control Experiments. IFAC/IFORS Conf. on Control Science and Technology for Development, Beijing, PR China.

Little, J., Mooler, C. (1985). PC-MATLAB User's Guide. The MathWorks, Inc., 124 Foxwood Rd., Portola Valley, CA 94025, USA.

Maier, G.E. (1984). Modula-2 Echtzeitbetriebssystem MODEB V2: Bedienungsanleitung und Installierung. Report No. 84-03, Dept. of Automatic Control, ETH Zürich.

Maier, G.E., Schaufelberger, W. (1985). A Program Structure Based on Exception Handling for Variable Structure Control Systems. 7th IFAC Conf. on Digital Computer Applications to Process Control, Vienna.

Mansour, M., Schaufelberger, W. (1981). Digital Computer Control Experiments in the Control Group of ETH Zürich. IFAC World Congress, Kyoto, Japan.

Mansour, M., Schaufelberger, W., Cellier, F.E., Maier G.E., Rimvall, M. (1984). The Use of Computers in the Education of Control Engineers at ETH Zürich. Europ. J. of Educ., 9, 135-151.

Schaufelberger, W., Sprecher, P., Wegmann, P. (1985). Echtzeitprogrammierung bei Automatisierungssystemen. Teubner Studienbücher Elektrotechnik.

# METHODOLOGY FOR LARGE-SCALE PROCESS CONTROL SYSTEM DESIGN

A. F. Konar

*Honeywell Technology Strategy Center, 1000 Boone Avenue North, Golden Valley, MN 55427, USA*

Abstract. This paper reviews the development and application of a specific computer-aided design methodology called "SIMKON" for large-scale dynamic process control systems. This methodology has been implemented in the SIMKON-IV software system. The set of algorithms used in SIMKON is based on state-of-the-art simulation and control technologies and is a spinoff from the well-known DIGIKON-IV, ADAPS and KONPACT programs developed for aerospace and defense applications by Honeywell over a period of time and includes about 50,000 FORTRAN-IV statements. The significance of this development is that it provides the user with the total system capability of developing alternative controller organizations and corresponding multiperformance figures for large systems under one unified interactive command structure. The interactivity with the user significantly improves the productivity of the user. The richness of its command repertoire and the advanced algorithms it uses for implementing its functions make SIMKON-IV one of the most complete and advanced system engineering tools for control applications in the process industry. Brief descriptions of this tool and its basic capabilities and application areas are given. In addition, typical inputs and outputs are illustrated.

Keywords. Process control; computer-aided system design; chemical variables control; industrial control; multivariable systems; control system analysis; control system synthesis.

## INTRODUCTION

Productivity is a major concern today for the health and survivability of industrial organizations. This is due to the rising cost of materials, labor and overhead, and extreme worldwide competition in the marketplace.

Industrial planners have turned their attention to automation to increase productivity and reduce cost. The fields of Robotics and Computer-Aided Design and Engineering (CAD/CAE) have become very active for industrial research and development in automation. In addition to these trends, in-house training activities have increased considerably to facilitate the use of a high degree of automation in industrial complexes.

Within the framework of automation, this paper reviews the author's continued work in the development and application of a methodology and a CAE tool for large-scale control system design. The focus here is on the system control functions. The automated design of system management functions such as redundancy management, fault monitoring and diagnostics, trajectory and set-point optimizations, etc., are beyond the scope of this paper.

An overview of the control system design process is presented to set the stage for a description of the CAD tools developed. This is followed by a brief description of the interactive software development and transfer procedure. The CAE package, SIMKON, is then described along with its capabilities and application areas. Three examples are presented to demonstrate an application of SIMKON to the process control system analysis/design. Finally, current activities are described.

## CONTROL SYSTEM DESIGN PROCESS

In general, a complex process control system analysis/design involves six steps: (1) Control system requirements/architecture definition, (2) modeling and identification of nonlinear process dynamics, (3) nominal operating point/trajectory definition, (4) linearization of process dynamics, (5) controller concept definition, and (6) controller optimization, tuning, and performance evaluation. These six steps are reviewed briefly to set the stage for the CAE software organizations.

### Control System Requirements/Architecture Definition

This step involves the mission statement and specification data for the controlled system. Most often this data includes the operating envelope and limits of the controlled system, specific command and disturbance inputs, their types, ranges, and expected responses, parameter uncertainty ranges in the process model, steady-state performance levels, and actuator activity and limits. Depending upon the nature of the specific control system, additional requirements may be imposed (i.e., maximum temperature, pressure, structural stress, and so on). These requirements are met by an appropriate system architecture.

Most often this architecture is in the form of multilevel and distributed control and may contain robust functional modules such as controllers, estimators, predictors, identifiers, optimizers, etc. The key requirement here is that these modules perform reliably over a wide range of operating conditions and application scenarios.

Figure 1 shows a block diagram of a typical system architecture for an advanced process control system. It consists of nine levels. The first level includes the process dynamics. Most often the elements of process subsystems dynamically interact with each other due to mass and energy transfer between them. The second level represents the equipment dynamics involved: compressors, pumps, etc. The third level includes the final control elements, such as valves, actuators, and sensors. A system composed of levels 1, 2, and 3 is called a process plant.

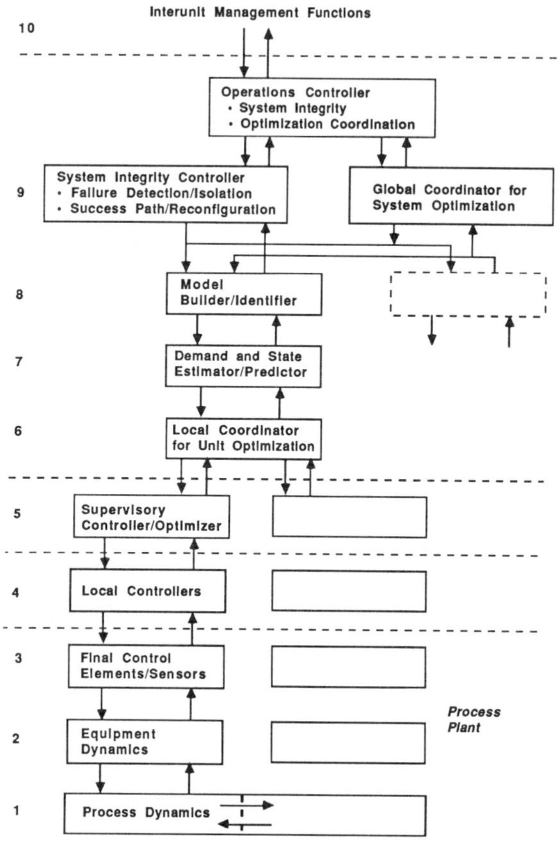


Fig. 1. Multilevel process control system architecture.

The fourth level is assigned to the local controllers, controlling the process variables such as temperature, flow, concentration, etc. The set points of the local controllers are driven by the supervisory controller at the next higher level. The supervisory controller depicted in the fifth level develops the optimal command profiles for batch processes or optimal operating points for continuous processes.

Allocation of local costs among the subsystems is performed by the local coordinator to optimize the unit operating cost.

The filtering of the measured plant signals and the estimation of the inferred plant and ambient system variables and their prediction into the future based on the data available up to the present time are carried out at the seventh level. This level consists of observers and linear and nonlinear Kalman filters of appropriate size. They contain the plant and ambient dynamic models in their structure. In addition, models of deterministic as well as stochastic disturbances into the plant are included. The model structures may be based on first principles or black box autoregressive moving average methods. These models are computed in the eighth level.

Because of the availability of the on-line model parameters of the load and the plant, plant status can be monitored. This is done in the ninth level. The current parameter estimates are compared with the normal operation parameters and any significant deviation is detected and isolated. The operator is informed, success paths to recover from the fault are suggested, and the control system is reconfigured for gracefully degraded operation until the failure is repaired.

This level also contains a global coordinator for the optimization of the total operating cost of the factory. Because of the interaction of the unit processes at the plant level, this coordinator adjusts the goals of the local optimizers on the lower level to achieve the best overall operating efficiency.

## Modeling and Identification of Nonlinear Process Dynamics

Accurate dynamic modeling of chemical process systems is by no means a trivial task. The processes are generally nonlinear, time varying, and have variable time delays and other non-minimum-phase effects. In addition they are often interactive and are perturbed by changes in feedstock and operating conditions. All of these factors make dynamic modeling and control somewhat different from that of typical electrical and mechanical systems where linear, time invariant stochastic models often give sufficient flexibility and accuracy.

The dynamics of a physical process are assumed to be characterized by the following set of differential/algebraic equations:

$$\frac{d}{dt}\left[\phi(x,\theta)\right] = \psi(\dot{x},y,x,u,\theta) \tag{1}$$

$$y = g(\dot{x},y,x,u,\theta) \tag{2}$$

$$r = h(\dot{x},y,x,u,\theta) \tag{3}$$

where:

$x$ = state vector of the process, $n_x$-vector

$y$ = algebraic/summing point variables, $n_y$-vector

$r$ = response/output variables, $n_r$-vector

$u$ = control/disturbance input variables, $n_u$-vector

$\theta$ = parameter vector of the process, $n_p$-vector

Application of the first principles to dynamic processes yields this form of equations (Konar, 1985; Luyben, 1973). For the dynamic simulation of the process, the equivalent form of Eq. (1) may be used:

$$\dot{x} = f(\dot{x},y,x,u,\theta) \tag{4}$$

The dynamic system parameters, such as the reaction rates and heat transfer coefficients in chemical process dynamics, are assumed to be represented either in the form of analytical expressions (i.e., y-variables) or in the form of experimental tables with multiple arguments.

The measured data history is used to extract model parameters, which may be done in batch or recursively in time. The parameter estimation function is accomplished by a repertoire of algorithms such as recursive and generalized least squares, extended, modified Kalman filters, maximum likelihood estimators, etc. Figure 2 shows a block diagram of a parameter estimator. The process model parameters and Kalman filter gains are changed on the basis of the likelihood function obtained from the residuals.

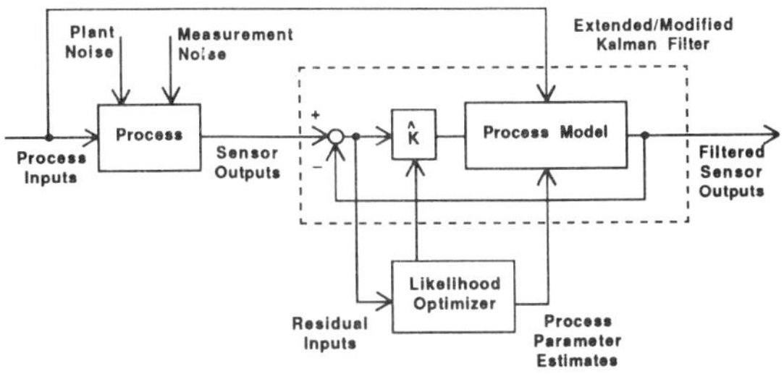


Fig. 2. Block diagram of a parameter estimator.

In process control systems there often may be significant time delays due to the distributed nature of the process dynamics, which normally are characterized by the partial differential equations. To accommodate these subsystems within the framework of ordinary differential equations, a mixture of lumped-parameter models with appropriate time delays is utilized. In this case, the process equations given by Eqs. (1), (2), and (3) are augmented with additional variables having delayed arguments, i.e.:

$$u_d = u(t-\tau_d),\ y_d = y(t-\tau_d),\ r_d = r(t-\tau_d) \qquad (5)$$

The software modules characterizing each algebraic and/or dynamic subsystem are then interconnected to form the total system simulation model, as shown in Figure 3. This approach provides great flexibility in forming very large dynamic systems.

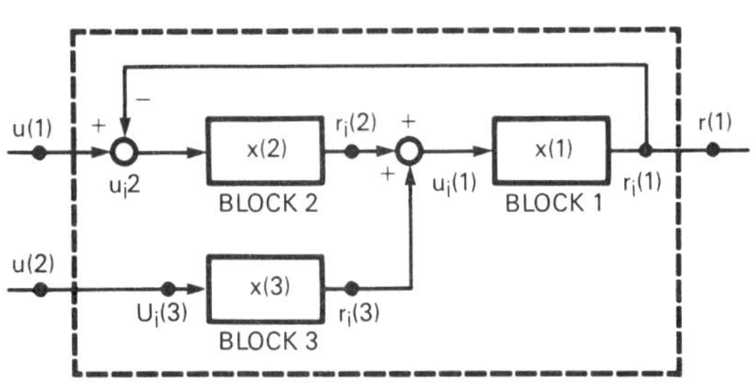


Fig. 3. Interconnection of subsystem blocks.

The interconnection equations are assumed to be in the following form:

$$u_i = Pr_i + Qu \qquad (6)$$

$$r = Rr_i + Su \qquad (7)$$

where:

$u_i$ = internal input vector
$r_i$ = internal output vector
$u$ = external input vector
$r$ = external output vector

and (P,Q,R,S) is referred to as the interconnection data of the system. Figure 4 shows a typical overall system interconnection diagram.

## Nominal Operating Point/Trajectory Definition

This step is most often concerned with the economics of process operations and involves a large-scale, nonlinear constrained-optimization task. Examples are the definition of temperature trajectory of an exothermic batch reactor, operating point of a distillation column, and so on. Two types of optimizations are carried out. In the dynamic optimization, the optimum profiles of the process variables are determined over a prescribed period:

$$[\dot{\bar{x}}(t),\ \bar{y}(t),\ \bar{x}(t),\ \bar{u}(t)],\ T_o \le t \le T_f \qquad (8)$$

In the static optimization, the optimal steady-state operating point is determined by:

$$(\bar{y},\bar{x},\bar{u}),\ \dot{x} = 0 \qquad (9)$$

In practice, the nominal trajectory definition step is normally separated from the controller design task. In the case of process control, it is known as the set-point optimization or supervisory control task. Optimal control theory and optimization theory are used to generate the nominal trajectory data and the nominal set-point commands to the control system (Ray and Szekely, 1973; Schuldt and Smith, 1971).

The integration of these functions for on-line variable trajectory generation and control is currently being pursued in aerospace scenarios (Comegys and Hanson, 1981; Jones, et al., 1981) and process control (Horn, 1978).

## Linearization of Process Dynamics

Designing a perturbation control system that will keep the dynamic process variables about the prescribed trajectory or an operating point under the influence of external disturbances, set-point changes, measurement noises, and system parameter uncertainties requires a perturbation model of the process.

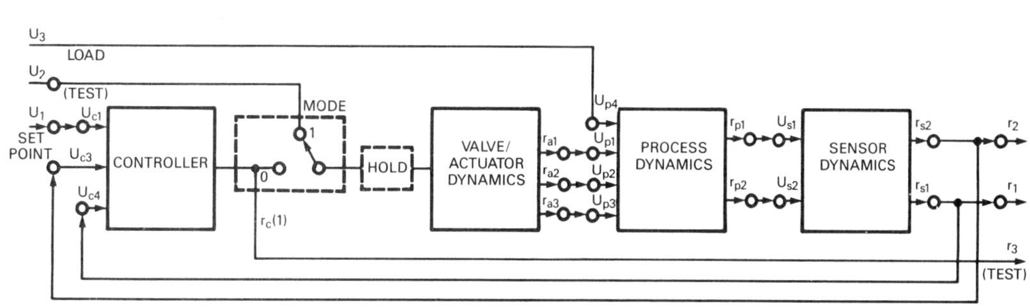


Fig. 4. Overall system block diagram.

The process dynamics given by Eqs. (2), (3), and (4) are linearized about the nominal trajectory defined by:

$$\dot{\bar{x}} = f(\dot{\bar{x}},\bar{y},\bar{x},\bar{u},\theta) \tag{10}$$

$$\bar{y} = g(\dot{\bar{x}},\bar{y},\bar{x},\bar{u},\theta) \tag{11}$$

$$\bar{r} = h(\dot{\bar{x}},\bar{y},\bar{x},\bar{u},\theta) \tag{12}$$

yielding:

$$\delta\dot{x} = F_{\dot{x}}\,\delta\dot{x} + F_y\,\delta y + F_x\,\delta x + F_u\,\delta u \tag{13}$$

$$\delta y = G_{\dot{x}}\,\delta\dot{x} + G_y\,\delta y + G_x\,\delta x + G_u\,\delta u \tag{14}$$

$$\delta r = H_{\dot{x}}\,\delta\dot{x} + H_y\,\delta y + H_x\,\delta x + H_u\,\delta u \tag{15}$$

The coefficient matrices in Eqs. (13), (14), and (15) are gradient matrices evaluated at the nominal point. These matrices are obtained either by analytical or numerical means. When a portion of the process variables is characterized by experimental tables, the numerical linearization becomes mandatory.

These equations subsequently are manipulated into the standard state-space form:

$$\dot{x} = A\,x + B\,u \tag{16}$$

$$r = C\,x + D\,u \tag{17}$$

where the set of four matrices (A,B,C,D) is called the quadruple data of the dynamic process (Konar, et al., 1976; Konar, et al., 1974). In general, the process quadruple is time varying. For control design, a subset of selected time points and the associated data are used. These finite points, referred to as operating conditions, represent the frozen-point linear process model within the operating envelope.

## Controller Concept Definition

The selection of controller concepts depends upon the nature of the process to be controlled, the effector and measurement points and qualities, the implementation considerations, and the controlled system performance requirements.

The synthesis of the actual control structure is often iterative and based on intuition and insight to select a feasible solution from the vast number of alternatives. Some algorithmic procedures to generate these alternatives were developed (Morari, et al., 1980; Govind and Powers, 1982).

The selection of the effectors (manipulated variables) and measurement points is also based on process knowledge. Algorithmic approaches for optimal sensor locations are given in Romagnoli and Stephanopoulos (1980).

Virtually all chemical processes are characterized by multiple inputs and outputs. In practice, the multivariable problem is decomposed into a series of single-input/single-output problems. To determine the best input/output pairings, several procedures were developed (Tung and Edgar, 1981).

For preliminary design, the optimal systems control in the form of a predictor and Kalman filter with state-estimated feedback as shown in Figure 5 provides a unified framework for design trade-off studies (Marshall, 1979; Kleinman, 1969).

Smith predictors, complementary control, self-tuning regulators and predictors, PID, feedforward, extended horizon, model algorithmic, dynamic matrix, internal model, two-port controllers and multivariable controllers are some of the frequently used concepts in the process control industry today (Stephanopoulos, 1984; Ogunnaike and Ray, 1979; Stephanopoulos and Huang, in press; Garcia and Morari, 1982; Joseph and Brosilow, 1978). Most often a control system must be modified to meet a set of steady-state requirements (i.e., temperature regulation for constant disturbance) by adding integrators, cross-feeds, and roll-off filters. If controllers are implemented using a digital processor, they are most often designed to operate in a multirate format due to different bandwidths of the process subsystems and finite word-length arithmetic operation (Konar, et al., 1981). At the end, the overall controller is characterized by transfer functions or differential (difference) equations and interconnections between them, as shown in Figure 3.

## Controller Optimization, Tuning, and Performance Evaluation

In the case of optimal controller configuration, automated tools have been developed for determining controller and estimator parameters over a period of time or for steady-state operations (Konar and Ward, 1972; Konar and Mahesh, 1977). For classical controller configurations, controller parameter tuning procedures are available (Luyben, 1973). Parameter tuning can also be accomplished by parameter optimization (Borow, et al., 1972).

Most often, design of complex processes requires a multiperformance evaluation capability to facilitate a quantitative analysis of all parameters that affect the design. Performance measures include: transient response, poles and zeros, singular values, open- and closed-loop frequency responses, stability margins, multivariable root locus, covariance response and power spectral density response. These responses are computed as functions of common design parameters such as gains, control filters/laws, noise, stochastic disturbances and operator inputs, multisample rates, word lengths, and process and

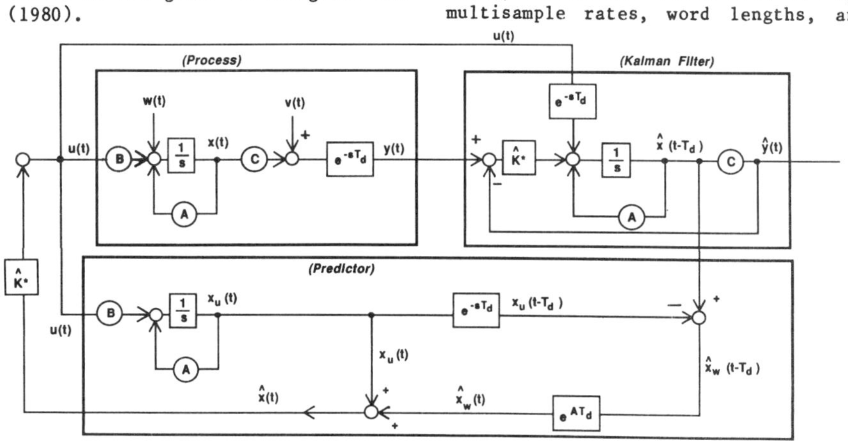


Fig. 5. Optimal feedback system for a process with transport delay.

computational delays. Final design is selected using the multiperformance data base generated within the operating envelope of the system (Konar, et al., 1974; Konar, 1973). The steps briefly presented here form the basis of the CAE software organization in process control.

## SOFTWARE DEVELOPMENT AND TRANSFER

Engineering software programs can be separated into two groups: a library of general-purpose subroutines and a set of structured programs with multiple analysis/design functions. The first group is used for solving one-shot problems. These programs provide maximum flexibility. The second group, often called software tools, is mainline oriented. These programs are equipped with a well-defined input/output interface for best use by nonprogrammer analysts. They are thoroughly tested and are expected to perform prescribed functions. There are many tools available, depending upon the need scenario as characterized in Figure 6.

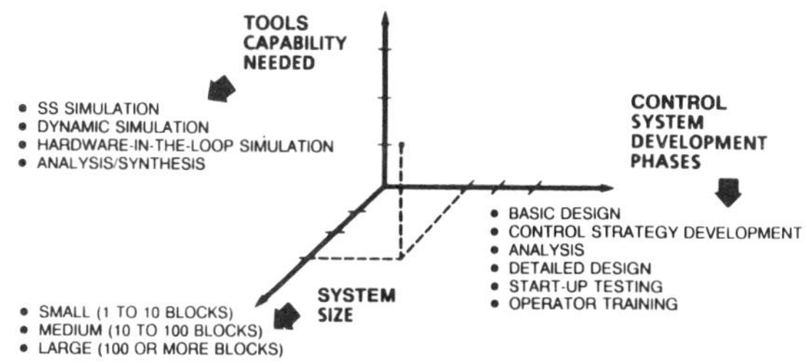


Fig. 6. Tools characterization space.

Their upgrading requires special teams. Because of their large size, documentation and users' manuals are essential for their operations. They require extensive resources to develop and maintain. The tool that is involved in this review belongs to this group.

The development of this class of software starts with establishing a set of software guidelines. A set of typical guidelines is given in Konar and Mahesh (1982). In parallel to this activity, the required analysis/design functions are spelled out. This step must draw from the working experience of senior designers/analysts and established design/analysis methods. The algorithms selected for the required functions must be based on the newest techniques available. The success and useful life of the tools developed depend greatly upon this continuous, long-range technology base development.

After individual sections of programs are developed and tested, the overall program is validated using a large-scale analysis/design test case. The reporting requirements must include a technical report summarizing the mathematical details of the algorithms and methodologies. Programs must be fully documented. This includes having program listings with comment cards explaining each major operation and argument. A users' manual must be prepared that gives typical demonstration examples.

To accommodate the transfer and use of these tools, on-site workshops for selected engineers must be conducted. The workshop periods and contents may vary. Workshops usually run 2 to 3 weeks. Typical contents of a CAE workshop for dynamic process modeling and control are shown in Figure 7.

These workshops not only train the users in the mechanics of interfacing with the tools, but more importantly explain the methodology of analysis and design and how and where the capabilities offered by the software are used for engineering efficiency and productivity.

## SIMULATION AND CONTROL PROGRAM (SIMKON)

The SIMKON Software System is structured to incorporate the methodology described above. SIMKON-IV is a knowledge-based system with: (1) models and data base; (2) problem-solving methods; (3) language understanding; and (4) image understanding. The set of algorithms used in SIMKON-IV is based on state-of-the-art simulation and control technologies as well as proven programs such as ADAPS, KONPACT and DIGIKON developed for aerospace and defense applications by Honeywell over a period of time (Konar, et al., 1974; Konar and Ward, 1972; Mahesh and Konar, 1983). Summaries of these programs are given in Konar and Mahesh (1982).

The software architecture and the functions implemented in SIMKON-IV and DIGIKON-IV are identical, the only difference being their dynamic model data bases; the former contains the industrial and chemical process models such as exothermic reactors, heat exchangers and so on, and the later houses the aerospace vehicle models such as the Space Shuttle, F-4 Tactical Fighters, etc.

**PART I. ANALYSIS (THEORY AND APPLICATIONS)**

A) DYNAMIC MODELING OR LARGE INDUSTRIAL SYSTEMS
- THERMAL/ENERGY SYSTEMS (NONLINEAR/LINEARIZED)
- PROCESS CONTROL SYSTEMS
- SIMPLE CONTROL LAWS MODELING (PID) DIGITAL ALGORITHMS/MODELING
- OVERAL SYSTEM MODELING
- MODEL REDUCTION PRINCIPLES

B) ANALYSIS TECHNIQUES
- POLES/ZEROS
- TRANSIENT RESPONSE
- FREQUENCY RESPONSE
- ROOT LOCUS
- RMS RESPONSE

**PART II. SYNTHESIS (THEORY AND APPLICATIONS)**
- STATIC OPTIMIZATION
- DYNAMIC OPTIMIZATION (TRAJECTORY CONTROL)
- MULTILEVEL OPTIMIZATION
- CLASSICAL SYNTHESIS METHODS (PID)
- QUADRATIC CONTROL AND ESTIMATION
- SYSTEM IDENTIFICATION CONCEPTS
- POLE PLACEMENT SYNTHESIS
- SINGULAR VALUES SYNTHESIS
- ADAPTIVE CONTROL
- CONTROL SYNTHESIS FOR SYSTEMS WITH LARGE TRANSPORT DELAYS

**PART III. SIMKON-V LAB**
- DESCRIPTION OF SIMKON-V
- SELECTED LAB PROBLEMS

Fig. 7. CAE workshop outline for dynamic process modeling and control.

DIGIKON represents an advanced, state-of-the-art tool for modeling and analyzing digital and continuous control systems. Since its first development in 1974 by Honeywell (Konar, et al., 1974), it has been constantly upgraded through various contracts from government and industry.

Interactive DIGIKON (DIGIKON-IV) is the most recent aerospace version, and it has been developed to fully utilize the interactive features of modern computers to reduce the engineer's time in modeling and analyzing continuous and digital control systems. It is currently operational on CDC Cyber computers (NOS), on the PRIME computer at the NASA Langley Research Center, and on the VAX system at Honeywell (Mahesh and Konar, 1983). Its process industry version (SIMKON-IV) is currently operational on the Honeywell MULTICS network.

The SIMKON Software System illustrated in Figure 8 uses the modern state-space approach for modeling dynamic control systems and analyzes them with respect to various performance measures. It facilitates a quantitative analysis of all parameters that affect performance and/or stability of control systems. Performance includes control and disturbance covariance response, transient response, and frequency response. Stability includes eigenvalues, root locus, gain and phase margins and singular values. The parameters include sample rate, word length, computational delays, multisample rates, control filters/laws, system bandwidth, measurement noise, and stochastic load and set-point inputs. The analysis capability of SIMKON can be used iteratively to tune control system design parameters. The optimal control and estimation capability can be used for initial benchmark designs. Dynamic modeling can be used to extract process models from the batch or on-line data.

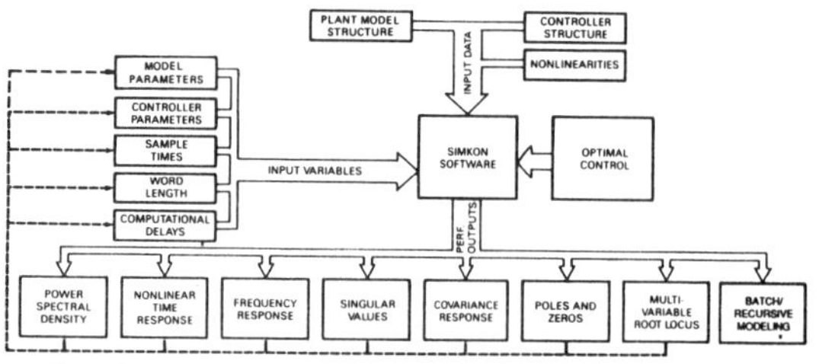


Fig. 8. SIMKON software system.

In general, an interactive command-based approach to computer-aided engineering enables the designer to interact with the computer directly, flexibly, and in apparent real time. The following facilities are usually provided:

- A graphics terminal;
- A software library of design, analysis and modeling programs;
- A high-level command language;
- Software for editing and macro command generation.

The terminal provides real-time computer access and control with effective graphic communication of design status, results, and trade-offs. It is usually augmented with hardcopy equipment and off-line printing facilities to provide permanent records of the communication. The subroutine library contains all the analysis/design algorithms and models needed for a specified class of engineering problems. The command language provides a simple means to invoke these high-level routines in any user-desired sequence without concern for input/output, data handling, internal data structures, or numerics. Finally, the editing and macro capability provides a way to define, save, and execute fixed sequences of user-specified commands. In effect, a user can create higher level commands suitable for his own design task, enhancing the power of the basic command language.

SIMKON employs a command-based approach for interaction between the user and the computer, providing an efficient use of the program by the experienced user. Help commands and prompting by the program for missing parameters provide sufficient assistance to the beginning user.

Maintenance of the modeling files is facilitated by means of a direct-access data base and the line editor developed for the SIMKON software.

Figure 9 illustrates the general organization of the SIMKON-IV program. The main executive handles the interaction between the user and the various programs. The various programs, in addition to communicating with the user, communicate directly with the various data files. The user can execute any program by typing in the appropriate command. Major commands are listed in Table 1. Data and/or options can be specified as arguments to the command. A complete list of command descriptions is given in Konar, et al., (1983). The main executive parses the command line, separating the command from the arguments, and passes the arguments to the corresponding program (as specified by the command).

The various programs, when called by the main executive, inspect the arguments to satisfy other data requirements or to set the option flags to appropriate values. Any missing parameter of additional data required by the program is prompted for directly through the terminal to the user. The user-entered data at the terminal is also read directly by the corresponding program. In addition, the program may need data from the appropriate data files to satisfy the data requirements. Once the computations/data manipulations are completed, the requested output is displayed on the terminal. In addition, the program writes any required data on appropriate data files. The execution control is then returned to the main executive, which prompts the user for the next command.

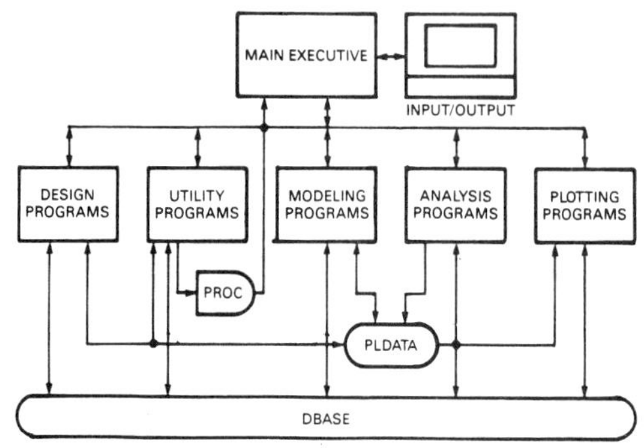


Fig. 9. Organization of SIMKON-IV program.

Table 1 List of SIMKON-IV Commands

| Type | Command | Description |
|---|---|---|
| Modeling | CDEM | Compute differential equation model |
| | CTFM | Compute transfer function model |
| | RQDM | Read quadruple data model |
| | CQDM | Compute quadruple data model |
| | CICM | Compute interconnection model |
| | CBDM | Compute block diagram model |
| | CCDM | Compute conditioned model |
| | CZT | Compute Z-transform model |
| | CTT | Compute Tustin transform model |
| | CMZT | Compute modified Z-transform model |
| | CZWT | Compute Z to W transform model |
| | CWZT | Compute W to Z transform model |
| | CWLT | Compute word length truncated model |
| | CVGM | Compute variable gain model |
| | CCM | Compute controller model |
| | CQSA | Compute quadruples for sensitivity analysis |
| | CNSM | Compute nonlinear simulation |
| | CMP | Compute model parameters |
| Analysis | CPZR | Compute pole zero response |
| | CFR | Compute frequency response |
| | CNFR | Compute nonlinear frequency response |
| | CLCA | Compute limit cycle analysis |
| | CTR | Compute time response |
| | CNTR | Compute nonlinear time response |
| | CRL | Compute root locus |
| | UDRL | Update root locus |
| | CGRL | Compute generalized root locus |
| | CNLC | Compute nonlinear covariance response |
| | CSVD | Compute singular value decomposition |
| | CPSD | Compute power spectral density |
| Design | CDNM | Compute design model |
| | CSFG | Compute state feedback gains |
| | CKEG | Compute Kalman estimator gains |
| | CSFS | Compute state feedback system |
| | COES | Compute optimal estimator system |
| | COS | Compute optimal system with estimator |
| | CCOV | Compute covariance analysis |
| | CCOR | Compute correlation analysis |
| | RSFG | Reset feedback gains |
| Plotting | PFR | Plot frequency response |
| | PTR | Plot time response |
| | PRL | Plot root locus |
| | PCR | Plot covariance response |
| | PSV | Plot singular values |
| | PSD | Plot power spectral density |
| Utility | EDIT, FN | Edit data file FN |
| | PRINT, FN | Print data file FN |
| | EXEC, FN | Executive data file FN |
| | HELP, CMD | Help with command CMD |
| | DELETE, FN | Delete data file FN |
| | LIST, DT | List data file names of type DT (DT can be PROC or HELP or FREQ or SIMU or ROOT) |
| | MSREAD | Read from procedure file JS into DBASE |
| | MSWRITE | Write from DBASE onto procedure file JS |
| | INIT | Initialize data base |
| | NAME, SYSN | Print name list data for system SYSN |
| | QUAD, SYSN | Print quadruple data for system SYSN |
| | RGPT | Read global plot title |
| | WGPT | Write global plot title |
| | FORMAT, OPTN | Define output format (OPTN can be batch or INTERACTIVE or NONE) |
| | BAUD, N | Change baud rate to N (default value for N is 1200) |
| | STATUS | Print status of global plot title, format, and baud rate |

The utility program facilitates storing command files and models on a direct-access data file. Once stored, this sequence of commands and models can be executed repeatedly. Also, the contents of any data file can be edited by the utility program. This provides the SIMKON-IV program with a unique capability to readily modify the system description and execute. The utility program also maintains a list of help files. Information on any specific command can be obtained by using the help command.

The modeling program obtains the quadruple data for each subsystem from physical descriptions. These descriptions can be in the form of differential (difference) equations, as given by Eqs. (13), (14), and (15), or continuous (discrete) transfer functions. The modeling program also performs parameter estimation with batch or recursive data. Arbitrary interconnections of the subsystems are allowed to form the overall control system. The decomposition of large dynamic systems into several subsystems and interconnections between them facilitates the modeling of large-scale, complex systems.

Transformations in s, z, and w planes are obtained using the system quadruples to accommodate sample rates, computational delays, and performance evaluation. The quadruple approach to internal computations results in high numerical accuracy and computing speed.

The analysis program evaluates the performance of the overall system. Basic capabilities are: poles and zeros, frequency response, singular values, covariance response, power spectral density and multivariable root locus and time response.

The multivariable root locus is a unique capability for sensitivity analysis of the system to changes in the system parameters. Covariance response and power spectral density analyses are used for evaluating the performance of the system to flow noise, stochastic ambient, and operator inputs.

The design program provides optimal controller and estimator synthesis for multivariable process control benchmark designs. It also performs parameter optimization based on the smart search process. Table 2 summarizes the SIMKON-IV capabilities.

## SIMKON-IV DEMONSTRATION EXAMPLE

SIMKON may be used in the following CAE areas:

- Integrated process design (control configured);
- Controller concept studies/parameter tuning prior to field test;
- Controller implementation checkout and validation;
- Detailed point design of complex control laws;
- Simulation of complex dynamics;
- Dynamic model building;
- Adaptive, filtering, prediction, and control.

Table 2 SIMKON-IV Capabilities

1. COUPLED, NONLINEAR AND LINEAR DIFFERENTIAL EQUATION INPUT
2. MULTIPLE TRANSFER FUNCTIONS INPUT
3. GENERALIZED INTERCONNECTIONS INPUT
4. MULTISAMPLE RATE CONTROL SYSTEM STRUCTURE
5. PROCESS TRANSPORT AND COMPUTATIONAL DELAYS
6. LARGE SYSTEMS
7. POLES, ZEROS, PSEUDO ZEROS, SINGULAR VALUES
8. FREQUENCY RESPONSE IN s,z,w PLANES
9. TRANSIENT RESPONSE AND NONLINEAR SIMULATION
10. RMS RESPONSE, POWER SPECTRAL DENSITY AND POWER, NONLINEAR COVARIANCE ANALYSIS
11. MULTIVARIABLE ROOT LOCUS, SENSITIVITY ANALYSIS
12. INTERACTIVE PARAMETER OPTIMIZATION
13. OPTIMAL CONTROL AND ESTIMATION SYNTHESIS
14. BATCH/RECURSIVE PARAMETER ESTIMATION

The following three examples illustrate typical SIMKON-IV files and outputs. Figure 10 shows a simple closed-loop system characterized by transfer functions. If the user wants to develop step and frequency responses for this system, he may build a file as shown in Figure 11. Statements 1 through 13 develop the model; statements 14 through 16 develop the responses. Figures 12 and 13 show, respectively, the step and frequency response outputs of the system when this file is executed.

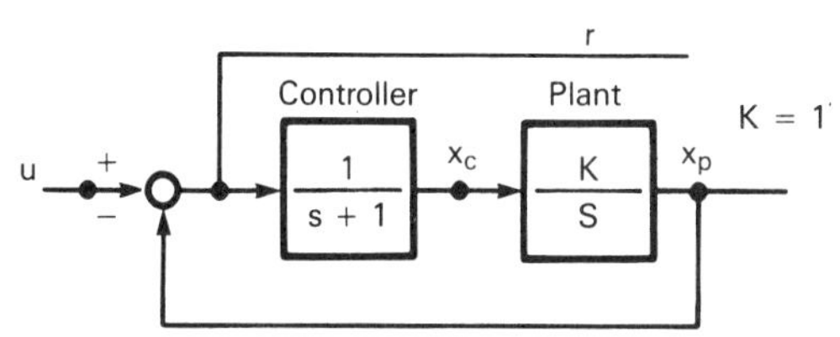


Fig. 10. Simple closed-loop system.

```
File = prob21    in USER    Data Base

 1  rgpt
 2  simple closed-loop system example
 3  ctfm(ex1,print)
 4  hs(1)=1/s
 5  hs(2)=1/s+1
 6  end
 7  ui(1)=ri(2)
 8  ui(2)=-ri(1)+u(1)
 9  r(1)=-ri(1)+u(1)
10  end
11  r1,output
12  u1,input
13  end
14  ctr(ex1,.05,.1,10,u1=.95)
15  ptr(p,output,input)
16  cfr(ex1,input,output,.1,10)

enter command
```

Fig. 11. Command file for simple closed-loop system.

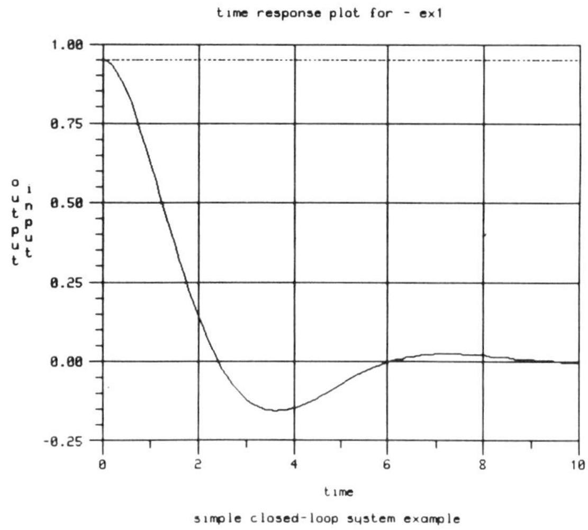


Fig. 12. Step response output.

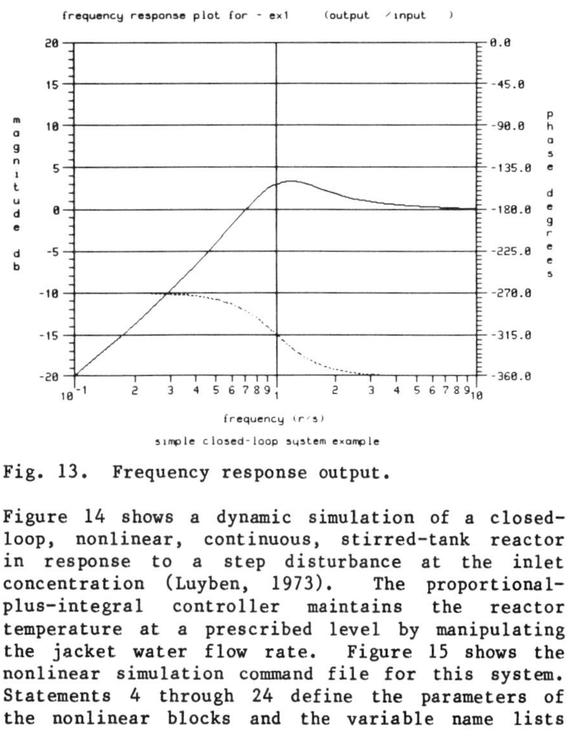


Fig. 13. Frequency response output.

Figure 14 shows a dynamic simulation of a closed-loop, nonlinear, continuous, stirred-tank reactor in response to a step disturbance at the inlet concentration (Luyben, 1973). The proportional-plus-integral controller maintains the reactor temperature at a prescribed level by manipulating the jacket water flow rate. Figure 15 shows the nonlinear simulation command file for this system. Statements 4 through 24 define the parameters of the nonlinear blocks and the variable name lists for these blocks. Statements 25 through 33 define the interconnections between the three blocks of the simulation; statement 34 provides initial conditions; statement 36 requests linearization of the CSTR at two time points; and statements 40 and 41 ask for linear quadruple data to be printed out. The rest of the statements are for the time plots of the variables. Statement 48 is for the cross-plot of the kinetic coefficient as a function of the reactor temperature. Figure 16 shows the inlet and reactor concentrations, and Figure 17 shows the inlet and reactor temperatures. Figure 18 shows the exothermic reaction rate, and Figure 19 shows the cross-plot of the kinetic coefficient as a function of the reactor temperature. The significance of this plot is that for small temperature swings, the kinetic coefficient is a linear function of the temperature, displaying the validity of local linearized models. Figure 20 shows the flow rates, and finally, Figure 21 shows the derivatives of the reactor states. Figure 22 shows the state diagram of the linearized CSTR dynamics. Figure 23 shows the poles and zeros of this linear system. Note that it is unstable as expected. Figure 24 shows the root locus of the closed-loop system with the loop gain varying. Observe that for gain in the neighborhood of 1.8, the closed loop is critically damped. Figure 25 shows the closed-loop poles and zeros, and Figure 26 shows the flow rates with this gain setting. It displays a critically damped behavior.

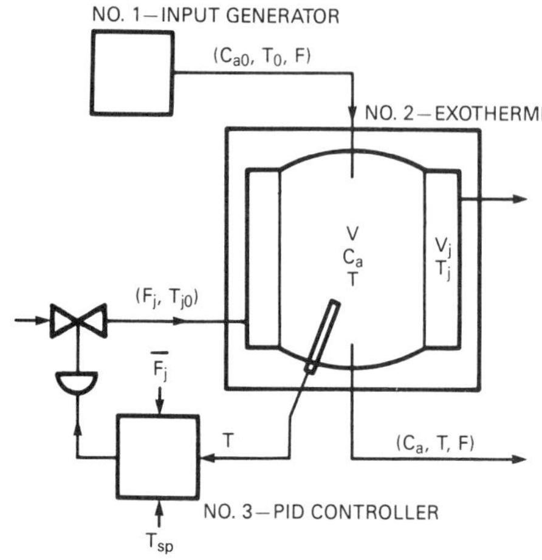


Fig, 14. CSTR control system block diagram.

```
        File = clsyspi  in USER   Data Base     02/25/86  1153.4

  1  rgpt
  2  non-linear cstr system
  3  cnsm(clsyspi,sim2,sim7,pic,.005,.01,8)
  4  step(.50,0,.1,2)
  5  end
  6  r1,ca0,moles/ft3
  7  end
  8  cp(n=1.0)
  9  end
 10  u1,ca0,moles/ft3
 11  u2,tr0,r
 12  u3,f,ft3/hr
 13  u4,tj0,r
 14  u5,fj,ft3/hr
 15  r1,ca,moles/ft3
 16  r2,tr,r
 17  r3,tj,r
 18  r4,k,/hr
 19  r5,qj,btu/hr
 20  r6,qg,btu/hr
clsyspi  c/r = continue, s=stop
 21  r7,cadot,moles/ft3 hr
 22  r8,trdot,r/hr
 23  r9,tjdot,r/hr
 24  end
 25  ui2(1)=ri1(1)
 26  ui2(2)=u(1)
 27  ui2(3)=u(2)
 28  ui2(4)=u(3)
 29  ui2(5)=ri3(1)
 30  ui3(1)=u(4)
 31  ui3(2)=u(5)
 32  ui3(3)=ri2(2)
 33  end
 34  ic(u1=530,u2=40,u3=530,u4=49.9,u5=600,x1=.245,x2=600,x3=594.6)
 35  end
 36  lin(sim7,.001,3)
 37  end
 38  end
 39  end
 40  quad,nlcs1
clsyspi  c/r = continue, s=stop
 41  quad,nlcs2
 42  mptr(p,ca,ca0)
 43  mptr(p,tr,tr0)
 44  mptr(p,tj,tj0)
 45  mptr(p,f,fj)
 46  mptr(p,qg,qj)
 47  mptr(p,k)
 48  mptr(cp,tr,k)
 49  mptr(p,cadot,trdot,tjdot)
 50  mptr(p,fint)

enter command
```

Fig. 15. SIMKON-IV nonlinear simulation command file.

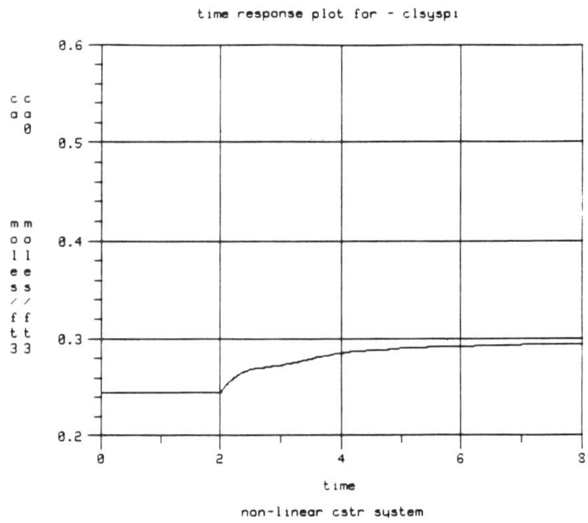


Fig. 16 Inlet and reactor concentrations.

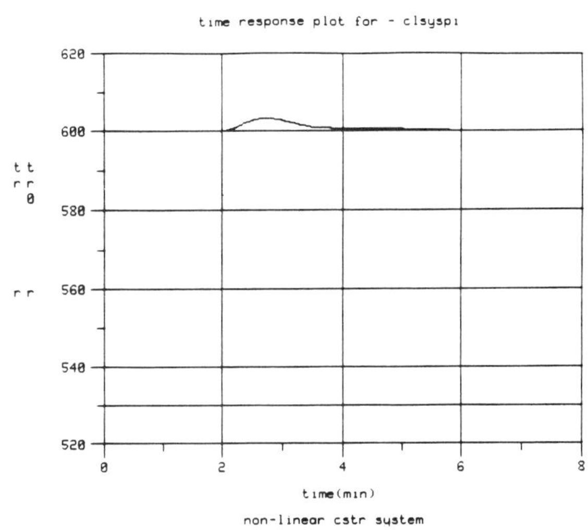


Fig. 17. Inlet and reactor temperatures.

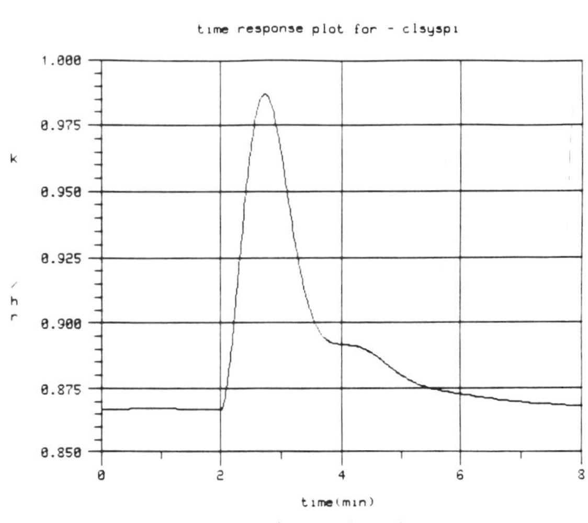


Fig. 18. Exothermic reaction rate.

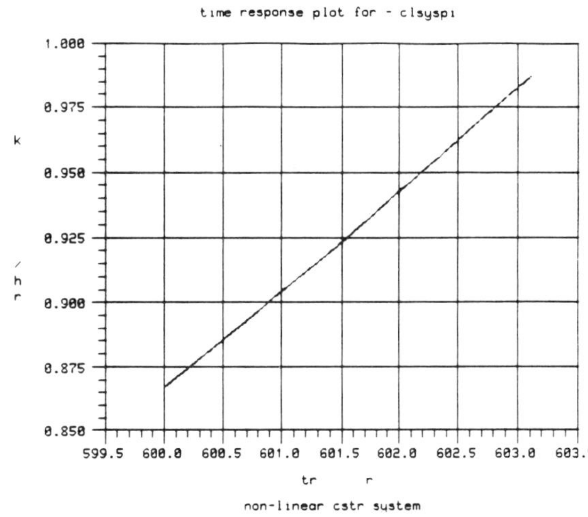


Fig. 19. Reaction rate versus reactor temperature.

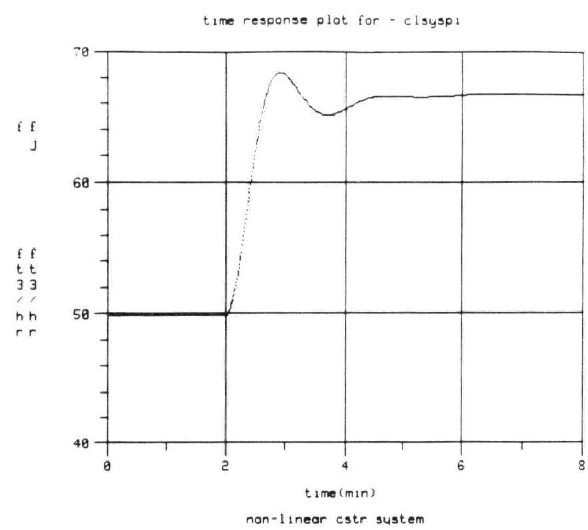


Fig. 20. Flow rates.

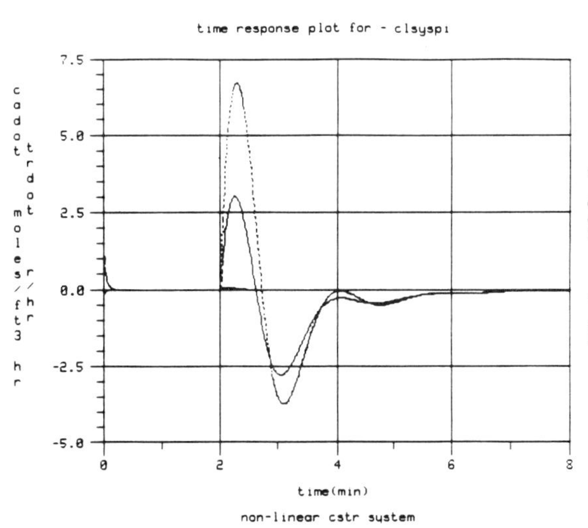


Fig. 21. Derivatives of the reactor states.

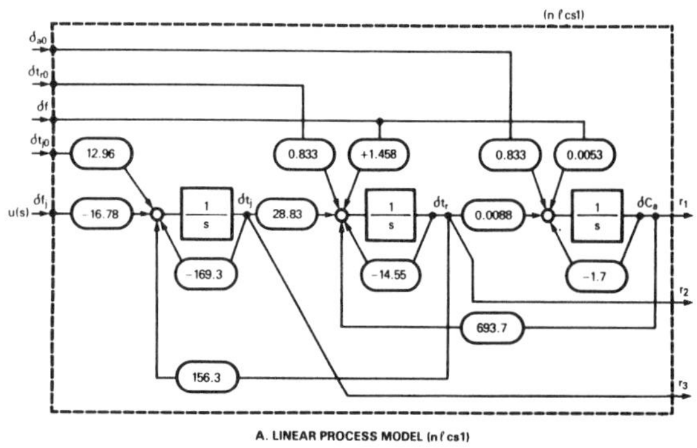


Fig. 22. Linearized CSTR dynamics state diagram.

```
*** s-plane poles of the system for - nlcs1   ***

 no     real          imag          damping       freq

  1   -188.1       0.0000E+00
  2   -.5320       0.0000E+00
  3    3.049       0.0000E+00

Press "page" c/r to continue

*** s-plane zeros for - nlcs1   (r2        /u5        ) ***

 no     real          imag          damping       freq

  1   -1.700       0.0000E+00

Press "page" c/r to continue
```

Fig. 23. Poles and zeros of the linearized CSTR.

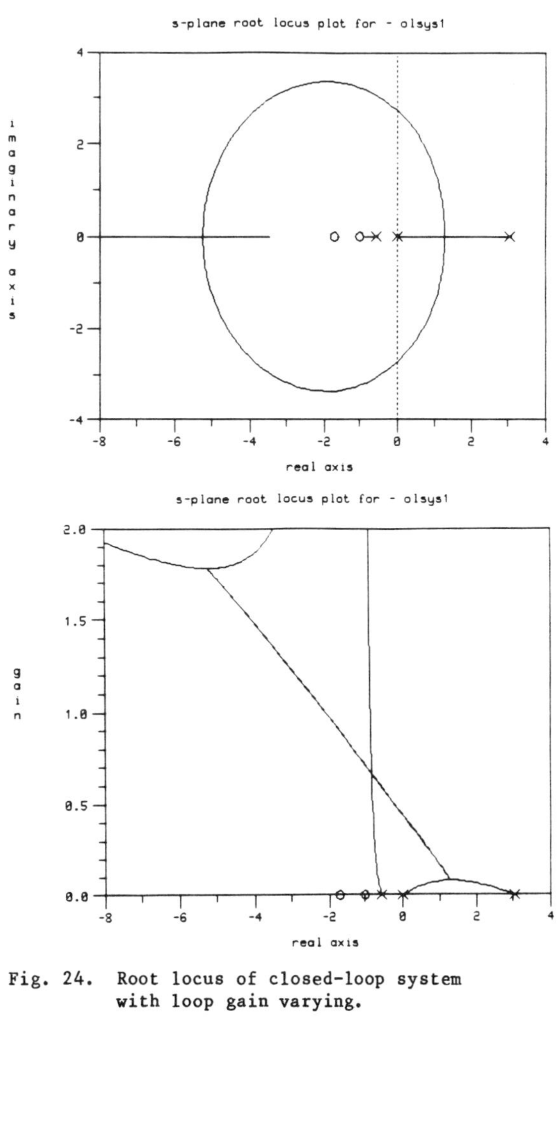


Fig. 24. Root locus of closed-loop system with loop gain varying.

```
*** s-plane poles of the system for - clsys11  ***

 no     real          imag          damping       freq

  1   -174.2       0.0000E+00
  2   -5.340       0.0000E+00
  3   -5.140       0.0000E+00
  4   -.8861       0.0000E+00

Press "page" c/r to continue

*** s-plane zeros for - clsys11 (r2        /u6        ) ***

 no     real          imag          damping       freq

  1   -1.700       0.0000E+00
  2   -1.000       0.0000E+00

Press "page" c/r to continue
```

Fig. 25. Closed-loop system poles and zeros.

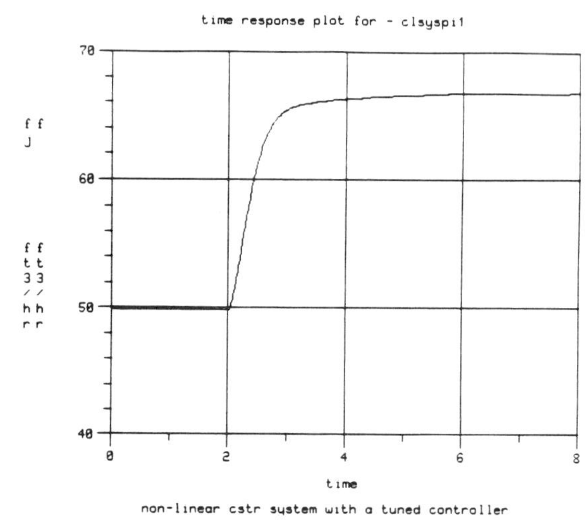


Fig. 26. Flow rates with the tuned controller.

The third example is on a simulation of an adaptive control system. This system is shown in Figure 27. It consists of three blocks: the instantaneous heat exchanger, the autotuning controller, and the set-point input generator.

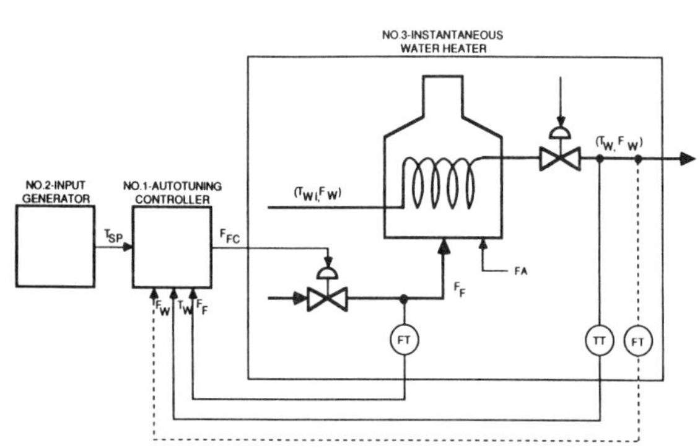


Fig. 27. Heat exchanger system block diagram.

The process model is assumed to be a first-order lag and transport delay:

$$G_p(s) = \frac{T_w}{Q} = \frac{K_p e^{-sT_d}}{T_p s+1} \tag{18}$$

The process gain $K_p$, the time constant $T_p$, and the transport delay $T_d$ are all functions of the hot water demand $F_w$. The water temperature is controlled by manipulating the fuel flow rate $F_f$.

The criteria of control is to have a dominant response with a given time constant and no overshoot. The internal model in the controller is taken to be:

$$G_m(s) = \frac{(\theta_{m2}+\theta_{m3}z^{-1}+\theta_{m4}z^{-2}+\theta_{m5}z^{-3}+\theta_{m6}z^{-4})}{(z-\theta_{m1})} \quad (19)$$

This represents the discrete heater transfer function model with a variable transport delay. The model parameters $\theta_{mi}$, i = 1,..6 are determined on-line by the parameter identifier, and the controller is tuned automatically.

For this simulation, the process parameters, the closed-loop system time constant, and the sample time are taken to be $K_p$ = 1 °C/liter/sec, $T_p$ = 10 sec, Td = 4 sec, Tcl = 2.5 sec, and $T_s$ = 1 sec. These yield the following discrete transfer function of the process:

$$G_p(z) = \frac{0.1\ z^{-4}}{(z - 0.9)} \quad (20)$$

which corresponds to:

$$\theta_{m1} = 0.9, \theta_{m2}=\theta_{m3}=\theta_{m4}=\theta_{m5}= 0, \text{ and } \theta_{m6} = 0.1 \quad (21)$$

Figure 28 shows the command file for the discrete nonlinear system simulation.

```
        File = htrsys1   in USER   Data Base     02/24/86  1457.4 cst

  1  ctfm(dheater,t=1)
  2  hs(1)=.1/s-.9
  3  hs(2)=1/s**4
  4  end
  5  ui(1)=ri(2)
  6  ui(2)=u(1)
  7  r(1)=ri(1)
  8  r(2)=u(1)
  9  end
 10  r1,tempw,deg c
 11  r2,fuelr,kcal/sec
 12  end
 13  rgpt
 14  PZP AutoTuner(tcl=2.5,th1=.9,th2=0.,th3=0.,th4=0.,th5=0.,th6=.1)
 15  cnsm(htrsys,sim6,sim2,dheater,1,1,50)
 16  cp(pi=10000.,thi1=.5,thi2=0.,thi3=0.,thi4=0.,thi5=0.,thi6=.5)
 17  cp(nd=1,nn=5,tcl=2.5)
 18  end
 19  r1,tempw,deg c
 20  r2,theta1m
htrsys1  c/r = continue, s=stop
 21  r3,theta2m
 22  r4,theta3m
 23  r5,theta4m
 24  r6,theta5m
 25  r7,theta6m
 26  r8,tempm,deg c
 27  r9,p11
 28  r10,p22
 29  r11,p33
 30  r12,p44
 31  r13,p55
 32  r14,p66
 33  r15,k1
 34  r16,tempsp,deg c
 35  r17,kn
 36  r18,temperr,deg c
 37  r19,yfb
 38  r20,fuelrc,kcal/sec
 39  end
 40  step(.99,0.,1.0,20.)
htrsys1  c/r = continue, s=stop
 41  end
 42  r1,tempspg,deg c
 43  end
 44  ui1(1)=ri3(1)
 45  ui1(2)=ri3(2)
 46  ui1(3)=ri2(1)
 47  ui3(1)=ri1(20)
 48  end
 49  ic(u1=1)
 50  end
 51  end
 52  ptr(p,tempw,tempm,tempspg)
 53  ptr(p,fuelrc,tempspg)
 54  ptr(p,theta1m,theta2m,theta3m)
 55  ptr(p,theta4m,theta5m,theta6m)

enter command
```

Fig. 28. Command file for adaptive water heater control system

Statements 1 through 12 define the discrete process model; statements 15 through 43 define the input parameters and name lists for the system blocks; and statements 44 through 47 define interconnection equations for the internal inputs.

Figure 29 shows the water temperature, model temperature, and set-point temperature input profiles. Figure 30 shows the fuel rate command and the set-point temperature input profiles.

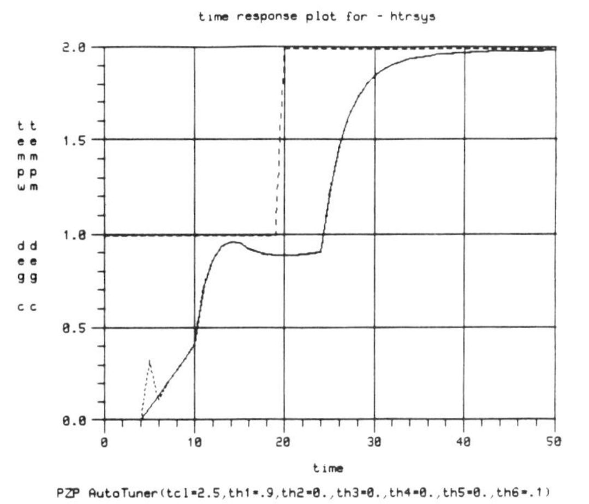


Fig. 29. Water temperature, model temperature, and set-point temperature input profiles.

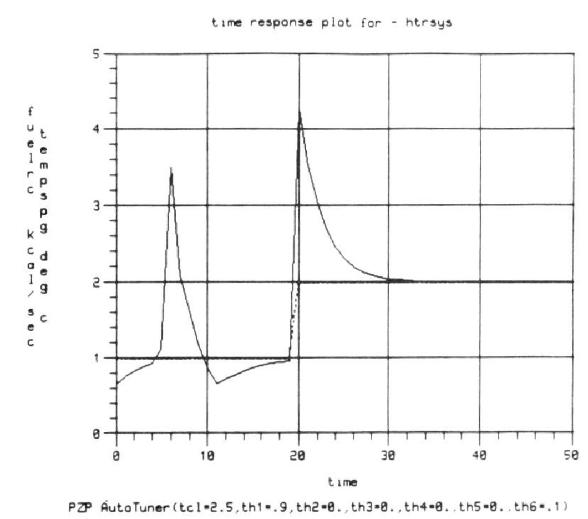


Fig. 30. Fuel rate command and set-point temperature input profiles.

Figures 31 and 32 show the model parameters determined from the sampled input/output data of the heater. Note that the model parameter values approach the correct values starting from the given initial condition. Also, the closed-loop temperature response after 20 seconds of learning time is excellent.

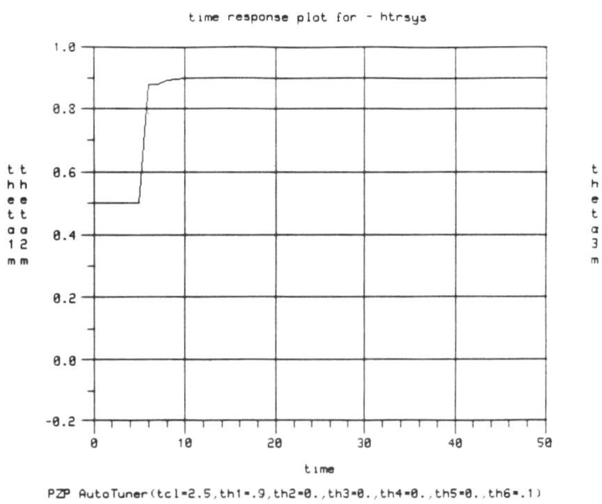


Fig. 31. Time response plot for th1m, th2m, th3m.

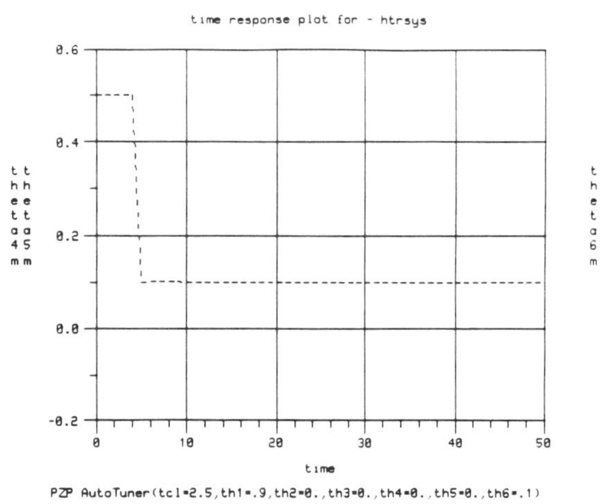


Fig. 32. Time response plot for th4m, th5m, th6m.

As Figures 11, 15, and 28 show, input of the dynamic system models to the SIMKON-IV system is relatively straightforward. Further enhancements are possible using the interactive graphics capabilities available in various engineering work stations. One such input permits direct dynamic system description via block diagram and is based on the Apollo work station. As shown in Figure 33, the Mentor graphics package is utilized to draw and extract the net list data for this system. Subsequently these net list data are utilized on the SIMKON-IV system to extract the modeling data.

## CURRENT ACTIVITIES

We are currently involved in three main activities: capability additions, user friendliness enhancement, and technology transfer to user organizations.

User friendliness is being enhanced by integrating the modeling input capability via an interactive graphics system with SIMKON.

For effective transfer of this tool, on-site training and technology transfer workshops are being offered to interested organizations.

## CONCLUSION

In this paper, an overview of a process control design procedure and the relevant interactive computer-aided design tool, SIMKON-IV, are

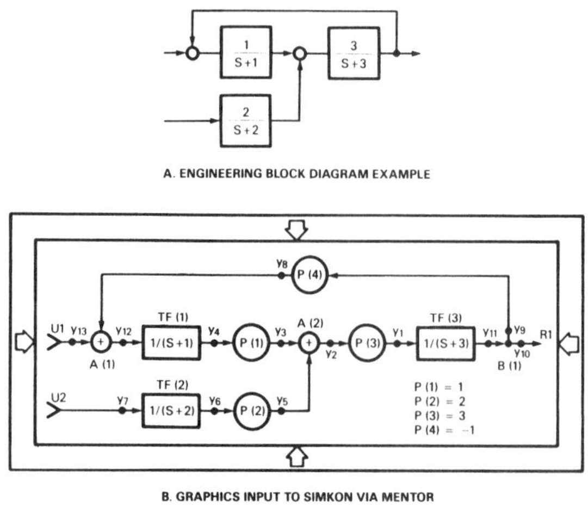


Fig. 33. Graphics input to SIMKON system.

represented. The significance of this development is that the user is provided with the total controller design capability under one unifying interactive command structure. The richness of its command repertoire and the advanced algorithms it uses for implementing its functions make SIMKON-IV one of the most complete and advanced system engineering tools for control applications in the process industry.

## ACKNOWLEDGEMENT

Several investigators have contributed to the results presented here. The author particularly would like to thank J.K. Mahesh and M.D. Ward for their continued efforts in implementing the interactive SIMKON programs. The author also would like to acknowledge fruitful dicussions with his colleagues, E.B. Lee, G. Stein, and C.B. Harwey.

## REFERENCES

Borow, M.S., et al. (1972). Navy digital flight control system development. Honeywell report 21857-FR GAP.

Comegys, G.L., and C.R. Hanson (1981). Integrated flight trajectory control development program. AFWAL-TR-81-3077.

Garcia, C.E., and M. Morari (1982). Internal model control. Ind. Eng. Chem. Process Des. Dev., 21.

Govind, R., and G.J. Powers (1982). Control system synthesis strategies. AIChE Journal, 28 (1).

Horn, B.C. (1978). On-line optimization of plant utilities. Chemical Engineering Progress, pp. 76-79.

Jones, A.L., et al. (1981). Integrated control design techniques. AFWAL-TR-81-3074.

Joseph, B., and C.B. Brosilow (1978). Inferential control of processes. AIChE Journal, 24 (3).

Kleinman, D.L. (1969). Optimal control of linear systems with time-delay and observation noise. IEEE Trans., Auto. Cont.

Konar, A.F. (1973). Precision weapon delivery. Trans. First NWC Symposium on the Application of Control Theory to Modern Weapon Systems.

Konar, A.F. (1985). Simulation and control workshop notes. Honeywell Technology Strategy Center.

Konar, A.F., E. Adali, and F. Tunali (1981). Multirate Multiloop Digital Control Systems Analysis. Technical University of Istanbul Press.

Konar, A.F., and J.K. Mahesh (1977). KONPACT program manual. Honeywell reports 77SRC42 and 77SRC43.

Konar, A.F., and J.K. Mahesh (1982). Computer-aided analysis and design of large-scale control systems. IFAC Workshop Proceedings, Ankara, Turkey.

Konar, A.F., J.K. Mahesh, and R.D. Jacobson (1983). SIMKON command description list. Honeywell Corporate Technology Center.

Konar, A.F., J.K. Mahesh, and B. Kizilos (1974). Digital flight control systems for tactical fighters. AFFDL-TR-73-119.

Konar, A.F., J.K. Mahesh, C.R. Stone, and M. Hank (1976). Active control synthesis for flexible vehicles. AFFDL-TR- 75-146, Vol. I, II, III.

Konar, A.F., and M.D. Ward (1972). Development of weapon delivery models and analysis programs. AFFDL-TR-71-123, Vol. I, II, III.

Luyben, W.L. (1973). Process Modeling, Simulation, and Control for Chemical Engineers. McGraw-Hill Co., New York.

Mahesh, J.K., and A.F. Konar (1983). Interactive flight control system analysis program. NASA contract report NAS 1-16438, Vol. I, II, III, IV.

Marshall, J.E. (1979). Control of time-delay systems. IEE Control Engineering, Series 10.

Morari, M., Yaman Arkun, and G. Stephanopoulos (1980). Studies in the synthesis of control structures for chemical processes. AIChE Journal, 26 (2).

Ogunnaike, B.A., and W.H. Ray (1979). Multivariable controller design for linear systems having multiple time delays. AIChE Journal, 25, No. .

Ray, H.W., and J. Szekely (1973). Process Optimization. John Wiley & Sons, New York.

Romagnoli, J.A., and G. Stephanopoulos (1980). On the rectification of measurement errors for complex chemical plants. Chem. Eng. Sci., 35 (5).

Schuldt, S.V., and F.B. Smith (1971). An application of quadratic performance synthesis techniques to a fluid cat cracker. J.A.C.C., Paper No. 3-E4.

Stephanopoulos, G. (1984). Chemical Process Control. Prentice Hall, New Jersey.

Stephanopoulos, G., and Hsiao-Ping Huang (in press). The 2-port control system. Chemical Eng. Sci.

Tung, L.S., and T.F. Edgar (1981). Analysis of ccontrol-output interactions in dynamic systems. AIChe, 27 (4).

# THE THREE WAVES OF AUTOMATIC CONTROL

†M. Cuénod

*Electrowatt Engineering Services Ltd., Zürich, Switzerland*

Abstract. This survey paper gives an overview on the 3 waves which can be recognized in the development of automatic control:

1st wave : Analog control loops.

2nd wave : Direct Digital Control (DDC) with a centralized process computer.

3rd wave : Distributed Control, using mini and microcomputers.

For each of these waves this paper sketches the main research topics and applications.

The conclusion points out the actual gap between the high capacity of the tools for control and the use which is made of these tools, and a challenging task: to reduce this gap.

## INTRODUCTION

I started my career as engineer in the large factory of Brown-Boveri in Baden, Switzerland, in the department of the mercury arc rectifiers, which has been completely cancelled now. One engineer who became afterwards one of the manager of BBC told me: "It is of no consequence for your career what your first job in a firm like BBC will be. Select your field of interest. Start to work in this field on your own. You will see it will become very soon your only field of activity". And he added: "Automatic control could be a good topic for you".

I started to learn the basis of what the automatic control theory was at that time. A few months later I was transferred to department for the development of control systems. I was associated with the development of the very first power system control, and, among others, I had the privilege to collaborate with Mr.Georges Darrieus, who developped the basic concepts of the load-frequency control. This start decided of the orientation of my career and of my special field of interest which is still the application of automatic control in energy systems, and specially in electrical power systems.

With some friends we founded the Swiss Association for Automatic Control, and I had the chance to persuade Professor Gerecke who was my PHD advisor to become its first President. Together we were at the Conference in Heidelberg in 1956, where the idea of the foundation of the International Federation of Automatic Control (IFAC) came up. Together we were in Versailles in 1958 for the foundation of IFAC. On the occasion of the first IFAC Congress in Moscow in 1960, professor Gerecke became vice-president, and later President of IFAC, and I became a member of its Executive Council as Treasurer during 20 years.

It gave me the opportunity to be deeply involved with very rapidly developping automatic control theory and applications. One privilege, at the end of my career, is to have been able to follow this evolution and to give me to-day the opportunity to make a survey of the 3 waves which can be recognized in the development of automatic control:

1st wave : Analog Control Loop.

2nd wave : Direct Digital Control (DDC), with a centralized process computer.

3rd wave : Distributed Control, using mini- and micro-computers.

I would also like to remind you the name of some of the pioneers of the automatic control theory and applications that I have had the privilege to know personally, and to learn from their research and development what I shall now share with you.

### 1st wave. Analog Control Loops.

This wave took place between 1945 and 1964, and is characterized by the analysis of the dynamic behaviour of the control loop, acting continuously, which is defined as the analog procedure.

The main purpose of this analysis was the determination of the stability conditions, and the comparison of the advantages and disadvantages of the different stability criteria, like Hurwitz, Nyquist, Leonhard, Bode, etc, and the implementation of stabilizing devices, like PID control elements.

Very quickly it was recognized that it was not enough to obtain a stable behaviour of the control loop. It was also required that the controlled variable be kept within prescribed limits with regards to its set point value, in spite of the perturbation acting on the controlled system. A link was established between the control theory and the signal transmission theory. The controlled system was identified to a filter between the fluctuation of the many perturbations acting on this system, and having in general a stochastic character, and the controlled variable to be kept constant. The behaviour of this filter was investigated by the use of the Laplace transform and of autocorrelation and spectral density analysis.

The determination of the transfer characteristics of the different elements of the control loops led to the development of identification methods based on the "black box" approach. Different procedures were developped to determine the dynamic characteristics of the transfer behaviour of this black box, based on the observation of the variations of the input and output variables of this black box, researches where Zypkin, Broida, Sage, Eykhoff and others brought significant contributions.

Very quickly also it was recognized that the hypothesis of the linearity of the control systems was only a first approximation and methods for the analysis of non-linear systems, like the Lliapunov method, were then developped.

It was equally recognised that a discrete action of some elements of the controlled loop was requested, which led to the use of the Z transform, defined by Jury, Regazzini and Franklin in the States, and of the finite automate theory developped by Professor Moisil from Rumania.

The main field of applications of this first wave of the automatic control process was related to continuous industrial process for example in cement plants, electricity production and the petrochemical industry.

Some remarkable results have been obtained in these fields. However, some limits were also recognized: the accuracy of the analog devices is limited; it is sometimes requested to change the structure of the control loops during the operation of the process, which is not easy with the technology of the analog control loops.

<u>2nd wave: Centralized control with DDC (Direct Digital Control).</u>

This wave can be situated between 1960 and 1980 and is linked to the tremendous hopes arising from the development of digital computers.

It was assumed at that time that men were unpredictable, uncertain elements which were to be eliminated as much as possible from the industrial process, and to be replaced by the process computer which was supposed to become the actual brain of the process. All informations related to the process should converge to this brain, which would then elaborate the control actions. The capacity and the cost of this brain were relatively high, and it seemed advisable to use it in time-sharing mode for the different tasks involved with the operation of a given process:

- the start and stop of the process.
- the optimal operation conditions for a given level of production.
- the monitoring and protection of this operation.

This second wave is characterized by the use of computers, not only as control tools, but as analysis tools, which led to the use of the state phase equations. This approach allows the extremalisation of an objective function, which often is the square value of the deviation of the step response. Many other optimisation methods have been implemented, like linear and dynamic programming, to which the names of Bellmann and Kalman have been associated.

These approaches also led to the development of estimations methods of the parameters of the state phase equations, as for instance the Kalman filter.

It must be recognized that the results of this 2nd wave were not as important as the hopes raised at the beginning of this wave. On one hand, the optimisation criteria were often selected, taking into account their ability for a mathematical formulation, but they did not correspond to the real needs encountered in practice.

On the other hand, the process computer was often considered by the operators as a mysterious and unpredictable black box. In 99 % of cases, this box acted satisfactorily, but in 1 % of unforeseen cases, the computer was not able to solve the problem. This led to a paper avalanche. The operators were puzzled and drowned by this avalanche and did not know how to restore the safe operation of the process.

In some cases of computer controlled cement plants, I saw the computer discreetly hidden under a cover, and the operation was performed manually by the operator, with unadequate means since the whole plant had been designed for a fully automated control.

It also appeared that a fully automated operation often did not correspond to the wishes of the operators whose presence and action were requested only for cases of unexpected disturbances, and for the maintenance of the plant.

With the exception of some special cases, a completely automated plant is an utopia. The centralized brain is a delicate device. When it is perturbed, the whole plant is affected. In fact, experience has shown that, often, in a completely centrally automated plant, the reliability and output of the process operation was reduced.

3rd wave: Distributed Control with mini and micro process computers.

This wave started in 1975 and is not yet over. It is characterized by the generalisation of the use of mini- and micro-computers, due to the tremendous reduction of the cost of the memory and of the data processing equipment. The reduction of those costs open the way to a distribution of the "intelligence" among the different phases of the process, and to the selection for each task of the most appropriate processor.

The actual trend is not so much to try to eliminate the operators, but rather to better assist them, and to give them more efficient tools, to be better informed about the behaviour of the process they are entrusted with, and to better intervene in case of emergency situations.

The accent is placed on the development of man-machine communication devices, with data logger and video screens giving the possibility to instore a dialog between the operators and the process via the process computer. This trend leads to the development of "convivial systems" given in principle by figure 1. The measurements of the state variables are transmitted on one hand to the control board, on the other hand to a simulator. Before acting on the process, the operators have the possibility to transmit the control orders to the simulator, which reacts and gives the possibility to see the results of the application of the control orders, and to analyse the future behaviour of the process.

Monitoring systems have been implemented to collect and process the different state variables of the process, and to transmit them to the different man-machine communication devices: digital measuring instruments, data logger or video screens. Special reference must be given to the "digital perturbograph": the main state variables are stored permanently, then deleted after a few minutes in a normal case of operation. In case of perturbations, this cancelling is suppressed. It is thus possible to be informed about the behaviour of the process before and during the perturbation.

Fig.2 gives the example of the monitoring and control system of a cement plant with distributed intelligence, namely with:
- 8 minicomputers.
- 4 programmable controllers.
- 4 video displays.
- 3 printers.
- 8 micro processors for the teletransmission.

One of the characteristics of this 3rd wave of automatic control is the recognition of fuzzy variables, defined according to the fuzzy sets theory developped by professor L.Zadeh in Berkeley, California and by professor A.Kaufmann of Grenoble and establishing the difference between certain variables, probable variables and fuzzy variables.

Another characteristic of this wave is the generalisation of the use of mini-and micro-computers as assistance in various fields:

-Computer Aided Design (CAD) gives the possibility to investigate different shapes, for instance in civil engineering and in modern electronic using printed circuits. This CAD involves also the determination of the optimal characteristics to be given to a project (capacity, intermediate storages, location etc), by simulation of its operations in many different possible cases.

-Computer Aided Creation (CAC) for the research of new ideas and new solutions, with the use of computer programs stimulating the creativity, combined with the methodology defined as "inventic" using invention teams and implying:
. a logic phase for the clear definition of the problem.
. an intuitive phase for the process of creation.
. a critic phase for the evaluation of the results obtained and sometimes a new definition of the problem.

-Computer Aided Decision, for the elaboration of a multi-criteria decision combining objective data and fuzzy data based on the concept of preference. This decision theory is becoming a huge field of research.

Fig.3 gives, in theory, an example of the use of an economic model permitting to evaluate the impact of political decisions and to associate the qualified public to this evaluation, with the help of television, through the feedback this public might give.

Computer-Aided Diagonostic using the approach of the expert-systems and of the artificial intelligence combining data banks and rules banks indicating how to handle the appropriate data, as a support for diagnostic, for instance in the medical or juridical fields.

Computer Aided Management (CAM) for the supervision of the implementation of project, as given by Fig.4 and involving:

1) a preliminary project:
. forecast of the needs to be fulfilled.
. analysis of the constraints to be taken into account (financial and time constraints, impact on the environment)
. definition of the project purpose.
. search of new ideas.
. preliminary design of the project.
. simulation of the project to forecast the results of the implementation, taking into account different evaluation criteria.

2) This preliminary project gives the basis for the main decision, which is the decision to implement, which involves the elaboration of the final project with:

. the final detailed design of the project.
. the implementation and commissioning of the project with computer aided programming and budgeting.
. the operation which checks if and how the needs are fulfilled, and sometimes the modification of the design and of the process operation.

We have already seen how the CAM can be used equally for the convivial control of an industrial process, as shown in principle by Fig.1.

Conclusion

The actual trend of the technology evolution raises among a large public certain fears against the possible outcome of an always less human and convivial world.

The development of the mini- and microcomputer technology makes it possible to recognize that these fears are not founded. It is more and more recognized that the aim of these tools is not to replace men, but to help them in their tasks, to support them and to make their activity easier and more efficient. The actual trend is to develop convivial systems allowing more man-machine communications, so that the operators become more and more able to judge and decide by themselves how to manage their process.

As stated by Erwin Töffler, the ability which will be more and more required is the ability to understand, to select, to create and to adapt. This evolution implies a promotion of the working conditions of the operators in charge of an industrial process.

The approaches of the classical automatic control theory of the 1st wave, and of the 2nd and 3rd waves are complementary. It is appropriate to apply them simultaneously to the same problems. The classical control theory gives a better understanding of the physical behaviour of the process. The modern control theory using state phase equations allows a deepest analysis and a better optimal setting of the control parameters.

The distinction between analog and digital systems becomes less obvious. In the same control loops some elements can be an analog system (for instance the process itself), and other elements can be digital, for instance for the transmission and the combination of the control signals.

The use of mini- and micro-computers is more and more generalized in technical applications as well as in non-technical fields. Does this application belong to "automatic control", or "operation research" or system analysis"? This question is irrelevant. The majority of the applications mentioned in this survey are handled in the different technical committees of IFAC. In many occasions, the question is put whether this Federation should change its name, but each time the idea has been given up. In fact the same concepts for analysis, comparison, selection, synthesis, using the resources of the modern informatic are applied in many fields of application, and there are few fields of human activity which have not been touched, one way or the other by this evolution. Even the arts are involved, as proved by the emergence of the synthetic computer music or of computer drawings.

It is striking to notice the world wide character of this evolution, as demonstrated by the world-wide attendance to this actual symposium.

The gap between theory and praxis was recognized some years ago. This gap seems to me now to be more between the capacity of the existing tools for the control and the process of the available information, and the practical use which is made of these tools, which more and more requests the ability of judgment, selection, analysis, decision on the part of the people in charge of these tools.

This evolution requests a higher level of the qualifications and know-how of the designer of a system as well as of the operators of these automatic controlled systems. The engineers are the initiators of the use of these new technologies. It is one of their responsibilities and tasks to train these designers and operators. It becomes ever more the duty of the engineers to become teachers and trainers.

The development of the mini- and micro-informatic is a challenge for the new generation of engineers. Taking into consideration my participation to the evolution of the automatic control technology during these past 40 years, I sometimes regret not to be able to start again my career as engineer, to help meet this challenge.

List of figures

Fig.1 Block diagram of the computer assisted control of an industrial process.

Fig.2 Example of the structure of a decentralized computer control system of a cement plant.

Fig.3 Principle of the on line simulation of economic systems as an aid for political decision.

Fig.4 Scheme of the computer aided procedure of the implementation of a project.

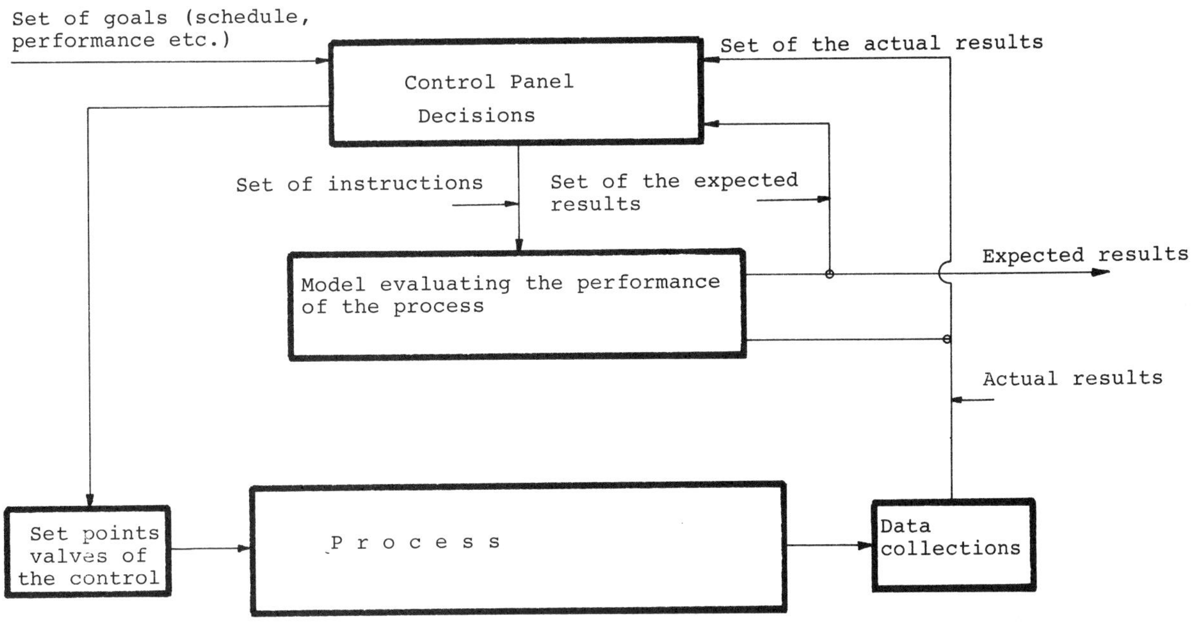


Fig.1. Block diagram of the computer assisted control of an industrial process

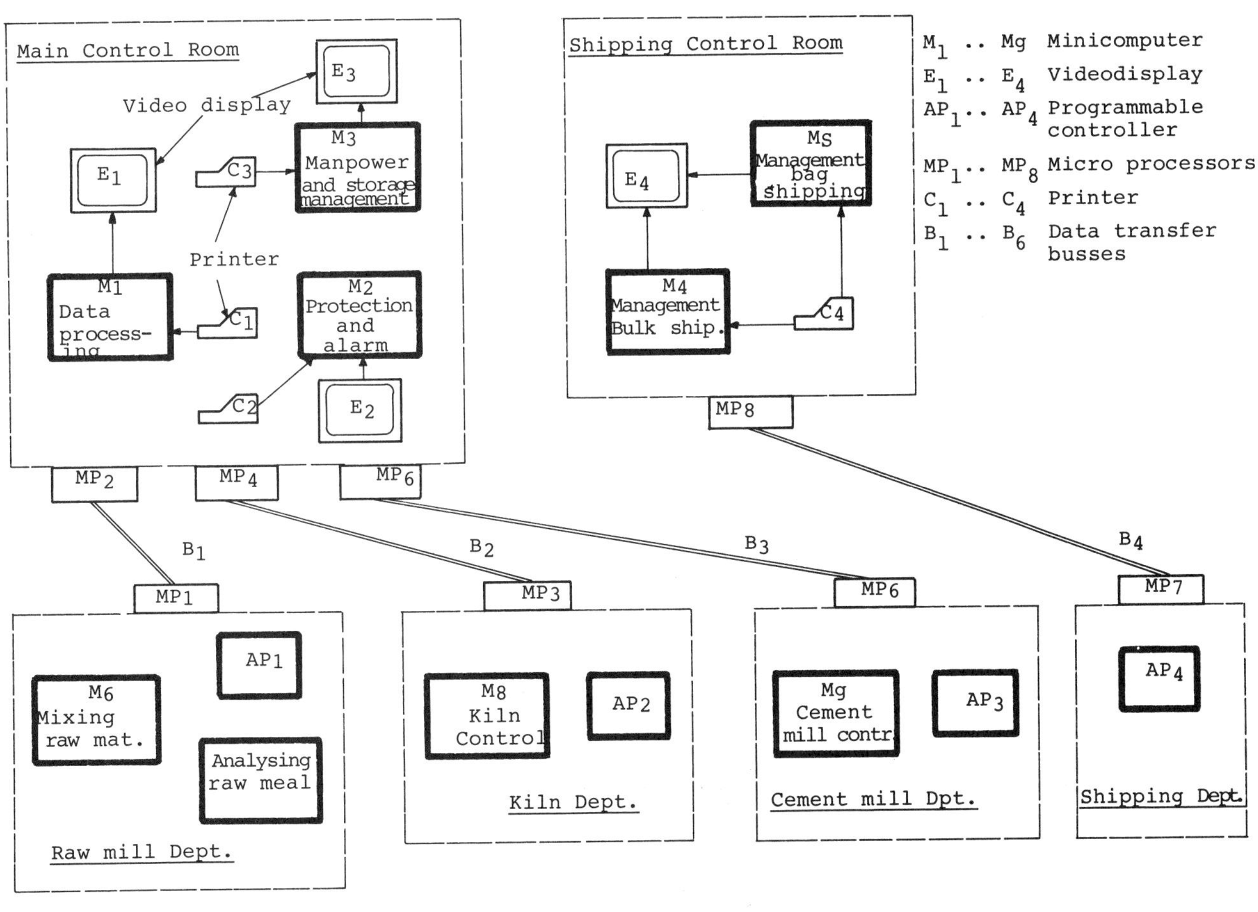


Fig.2 Example of the structure of a decentralized computer control system of a cement plant.

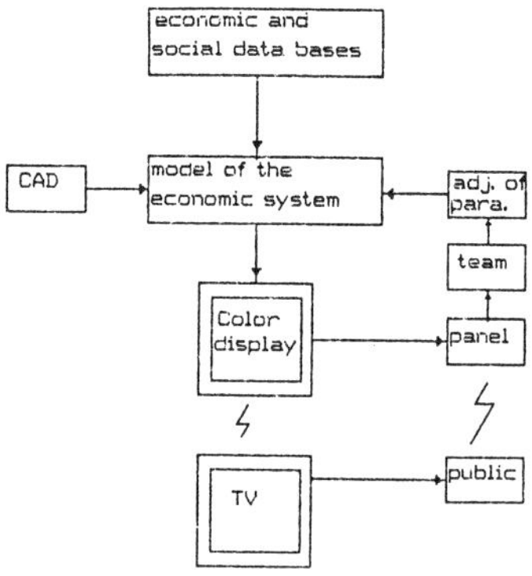


Fig.3. Computer aided political decisions

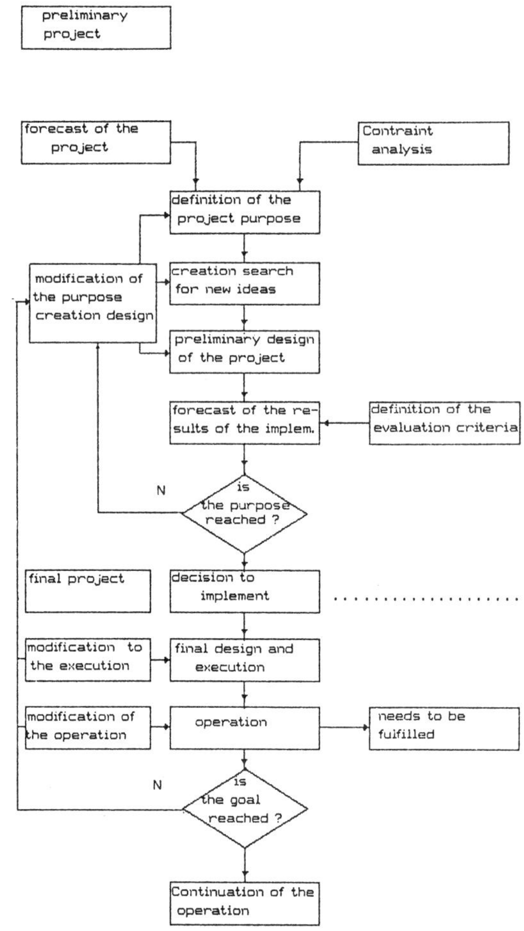


Fig.4. Scheme of the Computer Aided Procedure of the implementation of a project

# ENERGY MANAGEMENT FOR PROCESS INDUSTRY: HOW MUCH CONTROL IS NEEDED

**A. Kaya* and M. A. Keyes****

**Industrial Control and Instrumentation Laboratory, College of Engineering, The University of Akron, Akron, OH 44325, USA*
***Industrial Products Group, Babcock & Wilcox, 35 River Street, Chagrin Falls, OH 44022, USA*

Abstract. This paper provides an overall and unified knowledge on industrial energy management applicable for a broad base of industrial plants. Energy conservation and management opportunities are presented for various processes and equipment. Energy savings are discussed on particular power equipment such as: boiler, turbine, utility tie-line, and chiller which are common for process industry. Furthermore, few energy intensive processes are included to demonstrate the significant energy conservation opportunities. The methods for energy savings are described, which generally are: improved feedback controls; supervisory and optimizing controls; and coordinating controls. Considerations in implementing the savings by distributed controls including the hardware and software requirements are discussed. In conclusion, this paper is a survey of industrial energy management on fundamentals, methods, and implementation to reflect the current status of technology.

Keywords. Cogeneration, computer control, energy control, optimization, power management.

## INTRODUCTION AND OVERVIEW

Energy management, as the term is used today, is no less comprehensive in its scope than it was defined by Henri Foyal (1949). From 1948 until 1973 many energy management benefits were of little interest, even to large industrial energy users, due to the depressed price of energy. Today the situation has changed remarkably. The price of crude has rocketed from roughly $2.50 per barrel in 1973 to a spot price of over $40 per barrel (Keyes, 1980). Energy, its cost, and even more importantly its availability is a matter of individual concern, corporate survival, and national security. In this threatening environment, the need to manage our utilization and reduce our losses of energy is paramount. Some of the more direct benefits of energy management are: reduced costs, increased throughput, reduced cycle times, better feedstock utilization, reduced downtime, and lower pollution.

Accurate measurements or "sensors" are as key to the successful operation of a control system just as the senses are to the survival of a living organism. For this reason much of the major control system development over the past decade has focused on development of lower cost, reliable, and more accurate sensing devices.

Energy shortages caused some control problems or at least created challenges for control engineers. Some of them are: larger units, flow variations, varying feedstocks, lower quality feedstocks, new measurements, operating with small margin of safety and stability, varying costs, fluctuating market conditions, varying quality of fuel (Keyes and Kaya 1979). In general, three kinds of approaches are possible for an overall energy management program:

a) Energy generation: boilers (including black liquor recovery), electric power generation;
b) Energy transmission: pumps, fans and heaters of energy generation which are driven by steam or electricity as necessary; other auxiliaries for energy recovery, etc.; and,
c) Energy consumption: process area primarily for production (including by-products of energy sources such as black liquor and bark).

A coupling (interaction) exists between the production rate and energy availability in terms of reuse of process by-products such as black liquor and bark burning in pulp processing. Similar conditions exist in a petrochemical process in which by-product fuel gases are energy sources; however, those gases (e.g. ethylene) can be used independently. In the pulp and paper industry black liquor must be burned to recover chemicals for pulp processing. This explains the necessity of an overall plant energy management and coordination strategy. As for the control methods, three levels are considered in the order of hierarchy: on-line dedicated controls, supervisory controls, coordinating and planning (Kaya and Keyes, 1983).

On-line dedicated controls. These controls perform specific functions and are designed to be responsive to process transients or disturbances. Simply, they control designated variables optimally at the desired set point (set manually or automatically).

Supervisory controls. These functions perform optimization to allocate the loads (e.g. set points) of lower-level controls so that they may as a whole provide the most efficient overall operation. It is expected that a dedicated control may provide more efficient performance at the expense of lower efficiency of another one. The general idea here is to determine the set points so that the best performance of the aggregate operation is achieved.

Coordination and planning. Higher-level functions such as coordination between optimizing functions, under alternative energy production constraints and costs are performed at this level. The output of this level usually provides constraints and plans which must be satisified by the lower-level optimizing functions.

The general agreement is that the low level controls have: short control interval (sampling time), great deal of insight of process dynamics, local and dedicated, simple control criteria. At the same time high level controls tend to the other end of these as the level increases. The structure of multilevel controls including the controls (com mands) and measurement in between the levels are de scribed in Figure 1. A complete measurement and instrumentation system should first be available for any proper energy management system. Although measurements and dedicated controls belong to the lowest level of the hierarchy, they are essential for a successful energy management. The process area needs special attention; however, it varies based on industry and should be treated separately for each industry.

## ENERGY SAVING OPPORTUNITIES

There are numerous energy savings in process, equip ment and in control area. There are indicators of poor energy conservation such as: leaks, faulty traps, steam vented to atmosphere, uninsulated hot lines, oversized pumps, heat from waste gases not recovered, and smoking chimneys. These are commonly known by the industry.

Tight feedback control of processes are the single energy saving source with quick returns and minimum investments (Keyes and Kaya, 1979; Dukelow, 1977). A partial list of important energy savings due to feedback control is listed in Table 1. The energy savings usually provides 0.5 - 3 years payback periods for the investment. Energy management is more meaningful above the feedback control within the control hierarchy. This level of controls are supervisory and optimizing. The major energy savings are listed in Table 2. Although the savings by higher level controls are small in percent value, the absolute amount of energy saved is sufficiently high to warrant a payback period of 1 - 3 years.

TABLE 1: ENERGY SAVINGS BY FEEDBACK CONTROLS (Partial List)

1. Combustion controls of power and recovery boilers, process heater, soaking pits, lime kilns, etc.
2. Temperature controls of turbine condenser hot well, dryers, evaporators, compressor intake air.
3. Blow-down and soot blowing control.
4. Control of solids for increased throughput of the recovery boilers.
5. Temperature control to reduce cycle time of PVC rector.
6. Control of multi-stage flash evaporator and vapor compression cycle of desalination.
7. Control of octane of a catalytic reformer.

TABLE 2: ENERGY SAVINGS BY SUPERVISORY AND OPTIMIZING CONTROLS (Partial List)

1. Optimun load allocation of multi-service units: such as boilers, pumps, chillers working in parallel.
2. Turbine optimization to maximize the generation efficiency (by responding to plant steam demand and/or electric demand).
3. Utility tie-line control and management.
4. Soot blowing optimization.
5. Pulp and paper digester steam smoothing.
6. Chiller chilled water and condenser water optimizing temperature control.
7. Optimum heat recovery of condenser discharge.
8. Optimization of performance ratio, and compressor work per unit distillate in desalination.
9. Optimization of pressure in a distillation column.

## DESCRIPTION OF METHODOLOGY

Energy management requires instruments and calculations performed in real time. The savings in monetary value is more realistic measure than the energy itself, since fuel prices vary.

The efficiency of a boiler is higher for the lower loss of energy. Since fuel prices are known the fuel flow in $/hr can be known for a given output of the unit, such as kg/hr of steam output for a given fuel flow in $/hr. Several operating points of various $/hr versus kg/hr are captured to be a basis of curve fitting.

Similarly, the performance measurements of other units such as compressors, chillers, cooling towers turbines, and pumps, etc. can be determined. Table 3 indicates the units of performance measures of various energy conversion units. It appears that the increased efficiency of units are the first step towards energy savings. In the following, some of the feedback controls for improved efficiency is introduced.

### Feedback Controls

Combustion Control: Control of air/fuel ratio in a boiler can provide savings. Excessive fuel will cause unburned gases to escape through the stack which is a direct loss of energy. With excess air, on the other hand, the unused air is heated also causing a loss of energy.

TABLE 3: UNITS OF PERFORMANCE MEASURES OF EQUIPMENT

| Equipment | Units of Measurement |
|---|---|
| Boiler | Fuel input, $/hr. vs. output, kg. (of steam)/hr |
| Process Heater | Fuel input, $/hr vs. ouput, kJ (transferred)/hr |
| Chiller | Power input, $/hr. vs. output, Ton (or kJ/hr.) |
| Turbine | Steam input, $/hr. vs. ouput, KW (or kJ/hr.) |
| Cooling Tower | Power input, $/hr. vs. ouput, kg/hr (at std. conditions). |
| Pump | Power input, $/hr. vs. ouput, kg/hr |
| MSF Evaporator | Power input, $/hr. vs. product (distillate) kg/hr |

Further, as the excess air increases the temperature of the flue gases rise compounding this energy loss. Controlling the air/fuel ratio by the measurement of percent oxygen in stack gases provides accurate control and significant savings, especially in multi-fuel situations (Dukelow, 1977).

Combination of Opacity or Carbon Monoxide and Oxygen Sensor provide improved combustion control under certain conditions. Consider a spreader stoker boiler. Stokers among other idiosyncrasies have a habit of getting holes in the fuel bed on the grates. When this happens, the under grate air goes through the hole and not the fuel creating an air lean mixture in the bed even though total fuel/air ratio may be correct.

Fortunately, when this occurs, the smoke increases and an opacity measurement can be used to override the oxygen and increase air flow to the boiler. This supplementary measurement and control saves fuel and additionally insures compliance with stack opacity regulations. Furthermore, CO measurement (as an indicator of incomplete combustion) is used to optimize the set point of Oxygen to minimize the energy loss under varying operating conditions.

As another example of combustion, consider a recovery boiler. Sodium sulfide and sodium carbonate are recovered in smelt form for recycle. Additional benefits of precise combustion control over the increased steaming efficiency are incresed solids throughput by reduced excess air in case the mill is recovery constrained.

Blow-down control: There is an optimum value of conductivity (TDS) required by the boiler manufacturers. This value corresponds to a minimum overall loss of energy due to the blow-down and other long term losses such as efficiency loss due to scaling, increased maintenance and repair, shorter boiler life, loss of operation due to increased TDS. The necessity of controlling the conductivity is clear. The simple argument is that either excessive or no blow-down is both detrimental which leads into the necessity of an optimum value.

Turbine condenser (hot well) temperature control: Hot well temperature control provides savings on subcooling and preheating and a net increase on turbine output. The value of control point can be set by operators based on operating conditions and/ or by a supervisory adjustment of a supervisory logic (Derreberry, et al, 1985).

Control of split condensor for maximized heat recovery: Normal industrial plant hot water control systems supply hot water at constant temperature. Plant steam is normally used to maintain constant hot water supply temperature.

If the hot water supply temperature is allowed to vary while still meeting the heating demand, then the cost of both refrigeration and steam can be reduced. The energy saving strategy is based on utilizing the minimum hot water temperature necessary to meet the hot water heating demand. Reducing the hot water supply temperature in turn reduces the discharge temperature and pressure of the compressor (reducing the power input to the compressor). Cooling tower loads are reduced since condensor heat is partially dissipated by reheating the plant hot water supply. Steam use is also reduced since some of the hot water heating energy is recovered from the condenser (Enterline, et al, 1984, Kaya and Sommer, 1985).

Keyes and Kaya (1979) determined that for every minute trimmed from the cycle time by advanced control for a PVC plant provides significant savings due to increased throughput. Other examples are dye becks for textiles, smelters, steel mills, blast furnaces, etc. Keyes, et al, (1986) provided the methods to improve the efficiency of desalination plants by floating set point control of feedback loops. Szoke and Kennedy (1984) has reported over $1 million savings by improved Octane Control.

## Optimizing and Supervisory Controls

Supervisory functions at the second level determine the optimum set points of individual units (such as boiler and turbine loads) to assure an overall optimum operation. The major controls are listed in Table 2.

Load allocation of Parallel Units: A simple load allocation problem for parallel units is defined as to minimize the overall cost subject to satisfying the load demand and the operational constraints of each unit. It turns out that the necessary condition for optimum loading is that the incremental costs of each unit be equal (Kaya and Keyes, 1983). Optimum solution is obtained when the loads fall within the bounds of each unit. Otherwise the solution is suboptimal and the loads may be at the bounds.

Two different computational algorithms for optimization are currently used. The algorithms are:

1) Nonlinear constrained optimization. Several initial starting loads gave almost the same optimal load values due to the convexity properties of constraints and cost functions.

2) Increasing the load to the boiler with minimum $\Delta f/\Delta m$ value while decreasing the load of boiler with maximum $\Delta f/\Delta m$ value and so on until the incremental cost values $\Delta f/\Delta m$ are equal. There is a great saving on computational time. Comments on optimization methods and things to watch have been discussed by Kaya and Keyes (1983). This method can be applied to chillers, pumps, etc. working in parallel. On a demand change all unit loads are changed for a responsive system and subsequently incremental costs are equalized for optimization.

Turbine Optimization. Turbine optimization is very similar to the optimum load allocation problem. There are slight complications in real turbine systems since the energy from intermediate stages of extracting turbines can be utilized in several ways. The turbines may not all be fed from the same pressure header. The steam flows should be calculated in terms of $/hr rather than J/hr. Inflows should be assigned a value equal to the cost of steam from boilers. The extraction flows should be assigned a value equal to the cost of steam from boilers. The extraction flows should be assigned a $ value -- equal to their utility values (thus recognizing second law consequences of differing supplies). The generating turbines are primarily used for low-cost electric generation while the steam demand of the plant is satisfied. It may be possible that condensing flows are used to generate electricity only. This is the result of a higher level decision as "buy or generate".

A total turbine system optimization problem can be treated as a linear programming problem as discussed by Kaya and Keyes (1983). The generation-flow relations are derived from turbine performance curves under the assumed steady pressures at the headers (Hanson, 1978). Another simple approach is to assign the steam flows to extraction or exhaust conduits based on the principle discussed in multi-unit optimization. However, this approach, although computationally effective, still requires linear programming support upon appreciable changes in operating conditions.

Soot Blowing Optimization: Judicious soot blowing minimizes the cost of steam. Excessive soot blowing increases heat transfer and steam generation for a given fuel input. However, after the steam used for soot blowing is subtracted, the net results may be lower. That is, an optimum soot blowing policy should be defined, measured, and enforced. Soot blowing should be applied to the sections of boilers where it is necessary, rather than a simple time programmed cyclic regimen. Also, the black liquor recovery boilers in the Kraft pulp process represents the most lucrative potential application of soot blowing controls.

The measurements are needed for soot accumulation with respect to time and required energy to clean it. A soot blowing optimization problem is considered by Kaya (1981) and an example is worked out, for one section of a boiler based on the measurements. The optimization routine calculates: a) cost of steam, b) dirtiness of surfaces, c) required energy for cleaning, each with respect to time. The optimization routine provides the optimum soot blowing interval.

Chilled Water Temperature Optimization: Operation of chillers at elevated chilled water temperatures leaving the evaporator increases temperature and enthalpy of refrigerant entering the compressor. This , in turn, increases the cycle efficiency (coefficient of performance) of chillers. As a general rule, 1° C increase corresponds to two percent increase in efficiency. The control strategy for chilled water temperature optimization is based on maximizing water temperature while satisfying all users of chilled water. This is done by maintaining control valves to all users of chilled water at their maximum permissible open position.

A logic diagram for this control strategy using standard SAMA logic sumbology is given by Enterline et al, (1984).

Condenser Water Temperature Optimization: Operation of chillers at reduced condenser temperatures decreases the enthalpy of refrigerant exiting the compressor. This, increases the coefficient of performance. An improvement of two percent in efficiency for 1°C decrease in condenser temperature is expected.

The control objective is to keep the condenser water temperature as cool as possible, without wasting energy being supplied to cooling tower fans. A logic diagram for this control strategy (using standard SAMA logic symbology) is given by Kaya and Sommer (1985).

Keyes, Scheib and Kaya (1986) demonstrated that the performance ratio of a MSF evaporator is increased by floating the set points of control loops of regenerated feed temperature and evaporator vapor temperature. Furthermore, Keyes (1986) has shown that by selective controls the pressure in a distillation column can be lowered until the constraint is reached. The constraint is overhead cooling capacity in summer and reboiler capability in winter. This varying constraint improve the efficiency of column.

## Coordinating controls

These controls are over the supervisory and optimizing and set the operational limits for them. For example, cogeneration is a supervisory function over the use of extracting turbines to produce electricity while providing low pressure steam for other plant uses.

Cogeneration and Tie-Line Management: An industrial power plant will normally have two or more power generators. A plant's electrical power requirements can be satisfied from three sources of power: purchased power, power generated by extraction flow; and, power generated by condenser flow.

The usual philosophy is that the increased plant demand must be delivered by the source with minimum incremental cost $/kwh. In case demand decreases the reduction should be made from the generating source with the maximum incremental cost. There are cases when increase in generation or in purchase is not feasible. Then, the least significant loads should be shed. Tie-line management provides a guide for generation. At the same time, tie-line controls assures electric purchase by remaining within the contracted limit. Then, the necessary power generation is handled by turbine system. In case the plant operates on purchase mode, the power generation requirement is set free. If plant generation is designated, the load demand is controlled by the turbine system (Keyes and Kaya, 1983).

Supervisory Coordination of Chilled Water Temperature: As the chilled water supply temperature remains at its maximum allowable limit, the temperature difference $T_D$ (between supply and return) is reduced. This results in the increased flow to increase the overall heat transfer coefficient for the same load conditions. However, increased flow means increased pumping cost. An optimum trade-off exists between the increased cost of pumping and the decreased cost of refrigeration. Supervisory control adjusts the amount of chilled water temperature increase (Kaya and Sommer, 1985).

Plant Energy Management: A plant energy management is more than the extensive data gathering and monitoring. Multi-level control hierarchy of plant energy management is described in Figure 1, including a list of information communicated between various levels. There are other considerations which must be taken into account. Fuel have a high KJ/$ ratio is scarce. The load and fuel selection schedule of boilers must be constructed to prevent fuel unavailability. Boiler and fuel utilization limits can be set by higher-level heuristics to provide continuous operation. The industry has to meet a number of regulatory and operational data gathering requirements such as monitoring of pollution levels, efficiency limits, and energy inventories. Because of this, the literature in the energy management work includes various ancilliary functions: unit and overall performance evaluation and monitoring, trend analysis, logging and alarming, energy inventory summary, alternate fuel allocation, plant simulation, maintenance tracking, report generation, emergency operator guides, sequence of events monitoring, and economic analysis. A successful energy management requires the support of administration as well as engineering know-how (Kaya and Keyes, 1985).

## IMPLEMENTATION

There is much agreement among researchers and practitioners on the appropriate order of implementation of energy management systems. Implementation must start with low-level controls and logically build up to a higher level, leading eventually toward plant-wide energy management. Application guides for implementing of all levels of energy management are given in suppliers technical manuals (Bailey, 1986).

## Control Systems

Microprocessor-based control units are widely used at every level of control. Measurements are so far unchanged. Control functions of digital devices are mostly the same as their analog counterpart (PID, etc.). One other advantage of digital controls is their ability to communicate with other computers via databus. There are numerous workers who have reported their field experiences in applying computer systems to energy management. The equipment offered is a microprocessor-based distributed computer system. The functions performed are usually higher-level supervisory energy monitoring, and others such as reporting, alarming, etc. The trend is to introduce more flexible, lower cost distributed computer systems for energy management. Such systems can be retrofitted into the existing controls and instruments. The most complete system hardware for implementing the energy management is shown in Figure 2.

## CONCLUSIONS

It has been demonstrated that energy management can be achieved by multi-level control techniques and implemented by a distributed control system. Energy savings are substantial and additional side benefits are significant. An energy management audit determines the flexibility and the basis for further steps. The usual contents of an energy management audit report are given in Table 4.

TABLE 4: CONTENTS OF ENERGY MANAGEMENT AUDIT REPORT

1. Executive Summary
2. Summary of Savings and Sources
3. Description of Plant
4. Energy Savings in Percentages

5. Cost of Improvements of Savings and "ROI" for Each Step
6. Implementation Plan of Steps
7. Training
8. Support of Equipment and Software
9. Technical Literature:
    Articles
    Application Guides
    References

More and more advanced control implementation methods are being discovered for industrial plants. The benefits are substantial. As the trend continues, it will be justified to conduct energy management and product improvement audits to upgrade the plant operation periodically.

## REFERENCES


Bailey, (1986) "Energy Management Application Guides by NETWORK 90" Bailey Controls Co., Wickliffe, OH 44092 USA

Derreberry, B., et al., (1985), "Instrumentation and Control for Energy Efficient Cogeneration" Proceedings of ISA POWID Symposium, Vol. 28, pp. 1105 - 1116.

Dukalow, S.G., (1977), "Combustion Controls Save Money", Instruments and Control Systems,, November, pp. 35 - 41.

Enterline, L.L., Sommer, A.C., and Kaya, A., (1984) "Chiller Optimization by Distributed Control to Save Energy", ISA Transactions, Vol. 23, No. 2, 27 - 37.

Fayol, Henri, (1949), "General and Industrial Management", Sir Isaac Pitman & Sons, Ltd., London, p. 5.

Hanson, R., (1978), "Optimization in the Industrial Power Plant", TAPPI Engineering Conference Proceedings, pp. 602 - 614.

Kaya, A., (1978), "Industrial Energy Control: The Computer Takes Charge", IEEE Spectrum, Vol. 14, 0.48 - 52, July.

Kaya, A., (1981), "Optimization Methods in Soot Blowing for Energy Efficient Operation", Proceedings of Joint Automatic Control Conference, Vol. 1, No. TA-8D, USA

Kaya, A. and Keyes, M.A. , (1983), "Energy Management Technology in Pulp, Paper and Allied Industries", Automatica, Vol. 19, No. 2, pp. 111 - 130.

Keyes, M.A., and Kaya, A., (1983), "Plant Energy Management by Multi-level Coordination", Proceedings of 5th International IFAC Conference on Instrumentation and Automation . . .", PRP/5, Antwerp, Belgium, pp. 363 - 370.

Kaya, A., and Keyes, M.A., (1985), "Industrial Energy Management: Fundamentals and Implementation", Keynote address at Instrument Asia, Singapore.

Kaya, A., and Sommer, A., (1985), "Energy Management of Chillers by Multi-level Control and Optimization", Transactions Journal on Dynamic System Measurement and Control Vol. 107, pp. 270 - 277, Dec. 1985.

Keyes, M.A. (1980), "Energy Management", Proceedings of Workshop in Microprocessor Applications on Energy Management . . . ", (Editor: A. Kaya), The University of Akron, Akron, OH pp. 6 - 68.

Keyes, M.A. (1986), "Advanced Control Implementation with Distributed Controls", Proceedings of IFAC Workshop on Automatic Control in Petroleum , Petrochemical and Desalination Industries, Kuwait.

Keyes, M. and Kaya, A., (1979), "Instrumentation and Control Changes Brought About by The Energy Crisis", ASME Paper WA-79/En-7

Keyes, M.A., Scheib, T.J., and Kaya, A., (1986), "Multi-level Control and Optimization Methods for Desalination Plants", Proceedings of IFAC Workshop on Automatic Control in Petroleum, Petrochemical and Desalination Industries," Kuwait.

Szoke, R.W., Kennedy, J.P., (1984), "Optimizing Control of a Catalytic Reformer with Digital Instrumentation", NPRA Computer Conference (paper CC-84-117), October, 1984.


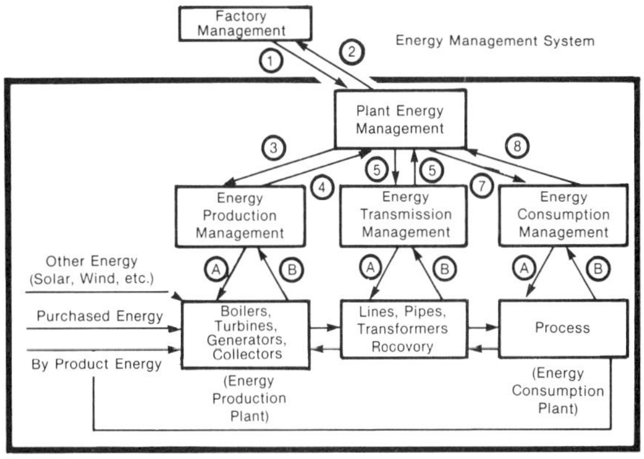


A. Command

B. Measurements
1. Purchased Energy
Environment Estimates
Production
2. Energy Cost/Product
In-Plant Energy
Readiness
3. Energy Demand
Energy Availability
Environmental Constraints
4. Energy Produced
Efficiency
Resources Utilized
5. Energy Inventory Scheduled
Maintenance
6. Energy Delivered, Received
Efficiency
Energy Recovered
7. Energy Availability
Production Demand
8. Energy Consumption/Production
Efficiency
By-Product Energy

*FIGURE 1 — Structure of Energy Management System for Industrial Plant*

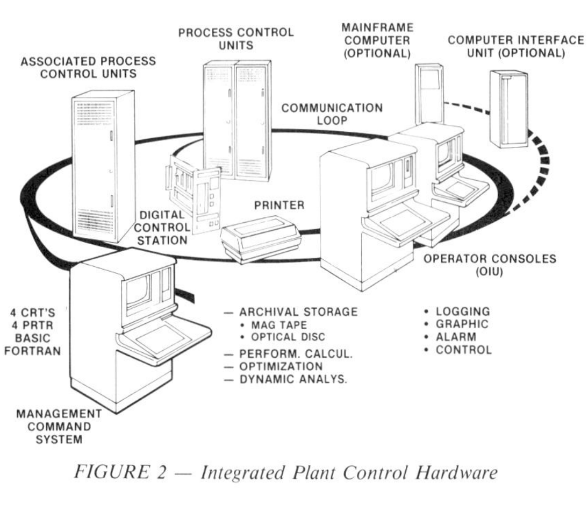


*FIGURE 2 — Integrated Plant Control Hardware*

# A MICROPROCESSOR-BASED MODEL REFERENCE ADAPTIVE CONTROL SYSTEM FOR HYDROELECTRIC GENERATING SET

**Wang Bailin**

*Department of Electrical Power Engineering, Huazhong University of Science and Technology, Wuhan, The People's Republic of China*

Abstract. This paper describes an adaptive speed governor of hydroelectric generating set, which can adjust its parameters automatically with the variation of operating conditions. An 8086 16-bit microprocessor supports the governor with the aid of special software. A new configuration of direct model reference adaptive control (MRAC) system is developed which can be used in non-minimum phase systems. The hydroelectric generating sets have wider stable ranges and higher frequency-following accuracy using the adaptive governors than those using conventional PID regulators.

Keywords. Adaptive control, Model reference adaptive control, Lyapunor stability, Computer control system.

## INTRODUCTION

Hydroelectric generating sets are the most important devices in a water power plant. The frequency and voltage of the electrical power supplied by them must be accurate and stable. The regulating system of the hydrolic turbine consists of turbine, penstock, generator, speed governor and so on. Mechanical governors were applied in the 40s and 50s, and electrohydraulic governors in the 60s and 70s. In 80s, microprocessor-based speed governors were introduced in some companies, such as Asea in Sweden, Neyrpic in Fance and Woodward in the U.S.A.

The conventional PID regulator whose parameters are constant, do not have good performances due to the non-linearity of the turbine and penstock and the change of the enviroment (water level, water pressure, network load, etc.). Some microprocessor-based speed governors (such as Asea's) employs new PID regulators whose parameters very with the variation of operating conditions in accordence with a parameter table designed off-line, but the design of the regulators requires sufficient data of the plant, which is difficult to determine.

We designed a year ago a new type of adaptive PID regulator (PID-MARC) applying model reference adaptive control theory and developed a prototype DONG FANG WT-II speed governor, which is supported by an 8086 16-bit microprocessor. Experiments showed that the governor possessed the desired adaptation function. It requires less data for the plant mathematical model, no extra excitation-signal, which is harmful to the power system.

## CONFIGURATION AND FUNCTIONS

The configuration of DONG FANG WT-II Speed Governor is shown in Fig.1. It consists of three main parts: (1) an 8086 16-bit microprocessor for frequency measurement, PID regulation, auto-tuning of the parameters, self-diagnosis, etc; (2) an alternative electronic PID regulator; and (3) a positioning device. The main functions of each part are given below.

### Monitor Program MONPC

The monitor program MONPC makes the 8086 microprocessor a powerful intelligent unit for process control. It directs the procedure counter to point to the system monitor program or to the application program; modifys parameters of the controller and reference input on-line through the key-board or potentiometers; displys the parameters of the controller and the variables of the controlled process by means of 7-segment LEDs; and gives error massages, etc.

### Basic Subroutine Library BSLPC

A special subroutine library for process control were designed, which includes subroutines for fixed point calculations of addition, subtraction, multiplication and division of integers with 32 bit accuracy and appropriate bounds; for solving differential equations by means of Runge-Kutta mehtod; for converting deciamal digits into hexadecimal digits and vice versa; for 12 bit ADC and DAC; and for digital filter, PID regulator and state regulator, etc.

### Self-Diagnosis and Treatment of Malfunctions SDTM

During the operation of the governor, the microprocessor makes diagnoses on the following items: procedure pointer errors, state of CPU, accuracy of ADC and errors of RAM. The processor makes corresponding treatment as soon as an abnormality is detected. In case the microprocessor should break down, the hydroelectric generating set would still operate safely and the processor could be fixed off-line when necessary.

### Frequency Measurement FMG and FMN

The sinusoidal signals of the generator or the network are converted into rectangular waves by the shaper, isolator and frequency dividor. The counter 8253, the "gate" of which is controlled by the rectangular wave, counts the input pulses (1228.8 KHz) which are proportinal to the cycle of the rectangular waves. At the sametime the rectangular wave is sent to the input port 8255 of the microprocessor to determine the appropriate time to read the data of the counter 8253. Then the frequency of the voltage of the generator or of the network can be calculated according to the

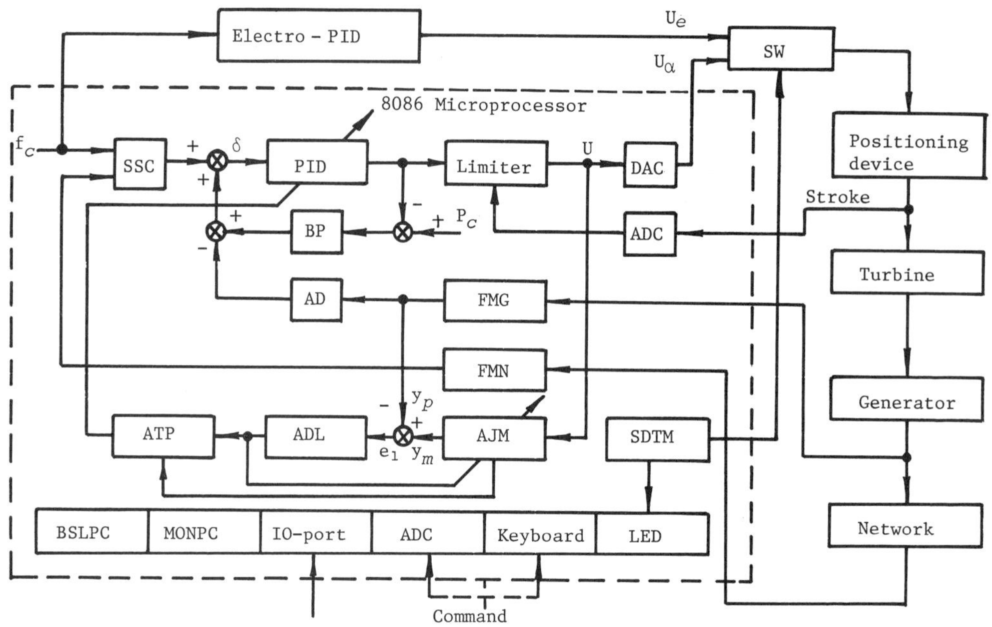


Fig.1 Block Diagram of the Adaptive Governor

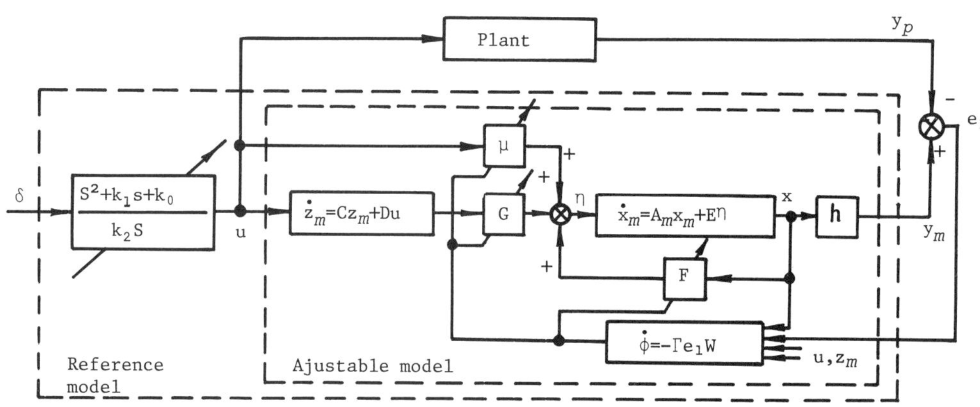


Fig.2 Adjustable Model and Reference Model

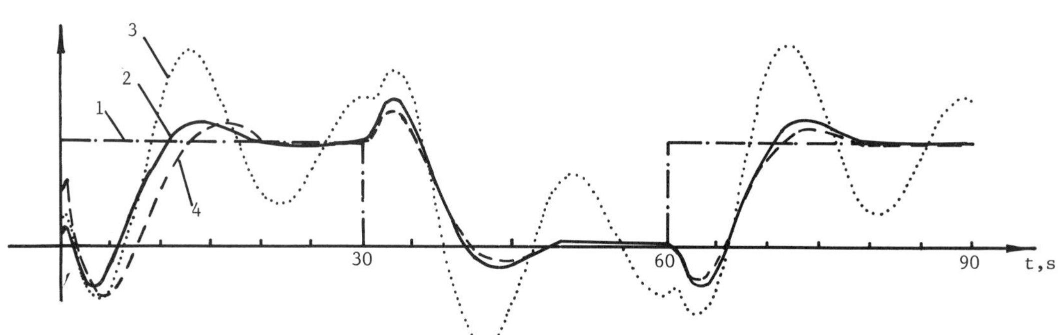


Fig.3 The result of Experiment

data by the 8086 CPU.

Auto-Tuning of PID Parameters ATP

If the parameters of the plant vary due to change of enviroment or conditions, the parameters of the PID regulator will be continuously adjusted till a matching with the parameters of the plant is reached. The auto-tuning loop consists of an adjustable model AJM, an adaption law ADL and an auto-tuning law of PID parameters ATP.

Other Units

The unit BP in Fig.1 is a mechanizm which gives the desired speed droop. Ad is an artifical insensitive area. SSC is a control unit of staring or stopping. The switch SW allows the control signal $U_d$ from the microprocessor to pass under normal operation or the control signal $U_e$ from the electronic regulator to pass in abnormality.

## ADAPTATION LAW

The configuration of MRAC system presented by Narandra and Valavani (1978) is successful and meaningful. Unfortunately, hydraulic regulating system can not apply the kind of MRAC directly, because it is a non-minimum phase system. A new configuration of MRAC is developed below which can be sued for minimum or non-minimum phase systems. In order to derive the configuration we give two lemmas at first.

Lemma 1. A single input-single output system characterized by a transfer function

$$\frac{Y_p(s)}{U(s)} = \frac{b_{n-1}s^{n-1} + \cdots + b_1 s + b_0}{s^n + a_{n-1}s^{n-1} + \cdots + a_1 s + a_0} \tag{1}$$

can be describec by a (2n-1)th order state equation

$$\begin{bmatrix} \dot{X}_p \\ \dot{Z}_p \end{bmatrix} = \begin{bmatrix} A_p & E\bar{B}_p \\ 0 & C \end{bmatrix} \begin{bmatrix} X_p \\ Z_p \end{bmatrix} + \begin{bmatrix} b_{n-1}E \\ D \end{bmatrix} u \tag{2a}$$

$$y_p = (h \quad 0) \begin{bmatrix} X_p \\ Z_p \end{bmatrix} \tag{2b}$$

where

$$A_p = \begin{bmatrix} 0 & 1 & \cdots & 0 \\ \cdots & \cdots & \cdots & \cdots \\ 0 & 0 & \cdots & 1 \\ -a_0 & -a_1 & \cdots & -a_{n-1} \end{bmatrix}_{n\times n}$$

$$E^T = (0 \ \cdots \ 0 \ \ 1)_{1\times n}$$

$$D^T = (0 \ \cdots \ 0 \ \ 1)_{1\times(n-1)}$$

$$h = (h_0, h_1, \cdots, h_{n-2}, 1)_{1\times n}$$

$$\bar{B}_p = (\bar{b}_0, \bar{b}_1, \cdots, \bar{b}_{n-2})_{1\times(n-1)}$$

$$\bar{b}_j = b_j - b_{n-1}h_j \qquad (j=0,1,\cdots,n-2)$$

$$C = \begin{bmatrix} 0 & 1 & \cdots & 0 \\ \cdots & \cdots & \cdots & \cdots \\ 0 & 0 & \cdots & 1 \\ -h_0 & -h_1 & \cdots & -h_{n-2} \end{bmatrix}$$

$$X_p^T = (x_1^p, x_2^p, \cdots, x_n^p)$$

$$Z_p^T = (z_1^p, z_2^p, \cdots, z_{n-1}^p)$$

Lemma 2. Let

$$h_e = (h,\ 0)\ , \qquad Be^T = (E^T, 0)$$

$$A_e = \begin{bmatrix} A_p & E\bar{B}_p \\ 0 & C \end{bmatrix}$$

then

$$h_e(sI-A_e)^{-1}B_e = h(sI-A_p)^{-1}E \tag{3}$$

The proofs of the lemmas are omitted. A theorem derived by means of the Lemmas is as follows.

Theorem. Consider a plant characterised by the eqn.(1), and giving a adjustable model

$$\begin{bmatrix} \dot{X}_m \\ \dot{Z}_m \end{bmatrix} = \begin{bmatrix} (A_m+EF) & EG \\ 0 & C \end{bmatrix} \begin{bmatrix} X_m \\ Z_m \end{bmatrix} + \begin{bmatrix} \mu E \\ D \end{bmatrix} u \tag{4a}$$

$$y_m = (h \quad 0) \begin{bmatrix} X_m \\ Z_m \end{bmatrix} \tag{4b}$$

where

$$A_m = \begin{bmatrix} 0 & 1 & \cdots & 0 \\ \cdots & \cdots & \cdots & \cdots \\ 0 & 0 & \cdots & 1 \\ -\alpha_0 & -\alpha_1 & \cdots & -\alpha_{n-1} \end{bmatrix}$$

$$G = (g_0, g_1, \cdots, g_{n-2})$$

$$F = (f_0, f_1, \cdots, f_{n-1})$$

$A_m$ is a given stable matrix, G, F and $\mu$ are adjustable gains, E, h are defined by Lemma 1. If function

$$W(s) = h(sI-A_p)^{-1}E$$

is strictly positive-real, and let

$$\dot{F}^T = -\Gamma_1 e_1 x_m \tag{5a}$$

$$\dot{G}^T = -\Gamma_2 e_1 z_m \tag{5b}$$

$$\dot{\mu} = -\Gamma_3 e_1 u \tag{5c}$$

then, we have

$$\lim_{t\to\infty} e_1(t) = \lim_{t\to\infty} [y_m(t) - y_p(t)] = 0 \tag{6}$$

where $\Gamma_1$, $\Gamma_2$ are nth, (n-1)th order positive definite matrixes respectively, $\Gamma_3$ is a positive number.

PROOF: Firstly, the adjustable model (4) has the same form as that of the plant model (2). The adjustable model matches perfectly with the model (2), only if

$$f_i^* = -a_i + \alpha_i \qquad (i=0,1,\cdots,n-1)$$

$$g_j^* = b_j - \mu^* h_j \qquad (j=0,1,\cdots,n-2)$$

$$\mu^* = b_{n-1}\ .$$

Secondly, let

$$F = F^* + \theta\ , \quad G = G^* + \psi\ , \quad \mu = \mu^* + \zeta$$

where $F^*$, $G^*$ and $\mu^*$ are perfectly matching parts of adjustable gains F, G and $\mu$ respectively; $\theta$, $\psi$, and $\zeta$ are non-matching parts of F, G and $\mu$ respectively, the adjustable model can be written as

$$\begin{bmatrix} \dot{X}_m \\ \dot{Z}_m \end{bmatrix} = \begin{bmatrix} (A_m+EF^*) & EG^* \\ 0 & C \end{bmatrix} \begin{bmatrix} X_m \\ Z_m \end{bmatrix} + \begin{bmatrix} \mu^* E \\ D \end{bmatrix} u + B_e\Phi W$$

where

$$\Phi = (\theta,\psi,\zeta) \;, \qquad W^T = (x_m^T, z_m^T, u) \;.$$

Let

$$e = \begin{bmatrix} X_m - X_p \\ Z_m - Z_p \end{bmatrix}, \qquad e_1 = y_m - y_p$$

then

$$\dot{e} = A_e e + B_e \Phi W \tag{7a}$$

$$\dot{e}_1 = h_e e \tag{7b}$$

Applying Lemma 2, yield

$$h_e(sI-A_e)^{-1}B_e = h(sI-A_p)^{-1}E$$

$$= \frac{s^{n-1} + h_{n-2}s^{n-2} + \cdots + h_1 s + h_0}{s^n + a_{n-1}s^{n-1} + \cdots + a_1 s + a_0} \;,$$

According to Kalman-Yacubovich theorem, there are positive define matrixes P and Q which satisfy equations

$$A_e^T P + PA_e = -Q$$

$$PB_e = h_e^T \;,$$

if function $h(sI-A_p)^{-1}E$ is strictly real positive. Now choose function

$$V = \frac{1}{2}\,(e^T Pe + \Phi\Gamma^{-1}\Phi^T)$$

then

$$\dot{V} = -\frac{1}{2}\, e^T Qe + \Phi e_1 W + \Phi\Gamma^{-1}\dot{\Phi}^T \;.$$

Let $\dot{\Phi} = -\Gamma e_1 W$

where $\Gamma$ is a given 2nd order positive define matrix, then

$$\dot{v} = -\frac{1}{2}\, e^T Qe \leqq 0$$

According to Lyapunov stability theory, we have eqn.(6), where

$$\Gamma = \mathrm{diag}(\Gamma_1, \Gamma_2, \Gamma_3) \;. \qquad \text{QED}$$

The adjustable model designed by the theorem is shown as Fig.2.

## ADJUSTMENT LAW OF PID PARAMETERS

The parameters of the PID regulator are designed on-line in accordance to the adjustable model. The criteria of the design could be optimal control, pole placement, etc. A simple and efficient method is employed in our governor. The linearization model of the hydroelectric generating set is characterized by the transfer function

$$\frac{Y_p(s)}{U(s)} = \frac{b_0 - b_1 s}{S^2 + a_1 S + a_0}$$

where $Y_p$ is frequency of generator, u is the output of the controller, and coefficients $a_0, a_1, b_0, b_1 > 0$. It is a non-minimum phase system. Its coefficients $a_0, a_1, b_0, b_1$ vary with the operation conditions.

Obviously, the transfer function of the adjustable model is

$$\frac{U_m(s)}{U(s)} = \frac{\tilde{b}_0 - \tilde{b}_1 s}{S^2 + \tilde{a}_1 S + \tilde{a}_0}$$

where $\tilde{a}_0 = \alpha_0 + f_0$, $\tilde{a}_1 = \alpha_1 + f_0$, $\tilde{b}_0 = g_0 + \mu h_0$, and $\tilde{b}_1 = g_1$.

The PID regulator can be characterized by a transfer function

$$W_c(s) = \frac{S^2 + k_1 S + k_0}{k_2 S}$$

The characteristic polynomial $S^2 + \tilde{a}_1 S + \tilde{a}_0$ is stable, in case the model and the adaptation law are properly designed. Because the values for the parameters of the adjustable model are exactly known, the coefficients $k_1$ and $k_0$ can be set as $k_1 = \tilde{a}_1$ and $k_0 = \tilde{a}_0$ consistently. Then, the transfer function of the reference model composed of PID regulator and adjustable model (see Fig.2) is

$$W_r(s) = \frac{\tilde{b}_0 - \tilde{b}_1 S}{\tilde{b}_0 + (k_2 - \tilde{b}_1)S} \;,$$

which is a 1st order dynamic unit. It is obvious that the adjustment law of $k_2$ can be easily obtained according to the ideal performance of process control.

## THE RESULTS OF EXPERIMENT

The adaptive speed governor has been used to control a test table which can simulate a real hydroelectric generating set and its enviroment. Fig.3 illustrates for the system the frequency following process of the generator with different controller. Curve 1 is the reference input. Curve 2 in the process when the system operates at the normal point ($a_0$=0.42, $a_1$=1.60) with a general (non-adaptive) PID regulator or an adaptive regulator. Curve 3 is the process, when the parameters of the plant become $a_0$=0.33, $a_1$=1.17 and the general PID regulator is applied. Curve 4 is the process, when the parameters of the plant become $a_0$=0.33, $a_1$=1.17 and the adaptive PID regulator is applied. The results show that the MRAC system always has good performance in spite of variation of the plant parameters.

## CONCLUSIONS

It has been shown that the microprocessor-based adaptive speed governor provides the following advantages: It requires less data of the plant mathematical model but possesses better performance under changing operation conditions and environment. The structure of the controller is more easily relizable. The algorithms for the adaptive mechanism is simple. The new configuration of the MRAC system belongs to the direct adaptation type, but it can be used in non-minimum phase systems. It may, therefore, be expected that an advanced hydraulic turbine speed governor will be available on the market in the near future.

## REFERENCES


Narendra, K.S. and L.S. Valavali (1978). Stable adaptive controller design — direct control. IEEE Tran. autom. Control, AC-23, 570-583.

Narendra, K.S., Y.H. Lin and L.S. Valavali (1980). Stable adaptive controller design, part II: proof of stability. IEEE Tran. autom. Control, AC-25, 440-448.

Landau, I.D. (1979). Adaptive Control — model reference approach, Marcel Dekker, New York.

# ADAPTIVE ROBUST SPEED REGULATOR FOR HYDROELECTRIC GENERATING SET

**Wang Hong**

*Department of Electrical Power Engineering,
Huazhong University of Science and Technology, Wuhan,
The People's Republic of China*

Abstract. This paper presents an adaptive robust speed regulator suitable for a hydroelectric generating set. By using lyapunov stability theory and E.J. Davison(1976) robust control theory, the control schemes and adaptive laws are obtained. The analysis of the dynamic properties of the whole speed regulating system is given and it is shown that the dynamic properties of the system is much improved by using the control schemes in the paper. Finally, a Z-8000 16bit microprocessor is used to realize the real-time simulation. The simulation results show that the dynamic preformance of an adaptive robust control system is better than the dynamic preformance of the normal and widely used speed regulating system in which a PID regulator is adopted. The response time is shortened by 30%, and the robustness of the system is increased.

Keywords. Adaptive control; robust control; speed control; stability; dynamic property.

## INTRODUCTION

In recent years, many control schemes have been developed which solve the control problem of an unknown plant. These control schemes may be classified into two main parts — adaptive control and robust control. Compared with robust control, adaptive control has witnessed some very important theoretical developemts and has been applied successfully in many industrial processes.

The main idea of robust control is to use a time-invariant controller for a plant with unknown constant parameters or slow-time-varying unknown parameters. This control scheme is simple in structure but is not suitable for a plant whose parameters change over a large range, such as the speed regulating system of hydroturbines and the voltage regulating system of synchronous machines. So it is necessary to study some methods which will improve the dynamic properties and enhance the adaptability and robustness of a robust control system.

## UNIFIED MODEL AND ITS ADAPTIVE CONTROL LAW

It is assumed that the plant is defined by the usual linear, constant coefficient dynamic equations

$$\dot{x}_n=A_n x_n+B_n V_n \tag{1}$$

$$y_n=C_n x_n \tag{2}$$

where $x_n$ is a state vector of dimension n, $V_n$ is a control vector of dimension m and $y_n$ is a measurable output vector of dimension m. $A_n$ and $B_n$ are unknown constant matrices, and $C_n$ is known matrix. The desired model is as follows:

$$\dot{x}_m=A_m x_m+B_m r_m \tag{3}$$

$$y_m=C_n x_m \tag{4}$$

where $x_m$ is an (n×1) state vector of the model, $r_m$ is an (m×1) input vector and $y_m$ is an (m×1) output vector of the model. $A_n$ is a stable matrix. The objective is to find control $V_n$ so that as $t\to\infty$, $e_m(=y_m-y_n)\to 0$ .

The realizable adaptive control law for the above objective is chosen as follows

$$V_n=L(Ky_n+r_m) \ . \tag{5}$$

Assume there exists L* and K* such that

$$A_n+B_nL^*K^*C_n=A_m \ , \tag{6}$$

$$B_nL^*=B_m \ . \tag{7}$$

Then the state error vector $e(=x_m-x)$ satisfies

$$\dot{e}=A_m e+B_m\Phi Z \tag{8}$$

$$e_m=C_n e \tag{9}$$

where $\Phi=[K^*-K,\ L^{-1}-L^{*-1}]$, $Z=[y_u^T,[L(Ky_u+r_m)]^T]^T$.

The adaptive updating law can be directly obtained from the following theorem.

Theorem.1
If there exists a positive definite matrix P such that

$$A_m^TP+PA_m^T=-I_n \tag{10}$$

$$C_n=B_m^TP$$

then the system described by

$$\dot{e}=A_m e+B_m\Phi Z \ , \tag{11}$$

$$e_m=C_n e \ . \tag{12}$$

$$\Phi=\Phi_p+\Phi_I \ , \tag{13}$$

$$\Phi_p=-T_1 e_m Z^T \ ; \quad (T_1\geqq 0) \ , \tag{14}$$

$$\dot{\Phi}_I=-T_2 e_m Z^T \ ; \quad (T_2>0) \tag{15}$$

is stable and

$$\lim_{t\to\infty} e=0 \qquad (16)$$

Proof: Consider the following quadratic function as a lyapunov function candidate for the above system:

$$V=\frac{1}{2}e^T Pe+\frac{1}{2}\text{trace}(\Phi_I^T T_2^{-1}\Phi_I) \ . \qquad (17)$$

On differentiating (17) with respect to time and using (11)-(15)

$$\dot{V}=-\frac{1}{2}e^T e-e^T PB_m T_1 B_m^T PeZ^T Z\leq 0 \qquad (18)$$

thus $\dot{V}$ is negative semidefinite and the system (11)-(15) is, therefore, stable. Furthermore, if $e\neq 0$, then $\dot{V}$ will be negative definite, and V decrease until $e=0$, so we have

$$\lim_{t\to\infty} e=0 \ . \qquad (19)$$

## ADAPTIVE ROBUST CONTROL SCHEME

Assume that the plant to be controlled can be expressed as

$$\dot{x}_p=A_p x_p+B_p u_p \qquad (20)$$

$$y_p=C_m x_p \qquad (21)$$

where $x_p$ is an $(n\times 1)$ state vector $u_p$ and $y_p$ are $(m\times 1)$ input and output vectors, and $A_p$ and $B_p$ are unknown.

The robust controller of this plant is composed of

a) Servo Compensator

$$\dot{\xi}=\Omega\xi+\theta e_p \qquad (22)$$

$$e_p=y_r-y_p \qquad (23)$$

where $y_r$ is the set point, and

b) Stabilizer Compensator

$$\dot{\eta}=\Lambda_0\eta+\Lambda_1\xi+\Lambda_2 u_p+\Lambda_3 y_p \ . \qquad (24)$$

The control vector is

$$u_p=K_1\xi+K_2\eta+K_3 y_p+Ly_r \ . \qquad (25)$$

In a normal robust control system, $K_i$ (i=1,2,3) and L is constant, this makes the dynamic properties of the system undesirable. Consider now the adaptive updating laws for these gain matrices.

The whole system can be described by

$$\begin{bmatrix}\dot{x}_p\\ \dot{\xi}\\ \dot{\eta}\end{bmatrix}=\begin{bmatrix}A_p & 0 & 0\\ -\theta C_m & \Omega & 0\\ \Lambda_3 C_m & \Lambda_1 & \Lambda_0\end{bmatrix}\begin{bmatrix}x_p\\ \xi\\ \eta\end{bmatrix}+\begin{bmatrix}B_p & 0 & 0\\ 0 & 0 & 0\\ \Lambda_2 & 0 & 0\end{bmatrix}\begin{bmatrix}u_p\\ u_1\\ u_2\end{bmatrix}$$

$$+\begin{bmatrix}0 & 0 & 0\\ \theta & 0 & 0\\ 0 & 0 & 0\end{bmatrix}\bar{y}_r \qquad (26)$$

where $u_1$ and $u_2$ are anxiliary input, $\bar{y}_r=[y_r,0,0]$. The output equation is

$$\bar{Y}_m=\begin{bmatrix}y_p\\ \xi\\ \eta\end{bmatrix}=\begin{bmatrix}C_m & 0 & 0\\ 0 & I & 0\\ 0 & 0 & I\end{bmatrix}\begin{bmatrix}x_p\\ \xi\\ \eta\end{bmatrix} \ . \qquad (27)$$

Let $x_n=[x_p^T,\xi^T,\eta^T]$, $V_n=[u_p^T u_1^T u_2^T]$, $y_n=\bar{Y}_m$ and let

$$A_n=\begin{bmatrix}A_p & 0 & 0\\ -\theta C_m & \Omega & 0\\ \Lambda_3 C_m & \Lambda_1 & \Lambda_0\end{bmatrix}, \quad B_n=\begin{bmatrix}B_p & 0 & 0\\ 0 & 0 & 0\\ \Lambda_2 & 0 & 0\end{bmatrix} \qquad (28)$$

$$C_n=\begin{bmatrix}C_m & 0 & 0\\ 0 & I & 0\\ 0 & 0 & I\end{bmatrix}, \quad E_n=\begin{bmatrix}0 & 0 & 0\\ \theta & 0 & 0\\ 0 & 0 & 0\end{bmatrix} \qquad (29)$$

then the equation (25) can be expressed as

$$\dot{x}_n=A_n x_n+B_n V_n+E_n\bar{y}_r \qquad (30)$$

$$y_n=C_n x_n \ . \qquad (31)$$

The model is chosen as

$$\dot{x}_m=A_m x_m+B_n\bar{y}_r+E_n\bar{y}_r \ , \qquad (32)$$

$$y_m=C_n x_m \ , \qquad (33)$$

then the control is the same as (5)

$$V_n=L(Ky_n+\bar{y}_r) \ . \qquad (34)$$

Using the results of the above section, we obtain the following adaptive apdating law for K and L

$$\dot{K}_I=T_2 e_m Z^T I_1 \qquad (35)$$

$$\dot{L}=LT_2 e_m Z^T I_2 L \qquad (36)$$

where $I_1$ and $I_2$ are sample matrices.

## APPLICATION TO THE SPEED REGULATOR OF A HYDROTURBINE

The model of the hydro-turbine to be controlled is expressed as

$$\dot{x}_p=\begin{bmatrix}-\lambda_1 & \lambda_3\\ 0 & -\lambda_2\end{bmatrix}x_p+\begin{bmatrix}b_1\\ b_2\end{bmatrix}u_p \qquad (37)$$

$$y_p=[1\ \ 0]x_p \ . \qquad (38)$$

where $x_p$ is an$(2\times 1)$ state vector, $y_p$ is the mesurable output. The servo compensator is chosen as

$$\dot{\xi}=-0.1\xi+e_p \qquad (39)$$

with $K_1(0)=1$, and the stabilizer compensator is as follows

$$\dot{\eta}=-2\eta+u_p-0.5y_p \qquad (40)$$

with $K_2(0)=-1$. The desirable model corresponding to (37)-(40) is

$$\dot{x}_m=\begin{bmatrix}-2 & 1.5 & 0 & 0\\ 0 & -1 & 0 & 0\\ -1 & 0 & -0.2 & 0\\ -1 & 0 & 0 & -4\end{bmatrix}x_m+\begin{bmatrix}-1.5\\ 2\\ 1\\ 0\end{bmatrix}y_r \qquad (41)$$

$$y_r=1(t) \ . \qquad (42)$$

$$y_m=\begin{bmatrix}1 & 0 & 0 & 0\\ 0 & 0 & 1 & 0\\ 0 & 0 & 0 & 1\end{bmatrix}x_m \ . \qquad (43)$$

The structure of the above system is shown in Fig. I

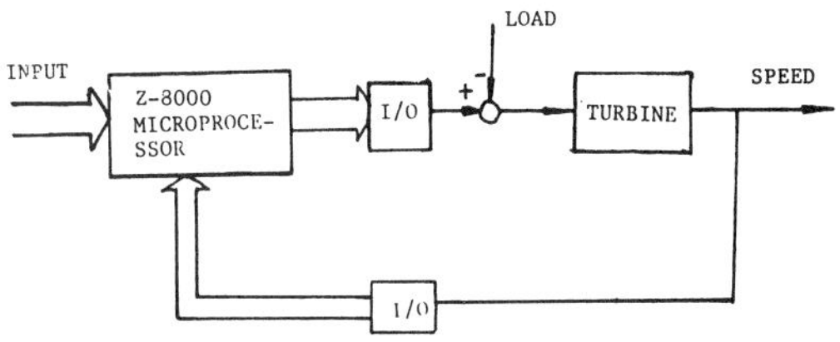


Fig. 1 Basic structure of adaptive robust speed regulator.

where a Z-8000 16bit microprocessor is used to implement the control scheme

Experimental Results

Using the control scheme described in the above section and the realization of Fig. 1, we can obtain the results about the set up of the speed of hydroaturbine which is shown in Fig. 2

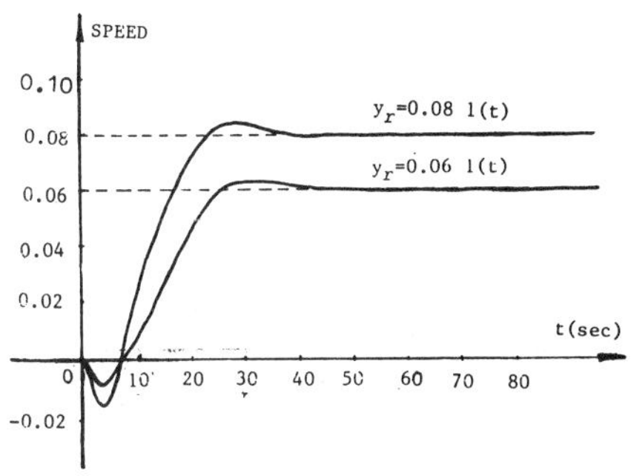


Fig. 2 The step response of the speed.

the response is better that that of a normal PID speed regulator.

## CONCLUSION

A new method for improving the dynamic properties of a normal robust control system has been obtained by combining the adaptive control method with a robust controller. The resultant control scheme can be implemented easily using a microprocessor. A large improvement of dynamic preformance has been obtained for the new control scheme as shown by one particular application.
A further problem to be solved is how to simplify the updating law.

## REFERENCES

Davison, E.J., Wang, S.H., (1975), On pole assignment in linear multivariable system using feedback, IEEE Trans. Automat. Control. Vol. AC-20, 516-518.

Davison, E.J., Goldenberg, A., (1975), The robust control of a general servomechanism problem: The servo compensator. Automatica. Vol.11, 461-471.

Narendra, K.S., Kudva, P., (1974), Stable adaptive schemes for systems identification and control-Part I, II. IEEE Trans. syst, Man, Cybern. Vol. SMC-4. 542-560.

Zames, G., Francis, B.A., (1981), A new approach to classical frequency methods: Feedback and minimax sensitivity. Proc. 20th Conf. Decision Contr., San Diego, CA, 867-874.

# NEW CONTROL ALGORITHMS BASED ON THE SLIDING MODE CONTROL METHODOLOGY

**S. Z. Sarptürk and O. Kaynak**

*Boğaziçi University, Bebek, Istanbul, Turkey*

Abstract. The use of the sliding mode control in conjunction with the full output feedback problem is investigated. New algorithms with adaptive nature based on the equivalent control concept are proposed to eliminate the high frequency noise introduced into the system by the well known discontinuous laws used in the sliding mode control methodology. A numerical example is discussed as illustration.

Keywords. Variable Structure Systems, Sliding Modes, Digital Control, Adaptive Control, Discrete Systems.

## INTRODUCTION

The sliding mode control systems, also known as variable structure control systems (VSS), are characterized by a discontinuous feedback control law which switches the structure of the system during the evolution of the state vector in order to maintain the state trajectories in a pre-defined subspace (Utkin, 1977).

By introducing the sliding modes into the system, it is possible to stabilize the system while achieving insensitivity to external disturbances and parametric deviations (Drazenovic, 1969). The basic philosophy of the sliding mode control can be summarized as follows:

Consider the following continuous-time system,

$$\dot{x}(t) = Ax(t) + Bu(t) + Dv(t) \tag{1}$$

where $x(t) \in R^n$; $u(t) \in R^m$ and (A,B) is a controllable pair. The upper and lower bounds for v(t) and for the elements of A and B are assumed to be known.

The objective of the sliding mode control methodology is to steer the states of the system into a (n-m) dimensional subspace $S \subset R^n$ and to maintain the subsequent motion of the state trajectories on this (n-m) dimensional manifold S, such that as $t \to \infty$ ; $x(t) \to 0$. The sliding manifold S can be defined as

$$S = N(C) = \bigcap_{i-1}^{m} s_i[x(t)] \tag{2}$$

where N(.) denotes the null space of (.) and $s_i[x(t)] = \{ x(t) : c^i x(t) = 0 \}$, $c^i$ being the i´th row of the real (m x n) matrix C.

The problem is to design a globally asymptotically stabilizing feedback control law $u(t) = u^*[x(t)]$ assuring the sliding motion on S. It can be shown that, once the manifold has been reached, the system behaves like an autonomous linear system of reduced order with poles arbitrarily located in the s-plane. It can also be shown that in this case the plant is insensitive to parameter changes and external disturbances (Utkin, 1977; Drazenovic, 1969; Bartolini, 1985; Ryan, 1983).

For the linear system of Eq. (1), u*(.) is obtained such that a suitable pair of functions $u^+[x(t)]$ and $u^-[x(t)]$ is chosen and the control is generated according to

$$u_i[x(t)] = \begin{cases} u_i^+[x(t)]; s_i[x(t)] > 0 \\ u_i^-[x(t)]; s_i[x(t)] < 0 \end{cases} \tag{3}$$

where $u_i(.)$ denotes the i´th input of the system and $s_i(.)$ the hyperplane attributed to the i´th input which satisfies the sliding mode condition

$$\dot{s}(x)s(x) < 0 \tag{4}$$

In the discrete time implementation of the sliding mode methodology, the sliding elements are replaced by a computing device which changes the structure of the system at discrete instants. This is in contrast with the continuous-time systems where the switching of the system structure can be made at any instant, as soon as the state trajectories cross the switching hyperplane. Since the control input is computed at discrete instants and applied to the system during the sampling interval, inevitably, a non-ideal sliding regime will appear.

The non-ideal sliding motion is the most important problem in the discrete-time sliding mode control systems since it introduces a big amount of high frequency noise into the system. Generally, if the system is of low-pass type, this noise may be at tolerable levels at the output of the system, however, the unmodelled fast modes existing in the system dynamics may be excited

In their form presented above, the sliding mode control systems requires the use the full state vector in their feedback control law. This is an undesirable feature because the availability of the states of the system is often restricted. Therefore the use of the output feedback in the control loop will be preferred. Here, the delayed outputs will be used as the "pseudo-states" of the system (Mosca, 1982). This naturally leads us to the ARMA type representation of the plant. However, the effect of the high frequency noise is more pronounced when full output feedback is used since the high frequency chatter is directly introduced to the output of the control system. If we want to use full output feedback in conjunction with the sliding mode control concept, special precautions should be taken as a remedy to the high frequency noise problem. A non-switching type control will obviously solve the problem. This fact will be discussed in the sequel and suitable control laws will be derived. These laws are based on the equivalent control concept introduced by Utkin (1977). The equivalent control is defined as the average value of the discontinuous control law which produces the sliding motion. If the sliding manifold is reached, the subsequent motion should continue on this manifold with the application of the equivalent control. But the existence of external disturbances and parametric deviations in the system makes impossible the computation of the exact value of the equivalent control, instead, only a nominal value can be computed. The application of this nominal value to the system will obviously cause a deviation of the trajectories from the desired manifold. Therefore the continuous control laws based on the equivalent control concept must compensate against the disturbing factors. As a result of this situation, the derived control laws have an adaptive nature. The first algorithm which will be presented below is a "corrective" algorithm and the deviations are compensated using the past data. The second algorithm utilizes a predictive scheme and the control input is computed according to a prediction error. We will call these algorithms C/A algorithm and P/A algorithm standing for corrective-adaptive and predictive-adaptive respectively. Both of them allow a direct determination of the error dynamics by a suitable choice of a polynomial defining the sliding surface. The saturation type nonlinearities can be considered automatically by these algorithms.

## PROBLEM FORMULATION FOR SISO SYSTEMS

Consider the following linear system represented in the ARMA form,

$$y(k+1) = A(q^{-1})y(k) + B(q^{-1})u(k) + D(q^{-1})v(k) \qquad (5)$$

where $q^{-1}$ is the unit delay operator. The polynomials A(.), B(.) and D(.) are of degree n,m and d respectively and are defined as follows:

$$A(q^{-1})=a_0+a_1q^{-1}+..+a_nq^{-n}$$
$$B(q^{-1})=b_0+b_1q^{-1}+..+b_mq^{-m}$$
$$D(q^{-1})=d_0+d_1q^{-1}+..+d_rq^{-r} \qquad (6)$$

where $m \leq n$ and $r \leq n$. It is possible to express $a_i$ and $b_i$ as,

$$a_i=\hat{a}_i+\delta a_i \quad ; i=0...n$$
$$b_i=\hat{b}_i+\delta b_i \quad ; i+0...m \qquad (7)$$

where $\hat{a}_i$, $\hat{b}_i$ denotes the expected values and $\delta a_i$, $\delta b_i$ the uncertainties. The sliding surface is defined as follows:

$$s(k+1) = C(q^{-1})y(k+1) = 0 \qquad (8)$$

where

$$C(q^{-1})=1+c_0q^{-1}+...+c_{n-1}q^{-n} \qquad (9)$$

Conformly to its definition, the equivalent control will be the control input which makes s(k+1) equals to zero. Clearly, with the definition of Eq. (8), if s(k+1) can be made zero for all k, the system dynamics will be governed by Eq. (8) without the influence of any external disturbances and internal uncertainties. For the stability of the control system, Eq. (9) should describe a stable polynomial i.e. all the roots of Eq. (9) must be inside the closed unit disc.

As a step in the derivation of the control law, initially, assume that A(.) and B(.) are known polynomials. In this case the equivalent control can be obtained from Eq. (5) and Eq. (8) as,

$$u=-(1/b_0)[F(q^{-1})y(k)+B_1(q^{-1})u(k)] \qquad (10)$$

where

$$B_1(q^{-1}) = B(q^{-1}) - b_0$$
$$C_1(q^{-1}) = q[C(q^{-1})-1]$$
$$F(q^{-1}) = A(q^{-1}) + C_1(q^{-1}) \qquad (11)$$

The application of the equivalent control obtained in Eq. (10) to the system yields the following output at the (k+1)´th instant as,

$$y(k+1) = -C_1(q^{-1})y(k)+D(q^{-1})v(k) \qquad (12)$$

In this case the value assumed by s(k+1) will be:

$$s(k+1) = D(q^{-1})v(k) \qquad (13)$$

As it can be seen from Eq. (13), if the system parameters are exactly known, the deviation of s(k+1) from zero shows the effect of the external disturbances acting on the system up to the k´th instant. This property can be used in order to derive a control algorithm which measures the effects of the external disturbances and corrects the input accordingly. The derivation of this algorithm is presented below.

## DERIVATION OF THE C/A ALGORITHM

Before proceeding further, a key assumption involving the external disturbance dynamics will be made and will be adopted throughout the work. This assumption is that the external disturbance dynamics is slow with respect to the sampling frequency utilized in the system. Therefore it is possible to write that,

$$D(q^{-1})v(k) \cong D(q^{-1})v(k-1) \qquad (14)$$

In order to compensate the effects of the external disturbances, let us add a scalar correction term denoted by w(k) into the control input obtained in Eq. (10). With the addition of this correction term to the equivalent control we obtain,

$$u(k)=-(1/b_0)[F(q^{-1})y(k)+B_1(q^{-1})u(k)+w(k)] \qquad (15)$$

The application of this control law to the system yields,

$$y(k+1)=-C_1(q^{-1})y(k)-w(k)+D(q^{-1})v(k) \qquad (16)$$

In this case s(k+1) becomes

$$s(k+1) = D(q^{-1})v(k) - w(k) \qquad (17)$$

In a similar way, at the previous sampling instant it is possible to express s(k) in terms of the previous correction term and the external disturbance as,

$$s(k) = D(q^{-1})v(k-1) - w(k-1) \quad (18)$$

If we solve Eq. (18) for the disturbance term, we obtain:

$$D(q^{-1})v(k-1) = s(k) + w(k-1) \quad (19)$$

where s(k) is a measurable value obtained from the system output and w(k-1) is a known value applied to the system. Therefore if we assume that the system parameters are exactly known,the effect of the external disturbances can be identified using Eq. (17). Conformly to the assumption made above, by replacing Eq. (19) into Eq. (17) and by letting s(k+1) equals to zero, we obtain the following recursive update equation for w(k) such that:

$$w(k) = w(k-1) + s(k) \quad (20)$$

In the presence of the exact knowledge of the plant parameters, the control given in Eq. (15) together with Eq. (20) will make the system asymptotically stable within the dynamics imposed by Eq. (10). This control will reject any slow disturbances from the system. Clearly the assumption that the perfect knowledge of the plant parameters are available is an excessive idealization which can never be reached in real world situations. It can be shown that, in the case of parametric deviations with known bounds, a suitable polynomial $C(q^{-1})$ can be found in order to place the poles of the closed loop system inside the unit circle. It can also be shown that the disturbance rejection property of the algorithgm is preserved in spite of the parametric uncertainties.

## Addition of a reference trajectory

The algorithm derived in the previous part drives the output of the system to zero. Since it is desired that the output reaches a certain steady-state value or tracks a reference trajectory, a reference input should be added into the algorithm. The addition of a reference can be achieved by a suitable coordinate transformation. Let r(k) denote the desired reference. In this case the output error can be expressed as,

$$e(k) = r(k) - y(k) \quad (21)$$

Here, r(k) may be a fixed reference value or a trajectory. If r(k) is a fixed reference value, the error dynamics can be arbitrarily determined by the choice of $C(q^{-1})$. On the other hand, if r(k) is a pre-determined trajectory, the error dynamics between the reference trajectory and the output can be arbitrarily determined. From Eq. (21), the sliding motion can be defined as,

$$s(k+1) = C(q^{-1})e(k+1) = 0 \quad (22)$$

Clearly, this definition allows us the determination of the error dynamics by the selection of $C(q^{-1})$. Since Eq. (22) can be expressed as follows;

$$s(k+1)=C(q^{-1})r(k+1)-y(k+1)-C(q^{-1})y(k) \quad (23)$$

from Eq. (5) and Eq. (15), the control input can be written as,

$$u(k)=-(1/b_0)[F(q^{-1})y(k)+B_1(q^{-1})u(k) -C(q^{-1})r(k+1)+w(k)] \quad (24)$$

By applying the control given in Eq. (24) to the system we obtain,

$$y(k+1)=-C_1(q^{-1})y(k)+C(q^{-1})r(k+1)-w(k)+ D(q^{-1})v(k) \quad (25)$$

therefore s(k+1) will be,

$$s(k+1) = w(k) - D(q^{-1})v(k) \quad (26)$$

which has the same form as Eq. (17). By following a similar way that we used in the derivation of Eq. (21), we obtain the following update equation for the correction term;

$$w(k) = w(k-1) - s(k) \quad (27)$$

Since the true values of the parameters may not exactly be known, only their estimated values can be used in the computation of the control law. A proof of the stability and disturbance rejection property can easily be made, however will not be presented here.

## Corollary

If the assumptions made in the derivation of the control law hold, the following algorithm will make the system asymptotically stable within the error dynamics described by Eq. (22) and any slow disturbance will be rejected from the system.

$$s(k)=C(q^{-1})[r(k)-y(k)]$$

$$w(k)=w(k-1)-s(k)$$

$$u(k)=-(1/\hat{b}_0)[\hat{F}(q^{-1})y(k)+\hat{B}_1(q^{-1})u(k) -C(q^{-1})r(k+1)+w(k)] \quad (28)$$

If the assumption made on the external disturbance dynamics does not hold, an error proportional to the derivative of the external disturbance will appear at the system output.

## DERIVATION OF THE P/A ALGORITHM

The aim of the C/A algorithm presented above was to identify the external disturbances acting on the system and to compensate their effects using a correction term added to the control input. However, the philosophy of the P/A algorithm is different. The objective is to predict the value of s(k+1) based on the past measurements and on the prediction error made at the k´th sampling instant. Then a control input is computed to make the predicted value of s(k+1) equal to zero.

In the derivation of the P/A algorithm it will be assumed that the measurement process is not corrupted with noise i.e. the measurement process is deterministic. This means that the uncertainties in the system dynamics comes from the parametric deviations and external disturbances. Therefore the predicted value of s(k+1) based on the measurements up to the k´th instant can be expressed as (Meditch, 1969),

$$\hat{s}(k+1|k) = \hat{y}(k+1|k) + C_1(q^{-1})y(k) \quad (29)$$

where $\hat{y}(k+1|k)$ is the predicted value of the system output at (k+1) and can be given as,

$$\hat{y}(k+1|k) = \hat{A}(q^{-1})y(k) + \hat{B}(q^{-1})u(k) \tag{30}$$

The prediction error which occurs at the k'th instant can be written as,

$$\tilde{s}(k|k-1) = s(k) - \hat{s}(k|k-1) \tag{31}$$

where s(.) denotes the actual value of s(.) obtained from the plant output. Now, by using Eq. (30) as an innovation term, the predicted value of s(k+1) can be corrected according to the prediction error at the k'th instant (Rouhani, 1982) such that;

$$\hat{s}(k+1|k,k-1) = \hat{s}(k+1|k) + \tilde{s}(k|k-1) \tag{32}$$

Here, our objective will be to obtain a suitable control law to render the left-hand side of Eq. (32) equal to zero. Since it is possible to express Eq. (32) as,

$$\hat{y}(k+1|k)+C_1(q^{-1})y(k)+\tilde{y}(k|k-1)=0 \tag{33}$$

where $\tilde{y}(k|k-1)$ is defined as follows:

$$\tilde{y}(k|k-1) = y(k) - \hat{y}(k|k-1) \tag{34}$$

From the definitions made above and from Eq. (33) the contrtol input can be computed as follows:

$$u(k)=-(1/\hat{b}_0)[\hat{F}(q^{-1})y(k)+\hat{B}_1(q^{-1})u(k) +\tilde{y}(k|k-1)] \tag{35}$$

The control law given by Eq. (35) will make the system output asymptotically stable within the dynamics described by Eq. (8). The algorithm rejects any external disturbances satisfying the assumption made on the disturbance dynamics. In the presence of parametric uncertainties, the P/A algorithm allows the use of the knowledge about the uncertainties through the innovation process.

In order to include a reference trajectory, the same method as the one used previously will be repeated. The definition of the output error was given in Eq. (21). Here, the predicted output error will be defined as,

$$\hat{e}(k+1|k) = r(k+1) - \hat{y}(k+1|k) \tag{36}$$

Therefore the predicted value of s(k+1) will be,

$$\hat{s}(k+1|k) = \hat{e}(k+1|k) + C_1(q^{-1})e(k) \tag{37}$$

since

$$\tilde{s}(k|k-1) = \tilde{e}(k|k-1) = -\tilde{y}(k|k-1) \tag{38}$$

The control law u(k) can be obtained as follows:

$$u(k)=-(1/\hat{b}_0)[\hat{F}(q^{-1})y(k)+\hat{B}_1(q^{-1})u(k) -C(q^{-1})r(k+1)+\tilde{y}(k|k-1)] \tag{39}$$

Simulations showed that the P/A control yields better results than the previous one. Better transient responses and faster convergence rates are observed. As expected, the P/A algorithm is less sensitive to parametric disturbances. The inclusion of the predictive scheme also allows us to consider the saturation type nonlinearities of the output. However, this algorithm requires more computational effort than the previous one.

Corollary

As long as the assumption on the disturbance dynamics holds, the following P/A algorithm will assure the asymptotic stability of the system within the dynamics described by Eq. (22):

$$\hat{y}(k|k-1)=\hat{A}(q^{-1})y(k-1)+\hat{B}(q^{-1})u(k-1) \tag{40}$$

$$u(k)=-(1/\hat{b}_0)[\hat{F}(q^{-1})y(k)+\hat{B}_1(q^{-1})u(k) -C(q^{-1})r(k+1)+\tilde{y}(k|k-1)] \tag{41}$$

## SIMULATION RESULTS

In the following numerical exemple ,the performance of the two algorithms derived above will be presented and a comparison will be made. Consider the following linear system;

$$y(k+1)=1.26y(k)-0.041y(k-1)-0.250y(k-2)+0.254u(k) +0.290u(k-1)+0.070u(k-2)+0.270d(k)+0.125d(k-1) +0.05d(k-2) \tag{42}$$

The poles of this system in the z-plane are located at (-0.378, 0.920, 0.718). We want to place the poles of the closed loop system at (0.5, 0.5). Therefore $C(q^{-1})$ will be,

$$C(q^{-1}) = 1 - q^{-1} + 0.25q^{-2} \tag{43}$$

It is assumed that the estimated values of the system parameters are as follows;

$$A(q^{-1})= 1.10 - 0.06q^{-1} -0.3q^{-2}$$

$$B(q^{-1})= 0.40 + 0.25q^{-1} \tag{44}$$

A step disturbance of amplitude 0.25 is applied at k=50 and r(k)=1 for all k.

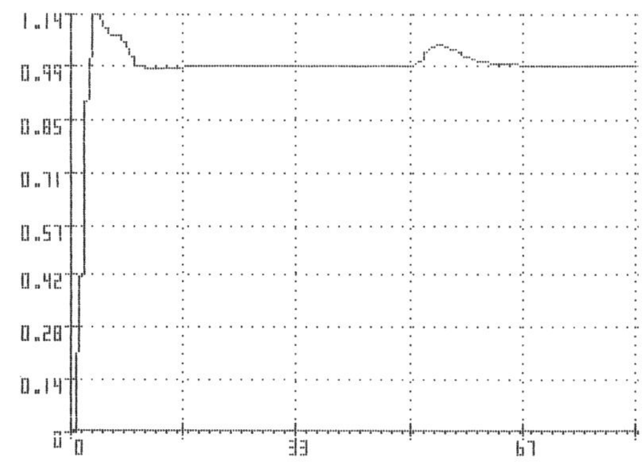


Fig. 1. Evolution of y(k) for the C/A algorithm.

Fig. 1 and Fig. 2 shows y(k) for C/A control and P/A control respectively. As it can be seen, both algorithms represent perfect disturbance rejection properties. However the P/A algorithm exhibits better transients. Fig. 3 and Fig. 4 shows the evolution of s(k)

## CONCLUSIONS

Two new control algorithms based on the sliding mode approach are represented above. The use of a non-switching control law which is uncommon to the classical VSS design eliminates the high frequency noise problem. Both algorithms reject any slow disturbances from the system. Simulations showed that the predictive algorithm is less sensitive to parametric disturbances. It is advisable to use the first algorithm in the cases where the knowledge of the system parameters is fairly good since it represents less computational complexity. The second algorithm gives better dynamic responses in the cases where there exist more ambiguity on the parameter values.

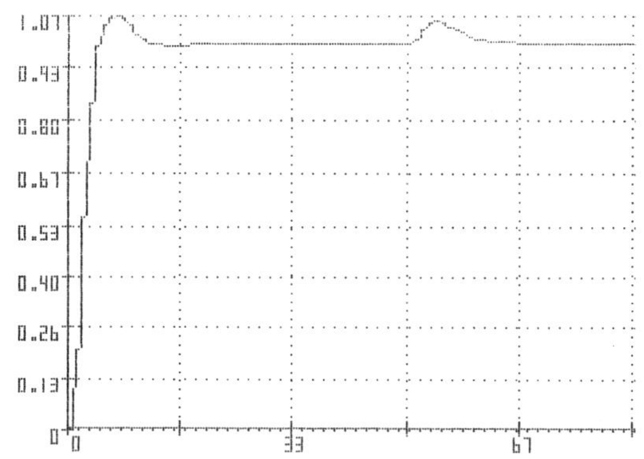


Fig. 2. Evolution of y(k) for the P/A algorithm.

As a concluding remark, we should note that the algorithms can be modified to incorporate them in model reference adaptive control schemes. It is also possible to use parameter updating mechanisms in conjunction with the control laws derived above to achieve better results. It is also possible to convert the algorithms in order to use them in multivariable control situations.

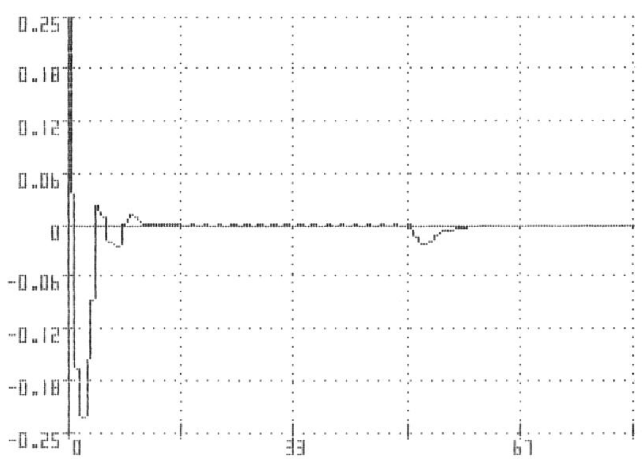


Fig. 3. Evolution of s(k) for the C/A algorithm.

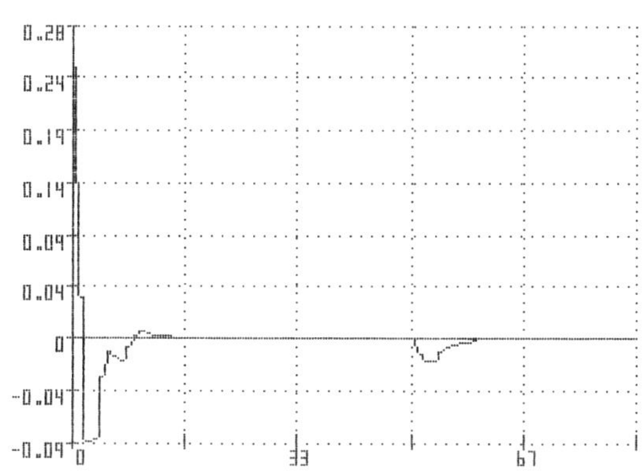


Fig. 4. Evolution of s(k) for the P/A algorithm.

## REFERENCES

Bartolini, T., and T. Zolezzi (1985). Variable structure systems nonlinear in the control law. IEEE Trans. Automat. Contr., AC-30, 681-684.

Drazenovic, B. (1969). The invariance conditions in variable structure systems. Automatica, 5, 287-295.

Itkis, U. (1976). Control systems of variable structure. New York, Wiley.

Meditch, J. S. (1969). Stochastic optimal linear estimation and control. New York, McGraw-Hill.

Mosca, E., G. Zappa and C. Manfredi (1982). Progress on multistep horizon self-tuners: The musmar approach. Ricerche di automatica, 13, 1, 85-105.

Rouhani, R., and R. K. Mehra (1982). Model algorithmic control (MAC); Basic theoretical properties. Automatica, 18, 401-414.

Ryan, E.P. (1983). A variable structure approach to feedback regulation of uncertain dynamical systems. Int. J. Control, 38, 1121-1134.

Utkin, V.I. (1977). Variable structure systems with sliding modes. IEEE Trans. Automat. Contr., AC-22, 212-222.

Young, K-K.D. (1978). Controller design for a manipulator using theory of variable structure systems. IEEE Trans. Sys. Man and Cyber., SMC-8, 101-109.

# A GENERATIVE MODEL FOR ASYNCHRONOUS FINITE-VALUED TIME SIGNALS AS APPLIED TO COMPUTER-BASED PROCESS CONTROL

**S. Y. Kwankam*, +M. E. Kaliski** and A. T. Miller****

**Ecole Nationale Superieure Polytechnique, Université de Yaounde, Yaounde, Cameroun*
***Northeastern University, Boston, MA 02115, USA*

Abstract. Computer-based process control is directly involved with the use of (asynchronous) finite-valued time signals. In this paper a generative model for such signals is proposed, one which allows both the direct imposition of physically-imposed signal contraints, and which allows for qualitative comparison of signal characteristics. Examples of the versatility of the model are given, and the relationship between this work and earlier work of the authors in asynchronous finite-state machine theory is explored.

Keywords. Sequential Machines, Automata Theory, Sequential Switching, Process Control, Computer Architecture

## INTRODUCTION

When computers are used in the process control environment, the plant outputs are often sampled asynchronously prior to being quantized, the sampling times determined by external events. The resulting signals that are processed by the controller belong the class of finite-valued asynchronous signals. This paper examines this family of signals, from the dual viewpoint of both generation and modeling.

At present, few analysis and synthesis methods are available for the evaluation and design of discontinuous feedback laws. This is particularly true for asynchronous systems, i.e. those for which the switching times of the input, output and/or state are not synchronized to a particular clock rate. The conventional theory of synchronous automata deals only with the <u>order</u> of switching and not the <u>times</u> of switching; it is not adequate to represent these asynchronous situations (Johnson, Kaliski, 1983).

This paper is concerned with asynchronous signals, and a special case, synchronous signals. A generative model for the production of asynchronous finite-valued time signals, introduced earlier by the authors (Kaliski, Kwankam, 1984), is enhanced to model such signals in a more general, fundamental way. This model will allow the direct imposition of physical contraints in an explicit, constructive fashion, and encompasses, as a special case, synchronous signals. It allows a basis for signal comparison and suitably restricted forms of it may be physically implemented to serve as self-contained signal generators.

## THE MODEL

The basic model considered is shown in Fig. 1. It consists of three parts, along with a real-time clock. The three parts are a <u>value-generator</u>, a <u>switching-time generator</u> and a <u>signal synthesizer</u>. The basic mode of operation of the system is as follows. The switching-time generator is initialized to a signal start time value, typically a non-negative real number, $t_o$, as is the real-time clock. The real-time clock generates a signal T(t) of slope equal to 1, with initial value $t_o$ (Fig. 2), which will be compared within the signal synthesizer to the output of the switching-time generator. Initially, these values are both equal to $t_o$, but in general the clock output will be less than the value of the switching-time generator output.

The value-generator is defined to operate over a finite domain, or alphabet, $A = \{ a_1, \ldots , a_n \}$, for some $n \geq 1$, and is initialized to a specific value in A, corresponding to the initial time value $t_o$. The switching-time generator

(+ This author's work has been supported, in part, by the United States Air Force Office of Scientific Research, AFSC, Contract F49620-82C-0080)

implements a mapping S: R x A --> R, using, as shown in Fig. 1, the current value-generator output and current switching-time generator output to generate the next switching-time generator output. This mapping might assume a variety of forms, allowing switching times to be uniformly spaced (the synchronous case), or recursively generated. We explore this further below. The output of the switching-time generator records the time of the next system "event". As will be explained shortly, when the real-time clock signal value becomes equal to this event time, an event occurs, causing the signal synthesizer to switch its signal output value to the value-generator output value appearing at its input terminals at the time of the event.

The value-generator itself implements a mapping V: R x A --> A, allowing both the current value-generator output and the current switching-time generator output to determine the next value-generator output. Again a variety of forms may exist for these maps.

The signal synthesizer generates the desired signal in real time (Fig. 3). The role of the signal marked "update" and the timing constraints inherent in these components are addressed next.

As seen in the detailed architecture of the synthesizer, the output signal produced is always a scaled unit step function (the box labeled step-function generator may be regarded as an idealized differentiator). The scaling amount is a value in the alphabet A and is stored in the "value store" portion of the block diagram. This value store, from a theoretical point of view, should be regarded as an idealized register capable of storing a value from the set $\{a_1, \ldots, a_n\}$ and holding this value until the input signal "change" goes from 0 to 1. (The signal "change" is assumed to go from 0 to 1 at the exact time that the system is switched on. The value initially present in the value-generator store, shown in Fig. 5, will be transferred into the signal synthesizer's value store at system start-up time).

At this point the stored value is instantaneously changed to $v_{new}$ and the output reflects this by immediately changing level from $v_{old}$ to $v_{new}$. The comparator, in ideal form, instantaneously compares the ramp input value, T(t), to the switching-time value of $s_{new}$. As soon as they become equal the comparator output becomes 1. It will immediately return to a 0 value thereafter as the value of T(t) immediately changes.

This idealized model is just that -- idealized. As soon as physical delays are imposed upon the implementation of both the value store and the comparator, as well as the functions V and S, they manifest themselves as imposed limitations on the speed at which signal switching can occur. In particular, $s_{new} = S(s_{old}, a_j)$ must differ from $s_{old}$ by some pre-determined amount for the system to function in a well-defined fashion. Further, the output signal generated will not be able to change at the value of $s_{new}$ specified. We address these issues in more detail later.

Figs. 4 and 5 detail the switching-time generator and value-generator, respectively. They are straightforward and require little additional comment.

## SOME PARTICULAR FORMS OF THE MODEL

### The Synchronous Case

We first note that this model can be used to generate arbitrary synchronous signals of clock rate $1/T_o$ as follows. Assume we wish to generate the signal with switching times 0, $T_o$, ..., $nT_o$, ... having value a(k) in A at time $kT_o$. Assuming idealized components, define S(s,v) to be equal to s + $T_o$, and define A(s,v) to be any map for which $A(nT_o,v) = a((n+1)T_o)$. Initialize the value-generator store to a(0) and initialize the switching-time store to t = 0, along with the real-time clock. Note that the map S depends only upon its first coordinate, as does the map A. This simplifies the structure of the signal generator significantly (Fig. 6) .

The issue of the effect of non-idealized components will be addressed in more general terms below.

### Generating Arbitrary Finite-Valued Signals

At the opposite extreme, the model proposed is capable of generating arbitrary finite-value time signals, and in effect, implements the <u>hybrid stack</u> described in the authors' work in asynchronous finite-state machines (Kwankam and colleagues, 1986). If we wish to generate a signal whose switching times are $t_o, \ldots, t_k, \ldots$, with value a(k) in A beginning at time $t_k$, we simply (i) initialize the value-generator to a(0) and the switching-time generator and clock to $t_o$, (ii) define S to be any extension of the map which sends $(t_k \cdot v)$ to $t_{k+1}$, and define V to be any extension of the map which sends $(t_k, v)$ to a(k+1), for k = 0,1,2,...

Note that here, too, the mappings S and V are dependent upon just the time-component, and that physical processing effects have not been considered.

### Recursively Generated Signals

In Johnson, Kaliski (1983), the issue of generating signal values recursively is addressed. That such a generative scheme can be implemented here is clear: we can basically allow V(s,v) to be viewed as a mapping on its second coordinate, allowing the current signal value to determine the next value in an iterative fashion. Similar remarks apply to the signal S(s,v). In fact, the general scheme introduced above defines $s_{new}$ iteratively in terms of $s_{old}$, but in both cases, employs monotone increasing maps to effect the iteration, since time always increases.

## LIMITATIONS IMPOSED IN THE PROCESS CONTROL ENVIRONMENT

The process control environment imposes a variety of constraints upon the control signals themselves. In the paragraphs below we discuss such constraints in general terms and show how they may be incorporated into our proposed generative model.

The most basic kind of constraint that must be satisified by any signal that is used in a control environment is that the signal dynamics must bear some relationship to the process dynamics. If the plant responds slowly to input stimuli, there is little gained by attempting to control the plant by a signal that varies too rapidly. In fact such rapid variation may be deleterious to overall system performance. There are two ways that such bounded signal variation may manifest itself. The first way is through the imposition of a minimum duration to each signal value, i.e. the existence of a time $T_{min}$ such that each signal piece persists for at least $T_{min}$ seconds. The second is through the imposition of a bound on the variation between the values of successive signal pieces, i.e. totally arbitary value changes between subsequent signal pieces are not allowed.

We can impose such constraints explicitly in our generative model. For the first constraint, we simply require that S obey a property of the form: $(S(s,v)-s) \geq T_{min}$, for all s and v. For the second constraint we assume that a distance measure dist: A --> R has been defined and we then require that $\text{dist}(V(s,v),v) \leq d_{max}$, where $d_{max}$ is the desired maximum "distance" between successive pieces.

## IMPLEMENTING THE GENERATOR IN THE PRESENCE OF PHYSICAL CONSTRAINTS

A few comments are in order concerning the physical implementation of a signal generator of the form described here. Physically, none of the components in our various figures can respond instantaneously to any signal changes. These physical limitations affect the model in two ways. The first, more obvious, limitation, is akin to that described in the preceding section: the duration of each signal piece must be long enough to allow all signals to propagate through the various blocks of the system. The second implication is more subtle: if the switching-time generator specifies a transition time in the output signal of $t = s_i$ then it will be physically impossible for the output signal to change at this time due to the physical delays in the system. Intuitively, if one agrees that the output time scale is shifted relative to the clock signal time scale by some fixed amount $T_{delay}$ then this problem can be circumvented. $T_{delay}$ must be large enough, however, to account for all physically-imposed delays of the type suggested above.

## QUALITATIVE MEASURES OF SIGNAL BEHAVIOR

Control theory is concerned, with characterizing the performance of controllers. Although performance criteria often involve such measures as the complexity of the controller, or of the sensors and actuators, one may well examine the nature of the control signals themselves. The generative scheme proposed here allows for such qualitative comparisons in a structured way.

We can compare finite-valued time signals by comparing the function S and V, for these are the only two "parameters" of the generator structure. For example, if the variations of S and V are "smaller" in one generator than in another, we might speculate that the first signal is better conditioned than the second. The process controller, then, might be both easier to implement and more reliable in the presence of perturbations.

## CONCLUSIONS

In this brief paper we have addressed the question of how one might generate finite-valued, generally asynchronous, time signals. A generative model was introduced and shown to encompass the full range of such signals, and, when suitably restricted, the class of synchronous signals. The role of the model in modeling physical processing constraints was examined, along with the modeling of physical constraints inherent in the implementation of the generator itself. The use of the generative scheme in qualitatively comparing process controller performance was explored.

## REFERENCES


Johnson, T.L., and Kaliski, M.E. (1983). Realization of asynchronous finite-state machines. Proc. 22nd IEEE Conf. on Decision and Control, San Antonio, TX.

Kaliski, M.E., and Kwankam, S.Y. (1984). Asynchronous real-time coders in the discrete control environment. Proc. 23rd IEEE Conf. on Decision and Control, Las Vegas, NV.

Kwankam, S.Y., Kaliski, M.E., Miller, A.T., and Johnson, T.L. (1986). Asynchronous finite state machines: simulations with imposed processing constraints. 1986 American Control Conf. Seattle, WA.

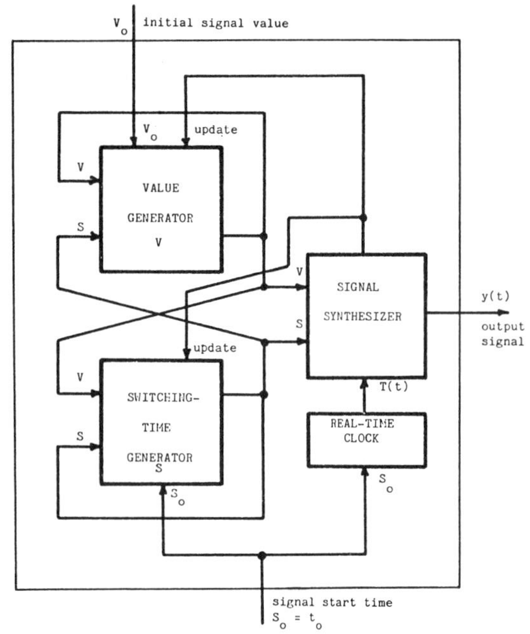


Fig. 1. The model

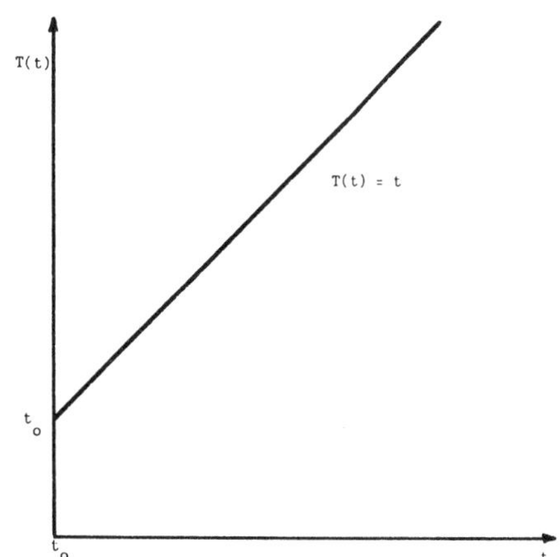


Fig. 2. Real-time clock signal

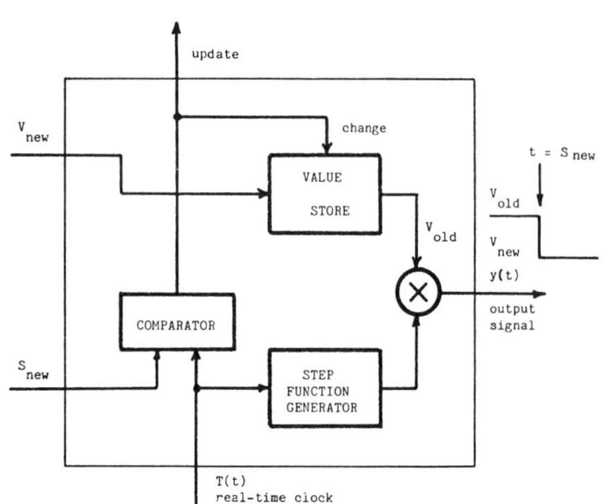


Fig. 3. Signal synthesizer

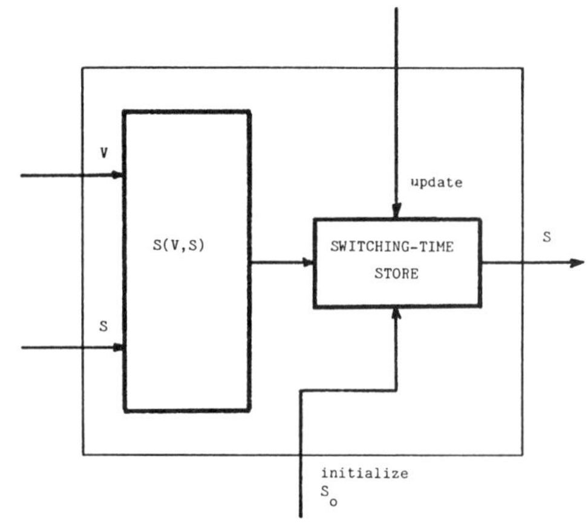


Fig. 4. Switching-time generator

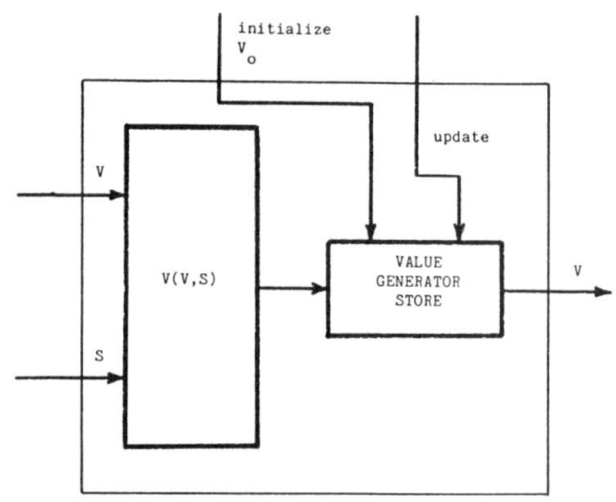


Fig. 5. Value generator

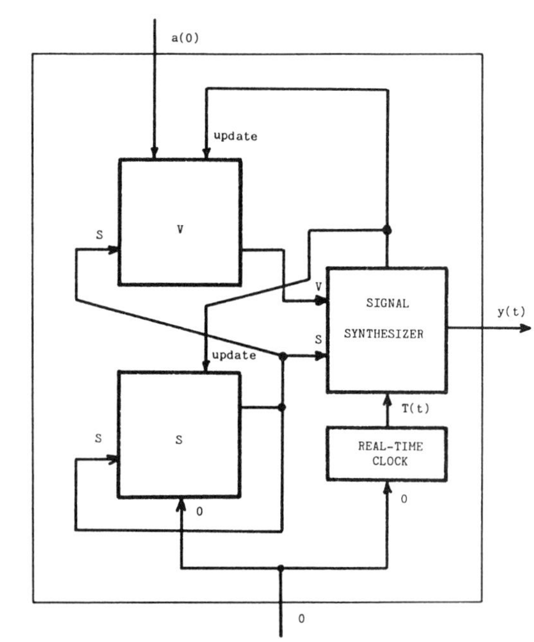


Fig. 6. Generating a Synchronous Signal

# ASSEMBLER GENERATED BINARY TEST SIGNALS FOR PROCESS IDENTIFICATION

**I. A. Henderson*, A. A. Ibrahim*, J. McGhee* and D. Sankowski****

**Department of Electronic and Electrical Engineering, University of Strathclyde, 204 George Street, Glasgow, U.K.*
***Instytut Elektroenergetyki, Polytechnika Lodzka, ul Gdanska, 176, 90–924 Lodz, Poland*

Abstract. Knowledge of the dynamical behaviour of many processes can assist in controlling plant efficiently in terms of power consumption and product quality. Binary test signals are used in the identification of electrical resistance furnaces which occur in many branches of manufacturing industry. Starting from a classification of test signals, the paper discusses the detail and special significance of binary sequences in process identification. The software involved in a microprocessor based instrument, which is designed to produce a wide range of binary test signals, is described. Both pseudo-random and multi-frequency binary sequences may be generated. Keyboard entry is an essential menu driven option and the choice of available multi-frequency binary sequences includes Van den Bos, Harris, Lodz and recently discovered Strathclyde base-band sequences. The use of assembler generated binary test signals permits wide flexibility of device function with minimum processing time from one hardware structure.

Keywords. Process control; identification; binary sequences; test signals; assembly programing.

## INTRODUCTION

Binary signals form an important class of signals which may be used as test signals in process control (Cusset and Mellichamp, 1975; Harris and Mellichamp, 1980; Plaskowski and Sankowski, 1984; Schwarzenbach and Gill, 1984). One obvious benefit is their ease of generation when compared with that of other signals. This is especially true when a comparison is made with analogue test signals. Although many final control elements cannot easily follow analogue signals, they can often transduce an electronic binary test signal into a corresponding low-level variation of an input to the process. A good example of the application of a binary test signal to such a process is the electrical resistance furnace (Plaskowski and Sankowski, 1984). Further applications of electro-heat are widespread and of significant importance (Davies, 1985; Lobodzinski and others, 1983; Michalski and Sankowski, 1977). However, these binary test signals have untapped potential for the identification of other widely differing systems in the manufacturing and process industries.

Both Pseudo-Random Binary Sequences, PRBS, and Multi-frequency Binary Sequences, MBS, have been successfully applied to process control systems. Apart from PRBS (Schwarzenbach and Gill, 1984), the published literature on MBS in process control originally mentions a set of Van den Bos sequences (Van den Bos, 1967) with the same statistics. In addition, there is a Lodz equivalent (Plaskowski and Sankowski, 1984) and two Harris MBS test signals (Harris and Mellichamp, 1980). It is important to realise that the number of potentially useful binary test signals is unlimited. In recent years, technological advances have made inexpensive micro-computers with a large memory available to tackle this problem. If the fundamental limits associated with the use of microprocessors in the production of these sequences are to be reached, then assembly language programing is required. This software engineering approach has been crucial in the recent advances which have been made in the proposal of new Strathclyde base-band sequences for use in the digital identification of systems (McGhee and Henderson, 1986; McGhee and others, 1986; Ibrahim, 1986).

In the identification of process control systems, the use of mini-computers and high level language programing has obscured the fundamental realities of digital identification of systems. As the zero crossovers of these sequences have a base code, it became imperative to re-examine the basis of this theory in the light of telecommunications and information theory. An awareness of a suitable process identification channel, which is shown in Fig. 1, is required. The nature of the signature of the system constitutes the input statistics to the channel, which dictate the choice of coded test signal suitable for digital identification. Naturally, any significant contaminating signal, due to disturbances in the power flow or influences from other sources, will cause distortion. This must affect the choice of test signal and the signal processing operations required to recover the signature of the process. These ideas are best expressed for the special case of digital identification by the Identification Channel Coding and the Identification Source Coding Theorems (McGhee and Henderson, 1986). As these theorems are quite general, their application has brought existing MBS under closer scrutiny. New high power per component, efficient Strathclyde base-band MBS test signals have also resulted. They offer a possibility of capturing a wide range of specifiable statistics dependent on those required from the signature of the process. A microprocessor based instrument with keyboard input (Ibrahim, 1986), which has been indispensible in this investigation, is still

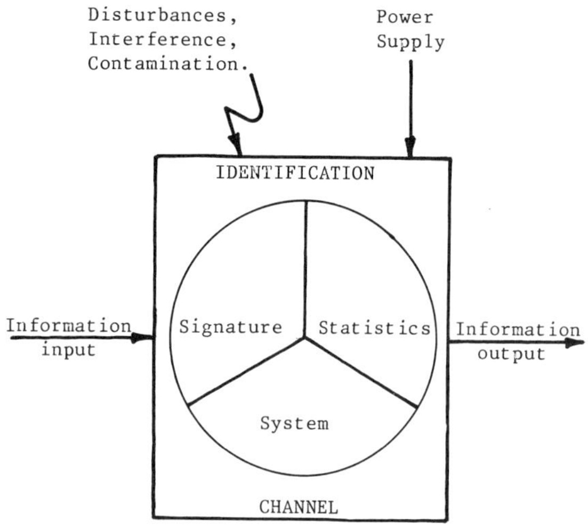


Fig. 1 A system identification channel

being actively used to examine implications of the two identification coding theorems for other binary coded test signals.

## TEST SIGNALS FOR IDENTIFICATION

Signals perform functions and possess structure. Symbols are the means by which information is conveyed. As signals can be defined as a group of symbols, the primary function of signals is their ability to carry information. Thus, the grouping of signals according to their properties is important in all information handling processes, such as in communication, measurement, control and signal processing. Signal structure, or shape, with time is determined by the manner in which the signal carries information. Time domain characterisation of signals indicates that they may be continuous or discretely sampled with amplitudes which are analogue, quantized or binary. Hence, there is scope for diversity in the choice of signals for identification purposes.

### General Aspects of Test Signals

External stimulation of systems, using test signals, has worthwhile advantages over normal operating inputs (Van den Bos, 1985). These benefits are:-

(i) They are accurately known and repeatable. Hence, it is not necessary to measure them.
(ii) Their shape or form may be chosen with comparative freedom.
(iii) They may be generated easily.
(iv) They are independent of other processes in the identification algorithms. This is important when correlation and cross-spectral analysis are used.

There are many ways of classifying signals depending upon the application. An emphasis, which is useful for identification but also for other areas, is shown in Fig. 2. As a full description is available elsewhere (McGhee and others, 1986) only the binary types of test signal will be considered here.

Binary signals form an important group of test signals. As they are based upon step functions they are easily generated. Several types of binary

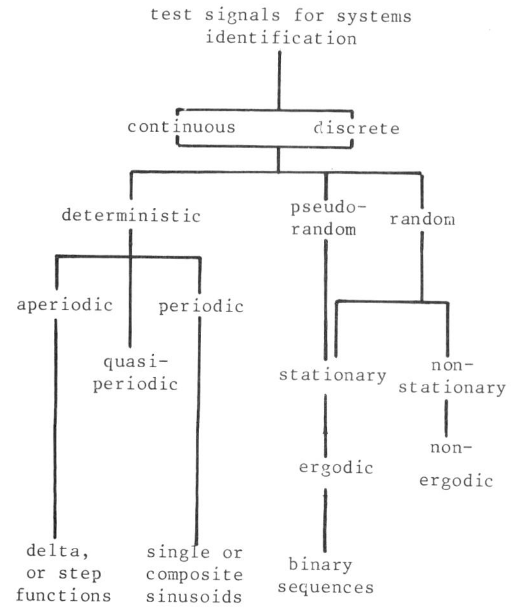


Fig. 2 Test signals for identification

test signal are known. These include (i) The maximal length binary sequences or PRBS (Briggs, Godfrey and Hammond, 1967; Davies, 1966) (ii) Multi-frequency Binary Sequences or MBS (Harris and Mellichamp, 1980; McGhee and Henderson, 1986; McGhee and others, 1986; Van den Bos, 1967).

From the structural point of view, these sequences may be considered as periodic and deterministic. From the point of view of an observer, it is also possible to consider them as random since the observer has no way of knowing the manner in which they are constituted.

### PRBS Test Signals

Maximum length binary sequences have been applied over a wide range of problems for many years. They now fill the same role in the application of correlation analysis to identification (Schwarzenbach and Gill, 1984) as the sinusoidal signal plays in the application of steady-state or harmonic analysis. The auto-correlation function of this sequence approximates to a "delta" function which is characterised by a narrow triangular function.

In addition to these properties, the PRBS is easy to process, synchronise and introduce into the process. Each PRBS test signal of period, $N = 2^n-1$ can be simply implemented by hardware. An n-stage feedback shift register with feedback to the first stage which consists of the modulo-2 sum of the output of the last stage and the output of one or more of the other stages is used (Briggs, Godfrey and Hammond, 1967; Davies, 1966). As a slow speed of computation is adequate in process identification, this logic circuit may be obtained by software engineering on a microprocessor development system (Ibrahim, 1986). Within the limit imposed by system memory for the generating format, this has the advantage of great flexibility in choice of sequence length.

PRBS test signals have two practical limitations. Their power is spread over all the frequency bandwidth components and off-line signal processing is required. Both of these disadnvantages are overcome with MBS test signals.

## MBS Test Signals

A primary requirement in identification using test signals is that the test signal should introduce no disturbance to the process set-point. Thus, binary sequences, which are to be used for identification, should have a mean value of zero. Hence, they belong to the group of binary sequences commonly referred to as maximum entropy (hereafter called max-ent) sequences since the number of logic "1" level excursions must equal the number of logic "0" level excursions.

Sequences of this type, which have been used for the identification of a number of different types of processes, are well reported. For convenience, it is beneficial to examine binary sequences in a normalised form. This assumes that the zero crossover base code excursions are +1v for logic level "1" and -1v for logic level "0". The normalised fundamental frequency is taken to be 1Hz. Table 1 is a collected summary of the parameters of existing MBS test signals for process identification.

## A CODING THEORY FOR IDENTIFICATION

If identification using coded test signals is regarded as a form of pulse code modulation, then the coded test signal acts as a carrier signal with a deterministic signature. The system to be identified writes its signature upon the probing test signal by inter-symbol interference. The differences between the identification channel and the data communication channel has led to coding theorems for the identification problem (McGhee and Henderson, 1986; McGhee and others, 1986). The IDENTIFICATION CHANNEL CODING THEOREM states that

"The most effective test signal for system identification will be source coded to arrange that the signal at the output of the system to be identified will have only those optimum statistical characteristics which match the identification channel".

The IDENTIFICATION SOURCE CODING THEOREM states that

"The optium test signal for the excitation of the system to be identified, must be source coded such that it is compact, but yet uniquely and instantaneously decodable, so that the response which it causes satisfies the aims for ouput signal statistics required by the Identification Channel Coding Theorem".

When this theory is combined with the theory of ideal filters, it becomes apparent that there is a fundamental limitation upon the minimum number of bits in a binary test sequence for identification purposes. Moreover, it can also be shown that this limiting minimum is unaffected by whether the identification is to obtain a non-parametric model in the time or frequency domain. Hence, it may be shown (McGhee and Henderson, 1986) that the minimum number of bits in a test sequence is given by

$$N > 2 + \left( \frac{\text{Observation Time}}{\text{Identification Period}} \right) \qquad (1)$$

Here, the observation time is the time required by an observer to capture the slow mode statistic of interest, whilst the identification rate is that Nyquist Rate required to identify the fastest mode of interest. The factor two arises as a result of non-stationary correlation statistics of a PRBS in the time domain or for signal processing after sampling when frequency domain models are the aim (Bellanger,1985; Ramirez, 1985). This limiting minimum is important from the point of view of economic generation of these binary test sequences.

## Strathclyde Base-Band MBS

The effectiveness of a test signal will normally

TABLE 1 Existing MBS Test Signals

| M.B.S. | Base Code | Component Amplitude, $E_n$ (n=harmonic number) |
|---|---|---|
| Van Den Bos Octave | $6^+,1^-,2,1,$ $11,6,2,14,3$ $1,6,6,2,3^-$ | $E_1$ =0.4821, $E_2$ =0.4832, $E_4$ =-0.5586, $E_8$ =0.4928, $E_{16}$=-0.4638,$E_{32}$=0.3979 |
| Lodz Octave | $5^+,2^-,14,7,$ $1,6,1,7,2,$ $1,7,6,2,3^-$ | $E_1$ =0.4509, $E_2$ =0.4705, $E_4$ =-0.6022, $E_8$ =0.5170, $E_{16}$=-0.3706,$E_{32}$=0.3979 |
| Strath-clyde Extended | $11^+,3^-,4,3,$ $2,1,3,2,1,$ $6,3,2,1,12,$ $1,1,4,2,2^+$ | $E_1$ =0.4588, $E_2$ =0.4519, $E_4$ =-0.3481, $E_8$ =-0.3687,$E_{12}$=-0.3383,$E_{16}$=0.3183, $E_{19}$=-0.3664,$E_{24}$=0.3428, $E_{32}$=-0.3183 |
| Harris Compact | $5^+,10^-,8,2,$ $8,29,7,11,$ $1,4,31,8,4^+$ | $E_1$ =-0.3156,$E_2$ =0.4170, $E_3$=0.3315, $E_4$ =-0.3962,$E_6$ =-0.4406,$E_8$=0.4269, $E_{12}$=0.4509, $E_{16}$=0.4412 |
| Harris Extended | $44^+,12^-,16,$ $14,7,4,14,$ $10,2,23,13,$ $6,4,46,6,5,$ $14,8,8^+$ | $E_1$ =0.4393, $E_2$ =0.4467, $E_4$ =0.3361, $E_8$ =-0.4058,$E_{12}$=-0.3488,$E_{16}$=0.3664, $E_{24}$=0.4095, $E_{32}$=-0.3321 |

TABLE 2 Strathclyde Base-Band MBS Test Signals

| Strathclyde M.B.S. | Base Code | Component Amplitude, $E_n$ (n=harmonic number) |
|---|---|---|
| Short Octave | $3^+,3^-,1,1^-$ | $E_1$ =0.7633, $E_2$ =0.6366, $E_4$=-0.6366 |
| S1-to-8 | $1^+,2^-,3,$ $1,3,5,1^+$ | $E_1$ =0.3965, $E_2$ =-0.3445,$E_3$ =-0.3882 $E_4$ =0.4502, $E_5$ =-0.3861,$E_6$ =0.2773, $E_7$ =0.5228, $E_8$ =0.6366 |
| Mid High Power Octave | $3^+,1^-,7,11,$ $5,3,1,1^-$ | $E_1$ =0.4232, $E_2$ =0.6111, $E_4$ =-0.7342 $E_8$ =0.5434 |
| Mid Level Power Octave | $3^+,1^-,7,4,1$ $7,4,3,1,1^-$ | $E_1$ =0.5038, $E_2$ =0.5231, $E_4$ =-0.5091 $E_8$ =0.6093 |
| Octave | $3^+,1^-,7,3,1,7$ $1,1,3,3,1,1^-$ | $E_1$ =0.5254, $E_2$ =0.5039, $E_4$ =-0.5091 $E_8$ =0.4775, $E_{16}$=-0.4775 |
| Compact | $1^+,3^-,4,7,$ $2,4,8,2,1^+$ | $E_1$ =-0.3402,$E_2$ =0.3621, $E_3$ =0.2777, $E_4$ =-0.4192,$E_6$ =-0.5290,$E_8$ =0.4502, $E_{12}$=0.5794, $E_{16}$=0.3183 |
| Octave Extended | $2^+,1^-,7,3,1,$ $3,1,4,1,1,$ $3,3,1,1^-$ | $E_1$ = 0.4513,$E_2$ =0.4391, $E_4$ =-0.4159 $E_8$ = 0.4115,$E_{15}$=0.5046, $E_{16}$=-0.3183 $E_{30}$=-0.2774 |
| Multi-Compact | $1^+,2^-,2,1,2$ $7,2,4,8,2,1^+$ | $E_1$ =-0.3298,$E_2$ =0.3997, $E_3$ =0.3478 $E_4$ =-0.3260,$E_6$ =-0.4645,$E_8$ =0.3842 $E_{12}$=0.3983, $E_{16}$=0.4775, $E_{18}$=0.3323 $E_{20}$=0.3290 |

be based upon its ability to extract or capture the identification statistics of interest from the process. However, a preliminary study gives some general requirements for MBS test signals. These are:-

(i) They should be capable of extracting statistics which satisfy the identification channel coding theorem.
(ii) They should be efficient in the utilisation of code.
(iii) They should be economical to generate.
(iv) They should posssess max-ent.
(v) They should possess a high proportion of total sequence energy in the MBS harmonics.
(vi) They should comply with other appropriate requirements which ensure that the code is in agreement with the identification source coding theorem.

Since there are $2^6=64$ and $2^8=256$ possible six and eight bit sequences respectively, it can be seen that the task of proposing new, max-ent sequences is non-trivial. There are two main aproaches to the formation of new MBS test signals (McGhee and Henderson, 1986). The first method commences from short bit lengths of say 6 or 8. Using the even time symmetry together with a combination of truth table entries and intuition, it is possible to develop an algorithm for the synthesis of new sequences. The second commences from known sequences and uses a computer search procedure. Although further consideration of these techniques is beyond the scope of the present contribution, a selection of Strathclyde base-band test signals is given in Table 2. An important feature of these sequences is their even or odd symmetry in time. If a sequence with either time characteristic is to be generated from internal memory, it is only necessary to store half the sequence in terms of its zero crossover code. Reading an even or odd half-sequence in one direction, need only be followed by reading the respective half-sequence in reverse order to produce the generating format (Ibrahim, 1986).

## ASSEMBLER GENERATION OF BINARY TEST SIGNALS

A priori knowledge of system requirements is normally required before a suitable binary sequence is chosen. In particular, the maximum allowable perturbation and an estimate of the slow and fast mode time constants of the process are required. These fix the magnitude of the binary sequence and the range of frequency in terms of the sequence bit duration/sampling period. Subsequently, a suitable test signal may be chosen with the following guidelines from the Identification Coding Theorems:-

(i) The test signal, whose number of bits should not be less than that given by equation (2), should excite all modes of interest in the process. Ideally, the total power should be concentrated in the dominant frequencies so that only essential information about the process dynamics is obtained.
(ii) The amplitude of the chosen sequence should neither drive the process out of its linear region nor alter the set point.
(iii) The lower the number of bits in the sequence the greater will be the time available for on-line signal processing and the smaller will be the memory required giving efficient use of storage.

### Software Engineering Implications

The principal benefit in designing with a microprocessor is the possibility for realising the widest possible fault tolerant (Domaradzki and others, 1986), functional flexibility from a single hardware structure. Whereas micro-computers with high-level software have widespread use, non-redundant machine code cannot be guaranteed because of the compilation process. On the other hand, microprocessor software design has a basic limitation which is epitomised in the continual compromise between efficient use of memory and speed of program execution (Leventhal and Saville, 1982). Software engineering applied to assembly language programing has an important part to play in the resolution of this design conflict. This is especially the case when microprocessors are applied in real-time engineering systems. A minimum execution time requires the use of "look-up" tables for consecutive storage of data in the simplest and normally most inefficient format. Thus, data distribution by a VIA to the engineering process or system is effected as quickly as possible. It is inevitable that this "look-up" table approach plays a major role in such applications.

A binary test signal generator design must be capable of generating all possible binary sequences. With a microprocessor based instrument, the ultimate limit for the bit time of a binary sequence, $\lambda$, is dictated by the microprocessor clock frequency. In practice, other additional fixed time penalties are introduced by the choice of analogue to digital convertor (ADC), digital to analogue convertor (DAC) and on-line signal processing since

$$\lambda = \lambda_{BS} + \lambda_{ADC} + \lambda_{SP} \qquad (2)$$

$\lambda_{BS}$, $\lambda_{ADC}$ and $\lambda_{SP}$ are the generation/DAC, ADC and on-line signal processing times associated with each bit of the binary sequence.

The initial specification will normally include the value of the fastest mode statistic of the process. Having minimised each of the individual parts of (2) and taken a decision regarding the data acquisition and distribution hardware, it is essential to use assembly language programing. This will ensure that the time required to produce each bit of the sequence and any on-line signal processing uses the minimum number of non-redundant instructions. As each instruction is related to the microprocessor clock time, it is possible to calculate the minimum bit time and hence the maximum specified fastest mode statistic which a particular binary sequence and processor may capture.

### Flow Diagram

Within the limits placed by both the available interface and microprocessor hardware, it is apparent that a common central software process is required to generate binary sequences no matter their form. Therefore, the software may be structured into the two main parts shown in the flow diagram of Fig. 3. The first part consists of initialising procedures whose outcome produces the sequence in a "look-up" table. This arranges that the second part starts from a common generating format which includes memory storage equal to the number of bits in the sequence. Taking into account level shifting, the data may be arranged so that the chosen amplitude is produced immediately at the output by means of the VIA and DAC. In addition, it is necessary to ensure the output voltage is zero at the start and finish of the test period. Hence, a user prompt for connection and disconnection may be required.

With the aid of suitable timing of the VIA, an

external interrupt may be generated. The bit duration must be independent of variations in time due to the different operations that take place during this interval. Each bit is directed to the process to be tested under the control of external interrupts. However, as it is only necessary to output each change of bit level, a suitable combination of interrupts and timeouts is used to produce the widest range of possible bit durations. In the case of the 6809 microprocessor, this range is from approximately 1ms to 30min. As previously stated, a fast limit of $\lambda$ = 1ms depends on the DAC, ADC conversion time and whether on-line signal processing is used.

Common Generating Format

This binary sequence initialisation section of the program permits a choice for the common generating format parameters which are sequence length, timing and amplitude data. PRBS and MBS with a keyboard entry option are possible while the span and resolution of binary sequence amplitude is fixed by the use of an appropriate DAC. From a priori knowledge of the process, the signal processing and the code of a suitable binary sequence may be specified. Subsequently, the bit time may be calculated. Since a wide choice is required, a flexible design allows the appropriate timeout and VIA timer counter values to be transferred to the common generating core of the program.

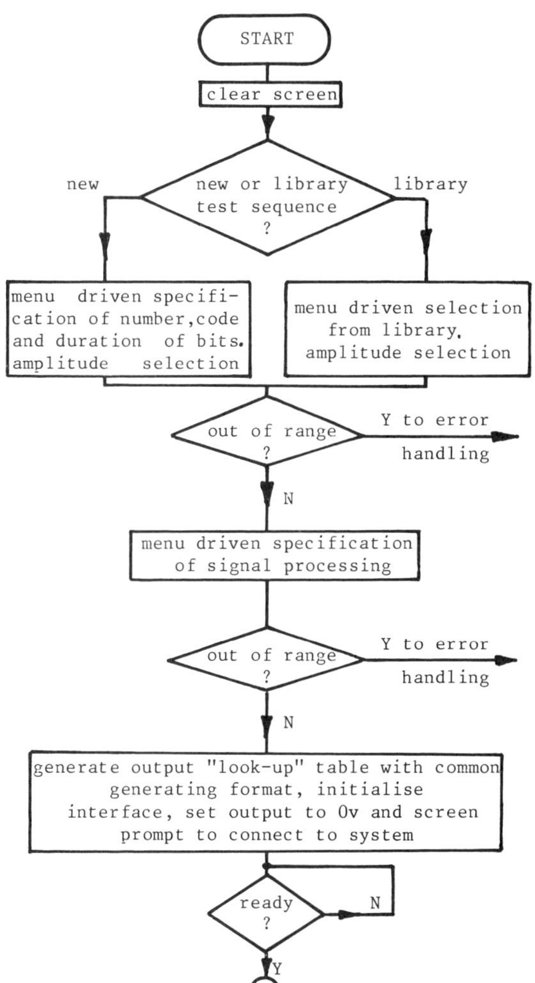


Fig. 3(a) Menu driven initialisation flow diagram

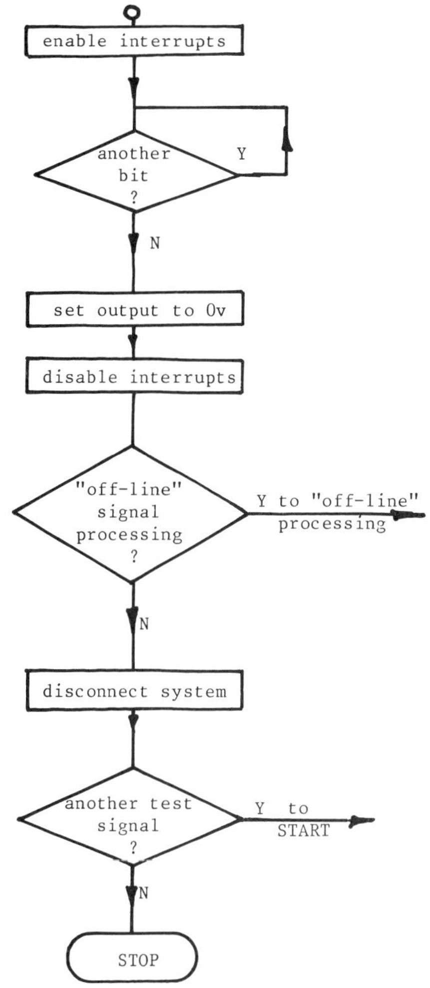


Fig. 3(b) Test signal generation flow diagram

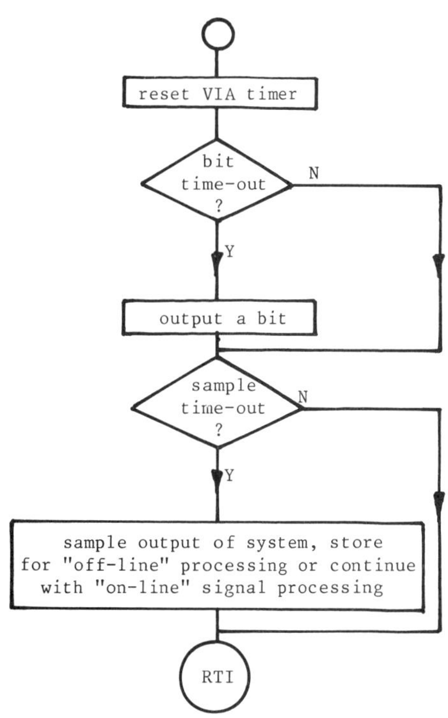


Fig. 3(c) Interrupt flow diagram

The output "look-up" table in its common generating format may be obtained from the following routes

(i) software logic,
(ii) conversion from efficient storage in internal memory,
(iii) a data base or library of binary sequences stored externally (disc or tape) and built up from keyboard entry.

As method (i) mimics the normal hardware circuit used to produce PRBS and uses very little internal memory, it is an obvious choice for PRBS test signals with a large sequence length. Initially, method (ii) was chosen to produce MBS because of the small number of known sequences involved. In retrospect, method (iii) uses the minimum internal memory. It offers the widest choice, since it places no limit in principle upon the number of binary test signals which may be made available. As the binary sequences may be saved and loaded in the common generating format, it is also quick and efficient.

It is appropriate that method (iii) be adopted in designs involving micro-computers where off-line memory facilities are available. Methods (i) and (ii) may be used if external memory facilities are not readily available as in most microprocessor developement systems which may be used in the initial design phase. In the latest portable version of this instrument, a Sinclair QL micro-computer has been used. It has been found to be ideal for this purpose with its large memory and micro-drive storage of all potential binary test signals. The use of superbasic with windows for the initialisation, and menu driven software makes the instrument "user friendly". An additional advantage is the fast generation and on-line signal processing possible using the M68008 microprocessor.

## CONCLUSION

A study of the literature on the digital identification of processes revealed a gap in its understanding and has produced identification coding theorems. These led to a software engineering approach to produce a suite of software packages to further the concepts involved with a practical implementation on a micro-computer. In particular, this involved the design of an instrument for the generation and storage of the large number of potential binary test signals that result from the investigation of the identification coding theorems.

The most effective structure splits the software for the instrument into two parts. A generating section, which requires the highest available speed of output from a common generating format, and an initialising, menu driven section which must be "user friendly". This leads to a high level language program for the main part of the software whilst assembly language programing is required for fast generation of the binary test signals. The storage problem for the large number of possible sequences is best resolved by use of external memory to produce a library of potential binary test signals for process identification.

An examination of existing MBS test signals with the new insight given by the identification coding theorems has produced the Strathclyde base-band MBS test signals. Other potential binary test signals are under investigation.

## ACKNOWLEDGEMENT

The authors gratefully acknowledge the accommodation and facilities mutually provided by both institutions, under their official exchange agreememt, for the conduct of this work.

## REFERENCES

Bellanger, M. (1985). Digital Processing of Signals, John Wiley and Sons, New York.

Briggs, P. A. N., Godfrey, K. R. and Hammond, P. H. (1967). Estimation of process dynamic characteristic by correlation methods, IFAC Symposium, Prague.

Cusset, B. F. and Mellichamp, D. A. (1975). On-line identification of process dynamics. A multi-frequency response method, Ind. Eng. Chem., Process Des. Dev., 14, 359-368.

Davies, E. J. (1985). Electroheat for industry, IEE Proc. Pt A, 132, 77-83.

Davies, W. D. T. (1966). Generation and properties of maximum length sequences, Control, 10, 302-304.

Domaradzki, P., Henderson, I. A., McGhee, J. and Szaniawski, W. (1986). Teaching fault tolerant techniques in measurement and instrumemtation courses, to be read at 8th IMEKO Colloquium on Higher Education, Warsaw, Poland.

Harris, S. L. and Mellichamp, D. A. (1980). On-line identification of process dynamics: Use of multi-frequency binary sequences, Ind. Eng. Chem., Process Des. Dev., 19, 166-174.

Ibrahim, A. A. (1986). Software for Digital Identification of Systems, MSc Thesis, University of Strathclyde, Glasgow, U.K.

Leventhal, L. A. and Saville, W. (1982). 6502 Assembly Language Subroutines, Osbourne / McGraw-Hill.

Lobodzinski, W., Orzylowski, M., Sankowski, D. and Szmanda, W. (1983). Diffusion furnace as multivariable temperature control system, (in Polish), Prace PIE, 24, No 88, 5-28.

McGhee, J. and Henderson, I. A. (1986). Test Signal Coding and Digital Identification of Systems, Internal Report, Dept. of Electronic and Electrical Engineering, University of Strathclyde, Glasgow, U.K.

McGhee, J., Henderson, I. A., Ibrahim, A. A. and Sankowski, D. (1986). Identification: Systems, Structures, Signals, Similarities, to be read at Systems Science IX, Wroclaw Technical University, Poland.

Michalski, L. and Sankowski, D. (1977). Automatic temperature control in electrical resistance furnaces, World Electrotechnical Congress, Moscow.

Plaskowski, A. and Sankowski, D. (1984). The use of multi-frequency binary sequences (M.B.S.) to on-line identification in electro-heat, IMACS European Simulation Meeting, Eger, Hungary, 285-296.

Ramirez, R. W. (1985). The FFT Fundamentals and Concepts, Prentice-Hall, New York.

Schwarzenbach, J. and Gill, K. F. (1984). System Modelling and Control, 2nd Ed., Edward Arnold, London.

Van den Bos, A. (1967). Construction of binary multi-frequency test signals, IFAC Symposium, Prague.

# A MULTIPROCESSOR ARCHITECTURE AND ITS APPLICATION TO ADAPTIVE CONTROL

F. Cesarini, G. Soda and G. Zappa

*Dipartimento di Sistemi e Informatica, Universita' di Firenze, via S. Marta 3, 50139 Firenze, Italy*

Abstract. The paper presents the use of a multimicroprocessor system in adaptive control applications. The system is a global memory multiprocessor with a single shared bus. Processes executed on different processing units exchange messages through the global memory. Adaptive control algorithms have been split into two processes which are executed in parallel and co-ordinate each other by means of suitable data exchanging. The effectiveness of this implementation has been outlined in the cases when the reduction of sampling period is the major concern. Experiments on laboratory process simulator have shown that a better steady state behaviour and/or a faster adaptivity can be obtained.

Keywords. Multiprocessor architectures; adaptive control; hybrid self-tuner.

## INTRODUCTION

The recent widespread development of integrated electronic makes the realization of powerful multiprocessor architectures possible at very advantageous prices. Complex problems can be solved by multiprocessor systems in a modular and efficient way through a set of cooperating processes executed in parallel.

A multi-microprocessor system has been implemented, simple and flexible enough to allow the investigation of multiprocessor software design approaches. The implementation was based on the massive use of standard, "off the shelf" hardware and software modules; only the parts specific to our goal have been developed. In this way, good modularity and flexibility in a rather inexpensive system have been obtained.

The effectiveness of this multiprocessor architecture has been demonstrated in a nonconventional implementation of an adaptive controller. More specifically the control and estimation processes, which constitute the main part of the adaptive control algorithm, are allocated on two different processing units and executed in parallel. Data (e.g. i/o variables versus parameters estimates) are exchanged through a global memory. The sampling period has been matched to the execution time of the control process, while the control parameters are updated at a lower rate. It is argued that this parallel implementation allows a lower sampling period, hence a better steady state behaviour, with respect to a conventional one, without reducing the adaptation capability. Laboratory experiments are reported which support this conjecture. In some sense, our implementation can be considered an approximation of the so called hybrid self-tuners, where the gains of a continuos-time controller are updated by a digital processor at discrete-time instants.

## HARDWARE ARCHITECTURE

The system is organized as a tightly-coupled, global memory multiprocessor machine having a single shared bus (Weitzman, 1980), see fig. 1.

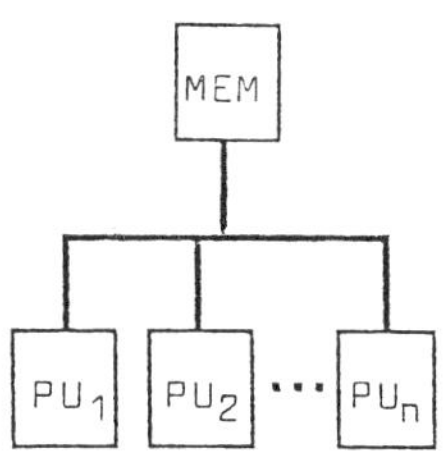


Figure 1 - Global memory multiprocessor architecture

No functional specialization of processing units is provided. A fully distributed executive, called EXEC, makes a virtual machine available whose processes allocated on different units are executed in parallel, while processes residing on the same unit are multi-programmed. The executive has been developed from a standard real-time multi-tasking operating system (MP/M-80). The reasons of this choice were both the use of the multitasking kernel for simplifying real time executive programming and the use of simple local resource interface obtainable through the I/O management system.

The system has a Direct Shared Bus (DSB) architecture. The shared bus is used only for global memory in order to get the maximum throughput in processor communication. The global shared bus (GBUS) is implemented as a Multibus subset which includes all the lines necessary for supporting global memory operations. The resulting architecture (Bagnoli, Gori and Soda, 1984) ,shown in fig.2, consists of a set of processing cells, implemented as Multibus systems, and interconnected by interface boards to the GBUS to obtain concurrent access to the global memory.

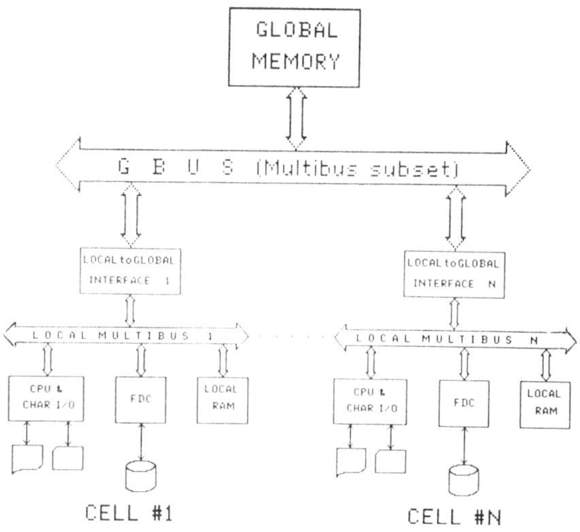


Figure 2 - System architecture.

The multiprocessor executive for a Global Memory Multiprocessor system without any functional cell specialization requires some adequate hardware features. The arbitration of the cell on the global bus must be accomplished by means of a strategy without priorities. The definition of semaphores allocated in global memory needs Read-Modify-Write instructions to avoid race conditions. An interprocessor interrupt mechanism is required in order to synchronize the operations of a distributed executive efficiently.

In our system all the problems connected with global access priority are solved by using a sequential, centralized arbiter which carries out our strategy. A global bus lock feature implements not-interruptable Test and Set sequences on global semaphores, as is done in the Multibus override feature. Interprocessor interrupts are generated by writing on an I/O port allocated on the local-to-global bus interface. The detailed architecture of our machine is shown in fig.3, and the not-standard modules we implemented are marked by a dashed line.

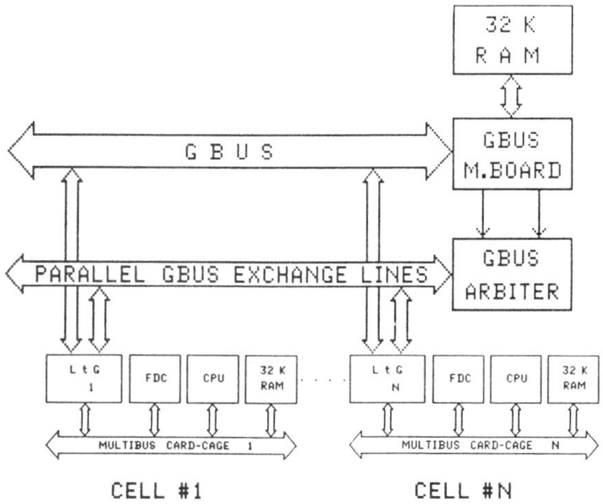


Figure 3 - System detailed architecture

The LOCAL-TO-GLOBAL INTERFACE (LtG) connects the local Multibus to the GBUS. It allows a cell to access to the global bus (and thus solves its conflicts contention with the other cells, it drives the GBUS properly, it supports the GBUS lock feature, and it generates interrupts to the local processor when requested by one of the other cells. The GBUS ARBITER accomplishes the centralized Gbus arbitration by means of a parallel resolution technique which uses a rotating priority strategy. The GBUS MOTHERBOARD only performs passive functions. Its main function is to bring the GBUS to a standard Multibus connector in order to implement the system global memory by means of a standard Multibus RAM board. It must also provide electrical termination for the Gbus lines in order to increase noise immunity and give maximum safety of communication.

In fig.3 the Parallel Gbus Exchange Lines and the Gbus are distinguished from each other to emphasize the function of the Gbus Arbiter in Gbus management, but they are not logically distinct from each other. This system is capable of containing up to eight processing cells. At present, two cells have been installed in the system. Each cell is a standard Multibus card-cage containing several Multibus-compatible boards. A cell is a really autonomous processing unit. It has a 8 bits AM 4006 CPU, a floating point processor which works on 32 bits floating point numbers (AM 9511 Arithmetic Processing Unit), a 32 K RAM local memory, a console and a disk. For our application, we added a standard 12 bits DA/AD converter (RTI-732 Analog Device) to one cell. The size of the global memory is 32K bytes.

## THE EXECUTIVE

The executive, called EXEC, (Bagnoli, Cesarini and Gori, 1984), is fully distributed, i.e. the processing units are not functionally specialized and the executive is fully replicated on each of them. The communication primitives are message passing primitives. This form of communication provides a high degree of abstraction. In fact, processes directly specify the process they want to communicate with and the structures which support communication are totally transparent to them. This produces a high-level process interaction scheme which simplifies application programming.

EXEC message passing primitives have the following syntactical form:

SEND(s,r,m) RECEIVE(s,r,m)

"s" and "r" denote the sender and receiver processes respectively and "m" is a pointer to the message.

The send and receive operations are based on a set of data structures allocated in the global memory. In order to achieve correct communication, the executive must handle all the synchronization problems. It must manage all the critical sections when accessing global data structures, it must control the running/blocked process state with regard to the primitives and it must cooperate with executives installed on the other units in order to accomplish inter-unit process activation.

The executive has two inputs: one for processes which invoke communication primitives and one for handling interprocessor interrupts from other units. This is shown in fig.4, which depicts a process interaction which requires an inter-unit activation. P1 and P2 are processes allocated on two different cells, Cell1 and Cell2 respectively.

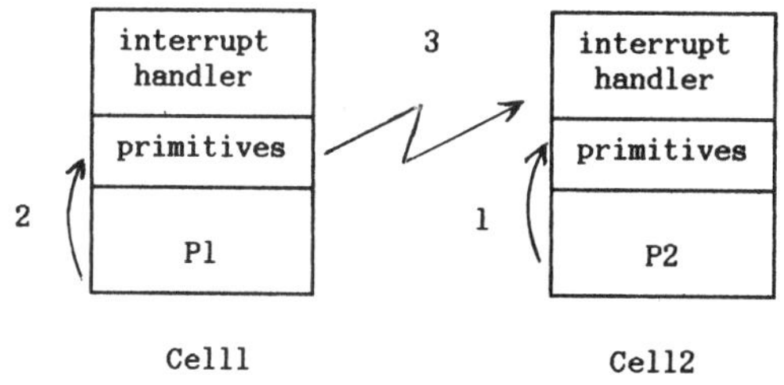


Figure 4 - Inter-unit process interaction

Interaction occurs in three phases: (1) P2 on Cell2 calls a RECEIVE from process P1. The executive residing on Cell2 inspects the data structures allocated in the global memory. Let us suppose that no message is available. Therefore P2 is blocked by the executive which marks this event in the global data structures. (2) P1 on Cell1 calls a SEND for P2. The executive residing on Cell1 stores the message in the global structures and realizes that P2 is waiting for it. (3) As a result, the executive on Cell1 must notify the availability of a message for P2 by an interrupt to Cell2. This interrupt allows the executive on Cell2 to awake P2. Exec operations are based on a set of primitives. An onion-skin view of the EXEC's internal structure is given in fig.5.

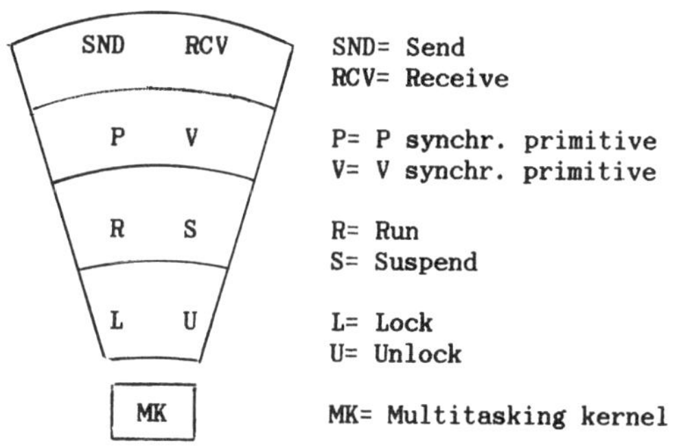


Figure 5 - EXEC global structure

User processes interact by using the high level SEND and RECEIVE communication primitives. The P and V synchronization primitives are employed to manage accesses to message buffers. The running/blocked state of processes during communication is handled by the RUN and SUSPEND primitives. Critical region accesses are handled by the LOCK and UNLOCK mutual exclusion primitives which operate on a set of global semaphores. For optimization purposes, we implemented a lock primitive which does not enter a busy-wait loop when the semaphore to be locked is busy, but retests the semaphore only after a programmable delay.

Three different versions of Executive can be loaded on the machine:

1) EXEC0: it supports static coupling without type replication,
2) EXEC1: it supports static coupling with type replication,
3) EXEC2: it supports dynamic coupling with type replication.

Type replication means that a process can be replicated on the same cell. This feature can be used when a certain service is heavily requested and risks to be a bottleneck for the system. The replication is transparent to the processes, i.e. a process sends a message to a certain type of process and does not know which process will actually receive it. Coupling refers to the capability of establishing a conversation between two processes. In case of static coupling, when the executive is installed it is necessary to declare the couples of processes which will interact. This declaration simplifies the management of global data structures used for the communication.



In this application we have used EXEC0 since it is faster than the other ones and is powerful enough for the needed requirements.

The executive has been tailored to MP/M-80 operating system in a modular way.(see fig. 6).

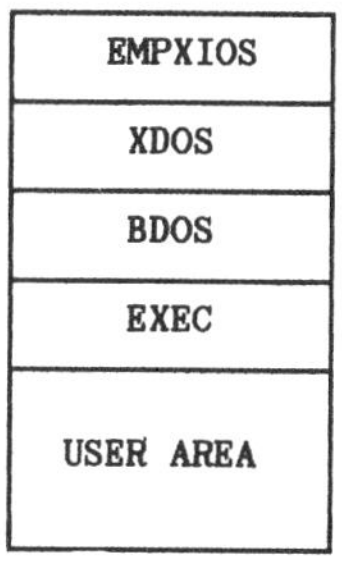


Figure 6 - Extended MP/M operating system

The EMPXIOS (Extended Multi-Processor XIOS) is a module which contains a Xios tailored to the local cell. It was extended by drivers which handle global bus locking, and interprocessor interrupt generation and servicing in order to support multiprocessor operations. The XDOS and BDOS modules are the standard MP/M kernel modules. The executive is implemented by a resident system process, named EXEC, which both extends the kernel functions and implements the communication primitives. The new kernel functions are the send/receive and lock/unlock primitives. The other primitives are local to the executive and transparent to the user processes. A maximum of 8 user processes can be allocated in the user area.

For the application we are going to describe, we have added to the Executive the primitive:

RECEIVENOWAIT(s,r,m).

The RECEIVENOWAIT primitive does not wait for a message; in case a message is not found, the primitive returns a suitable flag to the process.

The Executive is fully implemented in assembly language and requires about 2K bytes. We show in table 1 the execution time of the communication primitives in case the message length is 32 bytes. These times have been measured by means of a digital storage oscilloscope.

| primitive | execution time |
|---|---|
| SEND | 1.390 ms |
| RECEIVE | 1.390 ms |
| RECEIVENOWAIT: | |
| without mess. | 0.560 ms |
| with message | 1.390 ms |

Table 1 - Execution time of the communication primitives

## APPLICATION TO ADAPTIVE CONTROL

The multimicro architecture described above has been exploited for a parallel implementation of an adaptive controller. In order to explain its features let us briefly recall same basic facts of adaptive control theory.

Let us assume that the plant is decribed by

$$Y(s) = exp(-sT_d)G_1(s)U(s) + G_2(s)V(s) \qquad (1)$$

where U,Y,V are the Laplace transforms of the input $u(\tau)$, the output $y(\tau)$, and the disturbance $v(\tau)$ acting on the plant; $G_1$ and $G_2$ are rational transfer functions and $T_d$ is the i/o delay. It is well known (Astrom and Wittemmark, 1983), that the plant (1), equipped with A/D and D/A converters, is seen by a digital controller as a discrete-time system

$$A(q)y(t) = q^k B(q)u(t) + C(q)e(t) \qquad (2)$$

where $e(t)$ is a discrete-time "white noise"; $y(t)$ and $u(t)$ are the output and input values at the sampling instants $\tau = tT_s$, $t=0,1,\ldots$, $T_s$ being the sampling period, and $A,B,C$ are polynomials in the unit delay operator $q$. The integer i/o delay $k$ satisfies the following relation

$$T_d = (k-1)T_s + \delta k T_s \quad , \; 0 \le \delta k < 1 \qquad (3)$$

where $\delta k$ is the fractional delay. Notice that the coefficients of the polynomials $A$ and $C$ depend on $T_s$, while those of $B$ depend on both $T_s$ and $\delta$k. The controller design is based on the minimization of the quadratic cost functional

$$J = E\{[P(q)(y(t+k)-y^*(t)]^2 + [Q'(q)u(t)]^2\} \quad (4)$$

where $y^*$ is a\reference trajectory to be tracked, and $P,Q'$ are appropriate shaping filters of the closed loop transfer function. The control $u(t)$ which minimizes (4) is recursively generated by

$$G(q)u(t) = -F(q)y(t) = H(q)y^*(t) \qquad (5)$$

where

$$G := CQ+BE \; , \; Q := q_0Q'/b_0$$
$$H := PC \quad , \; PC = EA+q^kF \qquad (6)$$

$b_0$ and $q_0$ being the leading coefficients of $B,Q'$. The structure of the control law (5) is given by the triple $\delta G,\delta F,\delta H$ where $\delta$X denote the degree of X.

The *indirect* adaptive controller of the plant (1) based on (4) performs, during each sampling interval, the following tasks:

1') generation of the control signal according to (5);
2') computation of the control law coefficients according to (6);
3') updating of the parameter estimates of the discrete time model of the plant.

The alternative *direct* controller will be preferred for its computationally efficiency. It is based on the direct estimate of controller parameters. Let

$$\varphi(t) := P(y(t)-y^*(t-k)) + Q(u(t-k)) \qquad (7)$$

be an auxiliary output. Then it can be proved (Clark and Gawthrop, 1975) that if the estimate $\hat{\theta}(t)$ of the regression vector $\theta$ of the prediction model

$$\varphi_{t+k} = x(t)'\theta + \epsilon(t+k) \quad ,$$
$$x(t) := [y(t),y(t-1),..,u(t),u(t-1),..,-y^*(t-1),..]'$$
$$\theta = [f_0,f_1,\ldots,g_0,g_1,\ldots,h_0,h_1,\ldots]' \qquad (8)$$

converges to $\theta^*$, then the control law given implicity by setting

$$x(t)'\theta^* = 0$$

minimizes (4). Notice that the structure of $x$ and $\theta$ in (8) is matched to that of (5)Therefore the steps 1'-3' are replaced by:

1) generation of the control signal setting $x(t)'\theta(t-1) = 0$;
2) computation of $\varphi(t)$;
3) updating of the estimate $\theta(t)$ of $\theta$ in (8)

In order to show the potential advantages of a multimicro architecture, we shall focus our attention on the relations among the sampling time $T_s$ and the computing times $T_i$ required for the steps 1-3. Since in 1) and 2) only scalar products are performed, $T_1$ is proportional to the number of feedback gains n = dim $\theta$ and $T_2$ to $\delta P+\delta Q$. Conversely, for a recursive least square (RLS) implementation of 3), $T_3$ turns out to be proportional to $n^2$. Hence, in general,

$$T_3 >> T_1,T_2$$

and, in a standard implementation of an adaptive controller,

$$T_s > T_1 + T_2 + T_3 \qquad (9)$$

This constraint does not constitute a severe limitation in many applications. However there are cases, (plants with small time-constants), where the lower bound given by (9) may prevent the use of adaptive techniques. When this occurs, in order to reduce the sampling periods, the conventional implementation, hereafter denoted by a),must be replaced by:

b) Segmentation of the parameter estimation task (Sheirah and Malik, 1982);The task 3) is divided into $l_b$ subtasks to be executed sequentially. During a sampling interval only 1), 2) and one of the subtasks of 3) are executed. Hence feedback parameters are updated in $l_b$ sampling periods.

c) Use of a multiprocessor architecture; The control tasks 1),2) and the estimation task 3) are allocated on different processors. During each sampling period the first processor performs 1) and 2) and communicates i/o data to the other processor(s); controllers parameters are updated only when new estimates have been acquired by the the first processor.

d) A parallel execution of the whole algorithm on a dedicated architecture, e.g. systolic arrays;

Although the last alternative yields, potentially, the highest performance, it requires a complex hardware, equipped with several processing units, and special tools for software development.
The motivation for the alternatives b) and c) is that, if plant parameters undergo only slow changes, it is not necessary to update the control gains at each sampling instant. Hence, with a smaller $T_s$, a better noise rejection and tracking capability could be expected, without substantially reducing the adaptation speed. Bounds on $T_s$ and $l$ (number of sampling instants required for estimate updating) are now given by:

$$\text{case b)} \quad \begin{cases} T_s > T_1 + T_2 \\ l > T_3/(T_s - T_1 - T_2) \end{cases}$$

$$\text{case c)} \quad \begin{cases} T_s > T_1 + T_2 + T_C \\ l > (T_3 + T_C)/T_s \end{cases}$$

where $T_c$ is the time necessary for communicating between the processors. If $T_c << (T_1+T_3)$ the advantages of a multiprocessor architecture are evident: a quicker adaptation with almost the same $T_s$, or a shorter $T_s$ with the same adaptation rate.

## IMPLEMENTATION OF THE CONTROL ALGORITHM

The multiprocessor architecture described in the previous sections has been employed for the parallel implementation of the adaptive control algorithm, according to the strategy c). Two processing units (hereafter denoted as A and B) are required.

Remark. In order to reduce the delay between data acquisition and control generation, the computation of $u(t)$ has been split into two steps:

3a) compute the auxiliary input $z(t-1)$

$$z(t-1):= \sum_{i=1} (f_i y(t-i) + g_i u(t-i) - h_i y^*(t-i)) \quad (10)$$

3b) compute

$$u(t) = -(z(t-1) + f_0 y(t) - h_0 y^*(t))/g_0 \quad (11)$$

Therefore $z(t-1)$, not depending on the most recent data, is evaluated in the previous sampling period.

Unit A:
Data organization: local memory contains the following variables:

| | |
|---|---|
| i/o data | $y$, $y^*$, $u$; |
| regression vector | $x$; |
| auxiliary input and output | $z$ and $\varphi$; |
| feedback gain vector | $\theta$. |

At each sampling instant $t$ the following instructions are executed:

1) i/o data acquisition: $u(t-1), y(t), y^*(t)$ are loaded from the A/D converter;
2) Control signal generation: $u(t)$ is computed according to (11);
3) Sending of the control signal to the D/A converter.
4) Auxiliary output computation: $\varphi$ is computed according to (7)
5) i/o data communication: $\varphi$ and $x = [y(t-k),..,u(t-k),..,y^*(t-k)...]$ are sent to the global memory.
6) Regression vector updating: $x = [y(t-k+1),..u(t-k+1),..y^*(t-k+1),..]$
7) Feedback gain vector updating: test if a new feedback gain vector $\theta$ is available and, in the affirmative, receive it from the global memory;
8) Auxiliary input computation: $z(t)$ is computed according to (10).

Unit B
Data organization: the local memory contains the following variables:

| | |
|---|---|
| regression vector | $x$; |
| auxiliary output | $\varphi$; |
| feedback gain vector | $\theta$. |

The following instructions are executed:
1) Regression data acquisition: $\varphi$ and $x$ are transferred from global to local memory.
2) Feedback gain vector updating: $\theta$ is updated according to a UD factorization of the RLS algorithm.
3) Feedback gain vector comunication: $\theta$ is sent to the global memory.

Remark. Since $T_e$ is random, the parameter updating in the control algorithm does not take place at a costant rate. This allows a quicker adaptation.

## PERFORMANCES OF THE ADAPTIVE CONTROLLERS

Adaptive controller performances have been tested on a laboratory process simulator (PCS327 by Feedback), whose i/o transfer function is

$$G(s) = K_p \frac{1}{(s+0.01)^2}(1 + 1/sT_i)$$

Both the gain $K_p$ and the integrator time constant $T_i$ can be manually changed so that plant variations can be simulated. White noise has been added to the input so that, with reference to (1) we have $G_1 = G_2 = G$, and $T_d = 0$. When $T_i = \infty$, $G(s)$ has two coinciding real poles and the unit step response has a raising time of 33ms. In the cost functional (4) we have set $P=1$ and $Q'=1/2b_0$. A forgetting factor of 0.984 has been employed in the RLS algorithm. The reference signal is a square wave with period of 2s.

The control structure was chosen according to

$$u(t) = -[f_0 y(t) + f_1 y(t-1) + g_1 u(t-1) - h_0 y^*(t) - h_1 y^*(t-1)]/g_0 \quad (12)$$

which is matched to the plant only for $T_i=\infty$; hence (12) is a restricted complexity regulator.

A conventional implementation a) of the adaptive control algorithm has been compared to those based on the strategies b) and c) of the previous section. Since for the above cost-functional and control complexity we have $T_1 + T_2$ = 9+14 ms and $T_3$ = 70+80 ms, the following sampling intervals have been adopted:

$T_{sa}$= 100ms ;
$T_{sb}$= 32ms, of which 16ms for parameter updating,which implies $l_b$ = 6 (this is a good compromise between minimization of $T_{sb}$ and $l_b$;
$T_{sc}$= 15ms, i.e. $l_c$ = 5+6.

Notice that times required for data communication as given by Table 1, are inferior to algorithm execution times; hence a parallel implementation becomes, at least in principle, meaningful

| | strategy a) | | | strategy b) | | | strategy c) | | |
|---|---|---|---|---|---|---|---|---|---|
| $T_i$ | 50 | 25 | 10 | 50 | 25 | 10 | 50 | 25 | 10 |
| $t_r$ | 93 | 79 | 74 | 52 | 33 | 25 | 35 | 22 | 16 |
| $t_s$ | 300 | 500 | 800 | 400 | 130 | 140 | 70 | 54 | 68 |
| $M$ % | - | - | 8 | - | 3 | 21 | - | 3 | 16 |

Table 2 - Overshoot, raising and settling time of the control regulator for different strategies and integrator time constants

Comparisons carried out concern the steady state behaviour, both in term of response accuracy and noise rejection, and the adaptation speed to plant parameters changes. In Table 2, the raising time $t_r$(ms), the settling time $t_s$(ms) and the percent overshoot $M$(%) of the output response to a step reference signal are reported. They concern the strategies a), b), and c), and several values of the integrator time constant ($T_i$). All regulators have been optimally tuned according to (4).

In fig. 7 the time evolutions of the average (over 50 samples) value J of the cost functional are reported ($T_i$=∞) for the strategies b) (dashed line and c) continuous line.

Finally in Fig. (8) the plant output response to reference signal square waves is shown for all the strategies. The time scale is 250 samples/div. The integrator time costant has been suddendly changed from 25 to 10ms, so that adaptation speed can be evaluated.

From these experiments the following conclusion can be drawn:

i) Steady state performances of the tuned regulators improve (from a deterministic and stochastic point of view) when the sampling interval is reduced.

ii) The adaptation speed depends mainly on the data acquisition rate ($1/T_s l$) of the estimation process, while the sampling time does not appear to be a relevant parameter; i.e. a nonconventional implementation with a given sampling period $T_s$ and updating rate $l$, has the same adaptation speed of a conventional one with $T'_s = T_s l$. Therefore, the usefulness of a multiprocessor implementation of a non conventional strategy c) over b) are quite apparent.

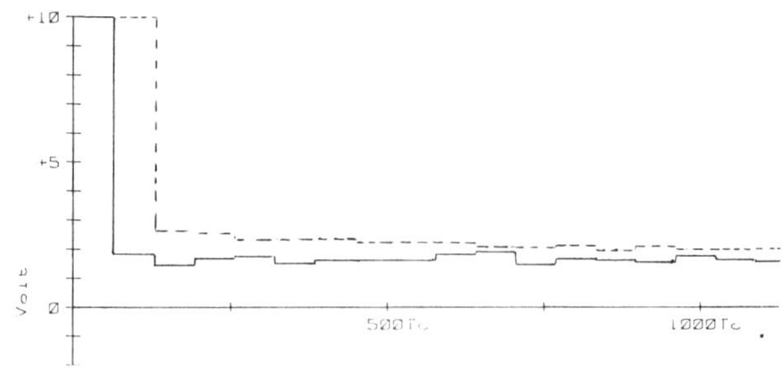


Figure 7 - Time evolution of the average value (over 50 samples) of the cost functional; dashed line: b); continuous line: c).

## CONCLUSIONS

The main hardware and software features of a low cost multimicro architecture have been described. Moreover a nonconventional implementation of an adaptive controller has been presented. This is based on the parallel execution of the control and estimation processes on different processing units at different rates. The advantages of this approach has been highlighted by laboratory experiments, in the cases where a conventional implementation does not allow a control sampling rate high enough.

## REFERENCES

Astrom, K. and B. Wittemmark (1983). Computer Controlled systems - Prentice-Hall.

Bagnoli, C., F. Cesarini and M. Gori (1984). EXEC: a distributed multiprocessor interprocess communication executive. Proc.MIMI 84, Bari, Italy, pp. 99-102.

Bagnoli, C., M. Gori and G. Soda (1984). Design and implementation of the GMMP multi-microprocessor system. Int. Journal of Mini and Microcomputers, 6, n. 2, pp.26-29.

Clark, D.W. and P.J. Gawthrop (1975). Self tuning Controller, Proc. IEEE, 22, n. 9, pp. 929-934.

Sheirah, M.A., and O.P.Malik (1982). Self-tuning microprocessor universal controller. IEEE Trans. Ind. Electr., IE-29, n.1.

Weitzman, C. (1980). Distributed Micro/minicomputer systems. Prentice Hall, New York.

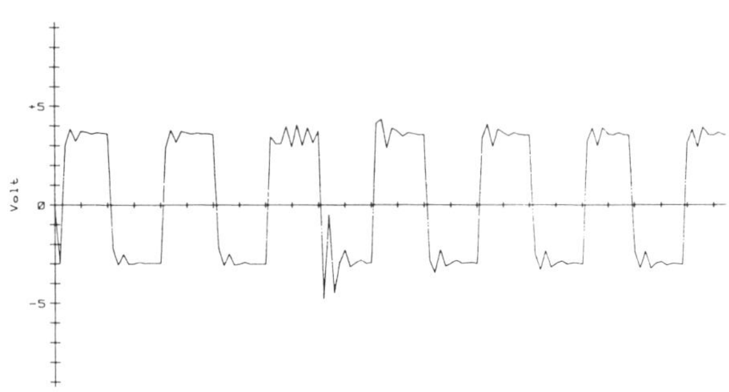


Strategy a)

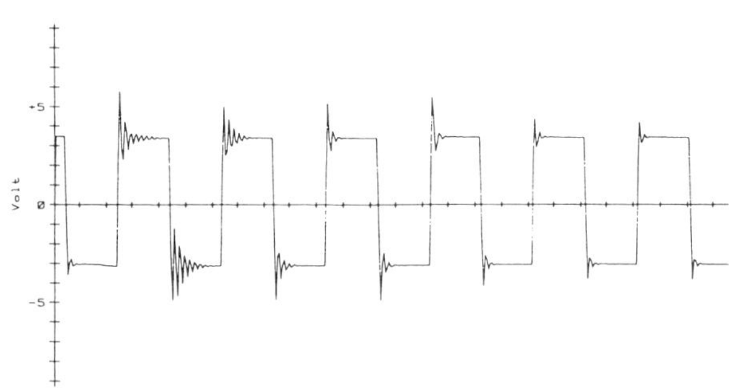


Strategy b)

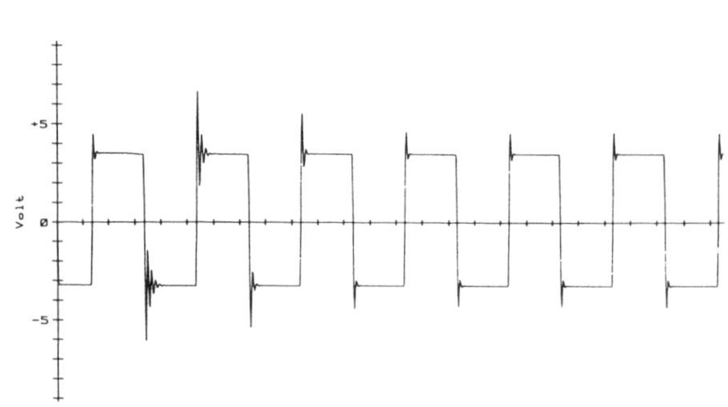


Strategy c)

Figure 8 - Plant output in presence of a sudden transition of $T_i$ from 25 to 10ms.

# MICROPROCESSOR IMPLEMENTED SUB-OPTIMUM SMITH CONTROLLERS FOR TEMPERATURE CONTROL

**C. Batur and J. Braun**

*Department of Mechanical Engineering, University of Akron, Akron, OH 44325, USA*

Abstract. The Smith dead time compensator is configured to operate with the Dahlin or the dead-beat controller. The innovation suggested in this paper is to avoid the estimation of the process dead time which is needed in the conventional Smith predictor scheme. This feature is believed to be a practical asset in real life applications where the estimation of the dead time may prove to be difficult. As expected, elimination of the dead time estimation increases the output variance. Further contribution of the paper is to show that the initially designed Smith compensator can always be made stable by placing a low pass filter in the feedback path and adjusting the bandwidth of the filter. The intuitive argument behind this approach is to make the overall system less sensitive to high frequency noise. Finally, these concepts are demonstrated on a pilot scale temperature control system.

Keywords. Smith dead time compensator; Dahlin control; Dead beat control; stability; temperature control.

## INTRODUCTION

The basic self-tuning regulator of (Astrom and Wittenmark,1973) and the self-tuning controller of (Clarke and Gawthrop,1975,1979; Clarke et.all, 1983) permit the process model and the controller to be updated for changes in the process dynamics. They explicitly assume that the process dead time is known. The pole placement controllers of ( Wellstead et.all.,1979a,b,1981) make the same assumption. All these algorithms can be modified to handle the dead time changes if the process dead time is also estimated on-line. Earlier attempts to estimate the dead time can be found in (Turtle and Phillipson,1971). They proposed a search technique to see if a given estimate can be assumed significantly different from zero. The number of zero coefficients are used to estimate the dead time. A confidence interval and a distribution function have to be assumed for the estimated parameters. (Astrom,1968; Unbehauen and Gohring, 1973). Furtermore this test is not recursive. The technique proposed by (Kurz and Goedecke,1981) estimates the dead time on-line at the expense of large computations. Pole placement algorithm of (Wellstead et.all,1979a) shares the same feature as far as the recursive estimate of the dead time is concerned. (Voger and Edgar,1982) have developed a dead time compensator for the Smith predictor control scheme operating with Dahlin controller. The approach does not require the explicit estimation of the dead time. This paper follows the same philosophy as far as the dead time estimation is concerned. However in addition to Dahlin controller, it also considers the dead beat controllers and proposes a practical procedure to stabilize the overall control system despite the inevitable modelling errors.

## PROCESS IDENTIFICATION

The stable process operating under stationary noise is assumed in the following form,

$$A(q^{-1})y(t) = q^{-k}B(q^{-1})u(t) + n(t) \quad (1)$$

where u(t),y(t) and n(t) represent the input, output and the noise respectively. The polynomial $A(q^{-1})$ has all the roots within the unit circle and defined as $A(q^{-1}) = 1-a_1q^{-1}-\ldots-a_nq^{-n}$. The process is assumed to be of minimum phase therefore the polynomial $B(q^{-1}) = b_1q^{-1} + \ldots + b_nq^{-n}$ is stable. The parameter k indicates the dead time index implying that the process has a total of (k+1)T dead time where T defines the sampling period.

The estimated process model will be assumed in the same form as (1) where the true parameters will be replaced by the estimated ones as

$$\hat{A}(q^{-1})y(t) = q^{-\hat{k}}\hat{B}(q^{-1})u(t) + \hat{n}(t) \quad (2)$$

A common way to avoid the explicit estimation of the process dead time k is to rewrite (1) as

$$A(q^{-1})y(t) = D(q^{-1})u(t) = n(t) \quad (3)$$

where $D(q^{-1}) = d_1q^{-1} + d_2q^{-2} + \ldots + d_{n+k}q^{-(n+k)}$

satisfying the following conditions

$$d_i=0 \ ; \ i=1,2,\ldots,k \quad \text{and} \quad d_i=b_{i-k} \ ; \ i=k+1,\ldots,k+n$$

Corresponding to (3), the estimated model can be written as

$$\hat{A}(q^{-1})y(t) = \hat{D}(q^{-1})u(t) + \hat{n}(t) \quad (4)$$

Obviously for the estimated parameters it can not be expected that the first k coefficients of $\hat{D}(q^{-1})$ will be identically zero. It therefore follows that the corresponding conditions on $\hat{D}(q^{-1})$ should be written as

$$\hat{d}_i \cong 0 \ ; \ i=1,2,\ldots,\hat{k} \quad \text{and} \quad \hat{d}_i=\hat{b}_{i-\hat{k}} \ ; \ i=\hat{k}+1,\ldots,\hat{k}+\hat{n}$$

This structure suggests that if the polynomial

$\hat{D}(q^{-1})$ is estimated then the number of first zero coefficients should indicate the process dead time. Since all the coefficients are estimates then a statistical procedure has to be performed to determine if a given coefficient is significantly different from zero. If the probability density function of the estimated parameters are known then by defining a somewhat subjective confidence interval level, statistical tests can be devised to determine if any coefficient is significantly different from zero (Astrom,1968; Unbehauen and Gohring,1973). However assuming certain probability density function on estimates imply that the probabilistic properties of the noise sequence are known. This may prove unrealistic for most industrial processes.
The above argument suggests that if the explicit dead time information is not needed in the controller design then instead of estimating $B(q^{-1})$ and k. separetly, one can only estimate the polynomial $D(q^{-1})$. The computational effort to be paid for this would be the increase in the number of estimated parameters from (n+1) (for B and k) to (n+k) (for D). However if the controller design requires the explicit knowledge of the process dead time this approach has to be further modified. This is discussed in the next section.

## CONTROLLER DESIGN

Consider the Smith predictor scheme to dead time compensation (Smith,1957 ; Chien et. all.,1984). Fig.1 shows the block diagram of the discrete-time Smith predictor. The process model (assumed to be correct in this figure) is in parallel to the actual process. The process and model outputs are compared to obtain the disturbance information. The feedback path contains a predictor to predict the future values of the noise. These predicted values are substracted from the reference input to minimize the effect of the noise on the process output. The controller in general is a function of the process model polynomials $A(q^{-1})$ and $B(q^{-1})$. Because of the properties of the Smith predictor scheme the controller does not require the explicit knowledge of the process dead time k. For the estimated parameters $\hat{A}(q^{-1})$, $\hat{B}(q^{-1})$ and $\hat{k}$ the Smith predictor scheme can be shown as in Fig.2. Where the polynomial $F(q^{-1})$ in the controller block diagram coresponds to a controller design for a system without dead time i.e. k=0. The predictor polynomial $H(q^{-1})$ is assumed to be a stable filter.

If the polynomial $\hat{B}(q^{-1})$ and $\hat{k}$ are available then the Smith predictor scheme can be formed easly. However this requires the explicit identification of the process dead time. If the approach suggested in the previous section is followed then instead of identifying $\hat{B}(q^{-1})$ and $\hat{k}$ separately one can estimate the polynomial $D(q^{-1})$ as described in the previous section. Referring to Fig.2 however this implies that the second term of the denominator of the controller will not available. To avoid the ambiguity associated with the use of $D(q^{-1})$, the polynomial $\hat{B}(q^{-1})$ in the controller will be replaced by the constant term $q^{-1}\hat{D}(1)$. Furthermore since $\hat{B}(q^{-1})$ and $\hat{k}$ are not explicitly available then the numerator of the model will be replaced by the same constant $q^{-1}\hat{D}(1)$. The rationale behind this modification is as follows. If the estimation errors are negligible then $\hat{D}(q^{-1})$ will be very close to $q^{-\hat{k}}\hat{B}(q^{-1})$ therefore the steady state properties of these polynomials defined by $\hat{D}(q^{-1})|_{q=1}$ and $q^{-\hat{k}}\hat{B}(q^{-1})|_{q=1}$ will be very close to each other i.e.

$$\hat{D}(1)\cong\hat{B}(1) \tag{5}$$

It therefore follows that this modification will not make a significant change in the steady state properties of the estimated model. A factor $q^{-1}$ has to be introduced in connection with $\hat{D}(1)$ because even under no dead time, k=0, the process and its model will have a dead time of one sample period due to the sampler in the discrete systems. Also note that replacing $q^{-\hat{k}}\hat{B}(q^{-1})$ by $q^{-1}\hat{D}(1)$ will avoid problems associated with the non-minimum phase systems. Even if the process is of minimum phase it is well known that (Clarke,1984) due to the sampling process the discrete model $\hat{B}(q^{-1})$ may contain roots outside the unit circle.

With these changes the Smith control system of Fig.2 takes the modified form shown in Fig.3. The classical feedback control version corresponding to Fig.3 can be obtained by block diagram manipulations and it is given in Fig.4. The controller transfer function is defined by the choice of controller polynomial $F(q^{-1})$. This is considered in the next section.

## CONTROL ALGORITHMS

Depending on the choice of the controller polynomial $F(q^{-1})$, several algorithms can be developed. Only three of those are considered here

a- Pole cancellation algorithm.

This algorithm corresponds to $F(q^{-1}) = K =$ constant. It is easy to see from Fig.3 that for very large K values the controller attemps to cancel all the poles of the process. However this solution will not be practical because of the large controller gain setting required.

b- Dahlin controller (Dahlin,1968).

This algorithm adjusts the step response to the set point with a user selected parameter (ν). The algorithm can be written in the following forms.

$$u(t) = \frac{\hat{A}(q^{-1})(1-e^{-\nu T})}{\hat{D}(1)(1-e^{-\nu T}q^{-1})}\,\zeta(t) \tag{6a}$$

$$u(t) = \frac{\hat{A}(q^{-1})(1-e^{-\nu T})}{\hat{D}(1)(1-q^{-1})}\,\xi(t) \tag{6b}$$

where (6a) corresponds to Fig.3 and (6b) corresponds to Fig.4. It is assumed that the predictor polynomial $H(q^{-1})$ is unity.

c- Dead beat controllers (Kurz and Isermann, 1977)

For $H(q^{-1}) = 1$ this controller takes the following forms

$$u(t) = \frac{\hat{A}(q^{-1})}{\hat{D}(1)(1-q^{-1})}\,\xi(t) \tag{7a}$$

$$u(t) = \frac{\hat{A}(q^{-1})}{\hat{D}(1)}\,\zeta(t) \tag{7b}$$

as before (7a and b) correspond to Fig.3 and 4 respectively. It can be seen from the comparison of (6) and (7) that the Dead beat controller is a special case for the Dahlin control algorithm and can be derived by substituting $\nu = \infty$ in Eqns. (6a,b). Furthermore comparison also shows that each algorithm has a built an free integrator to avoid steady state offset resulting from the step changes in the desired point. These sub-optimum Dahlin and Dead beat control algorithms will have practical implementation if they can be made stable even if there are large discrepencies between the process and the corresponding model. The next section discusses the stability problem.

## STABILITY OF THE SMITH PREDICTOR CONTROLLERS

The response of the Smith predictor control scheme is determined by not only the controller polynomial $F(q^{-1})$ but also by the process-model discrepancy and the predictor polynomial $H(q^{-1})$. This can be seen by writing the characteristic equation corresponding to Fig.3 as

$$A\hat{A}+AF\hat{D}(1)q^{-1}-HF[A\hat{D}(1)q^{-1}-q^{-k}\hat{A}B]=0 \qquad (8)$$

where for the sake of clarity the argument $(q^{-1})$ has been dropped from all polynomials. The characteristic equation shows that if the initial set-up for the Smith predictor control scheme is unstable then the operator has no tools to make the overall system stable. It may however be argued that if the Dahlin controller is employed then the adjustable parameter ($\nu$) of the Dahlin controller can be used to stabilize the system. The effect of this parameter can be analyzed by rewriting (8) for the Dahlin controller and for $H(q^{-1}) = 1$ as

$$A(1-q^{-1})\hat{D}(1) + \hat{A}(1-e^{\nu T})Bq^{-k} = 0 \qquad (9)$$

It is not obvious to see how the parameter $\nu$ will change the roots of the characteristic equation. An alternative approach is to manipulate the predictor polynomial in such a way that the roots of the characteristic equation can be guided to the inside of the unit circle. Since the feedback signal $X(t)$ of Fig.3 is a function of process disturbance and the process-model difference, the statistical properties of $X(t)$ are generally not known. Therefore sufficient information is not available to design an optimum predictor. Given this fact, a proper course would be to implement a low-pass filter in the feedback path to stop the high frequency component of the disturbance signal coming back to the controller. This filter is defined by the polynomial

$$H(q^{-1}) = \frac{1}{\gamma-(\gamma-1)q^{-1}} \quad 1\leqslant\gamma<\infty \qquad (10)$$

where the increase in $\gamma$ corresponds to the decrease in the cut-off frequency. If the Dahlin controller is used with this filter the characteristic equation becomes

$$[\hat{D}(1)(q^{n}A)q^{k+\hat{n}}e^{-\nu T}\Delta+(q^{n}B)(q^{\hat{n}}\hat{A})q^{2}(1-e^{-\nu T})]+$$

$$+\ \gamma\hat{D}(1)(q-e^{-\nu T})(q^{n}A)q^{\hat{n}+k}\Delta = 0 \qquad (11)$$

where $\Delta=q-1$. If this equation is viewed as a root locus equation with respect to parameter $\gamma$ then the zeros are located at $q=1$, $q=e^{-\nu T}$ and at the positions defined by the polynomial $(q^{n}A)$. Since $\nu$ and $T$ are positive quantities then all zeros are located within the unit circle with one exception namely the root at $q=1$. Since the number of zeros exceed that of poles by one there will be no zero at infinity. This means that if the root locus on the real axis approaches to $q=1$ point from the left, then the control system can always be made stable for large $\gamma$. This situation will occur if there are even number of poles to the right of $q=1$ point. To see if this is the case the following test is devised. If a polynomial $f(x)$ has even number of poles to the right of $x=1$ then it can be proved that the following sufficient condition is satisfied

$$\text{Sgn}[\lim_{x\to\infty} f(x)] = \text{Sgn}\ f(1) \qquad (12)$$

where Sgn is the signum function defined by

$$\text{Sgn}\ f(x) = 1 \quad \text{for } f(x)>0$$

$$\text{Sgn}\ f(x) = -1 \quad \text{for } f(x)<0$$

if (12) is applied to the poles of the characteristic equation (11) the sufficient condition for stability can be evaluated as

$$\hat{D}(1).B(1)>0 \qquad (13)$$

However since the process and its model are assumed to be stable then $A(1)>0, \hat{A}(1)>0$, combining (13) with those statements sufficiency condition can be established as

$$\frac{\hat{D}(1)}{\hat{A}(1)} \cdot \frac{B(1)}{A(1)} >0 \qquad (14)$$

This means that if the steady state gain of the model and the process have the sign then the Smith predictor controller can always be made stable by increasing $\gamma$ or equivalently by decreasing the cut-off frequency of the low pass filter.

## APPLICATION OF SMITH PREDICTOR WITH SUB-OPTIMUM DAHLIN CONTROLLER

A simple liquid heater system is used to prove the concepts outlined in the paper. 75W heater heats the fluid at the entrance. The fluid then follows a long unheated section of the heater. This is designed to introduce an artificial dead time into the process. The dead time is a function of flow rate and for nominal flow rates it is observed that it changes between 4 and 9 seconds. Because of this large change it would be difficult to estimate the process dead time on-line. Instrumentation details are given in Fig.5. The microprocessor used in the experiment is MC6502. The process disturbance is normally due to the measurement system in which a small thermistor faces the upstream flow. The air bubles and the steam pockets in the fluid are the main noise contributors.

Fig.6 shows the typical response of the modified Dahlin controller operating under the Smith predictor control scheme. The low pass filter is operating with $\gamma=1.01$ and the Dahlin controller is adjusted to give a step response with the adjustable parameter set to $\nu=0.1$ . Increasing $\nu$ further will unstabilize the system. However this can be counteracted by increasing $\gamma$. This is illustrated in Fig.7

## CONCLUSION

The technique proposed in this paper avoids the explicit identification of the process dead time. This feature is believed to be a practical asset in real life applications where the estimation of the dead time may be difficult. Further contribution of the paper is to manipulate the feedback path information of the Smith predictor scheme to insure the stability. This is achieved by implementing a low pass filter and decreasing the bandwidth until stability is obtained. It is proved that if the sign of the steady state gain of the process is known then the control system can always be stabilized.

## REFERENCES

Astrom,K.J., Wittenmark,B., (1980), Self-tuning Controllers Based on Pole-Zero Placement. Proc. IEE, No.3, pp 120-130.

Astrom,K.J., (1968). Lectures on the Identification Problem, Lund Inst. of Tech. Div. of Aut. Cont. Report, 6805.

Chien,I.L., Seborg,D.E., Mellichamp,D.A., (1985). A self-tuning controller for systems with unknown or varying time delays. Int.J. Control, Vol.42, No.4, 949-964.

Clarke,D.W., Gawthrop,B.A., (1975). Self-tuning Controller, Proc. IEE, Vol.122, No.9 pp. 929-934.
Clarke,D.W., Gawthrop,P.J., (1979). Self-tuning Control. Proc. IEE Vol.126, No.6, pp. 633-640.
Clarke,D.W., Hodgson,A.J.F., Tuffs,P.S., (1983). Offset Problem and K-Incremental Predictors in Self-tuning Control, IEE Proc. Vol.130, Pt.D, No.5 September.
Clarke,D.W., (1984). Self-tuning Control of Non-minimum Phase Systems. Automatica No.3, pp 501-517.
Dahlin,E.B., (1968). Designing and Tuning Digital Controllers, Instrument and Control Systems, Vol.42, pp.77-83, June.
Kurz,H., Isermann,R., (1977). Feedback Control Algorithms for Parameter Adaptive Control-Comparison and Identifiability aspects. JACC. WP.26.
Smith,D.J.M., (1959). A Controller to Overcome Dead Time. Instrument Society of America Journal, Vol.6, pp.28-33.
Turtle,D.P., Phillipson,P.H., (1971). Simultaneous Identification and Control. Automatica Vol.7, pp.445-452.
Vogel,E.F., Edgar,T.F., (1982). An Adaptive dead time compensator for process control. JACC., Arlington Virginia pp. TPS-E.
Unbehauen, Gohring,H., (1973). Application of Different Statistical Tests for the Determination of the Most Accurate Model in the Parameter Estimation. Proc. of 3rd IFAC Symp.on Idet. and Syst. Parameter Estimation, North Holland.
Wellstead,P.E., Prager,D., Zanker,P., (1979a). Pole Assignment Self-tuning Regulator, Proc. IEE, Vol.126, No.8, pp 781-787.
Wellstead,P.E., Edmunds,J.M., Prager,D., Zanker,P., (1979b). Self-tuning Pole-Zero Assignment Regulators, Int. J. Control, Vol. 30, No.1, pp.1-26.
Wellstead,P.E., Sanoff,S.P., (1981), Extended Self-tuning Algorithm, Int. J. Control, Vol.34, pp.433-455.

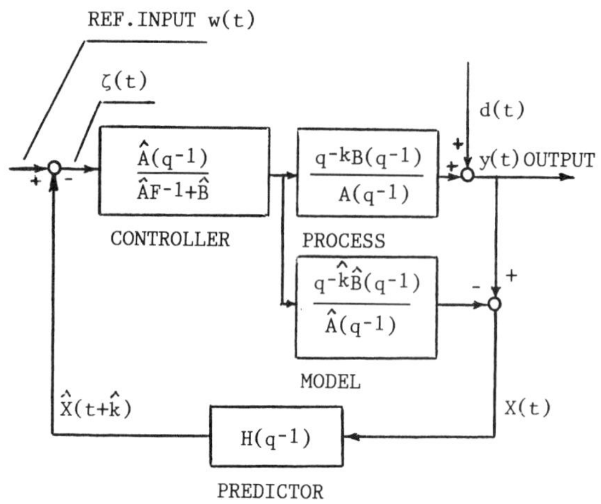


FIG.2. THE SMITH PREDICTOR SCHEME FOR THE ESTIMATED PARAMETERS.

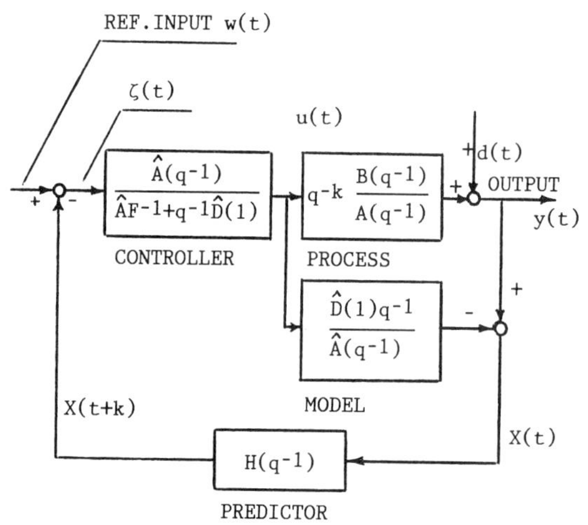


FIG.3 THE MODIFIED SMITH PREDICTOR SCHEME

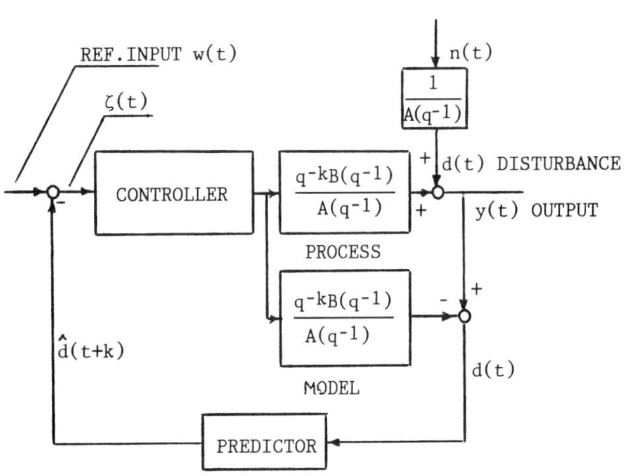


FIG.1 THE SMITH PREDICTOR SCHEME FOR THE IDEAL MODEL

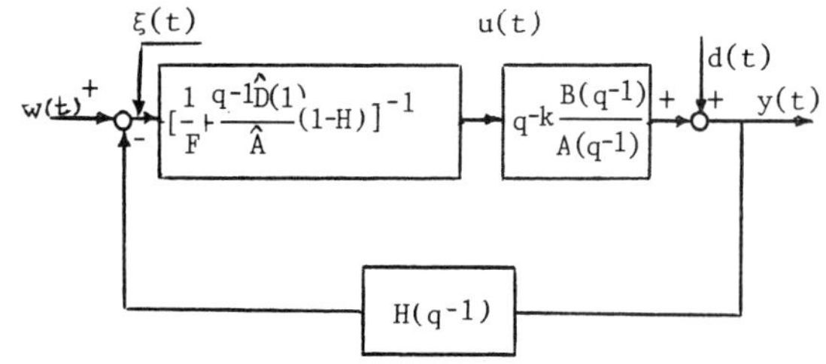


FIG.4 THE CLASSICAL FEEDBACK CONTROL VERSION OF THE MODIFIED SMITH PREDICTOR SCHEME

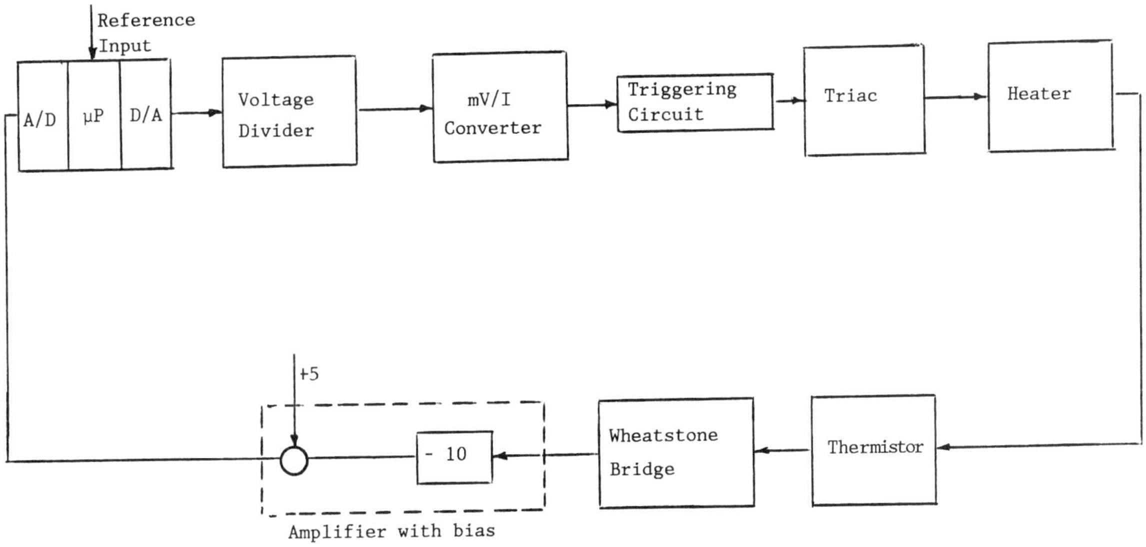


Fig.5 Instrumentation details

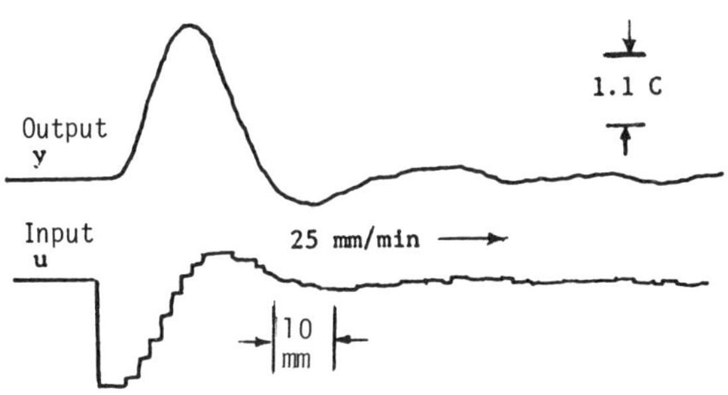


Fig.6 The response of Dahlin Controller operating under the Smith Control Scheme $\gamma$=0.01 $\nu$=0.1

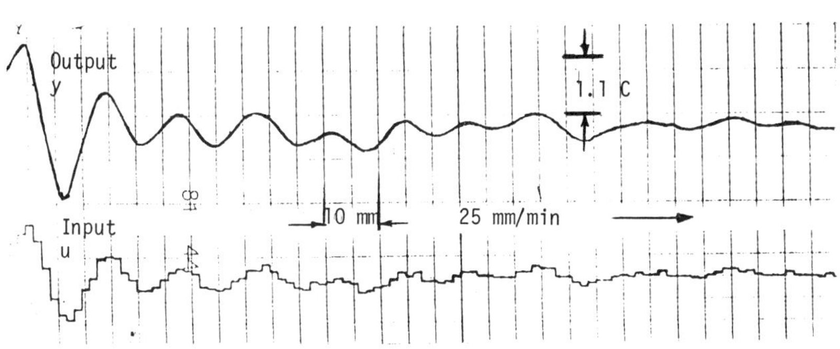


Fig.7 The response of Dahlin Controller operating under the Smith Control Scheme $\gamma$=0.5, $\nu$=0.08

# USING EXPERT SYSTEMS TO TROUBLESHOOT MICROPROCESSOR-BASED CONTROL SYSTEMS: AN EXPECTATION-BASED APPROACH*

**P. Schaefer** and H. A. Guvenir*****

***Martin Marietta Denver Aerospace, P.O. Box 179, M.S. 0427, Denver, CO 80201, USA*

****CWRU, Computer Engineering and Science, Cleveland, OH 44106, USA*

Abstract. The emergence of microprocessor-based controllers has allowed major improvements in control flexibility from simple overall system architectures, but has led to an increase in the difficulty of controller repair, due to the increased complexity of internal hardware. To alleviate this problem, therefore, it would be extremely beneficial to be able to package the necessary microcomputer expertise in a computer-based "Expert System," so that a technician without extensive microcomputer experience could effectively troubleshoot the control system. We describe an Expectation-based approach for such a troubleshooting expert system. In this approach, expert knowledge about cause-and-effect behavior and the implications of observations drives the inference mechanism to successful troubleshooting. The advantages of this approach are discussed, and an implemented system, FEXPR, is presented.

Keywords. Artificial Intelligence, Microprocessors, Controllers, System failure and recovery; Troubleshooting.

## INTRODUCTION

The diagnosis of microcomputer-based control systems is an important practical consideration in light of today's rapidly-advancing technology. These control systems often have a central role in an industrial or laboratory process, and down-time is therefore quite costly and to be avoided.

Because of the ever-increasing complexity of these control systems, however, it is difficult to find technicians familiar enough with microcomputer technology to do efficient troubleshooting. One conventional solution to the problem is to continually train and transport microprocessor experts from the controller manufacturer. This becomes less and less feasible through time because the rapidly-expanding technology requires higher and higher levels of expertise (Wilkinson, 1985). In addition to the increasing complexity, the troubleshooting problem is compounded because the technology itself is shifting from the more familiar analog hardware to digital techniques. What is needed, therefore, is means of providing the technician, who is knowledgable in the process being controlled, with an assistant expert in microcomputer technology. An effective tool for this purpose is available in the form of "Expert Systems," a class of Artificial Intelligence technology.

## EXPERT SYSTEMS APPLIED TO CONTROLLER TROUBLESHOOTING

Expert systems have for some time been applied to the task of diagnosis and troubleshooting in a variety of domains (Hayes-Roth, 1983). An Expert System is basically a computer implementation which contains the knowledge of a human expert about solving the problem at hand. During a process referred to as "Knowledge Engineering," the domain expert supplies his knowledge to a "Knowledge Engineer," who encodes it into the format required by the Expert System. Equipped with this knowledge, the Expert System should be able to perform the human expert's task within the boundaries covered in the Knowledge Engineering.

The most popular form of Expert System to date is the Production Rule based approach. In this kind of system, knowledge is encoded in condition/action pairs (Barr, 1981). A typical rule could be

IF <the power switch is ON> and
<there is no light from any LED>

THEN conclude that
<the problem is with the power supply>.

Unfortunately, there are several problems which result from using an expert system based on the condition/action approach. These problems result from the fact that the Expert System is not given an understanding of how the device works- it is provided only with appropriate troubleshooting actions to perform under various circumstances. This is not an important shortcoming in very simple problem domains; the knowledge engineer must simply ensure that enough questions are asked so that the system will never encounter a situation not covered by the rules.

In more complex domains, however, the lack of understanding in the condition/action system can be disastrous (Bonissone, 1985). It is clear that whether a certain rule applies in a given situation might be dependent on a great many other factors. These factors are often obvious to someone who understands the device being tested, but might easily be overlooked during knowledge engineering. For example, the power supply referred to in the example above might be connected to many different circuits.

*This work performed at the Center for Automation and Intelligent Systems Research, Case Western Reserve University.

Characteristics of those circuits might affect the validity of the rule. As the problem domain becomes more and more complex, therefore, the knowledge engineer would have to ask more and more questions to cover the special cases. Eventually, this process becomes completely infeasible.

Unfortunately, microcomputer-based controller troubleshooting is one such complex domain. A manufacturer usually wants to use some of the same circuitry or circuit modules in a wide variety of applications. Many different permutations of processors, A/D converters, control interfaces, communicatons cables, etc., might be installed at customer locations. Clearly, the problem of troubleshooting this wide array of possible systems is not a "limited problem domain," but requires a considerable amount of flexibility. To avoid the inherent shortcomings of the condition/action "shallow reasoning" expert-system approach, therefore, it is worthwhile to consider alternate expert-system formalizations. Ideally, it is now apparent, the controller-troubleshooting expert system should be given the expert's underlying understanding of how the components in the control system actually work, rather than only symptoms and conclusions. In this way, knowing how a particular system is configured, the expert system should be able to understand why incorrect behavior is occuring, in much the same way that the human expert would understand the problem. One approach to this "deep-reasoning" kind of expert system is the Expectation-based approach, which we now present.

## THE EXPECTATION-BASED APPROACH TO TROUBLESHOOTING

With an Expectation-based expert system, expert understanding of how components in a system behave is encoded, rather than symptoms and conclusions as in the conventional condition/action systems. The method is called "Expectation-based" because the expert's understanding is encoded as expectations about what will happen to a component when various inputs are present, and what is expected to be true about those inputs when observations about the component are made (Schaefer, 1986). This understanding knowledge is stored in the form of an Expectation Network.

The Expectation Network is the basic structure describing the understanding the expert system has of the controller. Its structure parallels that of the actual microcomputer-based controller. Each quantity, or "signal," of interest in the controller (data bus, LED, power-supply voltage, etc.) is a node in this network. Connecting these signals to each other are the expectation relations, or "rules," provided by the human expert. Each component, or "module," from which a controller can be built is described by a set of these expectation rules, which connect to all the inputs and outputs of that module. Figure 1 shows the generic organization of signals, modules, and rules. A simple example describing a Digital-to-Analog card is presented in Fig. 2.

An Expectation Network contains two kinds of expectation rules. "Causal rules" describe how the expert understands the cause-and-effect relationships among signals in the module. When the values of the signals input to these rules are known, the rules indicate what their output values are expected to be. For example, a causal rule could state

```
IF <the BOARD-SELECT signal goes high>
THEN <expect the value on the DATA-BUS to
      also be found at the
      input of the D/A chip>
```

The causal rules express similar knowledge to the knowledge in causal reasoning systems, a closely-related "deep-reasoning" approach in Artificial Intelligence (Bonissone, 1985). However, those systems usually require causal relationships to be at a "first-principles" level, at the level of transistors and gates. Our causal rules can be at whatever level the expert chooses to think about the module.

The second kind of rules in the Expectation-based approach are "evidence rules." These rules describe how the occurence of one or more signals leads the expert to expect the presense of other signals. An example evidence rule could be

```
IF <any of the LEDs is on>
THEN <expect +5 volts is present>
```

The Evidence Rules yield information about signal values which are not easily observable by the technician. If the assumption were made that the user could measure any signal, these rules would not be needed. However, it is clear that for efficient troubleshooting, an expert does indeed make use of this kind of expectation knowledge.

Using the Expectation Network of causal and evidence rules, all the processing necessary for controller troubleshooting can be performed. When some signal value needs to be tested, the Causal Network can be traced to indicate what test inputs should <u>cause</u> that signal to be observed. After the test input, the Evidence Network rules will indicate what <u>evidently</u> occured in reality. A fault is located when a module is found whose inputs and outputs no longer correspond to its causal rules. This indicates that the module did not respond correctly to the circuit conditions, and therefore should be replaced.

## THE FEXPR CONTROLLER TROUBLESHOOTING SYSTEM

To demonstrate the feasibility of this approach to troubleshooting expert systems, a program called FEXPR (Flexible EXPectation-based Reasoner) was implemented. It is written in VAX LISP and runs on a VAX 11/782. FEXPR has been demonstrated for troubleshooting a commercially-available 68000-based control system (Guvenir, 1985).

Description of the various modules is quite straightforward in FEXPR. Consider the example D/A card from figure 2. This module can be defined as

```
(defmodule D/A
  (write board-select data power) ;inputs
  (analog-out LEDs) ;output signals
  (latch latched-value)) ;internal signals
```

Any particular system to be analyzed can be "plugged together" from such modules by using a function PUT-MODULE. PUT-MODULE indicates the actual circuit points where one of the modules will be connected. In this way, zero or more of any module type can be included in the system. For the example, one could choose

```
(put-module D/A-BOARD1 D/A
  (pin-1 board-3-select data-bus power-bus)
  (pin-3 led-block-1)) ;outputs
```

which indicates which actual points on the controller this specific D/A-BOARD1 of type D/A will be connected. In this case, it will be to the board-3 position, as indicated by the select line, above.

Causal rules are used to describe the behavior ofa module. They can be defined using DEFRULE. One simple rule could be

```
(defrule latch-data D/A
  IF (high write)
     (high board-select)
     (+5volts power)
  THEN (same-bus-value latched-value data))
```

which means "If the write signal is high, the board select is high, and the power line has 5 volts, then the value on the latched-value bus will equal the data bus value."

Evidence rules are defined with DEFINFER:

```
(definfer power-good D/A
  IF (any-on LEDs)
  THEN (+5volts power))
```

which states that if any LEDs on the board are illuminated, 5 volts is evidently present on the power line. The predicates in these rules, such as "high," "any-on," "+5volts," etc., are predefined in FEXPR. The knowledge engineer may also define additional predicates by using a function called DEFPRED, if required by the particular application.

## INTERACTING WITH THE USER

When the entire system to undergo troubleshooting has been described to FEXPR as described above, the actual troubleshooting task can begin. The program assumes that the technician using it has some knowledge of the control system, and has a means of applying test inputs. As troubleshooting proceeds, the technician will be requested to apply various inputs and enter the observed measurements into the FEXPR console.

The first step in troubleshooting is to indicate to FEXPR what the faulty signal should have been. This is described in terms of the predicates discussed previously. A typical starting point for a trouble of the example D/A board might be

```
(dc-volts pin-3 5.3)
```

From this point, FEXPR searches back along the causal rules to find a set of test inputs that <u>should</u> cause a 5.3 volt signal to appear at pin-3. Assuming that a module with user-controlled read/write capabilities were also in the controller system, FEXPR might find that writing, say hexadecimal "4F" to board 3 should produce the desired 5.3 volts. The FEXPR prompt to the technician would then be

```
Please do the following:

  Try to WRITE 4F to BOARD-3.

press <RETURN> when ready to continue...
```

When this has been done, FEXPR searches through the evidence rules to see what measurements can be made to yield evidence about the desired signal. In this case, the desired signal can be measured directly. FEXPR then asks for the appropriate measurement.

```
What is the VOLTAGE of PIN-3?
```

Proceeding in this fashion, FEXPR will ask for actions and observations until a faulty module is found. The search path for the questions includes all the modules which should have contributed to the originally-desired output; in this example, 5.3 volts on pin-3. When there is evidence that the inputs to a module are as they should be, but that the outputs of that module violate the expectations of a causal rule, that module is suspected as faulty. In most cases tested, FEXPR was able to find evidence clearly locating the actual faults in the controller. In some cases, however, several potential faults were found which could not be unambiguously accepted or rejected from external measurements. In cases such as those, FEXPR simply yields a list of the suspects, which must be verified by hand.

## CONCLUSION

The implementation and demonstration of FEXPR has shown that Expectation-based reasoning is indeed a viable approach for computer troubleshooting of microcomputer-based controllers. Using only the causal and inference rules and the specific system configuration, Expectation-based reasoning can perform effective test generation and diagnosis, avoiding the need for a knowledge engineer to explicitly program those activities.

The greatest advantage of Expectation-based reasoning is the flexibility which results. After all the modules available from a manufacturer have been adequately described, the same program can troubleshoot any controller configuration installed in the field. Additionally, when updated controller modules or features are designed, the designer can simply write new causal and evidence rules to describe the new behavior at whatever functional level he desires. Expectation-based reasoning can proceed without reconsidering changes in the overall troubleshooting task. This is true even if new test entry or measurement points are provided on the controller, which could drastically alter the actual actions performed by the technician. Such features are not readily available from traditional computer troubleshooting systems.

One aspect of our Expectation-based reasoning approach which could be improved is that of statistical accuracy. In addition to understanding how modules function, an experienced expert knows, at least approximately, what types of failures are most likely and how likely they are. Our work currently does not address this issue. However, interesting work by others in related fields (Pearl, 1985) seems applicable to the Expectation Network. Another area for further consideration is the interface between the controller and the computer running the Expectation-based reasoning program. Currently, FEXPR does all of the interfacing via the human technician, who must make all measurements and perform all the test inputs. This is appropriate for troubleshooting previously-designed control systems. For future controllers, however, it would be interesting to study the costs and efficiency benefits of directly-connected sensors and test inputs. With such capabilities, it may be possible to significantly increase the autonomous troubleshooting capabilities of control systems without greatly increasing the cost.

ACKNOWLEDGEMENTS

Our thanks to the Center for Automation and Intelligent Systems Research, and to Professor Yoh-Han Pao, Director, for supporting our research on the FEXPR project, and to the several engineers who provided expert knowledge about Microcomputer controllers.

REFERENCES

Barr, Avron, and E.A. Feigenbaum (1981). Knowledge Representation. In The Handbook of Artificial Intelligence. William Kaufmann, Inc. pp. 190-199.

Bonissone, Pierro P., and Kimon P. Valavanis (1985). A Comparative Study of Different Approaches to Qualitative Physics Theories. Proceedings of the Second Conference on Artificial Intelligence Applications, 236-243.

Guvenir, H.A., and P.R. Schaefer (1985). An Expert System for Troubleshooting the Digital Control Systems. Center For Automation and Intelligent Systems Research, Case Western Reserve University.

Hayes-Roth, F., D.A. Waterman, and D.B. Lenat (1983). Building Expert Systems. Addison-Wesley, Reading, MA.

Pearl, Judea (1985). How To Do With Probabilities What People Say You Can't. Proceedings of the Second Conference on Artificial Intelligence Applications, 20-27.

Schaefer, P.R. (1986). A Flexible Approach to Expectation-Based Reasoning Systems. M.S. Thesis, Case Western Reserve University.

Wilkinson, A. Jesse (1985). MIND: An Inside Look at an Expert System for Electronic Diagnosis. IEEE Design and Test, 69-77.

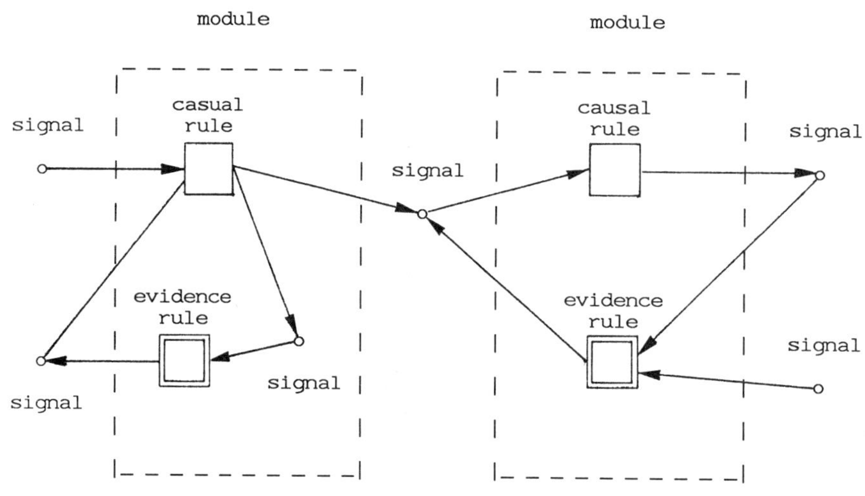


Fig. 1. Modules, Signals, Causal Rules, and Evidence Rules.

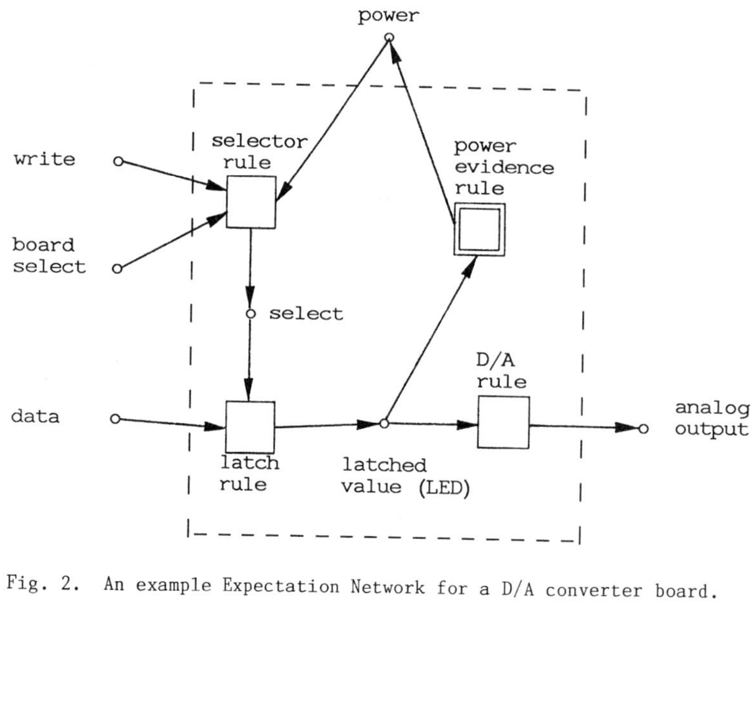


Fig. 2. An example Expectation Network for a D/A converter board.

# AN ASSOCIATIVE MEMORY BASED LEARNING CONTROL SCHEME WITH PI-CONTROLLER FOR SISO-NONLINEAR PROCESSES

**E. Ersü and S. Wienand**

*Isra Systemtechnik GmbH, Schöfferstraße 15, 6100 Darmstadt, Federal Republic of Germany*

Abstract. The paper discusses the real-time implementation of an associative-memory-based learning control scheme with PI-controllers for nonlinear processes. Starting with a pre-assumed PI-controller which only has to stabilize the process the controller parameters are optimized on-line by a predictive optimization. This optimization uses for prediction the model of the process stored in an associative memory which is also learned on-line. The situation-dependent optimized controller parameters are also stored in an associative memory. The concept is a modification of the LERNAS-system (Ersü,1984), which is also shortly described and compared to the system described here. Some experimental results with a nonlinear pH-control demonstrate the performance of the system.

Keywords. Adaptive control; Artifical intelligence; Learning systems; nonlinear control system; pH control; Predictive control; self-organising storage.

## INTRODUCTION

The fact, that most of the complex industrial processes, especially if they are uncertain or fully or partially unknown, have made self-organizing control schemes into an important research area in control theory. Since the beginning of the 1950's very active research has resulted in various forms of self-organizing control, e.g. adaptive, learning, fuzzy, linguistic, etc. An efficient implementation of these algorithms could only be realized by the rapid and revolutionary progress in microelectronics. A major part of recent industrial applications uses parameter adaptive control schemes by which a linear feedback law is on-line updated depending on an implicit or explicit identification of a linear model of the unknown process.

The learning control method LERNAS (Learning Control with Neuron-like Associative Memory Systems) represents a novel type of self-organizing control system which is motivated by the neurobiological and psychological research on the brain and the growth of the VLSI-technology as well. The main novelty of the method introduced by Ersü (1980, 1983) and discussed by Ersü and Tolle (1984) is that the concept uses a control action test and optimization through a 1-step ahead output predictive algorithm (similar to human problem solving) with both the predictive model of the unknown environment and the control strategy being represented by general mathematical (i.e. nonlinear) mappings. Both mappings are carried out by special neuron-like associative memory systems similar to information storage in neuronal networks.

The concept described in this paper is a modification of this method. The model for the process is the same and it is learned in the same way. The control-strategy, however, is different. The LERNAS concept described above assumes that a priori information about the process is hardly available. But in most cases this assumption is not valid. In general there exists sufficient a priori information to design a PI-controller which only stabilizes the process. Now instead of the controller-response to a given state the optimal parameter set for a given state of the process is learned. These parameters are optimized utilizing the predictive process model stored in an associative memory.

The advantages and disadvantages of this method with respect to the method described above will be demonstrated in this paper.

## LEARNING CONTROL CONCEPT

In classical system theory input-output descriptions are based on an assumed or predetermined mathematical structure, normally a set of differential equations. Replacement of these predetermined structures by input-output type of associative mappings leads to more general representations by (in general nonlinear) mathematical mappings between n-dimensional input vectors I and m-dimensional output vectors O:

$$M: \underline{I} \rightarrow \underline{O} \qquad (2.1)$$

Given a time-descrete representation of a time-invariant deterministic system

$$\hat{\underline{y}}(k+1)=f(\underline{\psi}(k), \underline{v}(k), \underline{u}(k)) \qquad (2.2)$$

with input vectors $\underline{u} \in U \subset R$

output vectors $\underline{y} \in Y \subset R$

measurable disturbance vector $\underline{v}$ $V \subset R^{n_v}$

state description

$$\underline{\psi}(k)=(y(k), \underline{y}(k-1),..., \underline{y}(k-\alpha)$$
$$\underline{u}(k-1),..., \underline{u}(k-\beta)$$
$$\underline{v}(k-1),..., \underline{v}(k-\gamma)$$

$\underline{\psi} \in \Psi$

scalars $\alpha$, $\beta$, $\gamma$

sampled time $t=k.T_s$

and sampling period $T_s$

an associative mapping for (2.2) can be defined by

$$S:(\underline{\psi}(t), \underline{v}(k), \underline{u}(k)) \rightarrow \underline{y}(k+1) \qquad (2.3)$$

which can be learned on-line via associative memorization. This procedure does not require any structural information about the process at hand except the scalars $\alpha$, $\beta$ and $\gamma$ representing the amount of history to be used in the model.

Associative memorization is a storage technique in which the location of the output information is addressed by the contents of the input information only (Kohonen, 1977).This type of information retrieval which is an outstanding feature of the associative neuronal network models avoids long searching procedures as they are generally necessary in conventional memory systems. The special memory system AMS (Associative Memory System) used for the implementation of the control concept is a Perceptron-like (Rosenblatt, 1962) distributed trainable associative memory system by which a real-valued function or mapping of several variables can be evaluated via simple content-addressing mechanisms rather than by complex mathematical operations.

Based on the concept of CMAC (Cerebellar Model Articulation Controller) proposed by Albus (1972, 1975) as a neuronal network model of the human cerebellar cortex AMS has been further modified and developed for real-time control purposes (Ersü, 1982). In AMS the information storage for an association pair ($\underline{I}_i$, $\underline{O}_i$) occurs by distributing the novel output information over a constant number ($r^*$) of memory locations, thus for any input information the output is always formed by a recollection of the information elements located in the corresponding (to $\underline{I}_i$) $r^*$ memory cells. Depending on the similarity of input information, e.g. the difference between two inputs $\|\underline{I}_i - \underline{I}_j\|$ the mapping procedure recalls corresponding sets of information elements which share the memory cells to an extent which is a function of the input similarity. This leads to one of the most important features of AMS called generalization, i.e. similar inputs tend to produce similar outputs. Mathematically it can be interpreted as a multi-dimensional interpolation in the neighbourhood of a trained point ($\underline{I}^o$, $\underline{O}^o$) illustrated in figure 1. The degree of generalization which is fundamental for on-line learning can be thus pretuned by the generalization variable $r^*$ and the quantization variable. As memory locations are shared by similar inputs the AMS procedure also reduces the amount of memory to a great extent.

## CONTROL STRATEGY

For simplification here only a single-variable process is considered. The method is nevertheless also applicable for multivariable-processes with some modifications.

The concept shown in fig. 2 incorporates two AMS-type memory systems, one for the predictive model of the unknown process

$$M: (\underline{\psi}_M(k), \underline{v}(k), u(k)) \rightarrow \hat{y}(k+1) \qquad (3.1.)$$

corresponding to $\underline{\psi}$ with some $\alpha_M$, $\beta_M$, $\gamma_M$ (instead of $\alpha$, $\beta$, $\gamma$)

and one for the control strategy

$$P: (\underline{\psi}_P(k), \underline{v}(k), y(k)) \rightarrow \underline{p}(k) \qquad (3.2.)$$

With setpoint value $w(k) \in Y$ and $\underline{\psi}_P$ in practice similar to $\underline{\psi}_M$.

$u(k)$ is calculated in a PI-control-algorithm:

$$u(k)=u(k-1)+q_0x_d(k)+q_1x_d(k-1) \qquad (3.3.)$$

Where $\underline{p}(k)=(q_o,q_1)^T$

The PI-algorithm was choosen because it is a robust algorithm with respect to model uncertainities. Further there are only two parameters to optimize.

The behaviour of the process is determined by a reference-model. This method results in a better learning-behaviour of the system because it always follows similar trajectories.

Assuming that

i. the unknown multivariable process at hand is a determinstic time-invariant or weakly time-variant BIBO-stable process,

ii. u,y and $\underline{v}$ are quantized and of finite subspaces,

iii. the overall goal

$$I_G=\sum_{k=0}^{k_e-1} L_G(y(k+1),\ w(k+1),\ u(k))\quad(3.4.)$$

can be substituted by a l-step ahead subgoal

$$I_S=\sum_{i=0}^{l} L_S(y(k+i),\ w(k+i),\ u(k+i-1))\quad(3.5)$$

IV. a parameterset $\underline{p}_V$ is known which stabilizes the process

V. the process is able to follow the model-trajectory

the algorithmic scheme in each sampling period is as follows:

1. the prediction-model is updated by the measured prediction-error $e(k)=y(k)-\hat{y}(k)$ where $\hat{y}(k)$ is the predicted value from step k-1.

2. an optimization scheme is activated. It tries to calculate the parameter-set $\underline{p}^*(k)$ which is the estimated optimal set for the subgoal $I_S(k)$. In fact only $\hat{I}_S(k)$ - the estimated subgoal - can be calculated, because for future $y(k+i)$ only the estimated value $\hat{y}(k+i)$ is available. Besides the whole prediction has to take place in the trained area of the model, because otherwise no information about the behaviour is available. That means that the resulting $\underline{p}^*(k)$ also depends on the information available in the process-model. The prediction-length l is adapted to the training-situation of the process-model in the boundaries

   $lmin < l < lmax$

   lmin is the shortest prediction which allows a critical evaluation of the parameter-set. lmax is the longest prediction which can assure estimated values which are to some extent accurate.

   There are three different possibilities for the starting point of the optimization:

   1. the initial parameter-set $\underline{p}_V$
   2. the old parameter-set $\underline{p}(k-1)$
   3. the parameter-set $\underline{p}^*(k)$ which is the preliminary optimal parameter-set out of the parameter-AMS.

   3. can only be used if the parameter-AMS is trained for the appropriate state of the process.

   2. can be used as long as this parameterset doesn't lead to an undesired behaviour of the process. This can be the case when the behaviour of the process in the actual state is (because of nonlinearities) very different to the behaviour in the state the previous parameter-set was originally optimized. To detect such a situation a supervisor-level was introduced.

   If none of the two parameter-sets is applicable the initial parameter-set $\underline{p}_V$ is used.

3. The parameter-set $\underline{p}^*(k)$ optimized by (2.) is memorized in parameter memory to be used as the best decision, making (2.) superfluous in the long range, and giving either an excellent optimization starting point or being used without further inclusion of the predictive learning loop after a user-defined point of time.

4. The last step is the calculation of the control-output $u^*(k)$. This is done via the PI-algorithm

   $$u^*(k)=u(k-1)+q_0x_d(k)+q_1x_d(k-1)$$

   where $(q_0,q_1)^T=\underline{p}(k)$.

   Again there are the three possibilities mentioned above for $\underline{p}(k)$.

   Before applying a control input to the process a supoptimal control input $\bar{u}(k)\neq u^*(k)$ with

   $$\|\bar{u}(k) - u(k)\| < \delta$$

   but

   $$\bar{u}(k)\notin G_T$$

   $G_T$ = trained area of model-AMS

   for some specified $\delta$ is determined.

   $\bar{u}(k)$ applied to the process excites it to further untrained information for the model AMS and, so, enlarges $G_T$. This exploratory procedure called active learning is speeding up the learning.

It should be pointed out that the fact that the old parameterset $\underline{p}(k-1)$ can be used for control is an advantage of this method in comparison to the original LERNAS-method. A parameterset in connection with a fixed controller-structure can provide a response to every state of the process. Depending on the nonlinearity of the process this will not be the optimal response in all cases, but in most cases it is similar to it.

This is not true if the u(k) are stored directly. Then the learned u is only a response to that specific state of the process.

With increasing learning-time the system can adapt the controller-parameters to the nonlinearities of the process. But, as the information "parameterset" is more global than the information "controller-output", it can not be adapted to such an extend. That means that the original LERNAS-system is better suited for very high nonlinear processes with fast changes in behaviour, whereas the described system is well suited for processes without these fast changes. In the latter case it has three major advantages:

1. It requires less memory, because a parameterset can be used for a greater area than a control-output and therefore it is not necessary to distinguish so many states of the process.
2. For the same reason it can learn more quickly: A parameterset once learned can be used in a greater area and there are less different parameter sets to learn than controller-outputs.
3. In the case of very high disturbances it leads to much better behaviour. The original LERNAS must learn the response to each disturbance-input. When this disturbance-input occurs the first time the reaction is slow, because the control-AMS has not yet learned a response to it. The system described here uses a PI-controller as backup. Therefore it can immediately react on a disturbance-input because the PI-controller "knows" an answer to it (even if it is not the optimal answer). It should be pointed out that in the case of smaller disturbances also the original LERNAS is able to react due to the ability of pluralization.

## SOME EXPERIMENTAL RESULTS

The described control-algorithm has been tested with a simulated nonlinear pH-control-process. The major nonlinearity of this process is the titration-curve (fig. 3). The system had to learn to change the pH-value periodically between 7 and 9 (the most nonlinear area) according to the given desired trajectory.

Fig. 4 shows the behaviour with the predetermined parameterset. It can easily be seen that the behaviour is not very good and that it depends on the set-point. In fig. 5 the behaviour in the region between 7 and 8 is shown after the system has followed three times the model-trajectory shown in fig. 4. In fig. 6 the same is shown after following the trajectory 15 times. Figure 7 shows the parameter-sets used in every step (the optimized parameters from the parameter-AMS) after 3 learning-cycles, fig. 8 after 15 learning-cycles.

The figures demonstrate that with increasing learning-time the system is able to follow the model-trajectory better and that it is able to adapt the parameters to the nonlinearity of the process.

The response to very high disturbance-inputs (up to 100 %) have also been tested. For this test a well trained parameter-AMS was used. Figure 9 shows that the system is able to achieve a good behaviour.

## CONCLUSION

A modification of the LERNAS-concept was described which uses associative memories to store the optimized parameters of a PI-controller. The parameters are optimized on-line with the help of a process-model which is also stored in an AMS and also learned on-line. The simulation shows that it is well suited for nonlinear processes with not too high nonlinearities. Only very few a priori-information about the process is needed. It is especially not necessary to have a mathematical model for the process.

The concept is at the moment very well applicable for SISO processes. The ongoing work will be concentrated on an extention to MIMO processes.

## REFERENCES

Albus, J.S., (1972). Theoretical and experimental aspects of a cerebellar model. PhD Thesis, Univ. of Maryland.

Albus, J.S., (1975). The cerebellar model articulation controller. Trans ASME, series G, Vol. 97, No. 3.

Ersü, E., (1980). Self-organizing control with associative memories. Internal Report at the Inst. of Control Engineering, Techn. Univ. of Darmstadt (in German).

Ersü, E. and Militzer, J., (1982). Software implementation of a neuron-like associative memory system for control applications. Proc. of the ISMM 8th Int. Symp. MIMI, Davos, Switzerland.

Ersü, E., (1983). On the application of associative neural network models to technical control problems. Proc. in Life Sciences: Localization and orientation in biology and engineerin ed. Varju and Schnitzler, Berlin, Heidelberg, New York.

Ersü, E. and Tolle, H., (1984). A New Concept For Learning Control Inspired by Brain Theory. Proc. of the IFAC 9th World Congress, Budapest/Hungary.

Ersü, E. and Militzer, J., (1984). Real-time implementation of an associative memory-based learning control-scheme for nonlinear multivariable process. Symposium "Application of multivariable system technique", Plymouth.

Kohonen, T., (1977). Associative Memory. Springer, Berlin, Heidelberg, New York.

Rosenblatt, F., (1962). Principles of Neurodynamics. Spartan Books, Washington.

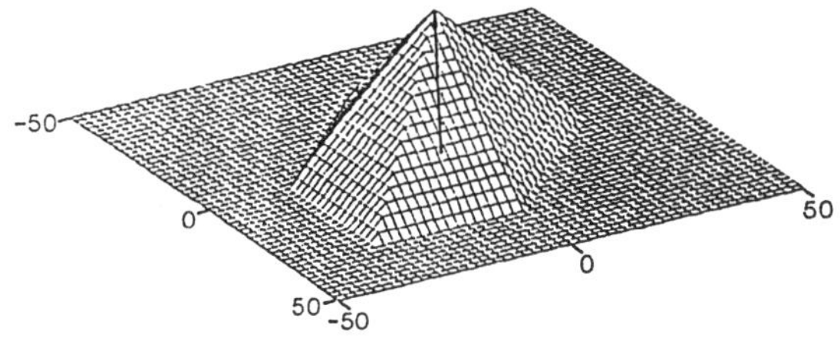


Fig. 1 Generalization demonstrated by training at the origin of the input plane of an AMS with two inputs and one output ($r^*=32$ and $\varepsilon=1$). The peak at the origin would be the AMS-response for $r^*=1$, i.e. no generalization, the case of conventional storage

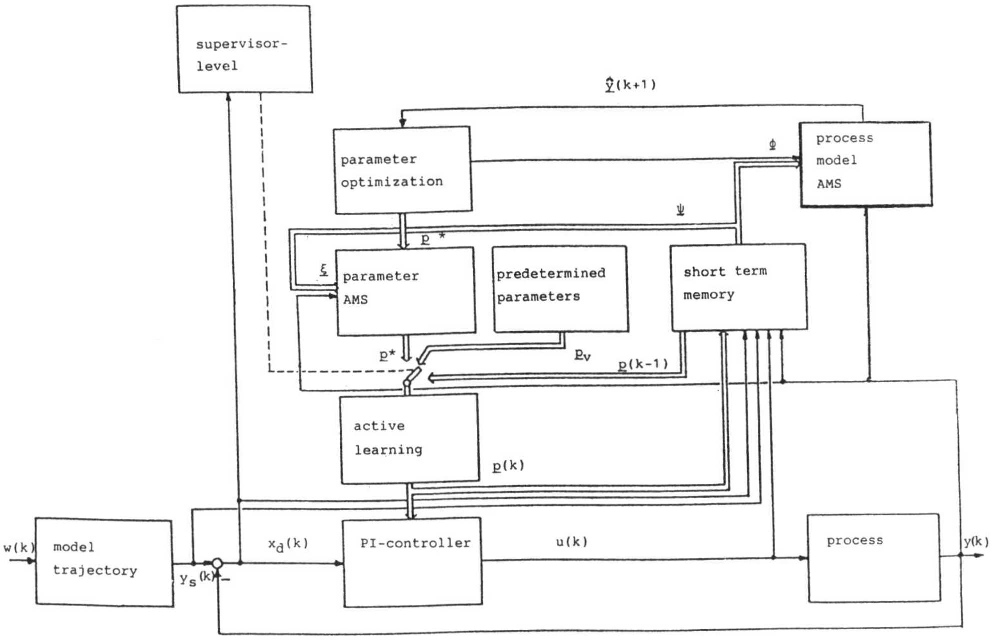


Fig. 2 Control-scheme

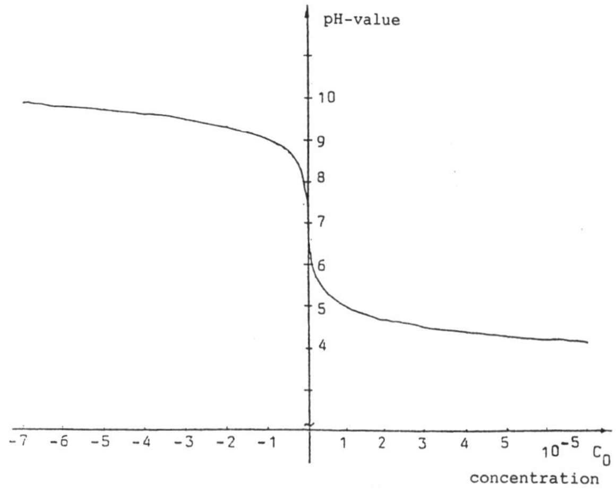


Fig. 3 Titration-curve

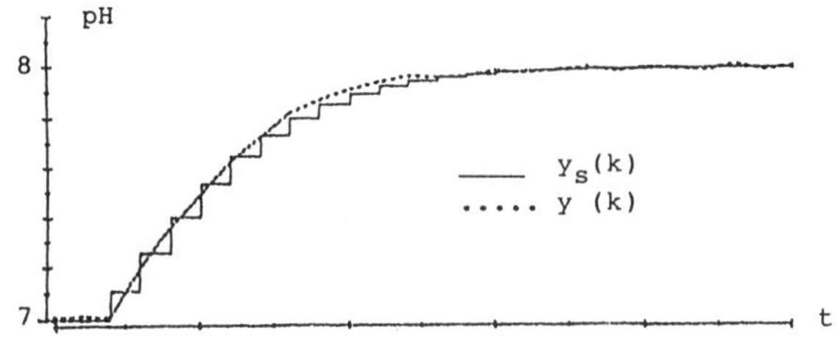


Fig. 6 Behaviour after 15 learning-cycles

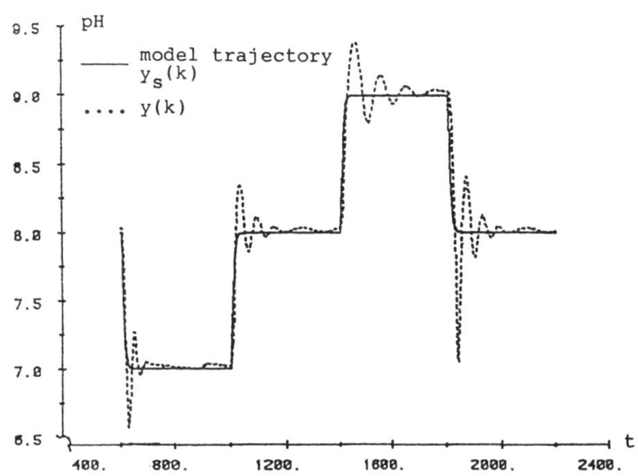


Fig. 4 Behaviour with initial parameterset

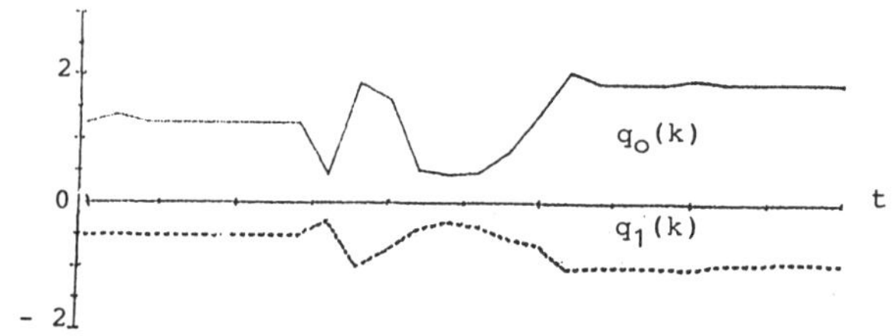


Fig. 7 Parameters after 3 learning-cycles

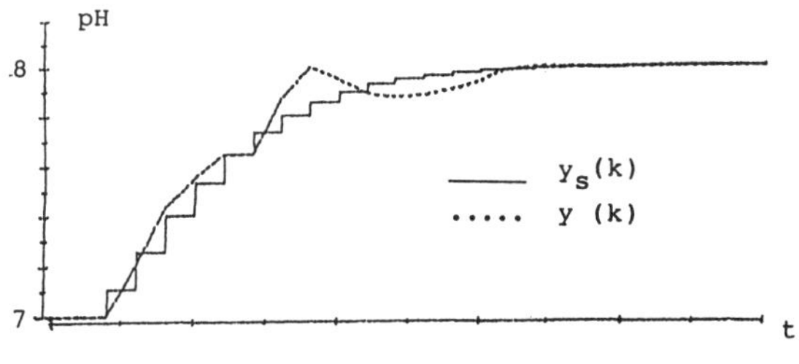


Fig. 5 Behaviour after 3 learning-cycles

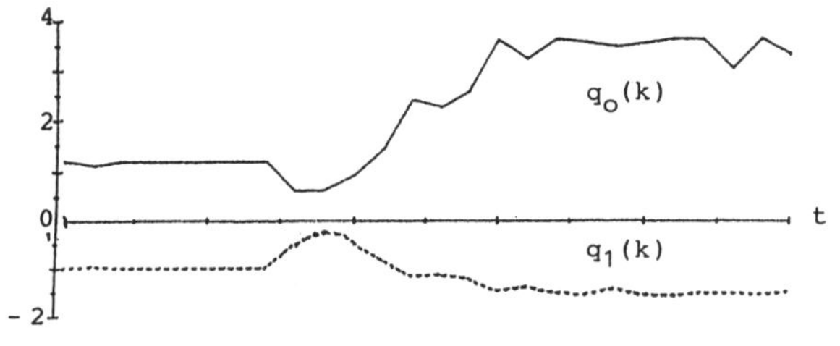


Fig. 8 Parameters after 15 learning-cycles

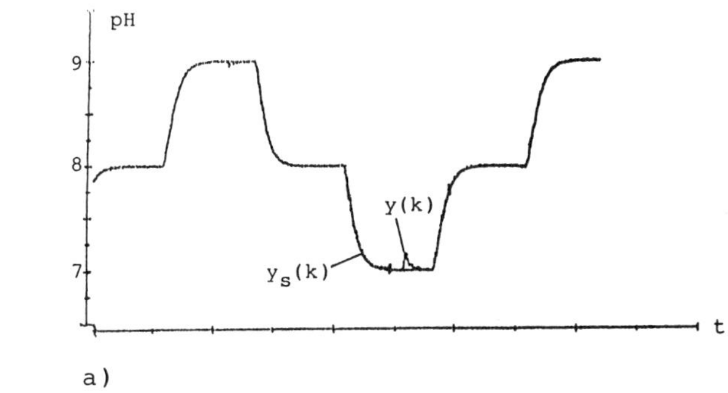


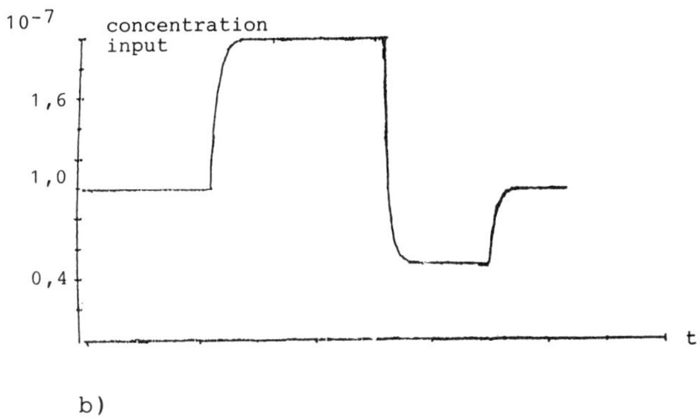


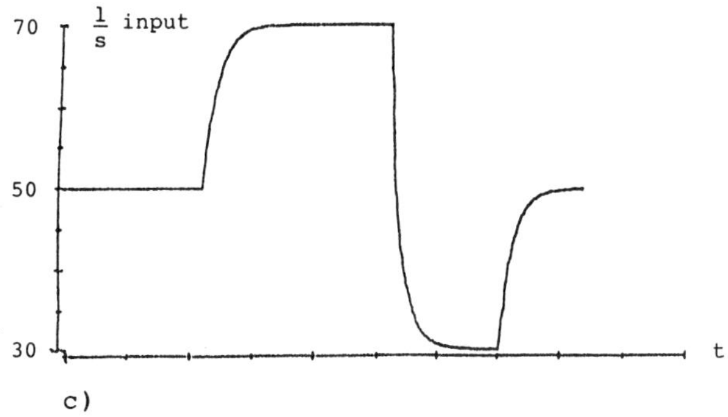


Fig. 9 Behaviour with input-disturbance

a) model-trajectory and output behaviour

b) flow-in concentration

c) flow-in mass

# APPLICATION OF MODERN NUMERICAL CONTROL TO AN INDUSTRIAL ROTARY CEMENT KILN

**J.-D. Marcuard and A. Meister**

*Control Laboratory, Swiss Federal Institute of Technology, CH 1015 Lausanne, Switzerland*

*Abstract :* *A digital control system for a rotary cement kiln has been developed and implemented on computer. The controller design is based on process identification results, obtained during open loop operation. Optimal kiln operation requires a very performant control system. The process control needs to ensure stable behaviour of the kiln in front of numerous and random system perturbations as well as to guide the process in optimal productivity conditions. Hierarchical and auto-adaptive control algorithms have been developed to adapt the controller the best to the production requirements. Actually, industrial application of the control system shows to work and is very well adapted to the nature of the process. Resulting improvements are very satisfying, mainly the remarkable energy consumption reductions.*

*Keywords :* *Cement industry, distributed parameter system, hierarchical structured control system, fault detection, target adaptive control, industrial application.*

## 1. INTRODUCTION

Cement manufacturing raw materials are very cheap. Production costs are dominated by energy consumption for cement elaboration and installation investments. Best productivity conditions demand an optimal process control to obtain the most favorable and desirable process behaviour.

Cement manufacturing contains following three parts :

- the raw material homogenisation
- the cement kiln process
- the clinker grinding process

The raw material homogenisation control problem was under consideration in an earlier project at our Control Laboratory and realised control strategies are succesfully applied since march 1984.

The burning phase of raw meal is the most important stage in the cement industries. Inside of the kiln, raw material temperature increases up to 1400 °C, essential to activate the chemical reactions for the obtention of clinker.

Prior to implementation of the control algorithm, the actual cement kiln was manually controlled by two operators. The kiln contained also some efficient conventional control loops. It has a capacity of 1900 tons of clinker per day. Human control isn't performant enough to manage all problems for such complex and perturbated processes as a cement kiln. In order to get an optimal clinker production hierarchical control algorithms have been developed and implemented on a PDP 11/23 PLUS. Illustrated below is the control structure (fig.1) which gives satisfying results for the industrial application. The implementation of the final control algorithm is in course.

## 2. DESCRIPTION OF THE ACTUAL KILN PROCESS

During the last years, rotary cement kilns obtained recognition for their favourable energy consumption. The actual kiln process can roughly be subdivised as follows :

- cyclone heat exchangers
- rotary cement kiln
- planetary coolers

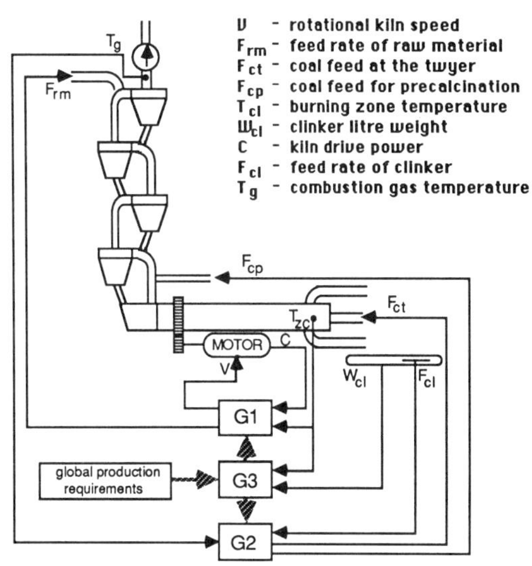


Fig. 1 : Flow sheet of the kiln process and of the implemented control loops.

The homogenised raw material passes across four cyclone heat exchangers, in the opposite direction of hot combustion gases. Besides preheating, the raw meal is already partly precalcinated, which requires a temperature of approximately 830 °C at the end of this phase. The raw material passes into the rotary kiln. The crossing of this section takes about 30 minutes, assisted by the low gradient and variable rotational speed of the kiln. In the burning zone, at the exit of the kiln, the material reaches temperatures of about 1400 $^{o}C$. High temperatures obtained by burning coal in the lower front end of the kiln are absolutely necessary in order to ensure the chemical reactions, transforming the raw meal to clinker. Finally the hot clinker is cooled down rapidly in planetary coolers to preserve the desired chemical properties before leaving the process. It is then grinded and stockpiled in vertical silos.

The output quality is represented by the free-lime content of the clinker (CaO which has not been captured in one of the cement compounds). As it is simpler to measure and excellently correlated with the free-lime content, the clinker density (represented by the litre weight $W_{cl}$) is actually determined.

## 3. IDENTIFICATION OF THE ROTARY CEMENT KILN

A cement kiln is a distributed parameter system. It has a highly complex dynamic behaviour due to the chemical reactions and the heat and mass transfers. Process control efficiency depends mainly on identification quality. Most controler designs require a quantitatif or qualitatif model describing the relationships between the control variables and the variables to be controlled. Furthermore, some criterion is required to define and specify the control problem. In the simple control system case, model inputs and outputs are directly known. In complex processes, such as the cement kiln process, there are a large number of inputs we can manipulate as well as measurements containing information about the actual state of the kiln. It is impossible to take into account all these variables to build a model. In some way, we have to select the essential variables from the point of view of the control problem. We have to find the best input variables to influence the process and on the other hand to ensure an optimal clinker production. In order to get the main information about the actual state of the kiln, we have to select the most representative measurements. One of the outputs of the model to be identified should include information about the clinker quality. While the clinker quality specified by its litre weight $W_{cl}$ can only be measured at the exit of the coolers (dead time about 50 minutes) and the hourly measurement, makes it impossible to use it as direct control variable.

In order to select the most important variables, we had discussions with the responsable stoff and we excited the process by manipulating multiple input variables during different experiment sessions. The temperature of the clinker in the burning zone $T_{cl}$ is linked closely to the output quality. Too little heat at the burning phase will mean that the material leaves with to high a proportion of free-lime. Too much heat means inefficiency and also an excessive burning zone temperature which results in a hard-to-grind clinker and probable damage to the kiln wall. Beside the quality, a desired clinker quantity is also required, it is represented by the measurement of the feed rate of clinker $F_{cl}$ at the exit of the coolers. It has been found, that the kiln drive power C is a very good indicator of the state of the kiln and also the main (indirect) perturbation variable. Distrubances, which are caused mainly by the formation and falling of blocking rings, are instantly indicated by this detector, as well as C measures also the variations in kiln loading. Ring formations have a very destabilsing influence on the dynamical behaviour of the process. Other perturbations, as variations of chemical proportions in the raw meal act upon this measurement. As a fourth control variable, we have chosen the combustion gas temperature $T_g$. It is a very good indicator for the quality of preheating and precalcination in the four cyclone heat exchangers.

As principal inputs, we have naturally the energy flow to the burning process ($F_{ct}$ - coal feed at the twyer, $F_{cp}$ - coal feed for precalcination), as well as the feed rate of raw material $F_{rm}$ entering into the kiln. The first considered variable must provide the burning of a good quality clinker, the second one a required production rate, and both need to yield an optimal efficiency of the industrial process. Another manipulated variable includes the kln rotational speed V, which controls the material residence time in the kiln. The kiln speed should be combined with the raw material rate, to maintain the internal loading of the drum and the heat transfer and chemical reactions characteristics.

Process identification during normal open loop manual control is not possible. Representative models are difficult to obtain, since the input variables are rarely excited. On the other hand, poor quality measurements of the output variables are varying and are strongly influenced by unmeasurable system perturbations. To get more information, parameter estimation using PRBS inputs was applied in november 1984. To computer transferred data was evaluated in two ways, by visual obervation of plots of significant variables and by closer statistical examination of the records. It has been found, that the settling of the system is shortest for the rotational speed of the kiln as input variable and it has a significant reaction on the burning zone temperature (fig. 2). The other input variables have slower and less revealing reactions on the state of the kiln, but are even as important as the speed for a stable behaviour of the process.

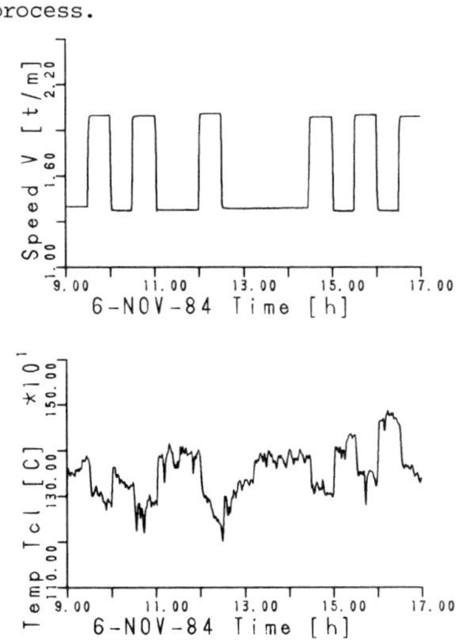


Fig. 2 : Input and output sequences during excitation procedure of the rotational kiln speed.

Model Structure :

Records transferred to the computer were used to identify the behaviour of the actual process. Recursive least square method with exponential forgetting factor was applied to estimate the unknown parameters of the mathematical model. The model structure corresponds to the ARMAX equation :

$$A\ Y(k+d) = B\ U(k) + C\ W(k) + sp \qquad (1)$$

with :

- k : discret time period normalised to the sample time $T_o$ ($t = k \cdot T_o$)
- d : dead time of the system (multiple absolute of the sampling period)
- Y : system output (burning zone temperature - $T_{cl}$)
- U : control input (rotational speed of the cement kiln - V)
- W : deterministic load disturbance (kiln drive power - C)
- sp : set point (constant, due to non-linearities of the system)

A, B and C are polynomial expressions of different orders of the discret time system in the backward shift form. Identification tests with the same model structure but with signal sequences of other input variables (raw meal feed rate $F_{rm}$, coal feed at the twyer $F_{ct}$) gave less significant results than the kiln speed.

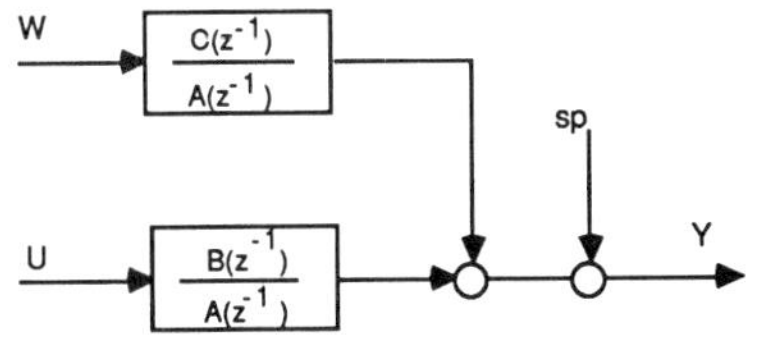


Fig. 3 : Structure of the identified model.

The determination of a discret time model is very important to quantify and to ratify the relationships between inputs and outputs of the actual system. On the other hand, different control strategies may be tested on the mathematical model before being implemented on the real process. Identification valuation has to be made to obtain information as to how the model describes the process. It explains which part in measurement information is reconstructable by knowing the input variables. The valuation in fig. 4 allows the comparision between the deterministic model output and the actual system measurement $T_{cl}$ and shows an excellent similarity between reality and the adaptive mathematical model.

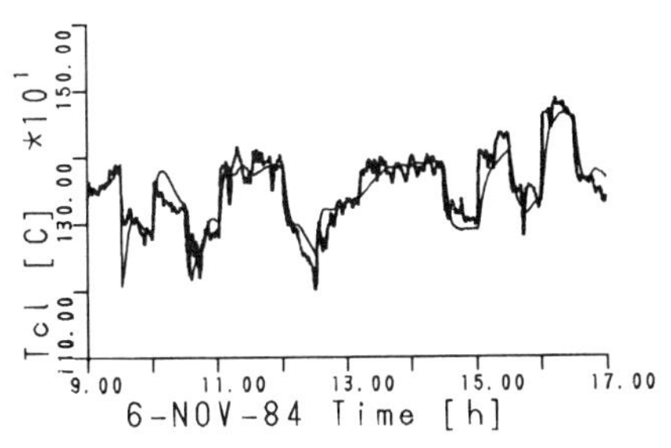


Fig. 4 : Valuation of the mathematical model ( ▬ system measurement, —— model output).

4. CONTROLLER DESIGN

In this section we discuss the control problem associated to the cement kiln process. The overall objective for producing clinker can roughly speaking be formulated as follows :

- produce a clinker of sufficient quality
- achieve a desired production rate
- minimize fuel consumption

It is quite difficult to fulfil all this objectives for multiple reasons. Ideal operation is often impossible due to serious disturbances caused mainly by the formation of rings, banking up of material and the calcining region to be starved. They are random and in function to the place where they occur, tolerated reaction time is very limited and demands interventions instantly. Other types of perturbations (coal energy rate fluctuations, chemical proportion varying in the raw meal), have greater time constants but are also very significant for the actual kiln state.

The control system has to command the cement kiln in ideal operation conditions. It has to compensate immediately for process troubles, due to perturbations. To get best control performances, a hierarchical control structure was chosen. It is possible by such a structure to correct fast disturbances to stabilize the dynamical kiln behaviour as well as to ensure best long-terme productivity conditions. Figure 1 illustrates the general controller design. Further explanations about the different control level are given below.

Control level G1

The regulator G1 maintain the burning zone temperature $T_{cl}$ around its setpoint value $T_{cl_o}$, generated by the target adaptive controller G3, by reaction on the input variables kiln speed V and raw meal feed $F_{rm}$ to the kiln. The control algorithms which are used are similar to those of a discret PID-controller :

$$\begin{aligned} u(k) &= u_1(k-1) + q_0\, e(k) + q_1\, e(k-1) \\ u_1(k) &= u_1(k-1) + q_2\, e(k) \end{aligned} \qquad (2)$$

dont :

- $q_0 = g(1 + T_D/T_0 + T_0/T_1)$
- $q_1 = -g\ T_D/T_0$
- $q_2 = g\ T_0/T_1$
- e : control deviation
- $T_0$ : sample time
- g : gain
- $T_1$ : integration time
- $T_D$ : derivation time

The controller parameters were ajusted on the basis of measured transient functions. They were reoptimized on-line according to the closed loop behaviour. As shown in fig. 2, settling time between kiln speed and burning zone temperature is very short. To obtain good control performances, a relatively small sample time $T_0$ = 1 min is used in the control procedure.

The raw material feed $F_{rm}$ is also manipulated by this PID regulator by linear combination with the kiln speed to maintain internal loading of the kiln.

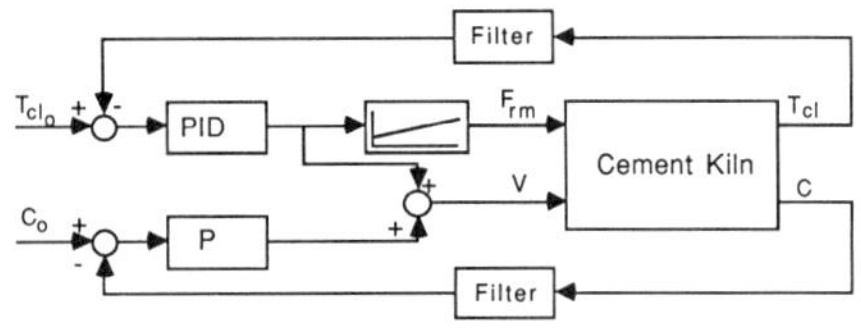


Fig. 5 : Feedback control system G1

Actual kiln speed is ajusted by a second superposed regulator with as control variable the kiln drive power W (filtred with a high time constant). This proportional controller aims at constant loading conditions inside of the kiln. Too much material in the drum produces lots of dust and so a low heat exchange efficiency with the hot combustion gases. In the opposite case, the system becomes very sensitive to distrubances due to its low thermic inertia.

## Control level G2

The control level G2 controls the production rate of clinker $F_{cl}$ and the temperature of the combustion gases $T_g$ at the exit. Due to the high dead time of the measurement $F_{cl}$, the manipulated raw meal feed $F_{rm}$ (filtred with a high time constant) is supervised by G2. Linear relations between this two variables are well known through the chemical transformations. The level G2 is controlled by two regulators with input variables the two coal feeds $F_{ct}$ and $F_{cp}$ to guarantee necessary burning conditions. Sampling periods for this two controllers are different.

At this level, adjustments of the coal at the twyer are most important to ensure the heat supply for burning clinker. The sample time is 60 minutes for investigations because the long settling times. The controller has to supervise the production rate of clinker. Varying heat supply influences the burning zone temperature, this one manipulates the production rate by action of the control level G1 ($F_{rm}$ as input variable).

A diagram of the implemented production rate controller is shown in fig. 6. The identification of a linear and auto-adaptive model, with as input and output variable the energy feed at the twyer and the clinker production rate, estimates namely variations in fuel quality. A fault detector supervises the actual kiln state and makes decision on the manipulated input. The goal of this supervision is to detect process changes and faults during normal operation and to take actions to avoid a damage of the process. In the actual case, application of fault detector is essential to prevent dangerous situations for the system i.e. mainly the superheating of the kiln.

The controller, which manipulates the coal feed for precalcination $F_{cp}$, acts on the process with a sample time of one minute (fig. 7). $F_{cp}$ depends lineary from the coal feed at the twyer to maintain the heat balance between this two energy sources. This linear combination is corrected by two superposed regulators. The first one controls the combustion gas temperature $T_g$ by proportional investigations to ensure good preheating conditions inside of the cyclone heat-exchangers. The second one acts as a feedforward regulator using the varying raw meal rate as a priori knowledge.

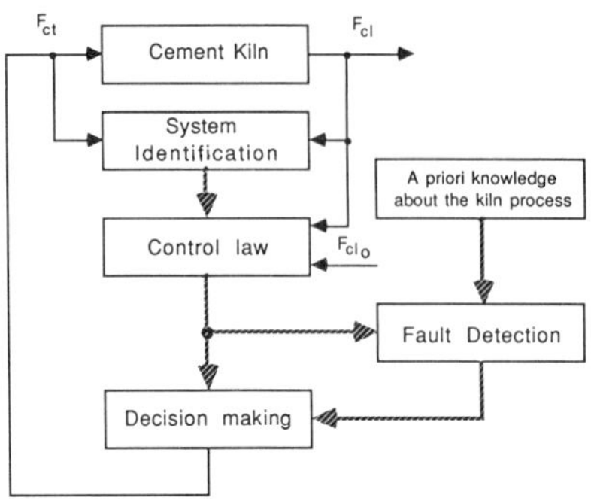


Fig. 6 : Generalized scheme of the auto-adaptive closed loop system with fault detector.

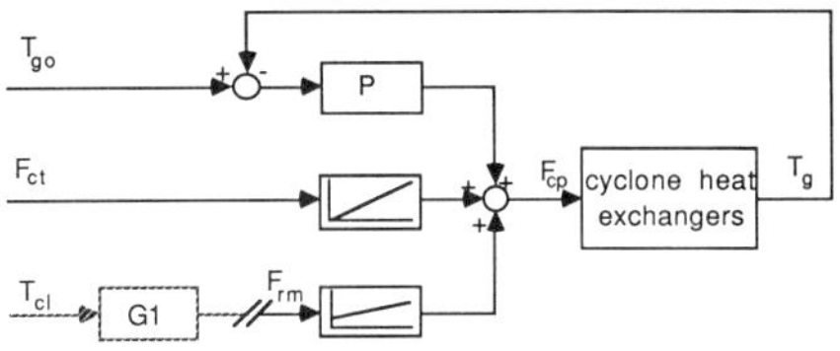


Fig. 7 : Control loop of the coal feed for precalcination $F_{cp}$ .

## Control level G3

The control level G3 assists the two lower levels. It generates the local setpoints of the controllers G1 and G2 in function of the actual kiln state and in order to get a required production rate of clinker.

To control the burning zone temperature represents only an auxilliary goal for producing good quality clinker. The output $T_{cl}$ is a very good indicator for the actual kiln state and also correlated to the clinker quality in the burning zone, but relations between this two measurements are not stationary for multiple reasons. Variations in chemical proportions of the raw meal and distorded and biased deviations of $T_{cl}$ demand adaptive performances in the burning zone. In the control system this has been solved by estimating a mathematical model between the two system outputs. As soon as there appears a new litre weight (introduced manually on computer), the target adaptive controller determines the actualized reference value $T_{cl_0}$. The litre weight setpoint $W_{cl_0}$ is used in calculating the corresponding setpoint value for the burning zone temperature by using eq. (3).

There are a lot of identification methods, applied algorithms (recurrent least square parameter fitting with variable forgetting factor) gives excellent results for this application. Due to the irregular arriving of litre weight measures, a static model combined with a dead time is identified in real time :

$$W_{cl}(t+d) \quad = \quad a\, T_{cl}(t) \, + \, b \qquad (3)$$

d : system dead time

Another task of this control level consists in guiding the adapted setpoint values along predefined trajectories to avoid too great variations for the manipulated inputs.

## 5. TEST RESULTS

The control system as described above was implemented on a cement kiln in octobre 1985. A PDP 11/23 PLUS is used as operating system. First results for the closed loop system with only the controller G1 in action were very encouraging. Thus a short time later, the control levels G2 and G3 were also implemented.

Figure 8 gives a good description of the actions of the control algorithms on the actual system. We may observe the reaction tendences of the different controller levels.

In fig. 8 we see the efficient functioning of the target adaptive controller. For nearly constant temperatures $T_{cl}$, litre weight increases and so changed burning conditions are detected. As consequence, the model as in eq. (3) is adapted and the setpoint value $T_{cl_o}$ modified. Also the production rate control by coal manipulations is visible in this graphics.

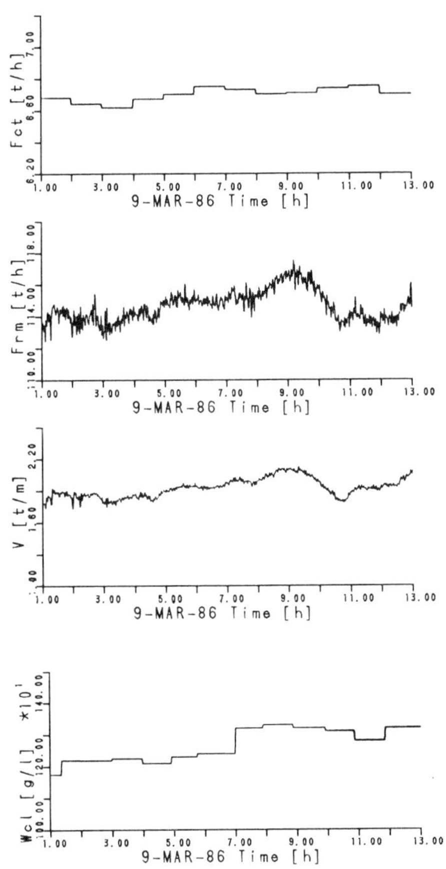


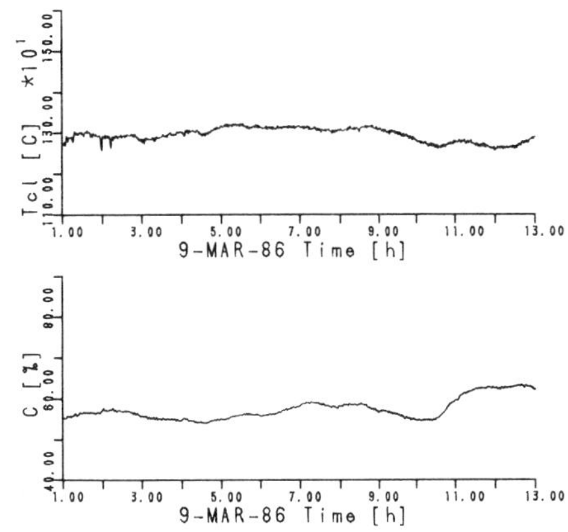


Fig. 8 : Closed loop behaviour of the kiln process (setpoint values : $W_{cl_o}$ = 1240 g/dm$^3$, $F_{rm_o}$ = 115.5 tons/hour)

Tests on the actual system have been pursued for several months. The controller characteristics were ajusted on the basis of obtained results. The implemented control system shows to work and is very well adapted to the nature of the process. Comparing to manual mode, automatic control makes the kiln behaviour much more regular. The influence of perturbations is remarkably reduced by the permanent supervision of the process. The overall objectives for optimal clinker production are achieved by this numerical control system. Resulting improvements are namely as follows :

### Clinker Quality $W_{cl}$ :

Clinker Quality has become much more regular thanks to the computer guided system, which allows to produce clinker of an admitted lower quality. The standard deviation of the litre weight measure has distinclty decreased and a lower setpoint value for the litre weight can thus be applied, allowing an economy of energy.

### Production Rate of Clinker $F_{cl}$ :

Actual results for the clinker production rate satisfyies entirely. During normal operation, the production rate follows its setpoint value rather close. When process changes occurs, detected by the fault detector, the clinker rate control becomes secondary to avoid dangers for the kiln system. This solution showed to work in this application. It represents a good compromise between setpoint following and system security.

### Economics on Heating :

Actual application of the described numerical control system implies an energy saving of about 2 % . This point needs to be pointed out when considering that the energy investments of such a process represents more than a half of the totality of production costs.

## 6. CONCLUSIONS

A hierarchical structured and auto-adaptive computer control system has been implemented on an industrial rotary cement kiln. Pre-examinations of the open loop behaviour were used for the controller design. The function of the control and its advantages were demonstrated during long test periods. The improvement of the operating performances are very remarkable, energy consumption of the plant has been reduced.

## REFERENCES

Åström, K.J., and B. Wittenmark (1984). Computer controlled systems. Prentice Hall.

Fortescue, T.R., L.S. Kershenbaum and B.E. Ydstie (1981). Implementation of self-tuning regulators with variable forgetting factors. Automatica, vol. 17, 831-835.

Isermann, R. (1971). Prozessidentifikation. Springer Verlag, Berlin, Heidelberg, New York.

Westerlund, T. (1981). A digital quality control system for an industrial dry process cement kiln. IEEE Trans. Automat. Contr., vol. 26, 885-890.

# SOFTWARE DEVELOPMENT AND ON LINE TESTS OF A MICROPROCESSOR BASED AUTOMATIC START UP SYSTEM FOR A FLUIDIZED BED LIGNITE COMBUSTOR

**H. Midoğlu* and E. Tulunay****

**Meko Microcomputer and Electronics Control Corporation, Paris Cad. 11–4, Kavaklidere, Ankara, Turkey*
***Electrical and Electronics Engineering Department, Middle East Technical University, Ankara, Turkey*

Abstract. A manual start up procedure has been developed for a 23 x 23 $cm^2$ fluidized bed lignite combustor which heats an 80 000 kcal/h conventional fire and smoke tube hot water boiler. Start up procedure is based on facts, rules of thumb and necessary knowledge concerning the fluidized bed combustion system that are collected and judged on by an expert. The procedure is basically composed of "if-then" rules. A microprocessor based controller is designed and constructed. A multitasking executive and a monitor program has been developed. This microcomputer is also used for the design and development of the control hardware and software. The fluidized bed lignite combustor has been started up semi-automatically, by manipulating the outputs of the computer upon operator commands entered on-line via the keyboard. A fully automatic start up is proposed to be possible and conclusions are derived for future developments.

Keywords. Fluidized bed combustion; automatic start up; heat systems; control engineering computer applications; microprocessors; multitasking; computer control.

## INTRODUCTION

Fluidization is a term used for defining the fluid-like behaviour of solid particles when subjected to a fluid flowing through them. The container of the solid particles, in which the fluidization process takes place is called the fluidized bed. In this application, sand and coal ash particles of 2-3 mm diameter are placed in a fluidized bed with 23 x 23 $cm^2$ net cross sectional area to form a bulk with a height of about 15 cm. The bed material resides on a perforated distributor plate with 1156 holes of 1.0 mm diameter. The fluidizing air is forced through the holes of the distributor and through the bed material. The bed material/fluidizing air combination exhibits fluid-like behaviour. At sufficiently high air velocities bubbles are formed and the sight is similar to a boiling liquid. In this combustion application, lignite particles with an average diameter of 3-5 mm act as fuel and the oxygen of the fluidizing air acts as a combustion agent. Fluidized bed is a process where gas and solid interaction is at a very high level. High combustion efficiencies can be achieved and sulphur dioxide emission can be minimized by adding limestone to the bed. Such favorable properties make fluidized bed an important process in modern combustion applications. Detailed information concerning the fluidization and fluidized combustion can be found in the literature (Howard, 1983; Kunii, 1969).

When starting the operation, the fluidized bed combustor has to be heated up to the operating temperature by auxiliary means. Liquified petroleum gas (LPG), is very suitable and can be ignited by a spark. As the bed temperature reaches 600-700$^oC$, coal feed to the bed can be started. LPG can then be turned off. A manual start up procedure has been developed (Mıdoğlu, 1985) for the 23 x 23 $cm^2$ fluidized bed lignite combustor which heats an 80 000 kcal/h conventional fire and smoke tube hot water boiler. Start up procedure is based on the facts, rules of thumb and necessary knowledge concerning the fluidized bed combustion system that are collected and judged on by an expert. The procedure is basically composed of "if-then" rules which follow a predetermined logical sequence. Thus the procedure developed may be adapted in future to construct a "rule based expert system". It may also be an interesting application of "fuzzy sets" in a fluidized combustion process control.

## HARDWARE OF THE SYSTEM

At present the manual start up procedure developed is implemented by using bivalent logic. A single card microcomputer based on 6502 microprocessor is designed and constructed which contains the contains the central processing unit (CPU), address decoding circuitry and memory. Additional printed circuit cards containing the keyboard and siplay interface circuitry, indicator and relay driving circuitry, 8-channel 8-bit analog to digital converter, input/output ports and power supply circuitry are also designed and constructed. These printed circuit cards use a common bus. They are rack mounted on a single back panel. Figure 1 illustrates the details of the hardware of the microcomputer constructed (Tulunay, Mıdoğlu 1985 a, b).

## SOFTWARE DEVELOPED

In this section, the basic structure of the softwere developed for realizing the start-up procedure is explained.

The entire program is written in 6502 assembly language and is logically divided into threemain parts:

1. The multitasking executive and the task library.
2. The monitor program.
3. The automatic start-up program.

The parameters for the start-up program can be examined and changed by the monitor program which handles and displays the command and data entries. Commands and simple test programs can be entered via the keyboard. Thus the monitor program is used for developing and debugging the automatic start-up program. On-line access and modification of the start-up parameters by the monitor program makes the system very versatile.

The Structure of the Multitasking Executive

Figure 2 illustrates the structure of the software developed. A real-time clock is used as an interval timer and it interrupts the CPU at 250 Hz. The task scheduler is executed each time an interrupt occurs. The scheduling algorithm executes one task for a fixed time, switches the processor to the next task for the same quantum period, and so on. Most of the tasks can satisfy their processing needs in one time slice. However, an overflow of the execution period does not cause any confliction. Tasks processes and that required a relatively long time must be divided into appropriate execution steps. Examples to the tasks and their functions are listed in Table 1.

Each task has an 8-byte register that contains the priority, starting address, execution step no, 16-bit status word, 8-bit "mail box" and the 8-bit checksum word. (see Fig.3.)

The processor performs the required task and then increments the "execution order number" namely EXECNO. Upon the next interrupt request it searches through the task registers to find the task that has a priority equal to the EXECNO. Finding the task, the processor executes it. Then it waits in an infinite loop for the next time slice. If no equal priority is hit until the ENDID (end identifier), then the EXECNO is reset to 1.

Note that the program has a dynamic structure since the priority of a task can be altered under program control, upon the receipt of a key depression for instance. Tasks are generally "turned off" by setting the priority to the hexadecimal number FF, because ENDID < #FF.

The task priorities are reorganised upon the entrance of the monitor and automatic start-up programs.

Although debugging an interrupt-dirven system is rather difficult, this dificulty has been overcome by using the structured modular programming method explained.

The Monitor Program

The monitor program is capable of:

1. Displaying and modifying the content of the entered address in the memory.
2. Modifying the contents of the processor registers.
3. Initiating the execution of the user programs and the programs in the ROM memory.
4. Setting break points in the user program.

The control of the system is transferred to the monitor program when the system is powered up or reset button is depressed. The monitor program then waits for the commands of the operator. The depression of the "start-up" key, transfers the control of the system to the automatic, start-up program.. But actually, the automatic start-up program has not been completely debugged yet. Since memory mapped I/O is used, by writing to or reading from a certain address in the memory, the operator can turn on or off the required output.

The fluidized bed lignite combustor has been started up semi-automatically, by only depressin a few keys, for several times. The start-up procedure has been further simplified and it has been a great success to start-up the combustor and obtain steady-state conditions in less than 5 minutes. The variation of the average bed temperature and the control actions of the operator are given in Fig.4.

The experimental observations and results are discussed in the next section.

## EXPERIMENTAL

Figure 4 is an actual average bed temperature versus time graph for one of the semi-automatic start up tests.

The valve position which controls the air flow rate is manipulated by the operator, by simply turning on/off the two related relay contact outputs of the microcomputer via the keyboard. Temperature is measured and displayed by the microcomputer continuosly and the operator gives decisions based on the increasing rate of the temperature. Spark ignitor, LPG, coal feeder are turned on or off by the operator via the keyboard.

Several start-up attempts have been carried out and all were successful. Note that the number of decisions and manipulations of the computer control outputs is kept at a minimum. The system rarely necessitates the rapid or continuous variation of the control outputs.

## RESULTS AND DISCUSSION

A manual start up procedure has been developed for the fluidized bed lignite combustor. This procedure has been simplified such that only the air flow rate is manipulated more than once, all of the other process variables are turned on or off at a certain instant of the start up process. This procedure is used by the operator working in front of the computer console. Several semi-automatic start-up tests have been performed which encourages the idea of a fully automated start up. After the development stage a simple and compact start-up controlled will be constructed.

## REFERENCES

Howard, J.R. (1983). Fluidized Beds, Combustion and Applications, Applied Science Publishers, London, New York.

Kunii, D. and O.Levenspiel (1969). Fluidization Engineering, John Wiley and Sons, Inc.

Mıdoğlu, H. (1985). Investigation of Automatic Start up Possibility for a Fluidized Bed Lignite Combustor, Master's Thesis, Electrical and Electronis Engineering Department, Middle East Technical University, Ankara, Turkey.

Tulunay, E. and H. Mıdoğlu (1985 a). Microprocessor based automatic start up system for combustion process. Preprints of the International Federation of Automatic Control/International Federation of Operations Research, Conference on Control Science and Technology for Development GSTD'85, Beijing, China, Vol.2, p.682-689.

Tulunay, E. and H.Mıdoğlu (1985 b). Experimental derivation of the automatic start up steps for a lignite combustor and a microprocessor based implementation. La Fluidisation, Actes 4 émes Journées Européennes Institut de Genie Chimique de Toulouse, Universite de Technologie de Compiegne, Toulouse, France, p.196-204.

TABLE 1. Example for the Tasks

| Task | Function |
|---|---|
| Supply Test | Checks if a power failure has occurred |
| Port Initialize | Inits all the I/O parts |
| Blink | Clears or restores the appropriate digit of the display buffer for cursor blinking |
| Display Mux | Outputs the contents of the display buffer in a multiplexed manner |
| Keyb Scan | Checks if a key depression has occurred |
| Key Check | Check if an invalid command or data entry |
| Key Handler | Accepts the key entry |
| Buzzer | Turns on or off the buzzer according to the requests of the other tasks |
| A/D Conversion | Performs the conversion of the analog signal of the requested channel |
| A/D Scaling | Converts the hexadecimal number to the required display formad |
| Monitor Init | Initiates the monitor program |
| Auto Init | Initiates the automatic start up program |

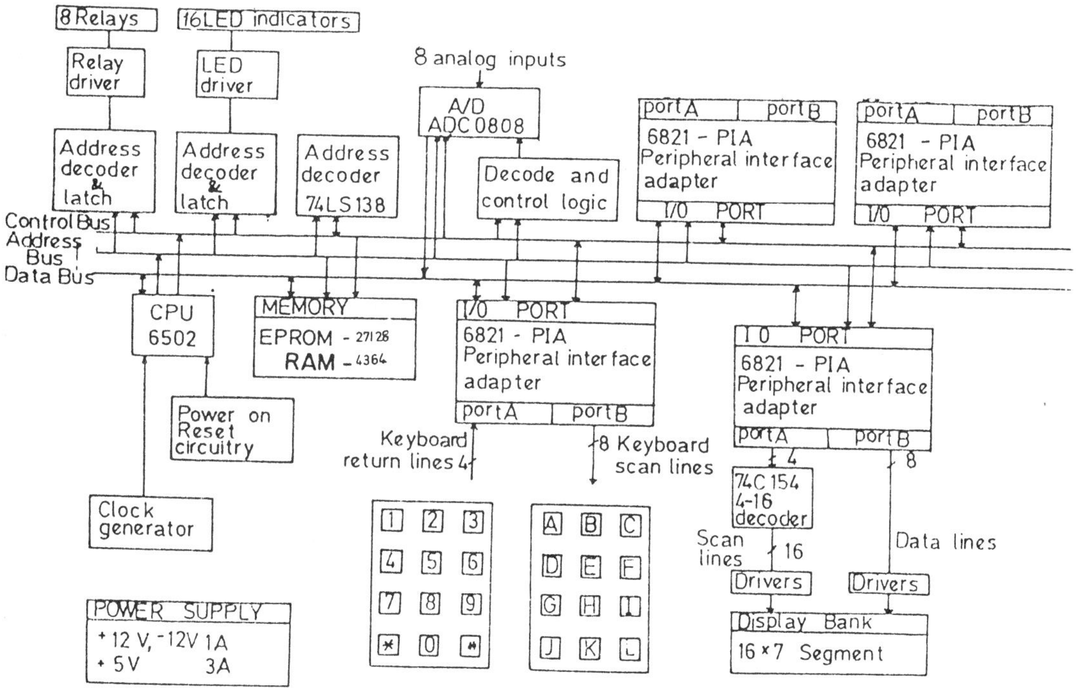


Fig .1 . Block diagram of the controller

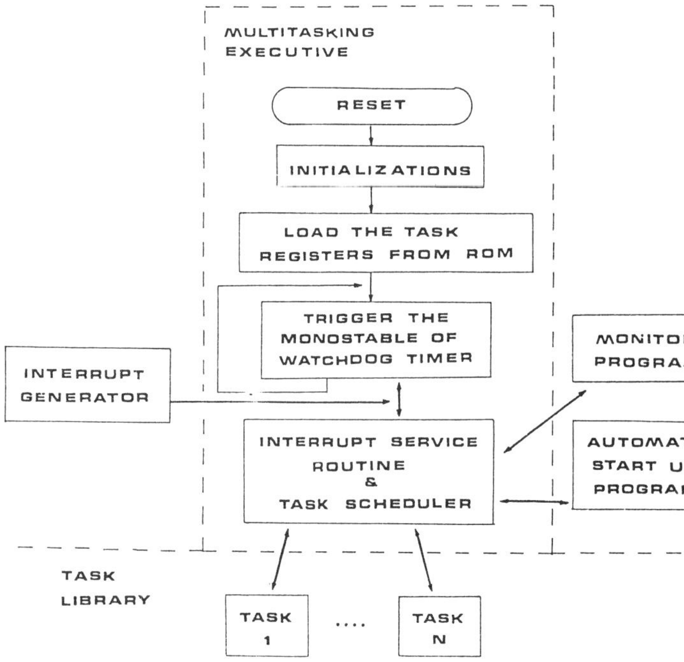


Fig.2 Basic structure of the software developed

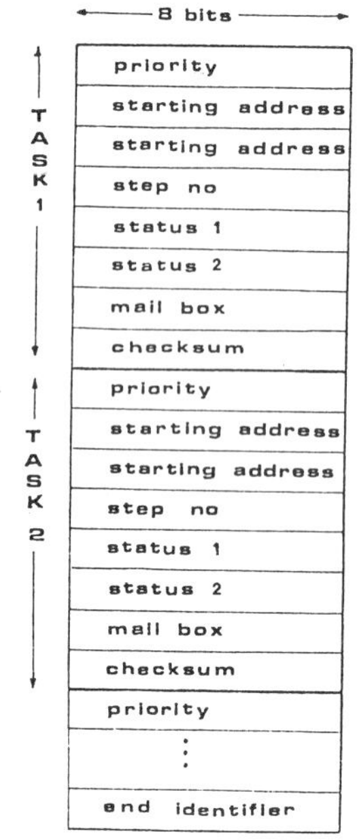


Fig.3 Task register structure in the physical memory

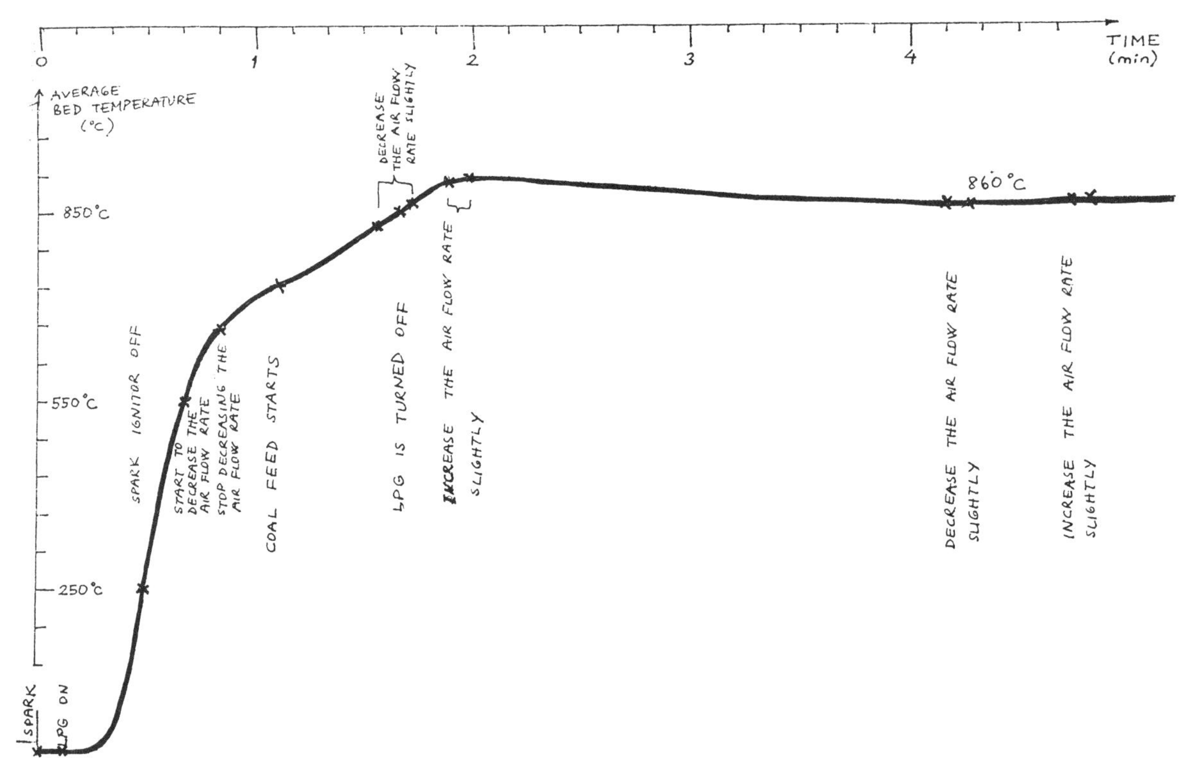


Fig.4 Average bed temperature variation during a semi-automatic start up

# COST-OPTIMIZED ENERGY CONSUMPTION IN SEWAGE TREATMENT PLANTS ON-LINE CONTROLLED BY DISTRIBUTED MICROPROCESSOR SYSTEMS

**G. Flotho**

*Department of Industrial Plants, Brown, Boveri & Cie A.G., P.O. Box 351, Mannheim 600, Federal Republic of Germany*

Abstract. Energy costs are the biggest operational costs while running a sewage treatment plant. A hierarchical microprocessor system on-line controls the plant and minimizes these costs in two ways. First is to shift energy demand into time zones, when energy is cheap. Second is to combine several energy sources and their portions of energy supply for achieving a cost minimum during the optimization period.
The strategy and the algorithms have to take into account different preconditions and restrictions of reality. However, the procedure is valid in general. An example for each of the two optimization ways is given below based on an existing plant. The results are discussed.

Keywords. Water pollution, Energy control, Optimization, Microprocessors, Hierarchical system, Process control, On-line operation.

## PRESENTATION OF THE PROBLEM

### General Structure of a Sewage Treatment Plant

Most of sewage treatment plants at the time being are divided into three main parts:

- mechanical part
- biological process
- sludge treatment.

The mechanical part consists of screen chambres, sandsettling tanks and presettling tanks. This part is to take most of the solid matter out of the sewages influent.
The biological process has two components, that together form an unit, the aeration and the secondary settling tanks. This part of the plant cleans the waste water from solved organic pollution by using microorganisms. Microorganisms need oxygen for life, this oxygen is brought into the water by blowing air into the tanks.
In both parts mentioned above you will get sludge as the result of treatment. This sludge contains a big amount of water. For dewatering the sludge treatment is used, whose main parts are the digestor tanks, sludgethickener and compression filters.
In the digestor tanks anaerobic microoganisms produce methane, the digestor gas, while decomposing the sludge.
Beside this there is a lot of auxiliary processes e.g. the neutralisation, desulphurising, waste water homogenisation. Furthermore you will find the controlroom, laboratories, offices and social buildings.
The whole plant is equipped with machinery like pumps, aircompressors, sludgescrapers. Heatings are used for the digestors and the buildings.
All this equipment needs energy and the main points of energy consumption are:

- air compressors
- inlet pumping station
- main pumping station
- primary sludge pumping station
- secondary sludge pumping station
- sludge heating
- heating of buildings.

The plant under consideration is used for treatment of both, community's and industry's waste water and an overall figure for that is:

380.000 inhabitant equivalents

Mostly there is a certain variety of energy sources for sewage treatment plants. Costs for each kind of energy are different and tariffs are a function of quantity and time. Therefore the followings may be considered as an example, the algorithm given later on will plantspecifically have to be modified, but is valid in general.

### Energy Optimization Problem

There are three sources for energy in this plant. The most important source is electrical energy, taken from the mains. Electrical energy costs depend on quantity and on three time zones, in which energy is consumed. Fig. 1.

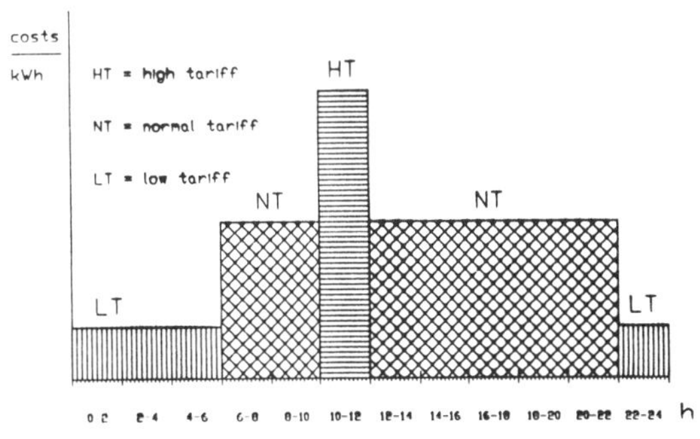


Fig. 1. Tarriff zones

Above all, due to the power supply contract, the maximum power is limited. A violation of this limit causes high penalty costs.
As a second source this plant may use natural gas. It is possible to use it for heating and to generate electricity on site by gasdiesel generators. To do this means having costs for natural gas, but for necessary machine maintenance as well.

The third and very interesting energy source is the digestor gas. It can be used in the same way as natural gas and the amount of gas production together with the three generators allows sufficient supply of energy demand for all important parts during a certain time. This is essential for emergency cases but also the cheapest way to meet energy needs of the plant. The optimization problem is to control the plant by achieving two aims

First : Minimize the energy demand in times of high energy costs

Second: Minimize the costs for energy supply.

## OPTIMIZATION STRATEGY

### Restrictions and Disturbancies

Every optimization strategy has to take into account the main purpose of any sewage treatment plant:

The purification of waste water to given minimum values of pollution reminders.

Especially the biological parts, aeration and digestors, will be heavily disturbed by changing their environmental conditions, e.g. quantities, temperatures, oxygen contents and some others. Best purification will be reached, if such conditions never change. However, you will have to change the conditions to a certain extend for energy optimization, but in such a range that good purification is obtained always. Some other main restrictions are given by pumping and aeration capacities, by the volumes of the basins, the gas storage and the gas engines.

Above that there will be "normal" disturbancies e.g. deviations from expected waste water quantity and grade of pollution during heavy rains and "unnormal" disturbancies, caused by defects anywhere in the equipment.

The optimization strategy has to consider the restrictions and has to be changed in case of disturbancies.

### Selection of Examples

The whole strategy is spread over following parts of the plant:

- primary settling tanks
- all pumping stations
- aeration tanks and air compressing
- secondary sludge recycling
- all heatings
- gas engines

In this paper two examples may be described in detail:

- control of the primary tanks, together with the main pumping station, which is part of trying to reach the first optimization aim
- control of the gas engines, which shows the way to the second aim

### Optimization of Energy Demand

For the selected example the algorithm is divided into steps:

- Prognosis of sewage influent
- Precalculation of setpoints for the main pumping station
- Calculation of buffer tank (primary settling tank) levels
- Correction of the main pumping setpoints in respect to actual influent

1 For abbreviations see list in Appendix A.

Prognosis of sewage influent. [1] The prognosis is based on measured data of the past. It is assumed (and later on verified) that the characteristics of the input are periodic. Period duration is one week. The data is stored in a table, Table 1

TABLE 1 Sewage Inflow $k_i$ during 2 hours

| | 0-2 | 2-4 | 4-6 | 6-8 | 8-10 | 10-12 | 12-14 | 14-16 | 16-18 | 18-20 | 20-22 | 22-24 h |
|---|---|---|---|---|---|---|---|---|---|---|---|---|
| MO | i=1<br>765 | 2<br>550 | 615 | 2070 | 4140 | 4370 | 4370 | 4410 | 3990 | 3945 | 3790 | 12<br>3470 |
| TU | 13<br>3200 | 2850 | 2480 | 3255 | 4225 | 4700 | 4540 | 4620 | 4220 | 4205 | 3915 | 24<br>3500 |
| WE | 3070 | 2830 | 2585 | 3395 | 4445 | 4535 | 4680 | 4825 | 4240 | 4315 | 3860 | 3540 |
| TH | 3175 | 2830 | 2610 | 3330 | 4400 | 4505 | 4405 | 4390 | 4085 | 3885 | 3750 | 3500 |
| FR | 3105 | 2855 | 2530 | 2945 | 4190 | 4420 | 4240 | 4305 | 3255 | 4030 | 3760 | 3355 |
| SA | 3085 | 2615 | 2445 | 2570 | 2885 | 2795 | 2795 | 2595 | 2190 | 1500 | 1295 | 1000 |
| SO | 73<br>745 | 495 | 495 | 535 | 1270 | 1475 | 1465 | 1000 | 1045 | 975 | 1025 | 84<br>970 |

while k is inlet quantity [$m^3$] during two hours. The table is updated every two hours this way:

If $k_{tab,old,i} / c_1 < k_{act,i} < k_{tab,old,i} * c_1$
then $k_{tab,new,i} = k_{tab\ old,i}*(1-c_2) + k_{act,i}*c_2$

If $k_{tab,old,i} / c_1 > k_{act,i} > k_{tab,old,i} * c_1$
then $k_{tab,\ new,i} = k_{tab,old,i}$

This method updates the table, following longlasting trends (f.i. seasons) and ignoring single events (f.i. heavy rains).

Precalculation of average setpoints for the main pumping station. Derived from the influent prognosis a table of setpoints $q_i$ is calculated rotatingly, Table 2, whose elements are:

$$q_i = \frac{\sum_{n=i-C_3/2}^{i+C_3/2-1} k_{tab,n}}{C_3} \quad [m^3].$$

TABLE 2 Quantity $q_i$ for Aeration Infeed

| | 0-2 | 2-4 | 4-6 | 6-8 | 8-10 | 10-12 | 12-14 | 14-16 | 16-18 | 18-20 | 20-22 | 22-24 h |
|---|---|---|---|---|---|---|---|---|---|---|---|---|
| MO | i=1<br>1585 | 2<br>1825 | 2110 | 2355 | 2600 | 2830 | 3040 | 3245 | 3435 | 3590 | 3690 | 12<br>3695 |
| TU | 13<br>3725 | 3730 | 3755 | 3775 | 3795 | 3805 | 3810 | 3800 | 3800 | 3805 | 3815 | 24<br>3835 |
| WE | 3820 | 3830 | 3845 | 3850 | 3860 | 3855 | 3855 | 3865 | 3865 | 3870 | 3865 | 3860 |
| TH | 3860 | 3835 | 3800 | 3785 | 3750 | 3740 | 3740 | 3735 | 3735 | 3730 | 3695 | 3680 |
| FR | 3670 | 3660 | 3650 | 3630 | 3645 | 3645 | 3635 | 3630 | 3610 | 3605 | 3570 | 3465 |
| SA | 3330 | 3210 | 3065 | 2925 | 2715 | 2510 | 2315 | 2120 | 1940 | 1780 | 1610 | 1475 |
| SO | 73<br>1365 | 1255 | 1120 | 1025 | 980 | 960 | 955 | 960 | 965 | 975 | 1100 | 84<br>1340 |

Calculation of buffer tank levels. Derived from the influent prognosis and the setpoint calculation the resulting levels in the buffer basin are calculated with starting level $l_{84} = 0$ at sunday midnight: Table 3.

$$l_i = \frac{\sum_{n=0}^{i} k_n - \sum_{n=0}^{i} q_n}{m * F * 1000} \quad [mm]$$

The calculations are repeated by changing $C_3$ until one element in the table reaches $level_{max}$ or $level_{min}$, Fig 2, i.e. buffer capacity is used and the aeration infeed is smoothened as much as possible, Fig. 3.

TABLE 3 Level $l_i$ in Presettling Tank

| | 0-2 | 2-4 | 4-6 | 6-8 | 8-10 | 10-12 | 12-14 | 14-16 | 16-18 | 18-20 | 20-22 | 22-24 h |
|---|---|---|---|---|---|---|---|---|---|---|---|---|
| MO | i=1 -820 | 2 -2095 | -3585 | min -3870 | -2330 | -790 | 535 | 1705 | 2260 | 2610 | 2715 | 12 2490 |
| TU | 13 1965 | 1080 | -195 | -720 | -290 | 600 | 1335 | 2155 | 2580 | 2980 | 3080 | 24 2745 |
| WE | 1995 | 995 | -270 | -725 | -170 | 510 | 1335 | 2295 | 2670 | max 3115 | 3115 | 2790 |
| TH | 2110 | 1100 | - 90 | -545 | 100 | 865 | 1530 | 2185 | 2535 | 2690 | 2745 | 2565 |
| FR | 2000 | 1200 | 80 | -610 | - 60 | 715 | 1320 | 1990 | 2235 | 2660 | 2850 | 2740 |
| SA | 2500 | 1905 | 1285 | 930 | 1095 | 1380 | 1865 | 2340 | 2585 | 2305 | 1995 | 1520 |
| SO | 73 895 | 135 | -495 | -985 | -695 | -180 | 325 | 365 | 445 | 445 | 370 | 84 0 |

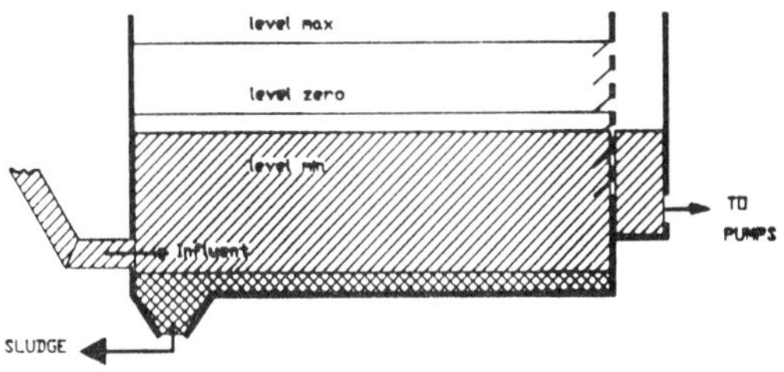


Fig. 2. Primary settling tank (buffer)

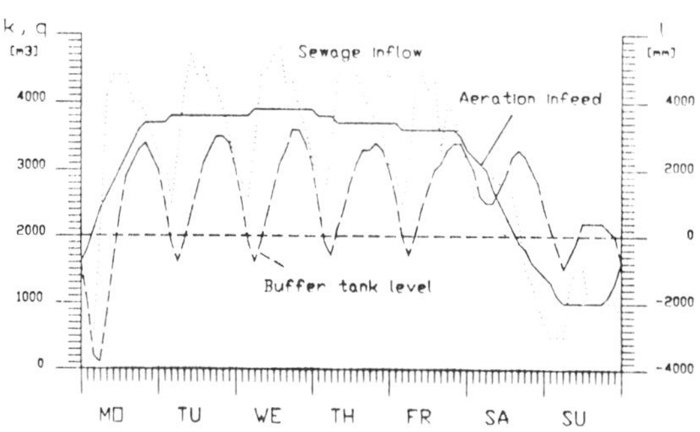


Fig. 3. Inlet flow, Aeration infeed, Buffer level

Correction of the setpoints. Up to this point, the pumping setpoints $q_i$ are based on the prognosis; if the prognosis is true, the pumps should run with these setpoints, obtaining most possible equal input into the aeration tanks, therefore and depending on that an equalizing and displacement of aircompressing and secondary sludge return, at last a displacement of power consumption.

Nota bene: Although control of aircompressing and sludge return depend like mentioned above, both will be optimized once again by additional algorithms, which are not part of this example.

Unfortunately there will be deviations from the prognosis during the day. Therefore the setpoints are corrected every 2 hours by the formula:

$$q_{cor,i} = q_i * \left[ \frac{\sum_{n=i-c_3/2}^{i-1} k_{act,n} + \sum_{n=i}^{i+c_3/2-1} k_{tab,n}}{12 * q_i} \right]^{1/4} * \left[ \frac{l_{act,i} - l_{tab,i-1} + l_{max} - l_{min}}{l_{max} - l_{min}} \right]$$

That can be considered a PD control algorithm.

Once again the setpoints are corrected every 10 minutes.

If:

$$\frac{k_i}{7.2 * c_1} < \text{input flow} < \frac{k_i * c_1}{7.2}$$

then

$$\text{flow setpoint} = \frac{q_{i,cor,new}}{7.2} = \frac{q_{i,old}}{7.2} + \left( \frac{q_{i,cor}}{7.2} - \frac{q_{i,old}}{7.2} \right) * c_4$$

Or if:

$$\frac{k_i}{7.2 * c_1} > \text{input flow} \qquad \frac{k_i * c_1}{7.2}$$

then

$$\text{flow setpoint} = \frac{q_{i,cor,new}}{7.2} = \frac{q_{i,old}}{7.2} + \left( \text{input flow} - \frac{q_{i,old}}{7.2} \right) * c_4$$

input flow = analogue value $\left[\frac{\text{liter}}{s}\right]$

$$\frac{m^3}{2h} = \frac{1000\ \text{liter}}{7200\ \text{sec.}} = \frac{1}{7,2}$$

(redimensioning factor)

Optimization of Energy Costs by combining Energy Sources

As shown in the example above, not only the total amount but also the time of energy consumption is influenced by different modes of running the plant. The said (first) optimization results in an equalizing of energy consumption. It is evident that the (second) algorithm, combining the energy sources for achieving lowest costs, may deequalize the consumption but due to the main restrictions, set by inlet quantity and the biological needs this deequalizing will lead to worse results in purification, which are not allowed.

Therefore the energy demand, caused by the first algorithm, is considered as preset and not fed back by the second. Based on this, the prognosis of electrical power demand, digestor gas production and sludge production are calculated.

The optimization interval is 24 hours.

Prognosis of the total electric energy demand. Once again a table is filled with actual values $W_i$ [kWh] from the past and then updated every two hours according to Table 4:

$$W_{i,new} = W_{i,old}\,(1-C_5) + W_{i,act} * C_5$$

TABLE 4 Electrical Energy Consumption $W_i$

| | 0-2 | 2-4 | 4-6 | 6-8 | 8-10 | 10-12 | 12-14 | 14-16 | 16-18 | 18-20 | 20-22 | 22-24 h |
|---|---|---|---|---|---|---|---|---|---|---|---|---|
| MO | i=1 1924 | 2 1656 | 1478 | 1325 | 1373 | 1414 | 1259 | 1252 | 1376 | 1270 | 1205 | 12 1170 |
| TU | 13 1146 | 1110 | 965 | 1057 | 1256 | 1340 | 1276 | 1481 | 1716 | 1738 | 1759 | 24 1908 |
| WE | 2083 | 2116 | 2025 | 1999 | 1969 | 2114 | 2144 | 2142 | 2115 | 2155 | 2189 | 2069 |
| TH | 2052 | 2038 | 1988 | 2135 | 2435 | 2407 | 2435 | 2438 | 2372 | 2454 | 2352 | 2445 |
| FR | 2416 | 2345 | 2216 | 2197 | 2396 | 2418 | 2421 | 2434 | 2424 | 2425 | 2458 | 2368 |
| SA | 2403 | 2417 | 2300 | 2263 | 2309 | 2168 | 2171 | 2191 | 2097 | 2088 | 2109 | 2094 |
| SO | 73 2093 | 2103 | 2093 | 2020 | 2092 | 2185 | 2128 | 2046 | 1971 | 1875 | 1885 | 84 1847 |
| | LT | | | NT | | HT | | | NT | | | LT |

Prognosis of gas and sludge production. Two tables with 7 elements each are filled, containing the daily averaged values of the past. Table 5 The tables are updated every day as follows:

$G_{j,new} = G_{j,old} * (1-C_5) + G_{j,act} * C_5$
$S_{j,new} = S_{j,old} * (1-C_5) + S_{j,act} * C_5$
$j = 1 \ldots 7$ = Monday ... Sunday

TABLE 5 Average Gas Production $G_j$
Sludge Production $S_j$

| | j = 1 | 2 | 3 | 4 | 5 | 6 | 7 |
|---|---|---|---|---|---|---|---|
| G | 3235 | 2969 | 3062 | 3137 | 3153 | 3053 | 3497 |
| S | 923 | 296 | 1509 | 1222 | 388 | 1153 | 1030 |
| | MO | TU | WE | TH | FR | SA | SU |

Prognosis of heat energy demand. This prognosis is based on:

$T_{dig}$ [°C] = average digestor tank temperature
$T_s$ [°C] = average sludge temperature in the prethickeners
$T_{air}$ [°C] = average air temperature by weather forecast

Therefore with:
H = heating energy demand per day [kWh]

$H = H_{dig} + H_{SB} + H_{MB}$
$H_{dig}$ [kWh] = heat for digestor tanks
$H_{SB}$ [kWh] = heat for social buildings, offices
$H_{MB}$ [kWh] = heat for machinery buildings

$$H_{dig} = S_j * C_6 (T_{set}-T_s) + o * C_7 (T_{set}-T_{air}) + o * C_8 (T_{set}-T_{dig})$$
only if > 0

$$H_{SB} = C_9 * (22 - T_{air})$$
only if > 0

$$H_{MB} = C_{10} * (12 - T_{air})$$
only if > 0

Total electric energy demand . This calculation is divided into three sections according to the time zones for low, normal and high tariffs.

$W_{total}$ [kWh] $= W_{Lt} + W_{NT} + W_{HT}$
$W_{LT} = \Sigma W_{i,\text{low tariff time}}$
$W_{NT} = \Sigma W_{i,\text{normal tariff time}}$
$W_{HT} = \Sigma W_{i,\text{high tariff time}}$

Electric energy demand $W_M$ from the mains is in general:

$W_M$ [kWh] $= W_{total} - W_G$
$W_G$ = el.energy by gas engines [kWh]
$W_G = n * P_{G,max} * (1 - x) * t$
$x = 0; 0.1; \ldots 1$

with
n = number of generators available (1 ... 3)
$P_{G,max}$ = maximum power of one generator [kW]

Therefore:
$W_{M,LT} = W_{LT} - n * P_{G,max} * (1-x) * t_{LT}$
$W_{M,NT} = W_{NT} - n * P_{G,max} * (1-y) * t_{NT}$
$W_{m,HT} = N_{HT} - n * P_{G,max} * (1-z) * t_{HT}$

x, y, z = current variables for the iteration

$t_{LT}$, ($t_{NT}$, $t_{HT}$) = duration time of low (normal, high) tariff zone
If x = y = z = 0, there is minimum el. energy consumption from the mains (max. self generation)
If x = y = z = 1, there is no self generation

Calculation of gas demand for engines.
$G_D = W_G * C_{12}$ [Nm³]

Calculation of natural gas demand for engines.
$G_{D,NE} = [W_G * C_{12} - ( G_j + G_s - C_{13})] * c_{14}$
only if > 0

Engines residual thermal energy
$H_E = W_G * C_{15}$

Natural gas demand for heating
$G_{D,NH} = [(H - H_E - G_j - G_D) * C_{16}] * C_{17}$
only if > 0

Cost function. The total energy costs, while running the plant, contain costs for:
- electric energy from the mains
- natural gas
- gas engine running costs

$$cost_{total} = cost_{el} + cost_{GN} + cost_E$$

This function has to be minimized.
In detail:

$cost_{el} = p_{el,LT} * W_{M,LT} + p_{el,NT} * W_{M,NT} + p_{el,HT} * W_{M,HT}$
$cost_{GN} = p_{NG} * (G_{D,NE} + G_{D,NH})$
$cost_E = p_E * (W_G)$

$cost_{total} = f (x,y,z) =$

$p_{el,LT} * [W_{LT}-n * P_{G,max} * (1-x) * t_{LT}]$
$+p_{el,NT} * [W_{NT}-n * P_{G,max} * (1-y) * t_{NT}]$
$+p_{el,HT} * [W_{HT}-n * P_{G,max} * (1-z) * t_{HT}]$
only if > 0 (el. costs)

$+ p_{NG} * c_{14} * c_{12} * n * P_{G,max} * (1-x) * t_{LT}$
$+ p_{NG} * c_{14} * c_{12} * n * P_{G,max} * (1-y) * t_{NT}$
$+ p_{NG} * c_{14} * c_{12} * n * P_{G,max} * (1-z) * t_{HT}$
only if > 0 (costs, if engines run by nat.gas only)

$- p_{NG} * c_{14} * (G_j + G_s - c_{13})$
(saving if replacing nat.gas by dig.gas)

$+ p_{NG} * c_{17} * H$
(heat by nat.gas only)

$- p_{NG} * c_{17} * c_{15} * n * P_{G,max} * (1-x) * t_{LT}$
$- p_{NG} * c_{17} * c_{15} * n * P_{G,max} * (1-y) * t_{NT}$
$- p_{NG} * c_{17} * c_{15} * n * P_{G,max} * (1-z) * t_{HT}$
only if > 0 (saving if using heat losses of engines)

$- p_{NG} * c_{17} * c_{16} * G_j$
(heating with all dig.gas available)

$+ p_{NG} * c_{17} * c_{16} * c_{12} * n * P_{G,max} * (1-x) * t_{LT}$
$+ p_{NG} * c_{17} * c_{16} * c_{12} * n * P_{G,max} * (1-y) * t_{NT}$
$+ p_{NG} * c_{17} * c_{16} * c_{12} * n * P_{G,max} * (1-z) * t_{HT}$
only if > 0 (available dig.gas is decreased if used for power generating)

$+ p_E * n * P_{G,max} * (1 - x) * t_{LT}$
$+ p_E * n * P_{G,max} * (1 - y) * t_{NT}$
$+ p_E * n * P_{G,max} * (1 - z) * t_{HT}$
(maintenance costs for engines)

Optimization. The algorithm combines the possible energy sources by calculating the cost matrix for the whole range of x, y and z in steps of 0.1 .
Not all elements are possible due to the regressions.
The combination of x, y, z with the lowest cost is the optimum. The value of x, y, z has a relation to the ratio of electrical energy consumption from the mains to self produced electrical energy and the necessary natural gas demand can be derived.

Calculation of setpoints. Having found the optimum values of x, y and z, the portions of electrical energy, natural gas energy and engine energy are known and will be spread to the elements $W_i$ for each time zone.
The setpoints for the control algorithm are:

$$P_{M,i} = W_{M,i} \; / \; h \qquad [kW]$$

Running the plant with this setpoints will lead to the cost minimum, if prognosis and reality are conform. Some deviations from the prognosis are normal and are met by spare capacity of the engines. But this will cause deviations from the cost minimum, of course.
Nota bene: The whole optimization shown above does not take into account the penalty costs when exeeding the allowed power consumption. To do this complicates the cost function but does not change it in general. In the realized software system this extension is implemented. That also contains a strategy for switching off energy consumers in an hierarchical order.

## REALIZATION AND EXPERIENCES

Control System Hardware.

The tasks mentioned above are performed by a decentralized, hierarchically structured control system. Fig. 4. At the lowest level there are analogue components controlling the actuators, getting actual values from the measurement equipment and handling them to the computer. In the middle level, the group control level, there are 6 microcomputers BBC-PROCONTROL I18
Here are some typical data:

- central processor PROCONTROL I 18
- 16 KB core memory
- watch dog timer
- process I/O
  with 200 to 700 digital inputs
  200 to 400 digital outputs

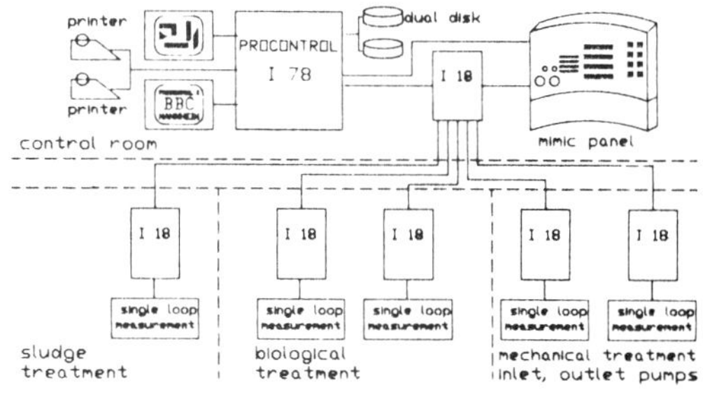


Fig. 4. Control System Configuration

The 6 microprocessors are linked as slaves to a mastercomputer BBC-PROCONTROL I 78 by a serial link.
The master consists of following main parts:

- central processor pdp 11/44
- 256 KByte core memory
- 2 disk drive units, 10 MByte each
- 1 CRT colour grafic display
- 1 CRT monochrome alpha-num display
- 2 printers
- serial links, 37.5 kBaud
- process I/O with 48 digital inputs
  96 digital outputs
  64 analogue inputs
  12 analogue outputs

Control System Software.

The software is written in a problem oriented high level language in a way that each system is able to run in stand-alone mode, controlling its dedicated part of the plant.
Furthermore the redundancy of the treatment is met by control system's redundancy. This makes the system safe in case of disturbancies.
At a basic level there are some main tasks for each system to do like:

- interlock and sequence control
- data aquisition and protocolling
- data communication, display of plant status.

That is state of the art and very similar in each system.
But above this software is implemented performing the tasks of energy optimization and, following the hierarchical structure, these tasks are spread over all system components.

Experiences.

For a period of more than two years the whole system runs without any problems, as far as the hardware and the software basic level is concerned. That is normal for an industrial designed, professional system and not to talk about in this paper.
But here are the main results of the energy optimization task:

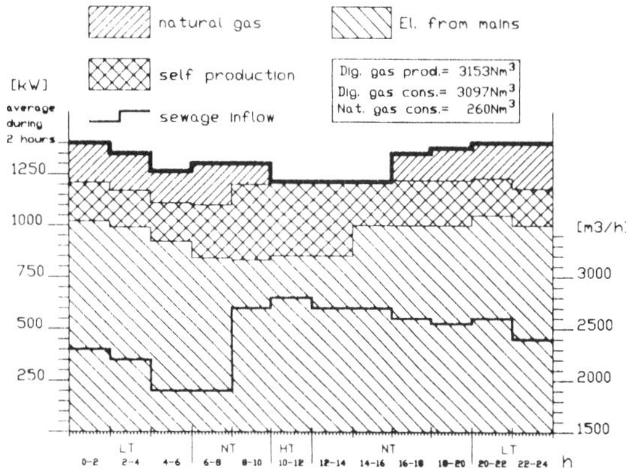


Fig. 5. Power Demand and Supply

The prognosis of influent and energy consumption is more exact than estimated before.

Therefore the control of the buffer basins, i.e. equalization of the aeration infeed allows to shift the treatment work, respectively the energy consumption into low tariff times.
In the first approach the equalization was too much perfect. Surprisingly the purification became worse and we had to learn, that

microorganisms have to be exponated to a certain stress-situation. We had to reduce the equalization a little bit.

Fig. 5. shows the total power demand during a typical day in March and the portions of external and self-produced supply.
Characteristics are:

- the digestor gas production is consumed nearly totally by generating electricity
- self generating has its maximum at times of high energy costs ( 8 a.m. to 2 p.m.)
- during the maximum inflow, power demand is at its minimum
- operating of gas engines is dispatched over the whole day to avoid too much natural gas consumption for heating.

As a summary you can say, that the optimization aims are achieved. But there are still some problems in adjusting the constants $c_i$ in the optimization algorithm which seem to be a function of waste water quality parameters.
To sort out these problems will be our task in the next future.

## REFERENCES


Lang, H. (1982). Energieoptimierung bei der Projektierung eines Klaerwerks. <u>Korrespondenz Abwasser</u>, Vol. 29, p.p. 399-403, Ges. Foerd. Abwassertechnik, St. Augustin, Germany.

Flotho, G., Scharpenberg, H. (1985), Leitsysteme in der Wasserwirtschaft. <u>Jahrbuch Wasser</u>, 1.st. ed. p.p. 152-156, Vulkan Verlag, Essen, Germany.


## APPENDIX A.

<u>List of Abbreviations in Order as used in the Text</u>

HT = high tariff
NT = normal tariff
LT = low tariff
h = hour
kWh = kilowatts * hour = electrical work
m = meter
k = inlet quantity [$m^3$]
$m^3$ = cubic meter
$m^2$ = square meter
i = running index (1, 2, ..., 84)
act = index for an actual measured value
q = quantity for aeration tank infeed [$m^3$]
$c_i$ = factors. All adjustable by operator
$c_1$ = sensivity factor, default = 1.1
$c_2$ = update factor, default = 0.05
$c_3$ = average interval (2, 4, ..., 42), default = 12
n = index
l = level in primary settling tank [m]
F = bottom square dimension [$m^2$]
n = number of tanks
min = minimum
max = maximum
cor = corrected
$c_4$ = correction factor, default = 0.05
W = work, electrical energy [kWh]
$c_5$ = sensivity factor, default = 0.05
G = average gas production per day [$Nm^3$]
$Nm^3$ = normalized $m^3$
j = index, 1, 2, .., 7, = monday .. sunday
S = average sludge production per day [$m^3$]
T = temperature [°C]
H = heat demand [kWh]
dig = digestor
SB = social buildings
MB = machinery buildings
$c_6$ = specific heat absorption of sludge [$\frac{kWh}{m^3 °C}$]
°C = degrees centigrade
$T_{set}$ = temperatur setpoint
o = number of digestor tanks
$c_7$ = surface heat loss by radiation [$\frac{kWh}{°C}$]
$c_8$ = sludge volume * specific heat absorption in digestor tank [$\frac{kWh}{°C}$]
$c_9$ = heat demand of social buildings [$\frac{kWh}{°C}$]
$c_{10}$ = heat demand of machinery buildings [$\frac{kWh}{°C}$]
$W_M$ = electrical work demand from the mains [kWh]
$W_G$ = electrical work produced by gas engines [kWh]
n = number of gas-engines available (1..3)
x,y,z = running indices (1, 0.9, ..., 0)
$P_{G,max}$ = maximum electrical power of one gas-engine [kW]
t = time
$t_{LT}$ = duration of low tariff zone
$t_{NT}$ = duration of normal tariff zone
$t_{HT}$ = duration of high tariff zone
$G_D$ = gas demand for engines [$Nm^3$]
$c_{12}$ = gas consumption per produced kWh [$\frac{Nm3}{kWh}$]
$c_{14}$ = factor in respect to higher energy content of natural gas
$c_{13}$ = minimum spare digestor gas in storage [$Nm^3$]
$c_S$ = gas in storage [$Nm^3$]
$H_E$ = residual thermal energy (heat losses) of engines [kWh]
$c_{15}$ = loss factor
$G_{D,NH}$ = natural gas demand for heat
$c_{16}$ = heat content of digestor gas
$c_{17}$ = factor in respect to higher heat content of natural gas
$cost_{GN}$ = costs for natural gas
$cost_E$ = running costs for engines (f.i. for maintenance)
$p_{el.}$ = price for electrical energy from the mains [money/kWh]
$p_{NG}$ = price for natural gas from supplier [money/kWh]
$p_E$ = price for running engines, empiric, [money/kWh]
$p_M$ = power from the mains [kW]

# DISTRIBUTED MICROPROCESSOR-BASED CONTROL SYSTEMS FOR FLUE GAS CLEANING

W. Latzel* and H. Kahle**

*Universität, GH Paderborn, Federal Republic of Germany
**Brown, Boveri & Cie AG, Mannheim, FRG

Abstract: Various microprocessor-based control systems are known today. They are either centralized, dedicated-centralized or decentralized in their functional structure of data acquisition and computation of control algorithms.

This paper presents a microprocessor-based, busoriented control system with freely selectable centralized, dedicated-centralized and decentralized functional structures for both data acquisition and computation of control algorithms within one overall system.

Demand in flexibility of functional structures is demonstrated in flue gas cleaning plants for coal-fired power plants. An analysis of the flue gas cleaning process is presented. The analysis shows that, with the exception of flue gas treatment itself, the following process stages of cleaning, product processing and end product treatment are separated from each other by intermediate storage containers. Resulting variable availability requirements for distinct process stages are demonstrated. These requirements are met by variable degrees of centralization within the functional structure of the control system.

A control system meeting the above requirements is presented. Some special system characteristics like closed-loop control via the data highway, modular system structure, event-driven communication are shown.

Keywords: flue gas cleaning; distributed control system; availability requirements; centralized functional structure; dedicated-centralized functional structure; decentralized functional structure.

## 1. ANALYSIS OF THE FLUE GAS CLEANING PROCESS

Flue gas cleaning processes for power plants belong to the state of the art. Several methods of process engineering have reached the status of technical application or have even gained substantial operational experience.

In nearly all used methods, appropriate chemical reagents are added to the flue gas after having left the boiler and the electrostatic filter. In this way the pollutants $SO_2$ and $NO_x$ are absorbed chemically and can be washed out of the flue gas stream /1/.

Out of approximately 20 methods available on the market, those methods using absorption liquid containing limestone or milk of lime have gained a market share of about 90%. The final end product resulting from these process engineering methods is commercially valuable gypsum. Because of the dominating market share, only these methods with gypsum as an end product will be considered furthermore.

The flue gas cleaning process as depicted in Fig. 1 can be subdivided into five stages:
- absorption
- flue gas reheating
- reagent preparation
- gypsum slurry processing
- sewage and end product treatment.

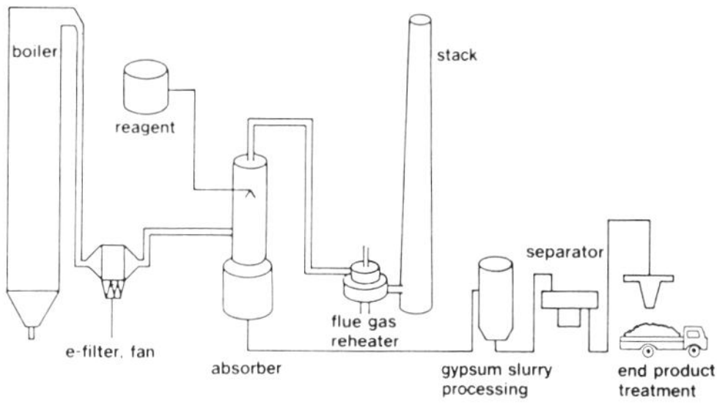


Fig.1 Outline of flue gas cleaning process

## 2. AVAILABILITY REQUIREMENTS FOR THE FLUE GAS CLEANING PROCESS

Among other criteria, the individual process stages differ in required availability level.

### 2.1. Availability requirements for distinct process stages

Process stages associated with the flue gas stream itself together with absorption and flue gas reheating functions must provide the same high availability as the power generation process itself.

According to applicable statutory regulations for air pollution control it is necessary to shut down or to drastically reduce production rate of the power plant in case these process stages go out of operation.

The stages of reagent preparation and gypsum slurry processing have to meet medium availability requirements, as they are equipped with mechanical storage containers.In order to bridge short disturbances of associated components, the volume of containers as well as the normally existing overdosing of absorbent within the storage containers can be utilized. Moreover, several components within the stages of reagent preparation, gypsum slurry processing as well as oxydation units are often installed in parallel /2/. In case of a component failure, these process stages can remain in operation by using parallel equipment.

Sewage and end product treatment stages call for the lowest availability requirements. Their mechanical storage containers provide such large buffer effects, that disturbances in these stages may last some hours, before the production rate of power generation must be reduced.

### 2.2. Availability requirements for the control system

To get a technically and commercially optimal solution, the different availability requirements for individual process stages must be imposed upon the control system as well. In case a uniform control system for the whole plant is desired, this system should possess inherent graded facilities to realize the full scale from high to low availability requirements.

Installing such a control system requires a division of the overall process into subfunctions of process control. Only with the resulting functional structure, the grade of centralization for the installed control system and hence the grade of required availability level can be exactly determined.

#### 2.2.1. Hierarchically structured control functions

As an example, Fig. 2.1 shows a section of the flue gas cleaning process with hierarchically structured control functions within process, group and individual control level. Clear delimitations, dependencies as well as redundancies of control functions are revealed. The required availability level can now be determined by combining a more or less large number of control functions in one hardware module. As a result, differently centralized structures for installed control system components are obtained.

#### 2.2.2. Various centralization in the control system for flue gas cleaning subprocesses

**Decentralized functional structure**

For components associated with the flue gas stream itself, high availability is required. This calls for a strictly decentralized and hierarchical structure of the control system (Fig. 2.2.a). This functional structure has successfully been used in power plant applications for more than 20 years. It comprises only one control function per hardware module in the individual control level as well as in the group control level. In case of overall automation, several group control functions are coordinated by a superimposed process control level.

Failure of modules used for process as well as group level control can be bridged by manual control actions in the next lower level of hierarchy until repair is accomplished. Only the degree of automation and comfort of handling is lowered during down time. Failure of a module used in the individual control level does not cause total shut-down of a plant, since those devices, which are decisive for the overall process performance are installed redundantly anyhow. For high availability requirements, mechanical redundancy is replicated in the installed control equipment in a consequent manner. Following this principle, failure of any control module does not affect overall plant availability.

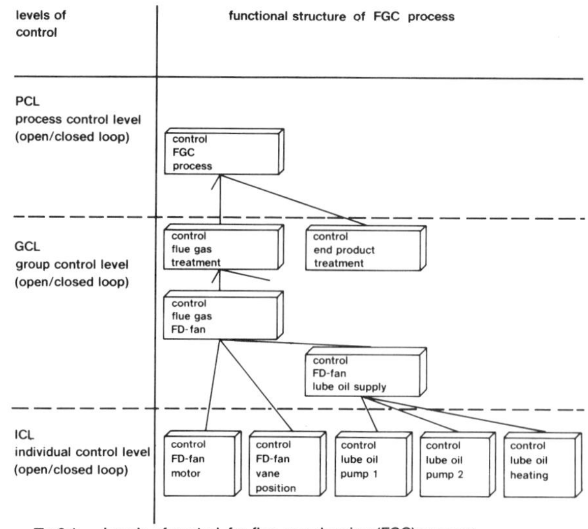


Fig.2.1 Levels of control for flue gas cleaning (FGC) process

| availability requirements | functional structure of control | | typical application example |
|---|---|---|---|
| a) high availability | GCL / ICL / devices | decentralized structure | flue gas fan |
| b) medium availability | GCL / ICL / devices | dedicated-centralized structure | reagent transport |
| c) low availability | GCL / ICL / devices | centralized structure | gypsum transport |

GCL = group control level
ICL = individual control level

Fig.2.2 Availability requirements and corresponding functional structure of control for flue gas cleaning (FGC) process

**Dedicated-centralized functional structure**

A medium level of availability is required for components of reagent preparation and gypsum slurry processing. Modules with allocated multiple control functions are used in group and individual control levels.

Concentration of several functions in one control module is only permissable within each technological coherent control function (Fig. 2.2.b). The hierarchical structure is maintained.

**Centralized functional structure**

Low availability is required for sewage and end product treatment. In this application, control functions in individual as well as in group control level are performed centrally in one common module (Fig. 2.2.c). Concentration of control functions in one control module is only limited by its maximum processing power.

All three functional structures offer an optimum for the ratio of availability to cost. The above mentioned considerations on differently centralized structures may be applied to all control functions like measurement, open and closed loop control.

## 3. PRESENTATION OF A CONTROL SYSTEM WITH FLEXIBLE FUNCTIONAL STRUCTURE FOR FLUE GAS CLEANING PLANTS

The control system PROCONTROL P, produced by Brown, Boveri & Cie AG, offers a selectable degree of centralization for realization of control functions. Before introducing those equipment components allowing this concept, a brief overview of the system configuration and types of control modules shall be given for better understanding.

### 3.1. PROCONTROL P system configuration

A generalized configuration of the system is shown in Fig. 3.1. The system can comprise up to 250 stations, which may be installed either locally decentralized throughout the whole plant site or locally centralized within the electronics room.

According to the assigned control functions, electronic modules for measurement, open and closed loop control, indication and annunciation may be installed in any mix and any number in every station. Modules within one station communicate with each other via the station bus.

All stations are connected with each other via the redundant remote bus. Both channels A and B of the remote buses are working non-reactively.

Up to eight redundant remote buses of 1.5 km length each may be coupled and laid out within the plant site. All process signals transmitted via a station bus are forwarded to the remote bus and are received by all connected stations in the same instant. Thus, all process signals within the overall system are available for further processing in all connected stations.

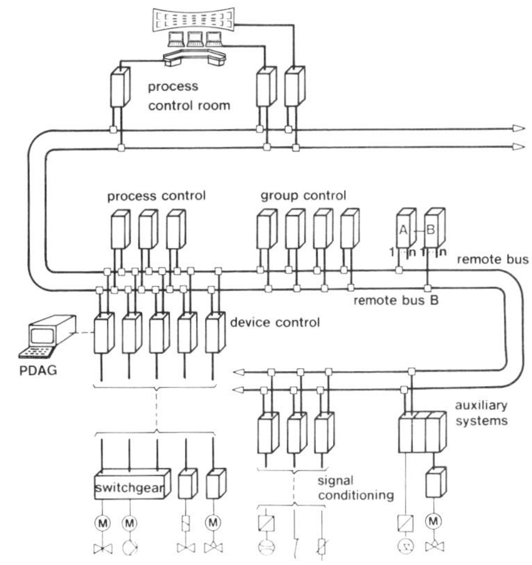


Fig.3.1 PROCONTROL P
System configuration

Use of a programming, diagnosing and indicating device (PDAG), which may be plugged in at any station, allows monitoring and substitution of all process signals on the remote bus. Equally, application-specific control algorithms may be read or written from or to any control module via the remote bus during full system operation.

Process monitoring and control may be performed either with CRT operator control stations or with conventional control instrumentation or a combination of both.

Depending on the type of process, a control system for a flue gas cleaning plant will consist of typically six to twelve stations, connected via the remote bus.

### 3.2. Structure of a PROCONTROL P station

A PROCONTROL P station is a cabinet for electronic equipment consisting of a maximum of four nineteen-inch racks. Each rack provides twenty-one slots for insertion of electronic modules.

Figure 3.2 depicts typical control modules. Nearly all modules are microprocessor-controlled. There are signal input and output modules as well as modules for device control (open and closed loop) with appropriate hardware for signal I/O to the process and (optionally) to a hardwired manual control station in the control room. Programmable processing modules contain application-specific control algorithms for use in process and group control level. In case of centralized or dedicated-centralized functional structures, processing modules perform individual control level functions as well. Connection to the CRT operator control station is achieved by the operator station coupler. The remote bus coupler connects the station bus to the remote bus and thus to all other stations and their modules throughout the system.

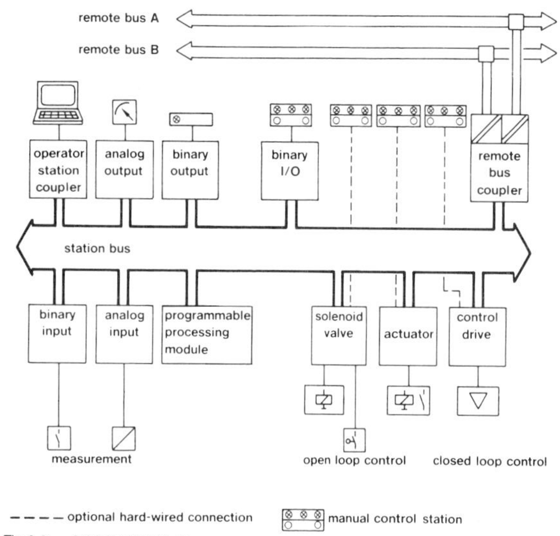


Fig.3.2 PROCONTROL P
Station bus with types of plug-in modules

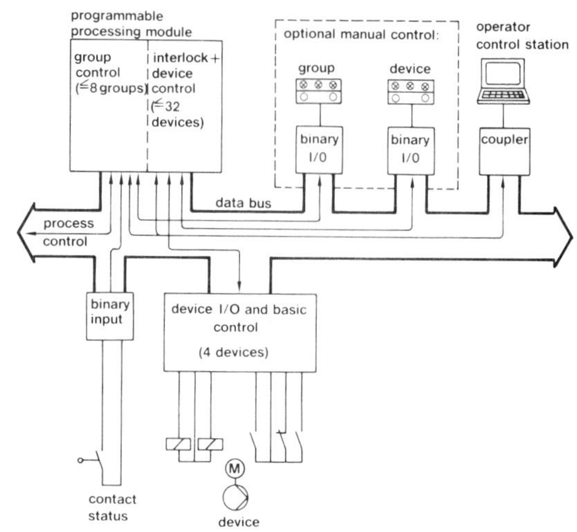


Fig.3.3 PROCONTROL P
Binary control, centralized functional structure

## 3.3. Realization of graded centralization

For the example of binary open loop control, the equipment facilities of PROCONTROL P to set up differently centralized structures are presented. In the following figures, station bus and remote bus are combined under the name of data bus.

### 3.3.1. Centralized functional structure

Several group and as many as possible device control functions together with associated interlocking logic are installed in the programmable processing module (Fig. 3.3.).

Coupling of devices is performed by simple device I/O and basic control modules. These modules have no microprocessor. The number of coupled devies is limited only by the maximal possible number of wires connectable to the module. Control commands by CRT operator control station within group and individual control level are directed to the programmable control module only. With additional cost, optional manual control stations for group or individual control level may be installed. Their communication with the programmable processing module is conducted via the data bus, because the large required number of contacts for parallel wiring cannot be installed on the processing module.

The centralized functional structure has the disadvantage to set out of operation several group and device control functions in case of a failed programmable processing module. Only device protection functions remain in operation. This structure has the advantage of lowest installation cost compared to the following variants.

### 3.3.2. Dedicated-centralized functional structure

Only one group control function and the hierarchically corresponding device control functions together with their inter-

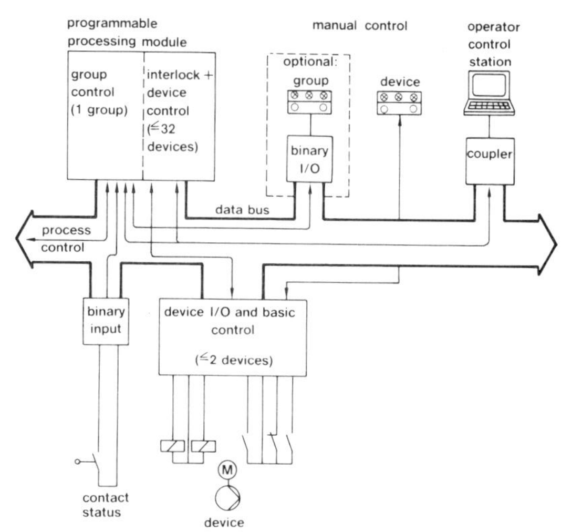


Fig.3.4 PROCONTROL P
Binary control, dedicated-centralized functional structure

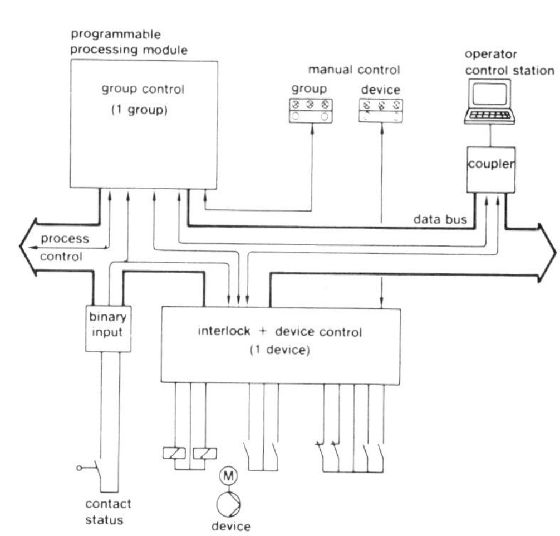


Fig.3.5 PROCONTROL P
Binary control, decentralized functional structure

locking logic are installed in the programmable processing module (Fig. 3.4).

Coupling of devices is performed by simple device I/O and basic control modules, comparable to those used in centralized functional structures. These modules are not equipped with microprocessors, however an additional hard-wired manual control station is installed in the control room.

Manual interference via CRT operator control station within group and individual control level is directed to the programmable processing module only, just like in the centralized functional structure. As an option with additional cost, a manual control station for group control can be installed. This function works in parallel to the CRT operator control station via the data bus.

The dedicated-centralized functional structure will limit consequences of a disturbance to a clearly predefinable subfunction within the hierarchy of overall control.Failure of the programmable processing module will affect only one group control function and its associated individual control functions together with interlocking logic. However, each individual device can still be controlled manually by use of the manual control station, overriding interlocking logic status.

### 3.3.3. Decentralized functional structure

For high availability requirements, a decentralized functional structure shown in Fig. 3.5 is chosen. Only one control function each within group and individual control level is installed on each hardware module. Every module is connected to a hard-wired manual control station. Manual interference via CRT operator control station can be directed to the programmable processing module as well as to the device control module.

Consequences of a failed programmable processing module are limited to the group control function only. Manual interference within the individual control level can be performed either via CRT operator control station or via manual control station. Failure of a device control module is not decisive for overall plant availability, since important devices are installed redundantly anyhow.

Compared to the previous structures, the decentralized, hierarchical functional structure provides a high overall plant availability. This high availability requires the comparatively highest instalation cost.

### 3.3.4. Physical location of control equipment installations

The majority of FGC plants is installed in already existing power plants. Changes in physical construction are only possible within narrow limits. Thus, only limited space is available for installation of additional control equipment. Fig. 3.6 shows two variants in allocation of control equipment. It is assumed in both cases, that control of the FGC plant has to be performed by the operator staff in the power plant control room.

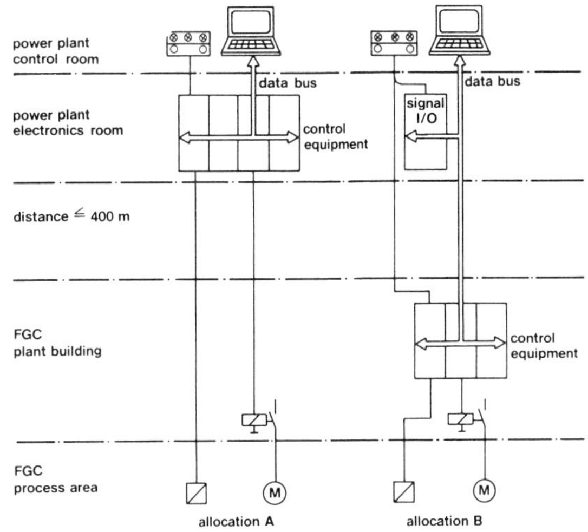


Fig.3.6 Various allocations for control equipment of flue gas cleaning process

Variant A assumes that there is still sufficient space for installation of the control equipment cabinets within the power plant electronics room. Signals for measurement and control are connected with multi-conductor cables to the FGC process area.

In variant B, electronic equipment cabinets are installed in the FGC plant building. In this case, the distance between the power plant electronics room and the FGC plant building is bridged by the remote bus. Since all process data are available on the remote bus, only stations with modules for signal I/O for wired push buttons, indicators and meters are set up in the power plant electronics room. There, the CRT operator control station is hooked to the control system in the same fashion.

The flexible structure of the presented control system allows a locally centralized or decentralized installation of the control equipment cabinets.

## 4. SPECIAL CHARACTERISTICS OF PROCONTROL P CONTROL SYSTEM

In this section, an overview of those system characteristics is given, which allow the flexibility of the system structure as described in section 3.

### 4.1. Source addressing and broadcasting with common acknowledge

Within the PROCONTROL P system, all process signals are transmitted in a digitized format to every station bus and remote bus. This task is autonomously performed by the system and need not be programmed. Signals are transmitted with messages. Messages contain mainly the binary coded process signal and the corresponding address. Systemwide, address and process signal represent a nonseparable entity according to the principle:

**"Systemwide, one signal is named identically " or "Systemwide, one message is addressed identically", respectively.**

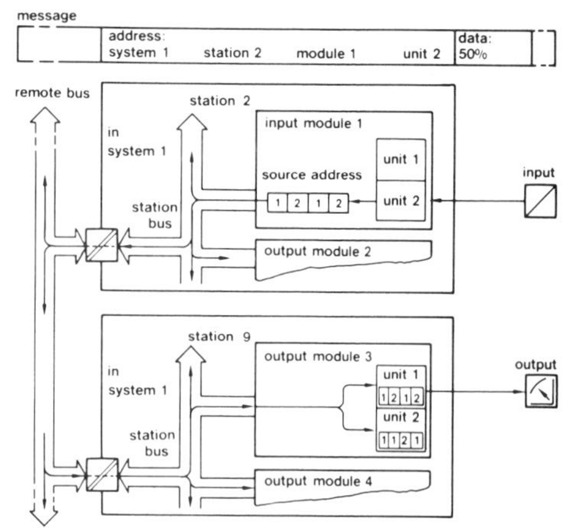


Fig.4.1 Principle of source addressing

The address is defined by that location, where the process signal is fed into the system. This location is the source for the process signal. The corresponding address is thus called source address. Figure 4.1 depicts in a simplified manner the principle of source addressing. In this example, an analog signal is fed into unit #2 of analog input module #1 in station #2 in system #1. The process signal with the address 1, 2, 1, 2 is transmitted over the station bus to the remote bus and furtheron to the buses of all other stations within the system. Transmission is thus performed corresponding to the broadcast principle:

**"During transmission from one source, all destinations are receiving".**

Every module provided with a receiving function compares the address of every newly received message with a list of application-specific addresses that is stored within its memory. This check decides on further processing of the received message in this module. In the above example, the address comparison matches on unit #1 on analog output module #3 in station #9 in system #1. The value of the transmitted process signal is stored in the output register and the signal is forwarded to the hard-wired panel meter.

The connection between analog input and analog output module is established simply by storing the signal's source address in the analog output module. With an equal programming procedure, any desired number of receiving modules may store and process the same message simultaneously. Connection between data source and one or several receiving modules is established as follows:

**"Only the receiving module decides on acceptance and further processing of messages."**

Principal advantages of this transmission scheme especially for real-time application within industrial control systems are shown in /3/ and /4/.

Every station can indicate a disturbed receipt of one or several messages by means of a convenient common acknowledge function, working in synchronism with message transmission. Missing common acknowledge provokes immediate retransmission:

**"A message is retransmitted to all receivers if at least one receiver does not acknowledge proper receipt."**

Use of the above mentioned principles allows the separation and distribution of one overall control function to nearly any number of functionally and locally distributed modules.

### 4.2. Data flow within the system

All functionally and locally distributed modules within the control system have to exchange process data in real-time. Guaranteing the necessary data flow within the system is of extreme importance for overall system performance. To ensure this, the presented control system uses a mix of cyclic and event-driven communication modes. According to momentary process signal dynamic performance, the system autonomously selects either cyclic or event-driven communication mode. Positive operational experience of this scheme in applications for overall control of coal-fired power plants is shown in /5/, /6/.

#### 4.2.1. Event-driven transmission

An event-driven transmission is stimulated by a change of value of any process signal within the control system. In this way, an excellent dynamic performance of process signal transmission is ensured throughout the whole system. Criteria for stimulation of a transmission are different for binary and analog process signals:

a) for binary process signals:
   An event-driven transmission is stimulated whenever a binary process signal changes state.

b) for analog process signals:
   An event-driven transmission is stimulated whenever the signal level of an analog process signal exceeds a preselectable increment compared to last transmission <u>and</u> after a preselectable inhibit-time has expired since last transmission.
   Typical values for increments are 0.4% of full scale and 50, 200 or 1000ms for inhibit-time.

Response times of typically 10ms for transmission in overall systems with more than 3.000 analog and 10.000 binary signals are obtained, /5/, /6/.

Event-driven transmission ensures the necessary response time for process signals.

#### 4.2.2. Cyclic transmission

The system PROCONTROL P performs event--driven transmissions with first priorty. However, a minimum portion of available bandwidth is reserved for cyclic transmission. During this mode of operation, the system itself activates all data sources to transmit their present value of process signal. This procedure is repeated continuously. In this fashion, all receiving modules within the system are updated with the latest value of process signal. Depending on overall system size, cyclic update of a given process signal is repeated in a scan of 1 to 6 seconds.

Control modules, distributed within the system, are continuously synchronized in this way. In addition, this mode of operation allows to cyclicly monitor the performance of all system components with the help of appropriate diagnostic provisions.

The cyclic mode provides the necessary robustness for transmission.

### 4.3. Closed loop control within the locally and functionally distributed control system

The principles shown in sections 4.1 and 4.2 allow the setup of a locally distributed closed loop control function with data transmission via station and remote bus (Fig. 4.2).

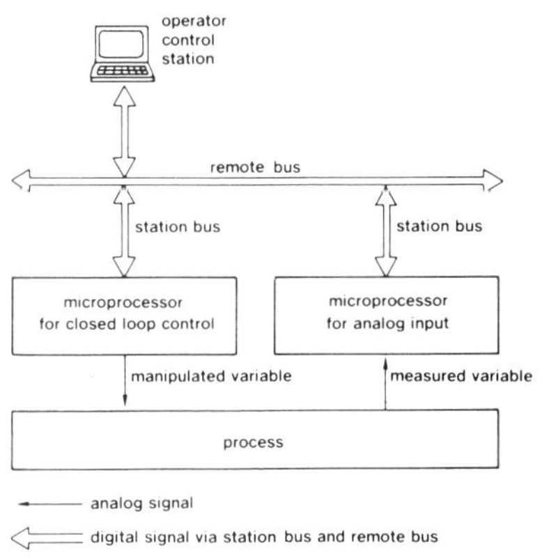


Fig.4.2 Closed loop control via station and remote bus in PROCONTROL P distributed control system

In /5/, the verification of stability for closed loop control functions is demonstrated in power plant applications.

This feature enables to set up all functions of control and instrumentation such as measurement, open and closed loop control, indication and annunciation with one homogeneous set of control equipment.

## 5. CONCLUSION

An analysis of the flue gas cleaning process shows variable degrees of availability requirements for certain process stages, while ensuring a high availability for the main power plant.

Variing availability requirements can be met by variable degrees of centralization in the setup of the presented control system. A technical and economical optimum will result. Any mix of centralized, dedicated-centralized and decentralized functional structures may be chosen.

The system PROCONTROL P has been successfully installed for control of industrial processes and power plants as well as in flue gas cleaning plants.

## 6. REFERENCES

/1/ Zimmermann, H. (1986). Konzept und Einsatzerfahrungen bei der Leittechnik von Rauchgasentschwefelungsanlagen. VDI-Berichte 589, VDI-Verlag, Düsseldorf. pp. 87-99.

/2/ Atzger, J. (1978). Entschwefelungsverfahren mit Endprodukt Gips - ein wirksamer und kostengünstiger Beitrag zur Reinhaltung der Luft, Tonindustriezeitung, 7, 413-418. Sprechsaal Verlag, Coburg.

/3/ Gueth, R.and T.Lalive, (1983). The distributed data flow aspect of industrial computer systems. In Proc. 5th IFAC Workshop on Distributed Computer Control Systems, Pergamon Press. pp. 1-8.

/4/ Gueth, R., J. Kriz and S. Zueger, (1985). Broadcasting source-addressed messages. In Proc. 5th Conference on Distributed Computing Systems, IEEE Computer Society Press. pp. 108-115.

/5/ Latzel, W. and H. Kahle (1984). Dynamik verteilter Prozeßleitsysteme, insbesondere für Regelungen. In GMR-Fachbericht Prozeßführung und Datenübermittlung, VDE-Verlag, Berlin. pp. 35-47.

/6/ Herrmann, R. and W. Körner (1986). Einsatz der Bus-Leittechnik in einem 750-MW-Kohlekraftwerk. In VDI-Berichte 589, VDI-Verlag, Düsseldorf. pp. 35-57.

# FAULT-TOLERANT MICROCOMPUTER CONTROL OF A HEAT EXCHANGER NETWORK

R. A. Martini* and A. Cinar

*Department of Chemical Engineering, Illinois Institute of Technology, Chicago, IL 60616, USA*

Abstract. As the use of digital computers in controlling chemical plants increases, loss of control due to computer failure becomes a major concern. A computer control system which is tolerant to faults of various degrees of severity was developed and experimentally tested using a heat exchanger network. Redundancy and diversity are created both in hardware and software.

The computer in the control loop generates new alternatives in handling failures of other devices in the control loop by creating functional redundancy. Consequently, tolerance to sensor failure is achieved by control law reconfiguration. The reassignment of control duties due to hardware, software, or sensor failure creates process upsets similar to temporary process disturbances.

Keywords. Fault-tolerant control systems; process control with microcomputers, heat exchanger control, time delay compensation.

## INTRODUCTION

The current trend in chemical plant control is to replace traditional analog controllers with digital controllers and computers. In chemical plants controlled by analog controllers failure tolerance is created by duplicating the control loops on the key variables. With computer control, all control loops are connected to one "controller" and its failure results in the loss of control of the plant. For creating failure tolerance, distributed control and multiple redundant computers are two approaches gaining acceptance in the industry. In distributed control where the computer is the supervisory unit and a local controller is used for each control loop. If there is a computer failure, the local controllers continue the control activity based on the latest information received from the computer. Tolerance to failure of local controllers can be implemented by duplicating local control loops. Hardware redundant computer systems consist of two or more identical hardware units running the same control program in parallel. When the active unit fails, the backup unit takes over the control duties. Since the units are identical, only hardware errors due to aging and malfunctions can be detected. In addition, these systems are unable to detect software errors due to software logic and design errors or software poisoning due to hardware faults.

Fault tolerance is the ability of a system to perform correctly in the presence of a limited number of faults (Frey, 1980; Rennels, 1984) and its extent depends on the redundancy and diversity built into the system. The complexity of the computer hardware and software, interface circuitry and sensors makes the elimination of all possible faults in the control system impossible. Therefore, to improve reliability, a computer system tolerant to various levels of hardware and software faults must be developed.

*Current affiliation: DOW Chemical Company, Midland, MI.

Hardware redundancy is accomplished by attaching one or more hardware units which, when failures occur, take the place of the failed unit and maintain the integrity of the system (Nieuwhof, 1978). To eliminate failures due to hardware logic design errors and common mode errors, hardware diversity is created by using different components with identical functions, such as different processors.

The fault tolerance of the system to software related failures must also be included in the system design. Software diversity is defined as the use of redundant program code which is generated in a different manner starting from one specification (Gmeiner and Voges, 1980) and is implemented by using different algorithms to perform the same function and comparing the results. Various levels of diverse programming can be implemented by using different programmers, different languages, and different computers. The probability of the same error existing in various implementations of the algorithm decreases as diversity increases. The outputs from diversely written but functionally identical modules can then be compared to reveal software errors. This eliminates the manual verification of each program. Often, the increase in cost in higher levels of software diversity can be offset by the reduction in verification costs.

With computer based systems, the destruction of the one to one correspondence between the processing units and the functional units shifts the redundancy of the system from hardware duplication to "functional redundancy" (Gilbert and Quirk, 1980). Preservation of the functional redundancy of the control activities becomes the focal issue, rather than the fault-tolerance of hardward or software alone. Furthermore, the computer may be used to create functional redundancy in other parts of the control loop. Functional redundancy can also be designed for the entire control system. Several control computers may be connected to a supervisory computer used for overall plant optimization. The control computers and the supervisory computer may be autonomously fault-tolerant by themselves.

Fault-tolerant controller design often ends at the I/O interface. However, failure of a sensor or an actuator can also cause a process to fail (McGill and Smith, 1984). The computation capability of the computer can be used to gain redundancy in sensors through the use of unlike components and analytic modelling (Van der Velde, 1984). Various control laws can be formulated for possible failures a priori and stored in the computer or algorithms for control system reconfiguration can be included within the control program.

The objective of this work is the construction and experimental testing of a fault-tolerant digital computer control system. Effects of hardware failure, software failure and sensor failure are tested. A heat exchanger network consisting of two heat exchangers in series and a fault-tolerant computer system with two microcomputers are used as the experimental system.

## THE FAULT-TOLERANT COMPUTER SYSTEM

### Computer Hardware and Software

In this study, as the dedicated process control computer a Falcon (SBC 11/21) single board computer (SBC) is used. The supervisory or host computer is an LSI 11/23 microcomputer. Both computers are manufactured by Digital Equipment Corporation (DEC). The SBC 11/21 contains the CPU, bootstrap PROMs, two serial I/O ports, and three bidirectional 8 line parallel I/O ports. In addition, the SBC system includes 64 KB of memory, an analog I/O board, an analog output board and a programmable real time clock. The host system has an LSI 11/23 CPU, 256KB of memory, analog I/O and output boards, a 4 line serial I/O board and a parallel I/O card as well as two floppy disks and a 10MB hard disk drive. The real clock is programmed to generate a signal at a rate of 1 Hz, so that the process measurements are read and control signals are sent out once every second. This configuration gives both the SBC and the Host the I/O hardward capable of controlling the process.

MicroPower/Pascal (MP/P) language (DEC, 1982) is used in the SBC. MP/P includes the software needed to operate the analog input/output operations and the real time clock. In the Host computer FORTRAN is used except in analog data adquisition and in Host/Target communications. A software package developed by Data Translations, DTLIB, handles analog data acquisition and some MACRO assembler programs have been developed for the Host/Target communications.

### System Configuration

The SBC is configured and programmed to read several process measurements, perform the control algorithm, and to send out control signals to the fina control elements. The SBC alco transmits the digital values of the process measurements and any change in the set points or control parameters to the Host over a serial communication line. At the same time, the Host is running a supervisory program that reads this serial information and displays the process measurements as well as any changes made to the control parameters on a local terminal. This continuous communication allows the Host to have all of the information needed to control the process.

Most importantly, the SBC also sends out two digital status lines from its parallel I/O ports. These signals are used as inputs (A and B) to the fault detection and control transfer circuit.

### Fault Detection and Control Transfer Circuit

The fault detection circuit is used to continuously monitor the SBC computer, determine its operating status, and send this status to the Host LSI 11/23 computer (Martini, 1985). The control transfer circuit is used to direct the measurement/control lines from the process to either the SBC or the Host, depending on the status of the SBC. A flow diagram of the fault detection and control transfer circuit is shown in Fig. 1. The circuits were designed to fail-safe so that control will be directed to the Host system. Optic isolators are used to keep any fault that develops on the fault detection and control transfer circuit board from damaging either computer.

The signals A and B are 2 parallel output lines from the SBC. During normal operation, the output line A should always be the complement of the output line B. In addition, after each A/D conversion, the output lines A and B are both complemented. When A equals B some major hardware fault has occurred, such as a power failure, or the software has been totally disabled. In the case where the software "hangs-up" or stops executing the control part of the program, the values of A and B will remain constant. They will still be complements, but they will not cycle. It will take a little longer than 2 clock cycles to realize that the signals A and B are constant. If a failure is detected, the fault detection circuit will latch the SBC STATUS line in the 'SBC FAILURE' state. This latch helps detect sporadic errors and will hold the value of the SBC STATUS line until the Host polls the fault detection circuit. Note that the Host can select to take over control by operator decision or programming logic and signal the control transfer circuit to direct the measurement/control lines to the Host. The SBC has no control over the fault detection or control transfer circuit.

The control transfer circuit consists of a set of relays which switch the connection of the lines from the process A/D interface to either the Host A/D or to the SBC A/D. The SBC STATUS line is used to control the connection of the relays (on - SBC, off - Host). Relays completely disconnect the lines from the faulty computer to the process. This keeps a faulty computer from generating spurious signals which could interfere with the signals sent from the controlling computer. In case the fault detection and control transfer circuit loses power, the relays will go to the unpowered state which will direct the measurement and control lines to the Host.

It is important to note that when the Host switches from supervisory to control mode it also passes on the last values of the control parameters. This allows the back-up control program to continue controlling the process at the same setpoints and with the same controller settings without any operator intervention.

### Functional Redundancy Against Sensor Failure

To include functional redundancy against sensor failure, it is proposed that in the event that a temperature sensor fails, the controller is reconfigured to use information from another temperature sensor. As an example, the failure of the outlet temperature sensor T2 from the second heat exchanger (HE2) is used. When this temperature sensor fails a downstream temperature sensor T3 will be used in its place. This sensor is located at the end of a 23 ft loop of uninsulated piping. The addition of a large velocity distance lag and heat losses to the surroundings makes the controller reconfiguration more complicated than the direct replacement of one measurement with another. Advanced control techniques must be used to compensate for time delays and heat losses (Holt and Morari, 1985; Jerome and Ray, 1985).

The first problem encountered in using sensor redundancy is detecting when the original sensor has failed. In our experimental system, the thermocouple amplification circuit is designed to output zero volts for an open thermocouple. To detect a shorted thermocouple, a test is made to check that the exit temperature from the downstream heat exchanger HE2 is always greater than the exit temperature from HE1. Even with steam valve S2 of HE2 closed, enough steam leaks past to elevate the water temperature a few degrees.

## Controller Configurations

Direct Sensor Replacement. Once the fault is detected, the controller is configured such that the auxiliary sensor, T3, directly replaces the original sensor, T2. This strategy would work well in a system where T2 and T3 are redundant sensors making the same measurement at the same point or in close proximity. However, since sensor T3 is downstream from the process, the measurement from T3 does not equal T2. The piping loop adds considerable velocity-distance lag to the measurement. This time delay causes the PI controller to become unstable, even with very conservative control parameters. Therefore, it is a bad assumption to assign T3 directly to T2.

Sensor Replacement and Time Delay Compensation. In this case, a General Multidelay Compensator (GMDC) (Jerome and Ray, 1985) designed by using the CONSYD CAD package (Morari and Ray, 1984) is added to the classic PI controller to remove the effects of the dead-time from the control loop. In order to realize the GMDC, the open loop response of T1, T2, andT3 to step changes in S1, S2, and flow disturbances are measured. From these responses, transfer function matrices (TFM) are modelled for the (T1,T2)system and the (T1,T3) system. When the sensor failure is detected, the information from T3 is used in place of that from T2 and the TFM constants and time delays from the (T1,T3) system replace those of the (T1,T2) system in the GMDC. This corrects for the additional time delay of the piping loop. In addition, the PI control parameters are replaced with those tuned for the (T1,T3) system with time delay compensation. Since there is some heat loss from the piping loop into the surroundings, the temperature T3 is a few degrees lower than T2. This would cause an offset between the setpoint value for T2 and its actual value.

Sensor Replacement and Compensation for Dead-time and Heat Loss. In order to make the reconfigured system perform as well as the original system, the offset in T2 must be eliminated. To accomplish this an inferential approach is used. The heat loss in the piping loop is modeled by examining the response of T3 to a change in T2. The model is then inverted to give an estimate of T2 based on a known T3. This estimate is then used as the measured variable in the dead-time compensated PI controller. Note that while the temperature T3 reflects the variations in the original variable T2, the time delay is still that of the T3 sensor.

# EXPERIMENTAL RESULTS

## Heat Exchanger Network

The fault tolerant control system was developed and tested using a heat exchanger network (Fig. 2) consisting of two pilot plant size shell and tube heat exchangers in series. Condensing steam on the shell side is used to heat water flowing in the tube side. The water inlet temperature, the water exit temperature from each exchanger (T1 and T2, respectively) a downstream water temperature at the end of a loop of piping 23 feet long (T3), and the supply steam temperature are measured by thermocouples, amplified, and transmitted to the computer. The water flow rate (FL) is measured using an electronic flow sensor/ indicator and is also sent to the computer. The water flow rate and the steam flow rate to each exchanger are regulated by pneumatic automatic control valves.

## Computer Hardware and Software Failure Recovery

Data was collected for simulations of SBC hardware and software failures for different operating conditions. The response due to total loss of control (Fig. 3) is compared with typical responses to hardware and software failures when the system is operating at steady-state (Fig. 4,5). The flow controller settings are $K_c$=0.95 and $T_I$= 3.0. The control parameters for both temperature control loops are: $K_c$=0.32 and $T_I$=7.0. First the water flowrate (FL) drops sharply and the outlet temperature T1 and T2 rise. The disturbance is caused because the water flow control valve is an air-to-open valve. In the brief time that neither computer is controlling the process, the valve stays in the closed position. As expected, first FL and then T1 and T2 return to their steady-state values. The deviations from the steady-state value for the software error are larger since it takes a little longer than twice the sample rate to detect a software error. The magnitude of the deviations of T1 and T2 dur to hardware and software failures are larger than the deviations caused by a typical flow disturbance, they are significantly more desirable than the response to the total loss of control.

## Sensor Failure Recovery

Alternate Control Configurations. In the experiments, the use of sensor T3 instead of T2 resulted in a closed-loop unstable system and a time delay compensator is needed (Fig. 6). Addition of the time delay compensator on the reconfigured control law stabilizes the controller using sensor T3 as well. The control settings are changed in the control law reconfiguration to $K_c$=0.25, $T_I$=7.0 for the T3 loop. Even though the response is stabilized the offset between T2 readings and its setpoint is unacceptable (Fig. 7). To correct for the offset, the piping loop was modelled as outlined and an expression to predict the value of T2 for a given T3 was obtained. This new configuration was then studied under the same step changes and disturbances. The system was stable under most cases and had a response similar to the original T2 controlled system (Figure 8).

## Controller Reconfiguration

A final set of experiments were conducted to observe the system behavior under on-line controller reconfiguration. Fig. 9a shows the effect of a T2 sensor failure and controller reconfiguration when the process is at steady-state. The transition has no noticeable effect on the outlet temperatures. Typical effects of a T2 failure and controller reconfiguration while the system was in transition to a new setpoint are shown in Figures 9b and 9c. In these tests a change in water flow rate (Fig. 9b) or temperature set point (Fig. 9c) is made and 10 sec later failure of sensor T2 is introduced. The results show that in general if the system with the T2 sensor is stable, the reconfigured system is also stable.

The functional redundancy achieved by controller reconfiguration against sensor failure is quite successful if the key features of the additional process characteristics are incorporated in the new control law. For the heat exchanger system, accounting for the extra dead-time and the heat loss were of critical value. Since this new controller is to be used until the original control loop is repaired some offset in the response may be tolerable. Consequently the new controller can be simplified. For the heat exchanger system, the use of the steady state gain alone for estimating T2 from T3 measurements resulted in the offsets shown in Fig. 9.

## CONCLUSIONS

The fault-tolerant computer control system described in this paper provides an added level of protection by building a hierarchy of computers in which the supervisory computers can also function as a control computer. The current trends in the chemical process industries are to have either distributed control or multiple identical processors. The configuration presented adds both hardware and software diversity and consequently increases fault-tolerance.

Experimental results illustrate that the successful transfer of control from one computer to the other due to hardware or software failure. Controller reconfiguration has proven to be an effective method to handle sensor failures. The time delay compensator designed enabled the stabilization of the system response with the reconfigured controller. With addition of fault-tolerance to the control system, a potential loss of control is reduced to control of a disturbance to the process. The magnitude and duration of the deviations from the set point values are related to the time needed to transfer control when a failure takes place.

## REFERENCES

DEC. (1982). MicroPower/Pascal User's Manuals Digital Equipment Corporation, Maynard, Mass.

Frey, H.H. (1980). Safety and reliability— their terms and models of complex systems, in R. Lauber (Ed.) Safety of Computer Control Systems, Pergamon Press/IFAC, 3-10.

Gilbert, M.H. and Quirk, W.J. (1980). Functional redundancy to achieve high reliability, in R. Lauber (Ed.) Safety of Computer Control Systems, Pergamon Press/IFAC, 59-64.

Gmeiner, L. and Voges, U. (1980). Software diversity in reactor protection systems: an experiment, in R. Lauber (Ed.) Safety of Computer Control Systems, Pergamon Press/IFAC, 75-79.

Holt, B.R. and Morari, M. (1985). Design of resilient processing plants V - the effect of deadtime on dynamic resilience, Chem. Eng. Sci., 40 (7) 1229-1237.

Jerome, N.F. and Ray, W.H. (1985). High performance multivariable control strategies for systems having time delays, submitted to AIChE J.

Martini, R.A. (1985). Fault-tolerant computer control of a heat exchanger network, MS Thesis, Illinois Institute of Technology, Chicago, IL.

McGill, W.F. and Smith, S.E. (1984). Fault-tolerance in continuous process control, IEEE Micro, 22-23, December 1984.

Morari, M. and Ray, W.H. (1984). CONSYD User's Manual, University of Wisconsin, Madison, WI.

Morari, M. and Ray, W.H. (1984). CONSYD User's Manual, University of Wisconsin, Madison, WI.

Nieuwhof, G.W.E. (1978). Standby redundancy system— two identical 100% units, Proceedings of the 1978 Canadian Reliability Symposium, Pergamon Press, 161-168.

Rennels, C.A. (1984). Fault-tolerant computing - concepts and examples, IEEE Tran. Comput., C-33(12) 1116-1129.

Vander Velde, W.E. (1984). Control system reconfiguration, Proc. 1984 ACC, San Diego, CA 1741-1745.

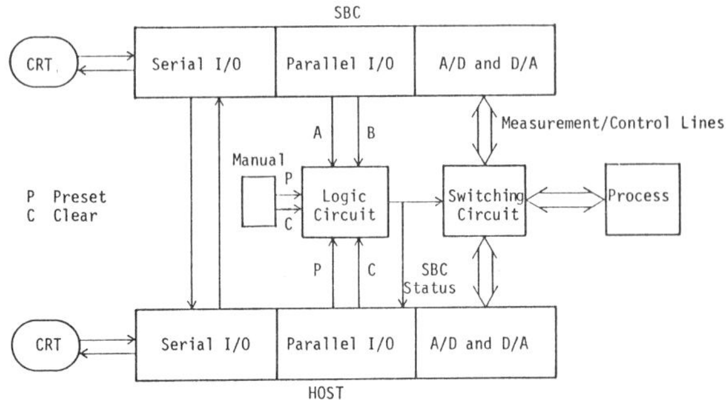


Figure 1. The Fault-Tolerant Computer System

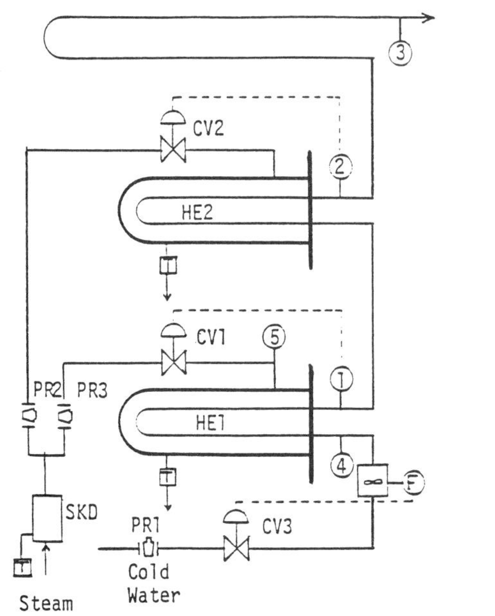


| | | | |
|---|---|---|---|
| F | Flow sensor | HE1 | Heat exchanger 1 |
| PR | Pressure regulator | HE2 | Heat exchanger 2 |
| SKD | Steam knock-out drum | T | Steam trap |
| CV | Control Valve | 1-5 | Temperature sensors |

Figure 2. The Heat Exchanger Network

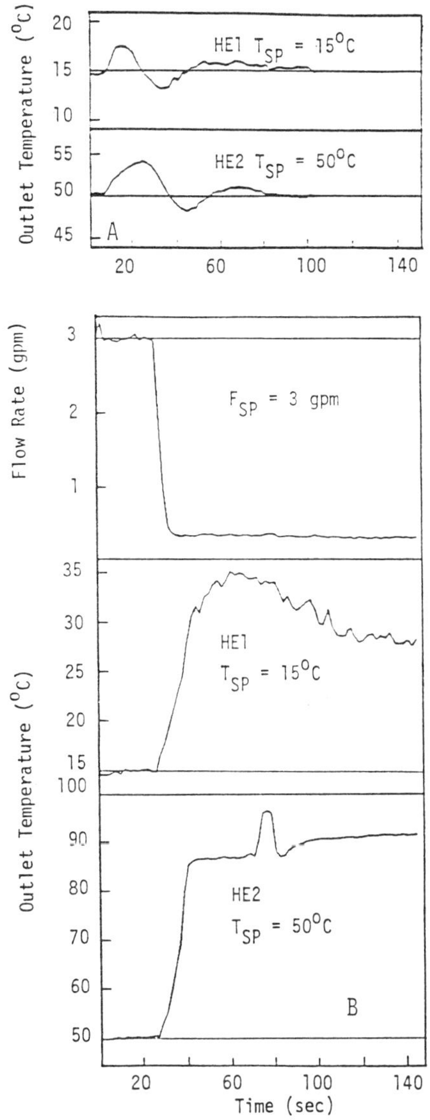


Figure 3. Process Response to a Change in FL from 3.0 to 3.5 gpm (A) and to Total Loss of Control (B)

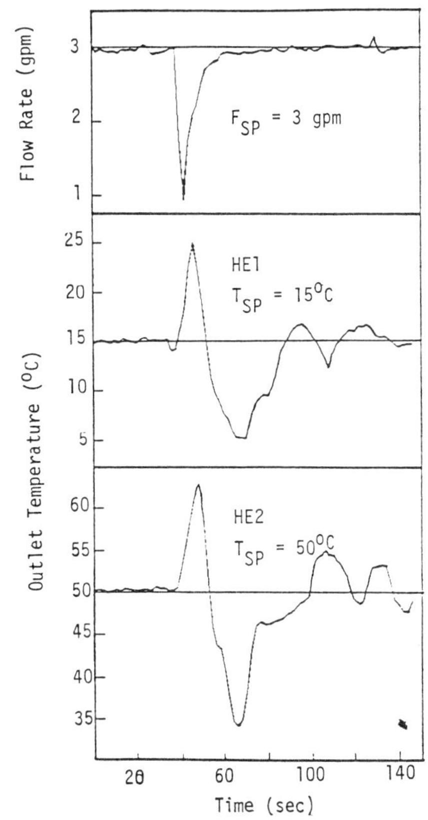


Figure 4. Process Response to SBC Hardware Failure

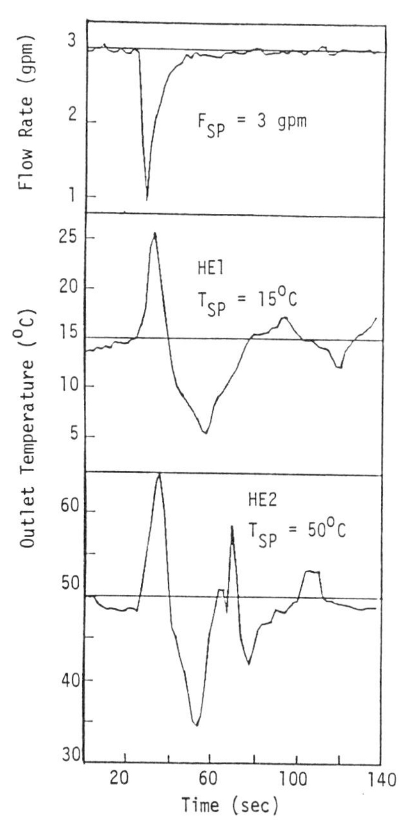


Figure 5. Process Response to SBC software Failure

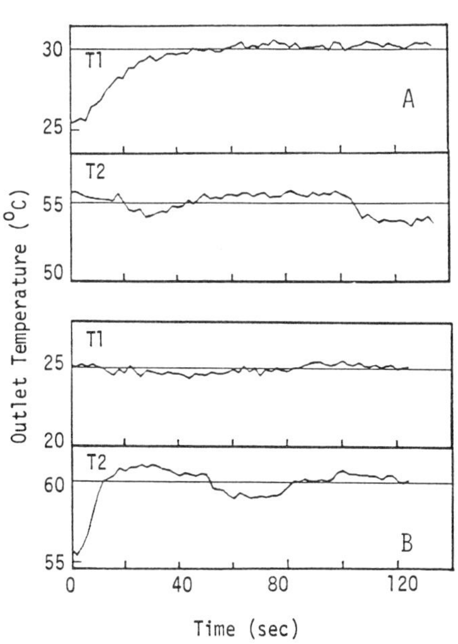


Figure 6. Process Response to a +5°C Step Change in T1 Set-point (A) and T2 Set-point (B); PI control with Time Delay Compensation.

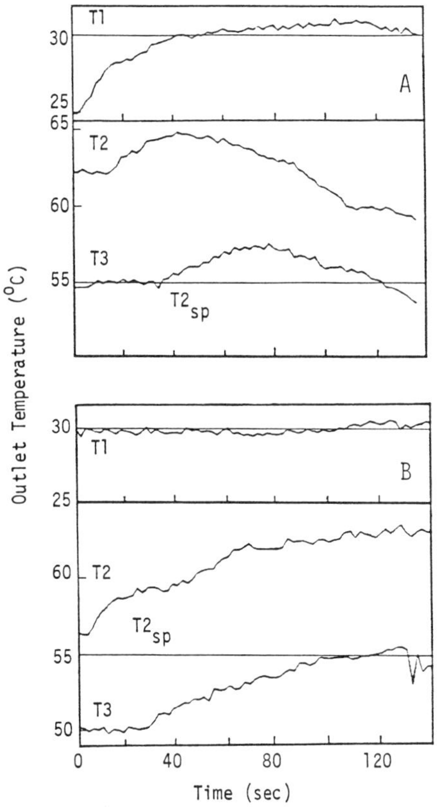


Figure 7. Process Response to a +5°C Step Change in T1 Set-point (A) and T2 Set-point (B) with T2 Sensor Failure; Controller Reconfiguration with Time Delay Compensation

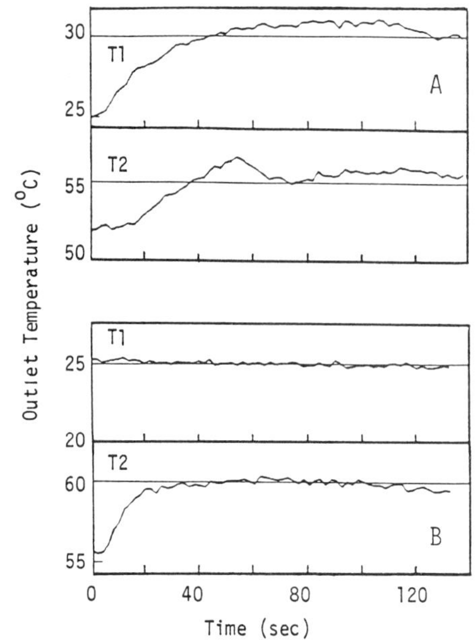


Figure 8. Process Response to a +5°C Step change in T1 Set-point (A) and T2 Set-point (B) with T2 Sensor Failure; Controller Reconfiguration with Time Delay and Heat Loss Compensation

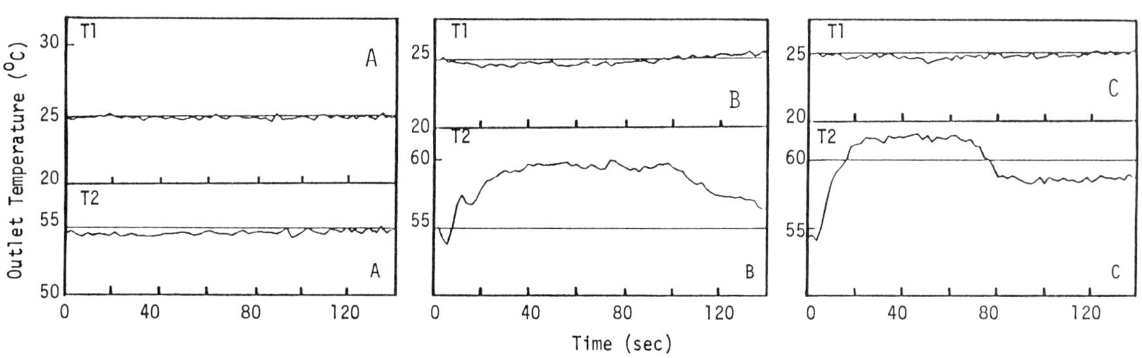


Figure 9. Process Response to a Controller Reconfiguration due to Failure of T2 Sensor while the Process is at Steady State (A), while in Transition to a New T2 Set-point (B) and while in Transition to a New FL Set-point

# MICROCOMPUTER APPLICATION IN LARGE POWER PLANTS

**B. Bachmann*, H. Cvetko* and M. Schurtenberger****

**Power Distribution Application,*
*Process Control and Instrumentation,*
*Engineering Department, BBC Brown, Boveri & Co., Baden*
***Department Sales and Engineering, H.E.P.P.,*
*BBC Brown, Boveri & Co., Zürich*

**Abstract** This paper briefly reviews the history of process control systems development. It presents a modern microprocessor based on a cable saving control and monitoring system with bus transmission. An example of the application of this system in a large Hydro Energy Power Unit is illustrated. Finally, the integrated control system which combines various control and protection functions completes this paper.

**Keyword List** Energy Control, Frequency Control, Hierarchial Systems, Integrated Plant Control, Man-Machine Systems, Microprocessors, Power Station Control, State Estimation.

## 1. INTRODUCTION

Hydro-electric power stations are an essential factor in power generation. They are readily used for covering base an peak loads. Although hydro-electric power stations can basically be considered as rather simple in comparison to thermal power plants, it should, however, be noted that, on account of their above-mentioned duties, ever increasing importance is attached to their availability and rapid operational readiness as well as to fast response to load variations. Moreover, hydro-electric power stations play a very important part in conjunction with the power system (frequency stability).

Today, microprocessors are reliable and safe enough to play a leading role in present-day control and monitoring systems. They are readily accepted by the customers owing to their possibilities of offering ease and great flexibility in their application.

We can start out from the fact that the use of microprocessors is nowadays considered as state of the art.

The present paper provides, to begin with, a short survey of the historical development from relay technology to programmable control and monitoring systems as described in the following chapter, pointing out its application in the field of hydro-electric power. A discussion and prospects for the so-called integrated control and monitoring systems of the future conclude the paper.

## 2. HISTORICAL DEVELOPMENT

An explosive development in the field of semiconductor technology has made it possible to provide standard control and monitoring systems for the complete range of power plants. This development also brought about an enormous improvement of the price-performance ratio, but at the same time caused worries about the formative processes involved with oncoming innovations.

Well-known manufacturers of control and monitoring systems have adopted a very cautious policy in introducing new technologies. The systems developed had to withstand comprehensive laboratory and field tests. It was only when sufficient operating experience had been gained that the systems were introduced in power plants. On one hand, this procedure permitted the manufacturers to be spared from serious setbacks and, on the other, to realize an adequate life span for one generation of control and monitoring systems. Figure 1 shows that, on an average, a technological innovation took place every six years in the past: the life-span of a generation being about ten years.

In retrospect, the following main statements can be made as regards to the introduction and possibilities of the individual generations:

### 2.1. Relay technology

For decades relay technology was the sole means of automating individual sub-processes of a power plant. Analog an binary control as well as monitoring functions always had to be performed by separate components on account of varying requirements.

### 2.2. Semiconductor technology (germanium,silicon)

With this technology, the electromechanical relays were replaced by logic functions such as AND/OR/memory and the like. It allowed to develop already more complex automation systems for analog and binary tasks. As in the case of the relay technology, the functions desired were accomplished by connecting logic functions in series or in parallel by means of wiring.

### 2.3 Integrated systems

On account of minimum space requirements, integrated systems have made it possible to combine repetitive standard functions (subroutines) into printed circuit board assemblies. User-specific requirements in addition to the normal standard functions were met by superimposing logic functions. This hybrid technology (fixed/wired program) already allowed the control and monitoring system structure to be matched very well with the given conditions of the power plant process. This is reflected in a decentralized, hierarchical structure of the control and monitoring system, which consists of unit, functional group and drive control levels. Moreover, it permitted free selection of the degree of plant automation, i.e. manual and automatic interventions became possible at all levels. Besides the structural improvements, dependability and reliability were considerably increased by introducing the active-1-technology and short-circuit-proof outputs.

### 2.4. Microprocessor technology

Introduction of the microprocessor technology for process automation was accompanied by a number of uncertainties. Memory requirements, processing times, complicated programming languages and much more had to be adapted to power plant-specific requirements in order to be accepted by the project engineers and ultimately by the power plant owner.

The first step into process control engineering was made with monitoring systems. For these subsystems, it was possible to make full use of the experience gained with telecontrol and supervisory netword control systems, thus leading to obvious advantages over conventional systems.

High-speed development of hardware as well as standardization of the programming languages soon led to centralized systems, designed to cover the complete range of requirements in the field of power plants. The last uncertainty was to know whether such complex systems comply with availability requirements. The most exacting was met by duplicating these complex centralized systems and using them as dual systems.

This design has not proved a success of long standing owing to centralization and the necessary computing time involved, to a high degree of software meshing, to difficult fault location and on account of high overall cost, the reasons why decentralized, transparent systems were called for.

Further development in the field of microprocessor technology and the resulting drop in prices made it possible to combine specific sub-racks into selfcontained, decentralized control and monitoring stations. In addition, bus systems with a high transmission speed were developed, which ensure serial data exchange between the control and monitoring stations.

With such a process control and monitoring system, configured on the basis of today's state of the art, the earlier mentioned disavantages of a centralized solution can be completely eliminated. Moreover, the use of bus systems contribute to drastically reduce the costs for plant cabling.

The configuration and technical data of such a modern, decentralized control and monitoring system with serial data transmission via bus systems is described in more detail in the following chapter.

## 3. A MODERN MICROPROCESSOR-BASED CONTROL AND MONITORING SYSTEM (1)

This chapter describes the BBC Brown Boveri control and Moniroring system PROCONTROL P, the youngest member of the PROCONTROL system family.

### 3.1. System overview

The BBC PROCONTROL P control and monitoring system for power plants performs all tasks required for process control, namely

- signal conversion, transmission and monitoring
- binary and analog control
- protection
- communication.

The PROCONTROL P system has a decentralized, hierarchical structure and is based on a digital, programmable control system. A bus-oriented data transmission and distribution system, consisting of a local and an intra-plant bus, is one of its main features. Figure 2 shows a typical configuration of this control and monitoring system (example in single-channel arrangement). The signal sources in the process are cabled conventionally to the nearest PROCONTROL P station (usually short cable links), where the signal is converted to digital form and brought onto the local bus which connects all modules belonging to the same station (input, output and processing modules).

The individual stations are interconnected via the intra-plant bus. Thus, each item of information generated in the system is available to all stations connected to the intra-plant bus. The intra-plant bus and the bus/station coupling modules can be selectively executed in redundant two-channel arrangement. The stations can be planned, erected and tested independently of each other and of the intra-plant bus line.

### 3.2 Module overview

To perform the tasks assigned to it, the PROCONTROL P system is equipped with modules for the following functions:

- Processing of measured values / Binary and analog control
- Data transmission
- Communication between man and the control and monitoring system

The modules used for the first functions are equipped with microprocessors.

Any one of the above-mentioned functions can be performed by plugging the corresponding module into the stations. Moreover, each station is equipped with basic modules for executing data traffic, for fault diagnosis and for power supply.

Figure 3 shows the typical configuration of a station with the different types of modules. This diagram clearly shows the modular configuration within a station, i.e. the selective assignment of different functions to particular modules.

All modules contain function-supervision facilities, and processing modules in particular are capable of self-diagnosis.

The basic modules (see fig. 3) are:

- power supply modules
- bus access modules and local/intra-plant bus coupler: It is possible to retain the redundant signal transmission structure with these two latter modules and a doubled intra-plant bus
- bus traffic director: Controls the data traffic on the local bus
- service case: Can be connected to the local bus via socket and allows to change the program of the processing modules and to simulate signals

Input modules (see fig. 3):

- Binary and analog input modules are available for the signal input: contacts, transducers, resistance temperature detectors and thermocouples can be connected

Output modules (see fig. 3):

For the output of signals, both binary and analog output modules are available. The output signales are of the form 0/24 V or 0/4-20 mA.

Processing modules (see fig. 3):

For data processing, modules are available whose specific functions are specified by programming.

### 3.3. Data transmission

The intra-plant bus which connects the different stations of the control system, consists of a coaxial cable. The serial signal telegrams on this bus can transmit one single byte of information. Figure 4 shows a typical data telegram of medium duration of 60 μsec.

Address and data have different voltage polarity. In the station designed by the telegram address, the intra-plant/local bus coupler takes the data from the intra-plant bus, converts it to a different bit sequence and makes it available on the station local bus. On the local bus, data is transmitted simultaneously with two polarities, as true data and inverted data. This feature represents the high data transmission reliability within the station. The traffic on the local bus is governed by the bus traffic director. The cycle time on the local bus can be selected to be 5 up to 20 msec.

### 3.4 Input, signal conversion, output

Analog and binary signal input and signal conversion takes place in the stations in close proximity to the process, once and for all modules involved in further processing. Stations with input modules are located near the process, in order to keep conventional cabling as short as possible and to profit as much as possible from the advantages of the bus transmission system.

The input modules perform the following tasks: Supply of power to transmitters and transducers, signal verification, digitalization and parallel/serial conversion of measured values, address allocation, potential separation, measurement range adaptation and linearization if required, filtering of analog signals and binary signal delay.

The ouput modules perform the following functions: Conversion of the serial digital values from the local bus into parallel digital analog signals for output, potential separation and protection against external influences when required.

The processing modules assumes among other things the tasks of flow correction, limit-value formation and other higher-order signal processing tasks.

### 3.5 Binary and analog control

If we speak of "analog control" in this section, we understand this to be a closed-loop feedback control circuit which is capable of setting the process control to a given target value and of keeping it in this state also after disturbances (e.g. PI, PID controllers). "Binary control" is understood to be the normal issuance of commands and the supervision of their execution.

Analysis of the power plant process indicates the advisability of dividing the complex overall process into individual sub-processes having clearly defined process functions. This provides insight into the interdependency of the sub-processes and their dependency upon associated individual subordinate functions. The division of the process into clearly defined groups results in a horizontally and vertically organized process structure. Whereas the horizontal structure is determined by the number of drives and actuators, in the vertical direction the process structure consists always of three basic hierarchical levels, these being the

- individual control level
- group control level
- process control level

The structure of the PROCONTROL P binary and analog control system is built-up exactly in accordance with this hierarchical structure.

The task of the individual control system is to control and monitor a particular drive (ON/OFF, position control etc.). This includes the realization of protection interlocks.

On the group control level usually the lowest level of automation is realized. The purpose of group control is to activate the individual controls according to a specified program, as a function of events or on a chronological and sequential basis so that the task of the entire group is performed. Another task of the group control is to control a group of drives, say a main drive with its auxiliary drives. In the analog control field, it is the task of the group control systems to form the reference variables for the individual analog control devices. However, in the PROCONTROL P system it is possible to integrate formation of reference value and related individual position or speed control in one module. (Figure 5 shows a typical PROCONTROL P tier).

At the process control level, the binary control represents the top automation level and controls the groups via group control systems, following a fixed program and according to a particular time schedule and sequence, so that the task of the overall process is performed. The process control level analog control forms the reference variables for the group analog controls. Its task is to adhere to the process variables given by the operator by hand setting or prescribed in a time schedule or received from a higher order control system which controls several power plants.

At all three control levels, programmable modules are used. Their inherent program developed from normal functional diagrams, is written in the program language PROCONTROL P 10 (DIN 40700).

A standard routine is assigned to each electrical symbol that may appear in a functional diagram. These routines are called by an instruction code, followed by one or more addresses. These addresses indicate the routine where to get the inputs and where to deposit the outputs. Figure 6 shows some examples of electrical symbols and corresponding instruction codes.

### 3.6. Man/machine communication

The PROCONTROL P system offers high-standard facilities to represent the entire process or parts of the process on colour screens, supporting the operator to a maximum extent with prepared information, in order to enable him to get an immediate overview over the process situation and to take the necessary decisions quickly. Also other output means are available (printer, board, registrator). The screen pictures can be modified any time by the operator to adapt them to his personal wishes and habits. The displays are organized such that the operator never looses the overview over the plant, even if he is presently occupied with a plant detail.

For trouble-free testing at works, assembly, commissioning at site and maintenance of the PROCONTROL P control and monitoring system a variety of equipment is available: Servicing, test and simulation units; programming units for the programming of processing modules; diagnosis station for localization and analysis of disturbances. The diagnosis station is the central working station for the dialog with the entire control and monitoring system. It performs the following functions:

Continuous: - Indication and logging of all alarm annunciations within the system in clear text.

On request:
- Indication and/or logging of any local bus signals
- Indication, remote adjustment and simultaneous logging of analog control parameters.
- Indication and/or logging of internal operational values of the processing modules
- Indication and/or logging of the software marshalling links in the local/intra-plant bus couplers
- Recording of any selectable binary or analog values
- Recording of software marshalling links on tape

Based on these functions, the diagnosis station is the suitable instrument with which analog control circuits can be optimized (even during operation) and to analyse disturbances. Modification of program instructions in the processing modules (i.e. changing the electrical functional diagram) as well as simulation of signals can be easily performed with the above-mentioned equipment.

## 4. APPLICATION OF PROGRAMMABLE CONTROL AND MONITORING SYSTEMS IN HYDRO-ELECTRIC POWER STATIONS

### 4.1. Hardware structure

A possible PROCONTROL P control structure for a large hydro-electric power plant is illustrated in figure 7. Consideration was given to overall availability in that a decentralized dual intra-plant bus system was chosen for the data traffic in the power plant. On account of short distances to the power plant process, the arrangement of the decentralized control and monitoring stations is geographically centralized for each power unit and its auxiliaries.

The control room is equipped with a central computer-based main control desk, capable of communicating with the process. As stand-by, provision can (optionally) be made for conventional, decentralized secondary control desks which allow the power plant to be managed in the event of a computer failure. Instead of these conventional operating facilities, a redundant, completely independent and computer-based control desk could be provided as stand-by control desk.

### 4.2. Main control and monitoring functions

The control and monitoring functions can be divided into two groups, i.e.:

- Process-specific function blocks
- Management-specific function blocks.

The process-specific function blocks cover, above all, the functions which are fundamentally important for operating the individual power generating unit, its switchgear and auxiliaries. On the whole, these are defined by the machine and switchgear builders and briefly surveyed in figure 8.

The management-specific function blocks include the functions which are of importance for the economy and optimization of one or several power plants. To this end, the task assignment between the higher-order power system control centre and power plant control centre are decisive in every single case. While in the past virtually all higher-order strategies have been computed in the network control centres, there is nowadays an obvious tendency towards a transfer of certain intelligence to the power plant.

This trend of delegating the responsibilities permits the following management-specific functions to be derived:

Joint control of active and reactive power

The primary task of the higher-order unit control equipment is to divide a set-point value for the total generated output equally between the jointly controlled units in the power plant. Such equipment offers the following advantages:

- The network control centre can regard a power plant with several units as a single power generator and dictate a set-point for the overall power plant without considering the individual loading of the machines.
- When the loads of several units feeding a busbar in parallel are shed, no problems arise regarding transients in the active and reactive power range, since the set-point is the same for each unit.
- Load surges are divided equally between the machines so that no single unit has to be disconnected due to overload.

Power plant commitment

The strategy for power generation in a power plant is normally planned by the power system control centre over a 24-hour period. This period can be divided into sections according to the set-points and the generator duty program determined accordingly. The benefits of such a function are as follows:

- Spinning reserves are available for individual or joint-control operation.
- Start-up and shut-down phenomena are reduced, resulting in less material wear.

Water level control for cascade power plants

In the case of cascade power plants, each load change has an immediate effect upon the water level of the individual cascade sections. In order to obtain a tolerable water level, these plants are therefore usually operated in the uneconomic, rigid base-load range. To economically utilize the available resources, a cascade control can be installed with a view to achieving the following objectives:

- Simplified plant management
- Levels maintained within tolerable limits
- Improved efficiency.

The above list of power application software functions is meant to demonstrate how the tasks are delegated by the network control centre. It could be completed with other functions such as network analyses, energy management computations, diagnostic functions and the like.

The presented PROCONTROL P process control and monitoring system has all the hardware components to accomplish the above-mentioned process. Management of specific tasks in a reliable and transparent manner according to a standard concept can also be accomplished. Fully developed and easily applicable standard software modules are available for repetitive standard functions.

## Outlook

This paper demonstrates that microprocessor controlled systems introduced into the market are at today's level in technique. Meanwhile more and more functions are being integrated. For this purpose, one has to combine different basic systems, such as:

- Procontrol P13/42 (as described in this paper)
- PHSC (Programmable High Speed Controller)
- BECOS 10 Plus (MMC and Remote Terminal Unit)

The appropriate combination of these subsystems leads to an integrated distributed computerized station control system for any power system application with serial data transmission and with the capability of different degrees of integration of functions on it. It is based on the well proved subsystems P13/42, PHSC, BECOS 10 Plus having been in service for many years.

With these 3 subsystems one covers standard control functions (P13) as well as fast or very fast functions (PHSC) such as protection and/or centralized functions (BECOS 10 plus). The data transfer is realized by local bus (P13) and by intra plant bus (P42).

Figure 8 shows the functionality of the integrated control system to which it can be applied

- Hydro Power Plants
- H.V. Substations (AIS and GIS)
- Static - Var. Systems and Converters

The practical experiences are positive and all the advantages of such a concept will enable it to establish a strong position in the market. The functionality, the availability, the comfort, the optimal separate HW and SW, the possibilities for extensions, modifications or adaptions of functions even during the commissioning phase, the significant reduction of cables due to the bus-system, etc. will guarantee its future and its success.

## Bibliography


(1) PROCONTROL P13/42. Digital, Decentralized Control and Monitoring System. Publication BC, Nr. CH-T 100 424 E

(2) PROCONTROL P - a Control and Monitoring System. BROWN BOVERI REVIEW 71, 1984.

(3) Control and Monitoring Systems for H.E.P.P. - an integrated part of Supervising Network Control. BROWN BOVERI REVIEW 9/10, 1984.


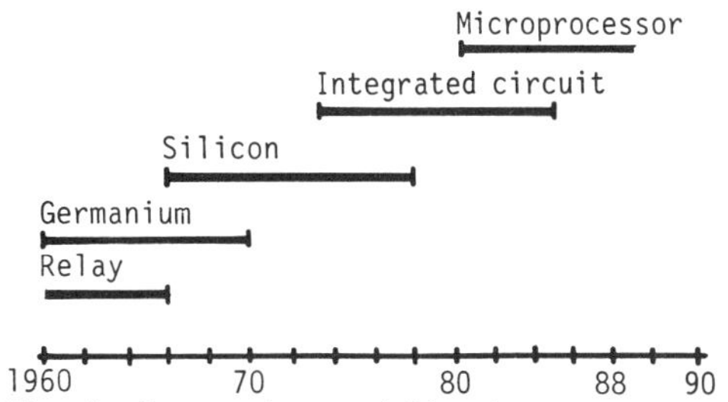

Fig. 1: Innovations and Time Span

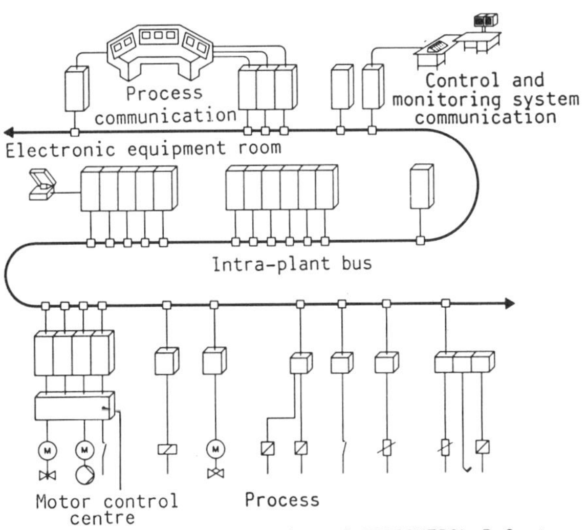

Fig. 2: Typical Configuration of PROCONTROL P System

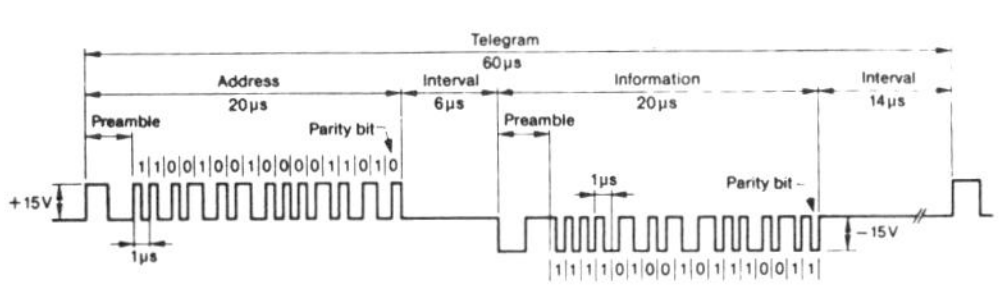

Fig. 4: Data Telegramm
LK=Clock, AD=Adress
DN=True Data, DI=Inverted Data

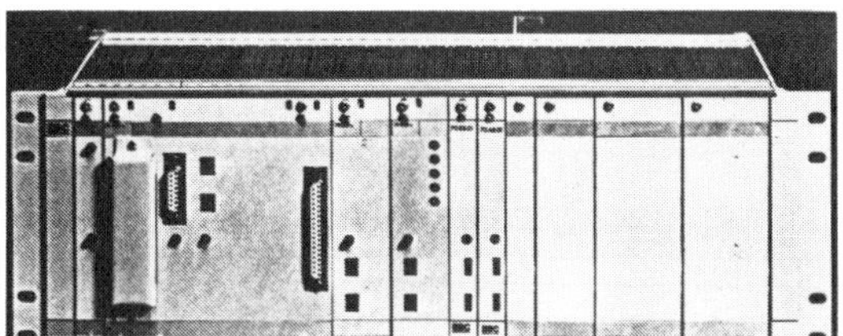
Fig. 5: PROCONTROL P tier

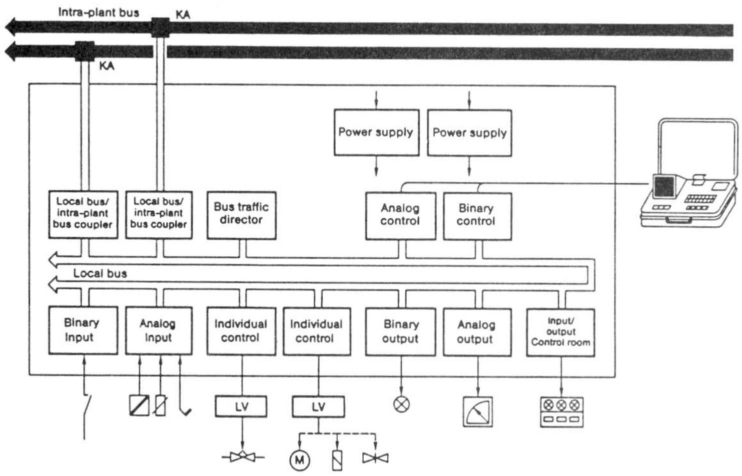


Fig. 3: Typical PROCONTROL P station
KA=Bus access coupler
LV=Motor control centre

| Symbol | Name | Instruction code |
|---|---|---|
| E01–X, E02–Y, X/Y, –A01 | Divider<br>$A01 = \frac{E01}{E02}$ | C0A3 (Call-in function)<br>401B (Allocation to input E01)<br>401C (Allocation to input E02)<br>A032 (Allocation to output A01) |
| E01, E02, ENN, &, –A01 | And<br>A01=E01·E02·....ENN | C0B1 (Call-in function)<br>23C3 (Allocation to input E01)<br>4912 (Allocation to input E02)<br>?<br>4F12 (Allocation to input ENN)<br>A333 (Allocation to output A01) |
| PD, E01–+, E02–KP, E03–KD, E04–T, E05–R, +–A01 | Differentiator<br>$F(s)=KP+KD\frac{T\cdot s}{1+T\cdot s}$ | C0C8 (Call-in function)<br>409A (Allocation to input E01)<br>D940 (Allocation to proportional action factor KP)<br>D480 (Allocation to derivative amplification KD)<br>DA30 (Allocation to derivative time constant T)<br>23C1 (Allocation to reset)<br>B09B (Allocation to output A01 |

Fig. 6: Electrical Symbols and PROCONTROL P10 Programming Language

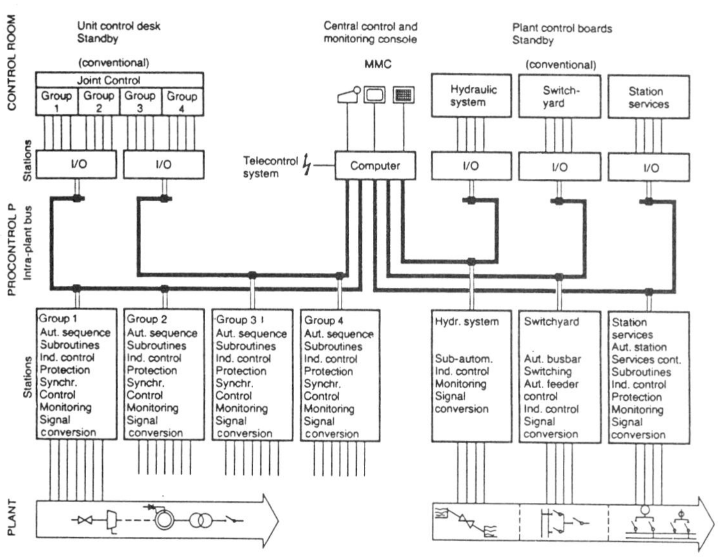


Fig. 7: Hardware Structure for a Large H.E.P.P.

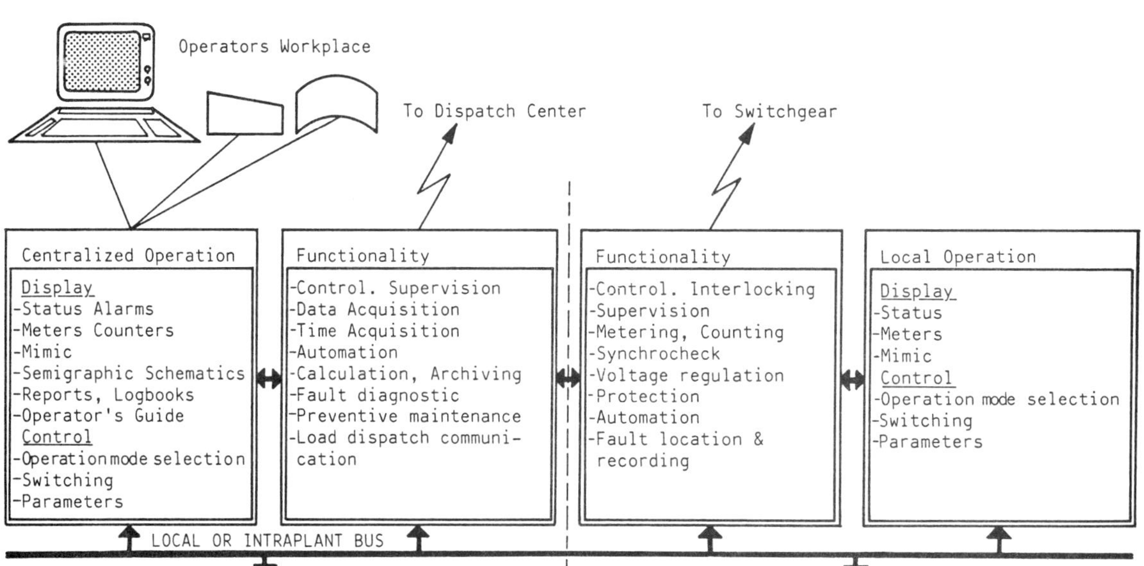


Fig. 8: ISC System Functionality and Operation

# MICROCOMPUTER AND COMPUTER APPLICATIONS IN THE HYDRO-ELECTRIC POWER PLANTS KARAKAYA AND ATATURK

†M. Cuénod*, H. Regli** and M. Schurtenberger***

*Electromechanical Equipment Atatürk, H.E.P.P. Elektrowatt Engineering Services Ltd., Ankara, Turkey
**Section Power Generation and Distribution, Elektrowatt Engineering Services Ltd., Zurich, Switzerland
***Department Sales and Engineering, H.E.P.P., BBC Brown, Boveri & Co., Zurich, Switzerland

1. Introduction
2. Process sequence control system
3. Process operation control system
4. Protection systems
5. Supervisory system
6. Automatic transfer/load shedding system
7. Conclusion
Bibliography

ABSTRACT

The paper discribes the control and monitoring systems of Turkey's two biggest hydro-electric power plants, Karakaya and Atatürk. In these systems micro-computers and computers are used for selected particular functions and tasks, distributed in the different parts of the plants.

KEYWORDS

Alarm systems, Analog computer control, Control equipment, Controllers, Digital control, Frequency control, Load regulation, Microprocessors, Motor control, On-line operation, Power control, Stability, Voltage control.

## 1. INTRODUCTION

The Karakaya and Atatürk hydro-electric power plants are among the most important hydro-electric schemes in the world.

According to the survey of the major dams of the world prepared in July 1985 [1], the Atatürk dam will be the 6th world largest volume dam, and its output will be the 15th largest of the world hydro-electric plants. With the exception of the Soviet Union, it is more important than any other hydro-electric plant in Europe.

With a total capacity of 4.200 MW, the output of these two plants is about half of the actual peak load of the turkish power system, which points out the importance of these plants for Turkey.

Figure 1 shows the location of these two plants in the eastern part of the TEK grid (Electric Authority of Turkey). Figure 2 shows the cascaded implementation of these two plants on the Euphrates River.

Figures 3 and 4 give the general layout of these two plants. Due to the difference of their topographical environment, their layout is very different: the Karakaya dam is a concrete gravity arch located in a gorge of the Euphrates River, with an overflow spillway and a power house integrated in the dam [3,4].

The Atatürk dam is a rock filled type with a central clay core, with the 8 penstocks and the power plant on the right bank and the spillway and diversion tunnel on the left bank of the river [2,5].

Table 1 gives a survey of the main technical data of these two plants.

TABLE 1 Main technical data

| | Karakaya | Atatürk |
|---|---|---|
| Storage lake length | 166 km | 180 km |
| Catchment area | 16450 $km^2$ | 11800 $km^2$ |
| Total storage | 9586 $Mm^3$ | 48700 $Mm^3$ |
| Useful storage | 5580 $Mm^3$ | 19300 $Mm^3$ |
| Net head | 142 m | 151 m |
| Rated discharge | 1182 $m^3/s$ | 1746 $m^3/s$ |
| Installed capacity | 1800 MW | 2400 MW |
| Number of units | 6 | 8 |

Figure 5 gives the general configuration of the control and monitoring system of the Karakaya and Atatürk plants, which is a hierarchical system with the 4 following levels of control:

For the units:
- direct manual control of certain auxiliaires
- local manual and automatic control from the unit control room, located in the power house near to the different units
- centralized control from the main control room
- remote control from the load dispatching center of Ankara (future)

For the switchyard equipments:
- direct control at the h.v. apparatus
- remote control from the local control/marshalling buildings in the bays of the switchyard
- remote control from the switchyard control room
- remote control from the main control room near the power house
- remote control from the load dispatching center of Ankara (future)

We shall now consider the different tasks of the automatic control, with the relevant control systems, namely:
- the process sequence control system
- the process operation control system
- the protection system
- the supervisory system

The systems described in this paper are - if not mentioned differently - used for both plants.

## 2. PROCESS SEQUENCE CONTROL SYSTEM

This system involves the automatic connection of all the operations required for the start-up and shutdown of the main units with their auxiliaries, in Atatürk plant also for the auxiliary hydraulic unit.

This automatic function is performed in Karakaya by the BBC solid state control system SECONTIC K, in Atatürk plant by the BBC microprocessor based control system PROCONTROL P.

Figure 6 shows the block diagram of this process sequence controller. This hierarchical control has the 3 following levels of control:
- Unit sequence control for the overall control of the complete unit.
- Function group sequence control related to the different sections of the control actions.
- Drive control for the control of the different unit auxiliaries.

The control is subject to different criteria, such as starting criteria, stepping criteria, monitoring criteria, overrun criteria, and the commands are amplified and transmitted to the different actuators of the process.

A feedback of the process action is obtained through data acquisition and data processing. This feedback is used to check, at the different levels, if the control actions have been correctly performed.

The obtained analog and digital data are also transmitted to the control room to inform the operator with the help of indicators and recorders in the alarm and mimic boards.

The data are then also picked up by the BECOS 10 supervisory system, to be displayed on printers or video-displays (see section 5).

The BBC SECONTIC K sequence control system used at Karakaya plant is composed of the BBC electronic control system DECONTIC K (Decentralized Control Automatic) and is a hard wired system with diode plug-in matrices for easy programming [6,7].

The BBC PROCONTROL P sequence automatic of Atatürk plant is a microprocessor based, fully programmed control system, which is described in detail in [19].

The common automatic synchronizing equipment located in the switchyard control room will close the 380 kV circuit breakers of each unit in the switchyard, with a remote adjustment of the speed and voltage. The selection of the unit to be synchronized is made with a synchronizing switch either in the main control room or in the control room of the switchyard [8].

## 3. PROCESS OPERATION CONTROL SYSTEM

This system involves all the equipment necessary to keep the operation of the power plant in a given state and between given limits of the controlled variables, namely:
- frequency,
- voltage,
- active and reactive power.

This equipment is made of:
- speed/active power control system,
- voltage/reactive power control system,
- automatic transfer/load shedding system of the station services (see section 6),
- plant joint active and reactive power control system.

Figure 7 shows the functional diagram of the speed control. The speed governor is a solid state PID controller which issues the speed control action, taking into account the measurement of the speed n and the active power P, and the adjusted values of the different control parameters [9].

Figure 8 gives the functional diagram of the voltage control : A static excitation system is used with thyristor rectifier supplied from the generator terminals [10].

The voltage controller is a BBC UNITROL solid state PID controller acting on the converter with the requested excitation build-up, desexcitation facilities and adjustable over- and under-excitation limiters, according to the following criteria:
- stator current too high,
- rotor current too high,
- rotor angle too high.

To improve the stability of the electricity transmission to the western part of Turkey also during oscillations of the active power, a slip stabilization is foreseen, taking into account the variation of the active power produced by the generator [11 to 14].

Due to the high value of installed power and the availability of stored energy, the plants of Karakaya and Atatürk will participate in the secondary frequency control of the turkish power system. For that purpose, the joint controller can be used either as total active power controller or else as secondary frequency controller, with the following actions:
- to control that the total power will be produced by the plant according to the following control characteristic E :

$$E = \Delta P + K\Delta f$$

- to equalize the output of the units participating in the control (load sharing).

Figure 9 gives the functional diagram of this joint control:

- the sum of the output $P = P_{u1} + P_{u2} + P_{u3} + \ldots$ and the frequency are compared with their target values $P_o$ and $f_o$
- the control error $\Delta f = f - f_o$ is weighted with the system bias $K$ and added to the control error $\Delta P = P - P_o$
- the control error $E = \Delta P + K\Delta f$ is transmitted as input to the PI controller.
- the output signal $Y_M$ is characterized by the following equation:

$$Y_M = K_p * E + 1/T_N * \int E\,dt$$

with $K_p$ = proportional constant
$T_N$ = integral action time

The joint controller output $Y_M$ is transmitted to all the turbine controllers as a reference value, so that all machines can be operated with identical power and speed droop. This controller output is also stored in a digital memory.

A similar joint controller is also foreseen for the reactive power production.

## 4. PROTECTION SYSTEMS

There are two main groups of protection functions provided at Karakaya and Atatürk power stations:
- The "classical" protection with electromechanical and electronic protection relays [15], oscilloperturbographs [16],
- The process protection provided by the control and monitoring system which permanently supervises the criteria allowing the process to run and immediately provokes safety switchings in case one or more criteria should no more be fulfilled.

At Atatürk power plant, the latter function will be partly executed in the microprocessor based BBC PROCONTROL P system, which is described in the paper [19].

## 5. SUPERVISORY SYSTEM

Figure 5 shows the principle of the supervisory system BBC BECOS 10 which provides the following main facilities [17,18]:

1) Mimic and alarm boards:
All switching apparatus of the h.v. switchyard are represented in the mimic board by a symbol providing two LEDs for the ON and OFF positions respectively. These LEDs are permanently updated on the basis of the indications provided by the supervisory system. Feeder identification and alarm signals are indicated by lamp windows flashing until acknowledgement. Measured values are displayed on analog and digital instruments.

2) Logging:
Tables providing information about the state of the plant are given on alphanumeric displays of video terminals. The state of the individual items are written out and updated in the "object status tables". All persisting alarms are listed in the "alarm tables" and displayed on a separate alarm board.

The logs are used for documentation of the operation sequence, for error analysis, for chronological monitoring of the behaviour of the process variables and the operations performed, as well as for recording the current system status on data loggers. These functions correspond to the following types of logs:
- "event logs", containing all the operators actions, status changes, limit overruns in the order they have taken place,
- "alarm logs", containing all alarm conditions,
- "measurand logs", containing a group of instantaneous measurands and calculated values.

The BBC BECOS 10 software package running on the supervisory computers is working in real time, with the operating system RUMOR to drive the hardware equipment.

The supervisory system is divided into the master station and different remote stations:
- the unit supervisors,
- the station services supervisor,
- the switchyard supervisor.

For safety reasons, two computers are provided. One of them is in on-line mode, executing the process tasks. The other one is in hot stand-by mode and also available for program development. The computers can be interchanged.

The following peripherals are connected to the supervisory system:
- MOS memory
- system typewriter
- floppy disc (as loading device)
- disc storage
- data loggers
- video terminals

The BECOS 10 also has the function of data exchange between the power plant and the switchyard (3-4 km). For this important task, a separate back-up transmission system is provided.

3) Hydraulic equipment control board:
Due to the importance of the hydro-mechanical measurements for the operation of the plants, a special hydraulic control board gives a survey:
- of the different heads of the reservoirs and of the tail race,
- of the discharges through units, over spillway and through bottom outlet.
- the spillway and bottom outlet gates openings.

Some of these values cannot be measured directly, but are computed from other physical values by a separate computer with specialized software for hydraulic calculations.(Fig. 10)

## 6. AUTOMATIC TRANSFER/LOAD SHEDDING SYSTEM

Each station of the 6,3/0,4 kV station services network is of redundant design. In case of outage of power supply from the 154 kV network or outage of one or more components (transformers, cables, busbars), this system automatically performs the necessary switchings in order to restore the supply to the network downstream of the faulty component(s). It also decides about the start-up of the emergency diesel units and outputs the necessary orders. In case of diesel operation, it automatically drops all consumers of lowest priority in order to avoid overloading of the diesels. Upon return of lost supplies, the AT/ALSS automatically returns the station services to predefined "normal" network configuration.

The AT/ALSS is built-up of a fully programmed control system (BBC PROCONTROL P, see [19]), which allows immediate change of the stored program. Thus, the "normal" and "faulty" network configurations can be changed at anytime by the operator in special situations like later system expansion or component outage of long duration. Also other parameters of the AT/ALSS can be changed by programming.

## 7. CONCLUSION

The Karakaya and Atatürk schemes can be considered as an example for the application of several computers and microcomputers distributed in the different parts of these plants, and for different individual control functions. The purpose of this control and monitoring system is not so much to replace the operators, but to relieve them from pure routine works. It is designed to assist them, to make them more available to exercise their judgement and initiative, to facilitate the man/machine communication and to increase the reliability of the plant operation.

In case of emergency, the origin of any perturbation can be identified at once, so that the normal operation can be restored very quickly. Thus, the global availability of the plant will be increased by a reduction of its outage periods and of the time required for diagnosis of the failure.

Back-up systems are foreseen to increase the reliability of the plant, protection systems are installed in order to prevent serious damages and through this also improve the availability of these power plants.

The plants of Karakaya and Atatürk will - through the use of modern technology - contribute significantly to an increase in the security of operation of the Turkish power system.

## LIST OF FIGURES


Fig. 1 Development of the 420 kV Turkish Grid (planning stage in 1990)

Fig. 2 Development of Lower Firat Longitudinal Section

Fig. 3 General Layout of the Karakaya Plant

Fig. 4 General Layout of the Atatürk Plant

Fig. 5 General Configuration of the Control and Monitoring System of the Karakaya and Atatürk Plants

Fig. 6 Block Diagram of the Process Sequence Controller

Fig. 7 Functional Diagram of the Speed Control

Fig. 8 Functional Diagram of the Voltage Control

Fig. 9 Functional Diagram of the Joint Control

Fig. 10 Hydraulic Board


## BIBLIOGRAPHY


1. Major Dams of the World 1985. Water Power and Dam Constructions July 1985, pp.61 to 67

2. Atatürk Dam and H.E.P.P. Modern Power Systems Magazine, London, March 1985

3. W.Howald, J.Fornasieri, W.Pantli. Karakaya, a Hydro-electric Power Plant in Turkey. Brown Boveri Review, February 1980

4. M.Quénod, R.Markulin. Tendances actuelles dans le développement des systèmes de conduite de centrales hydro-électriques. Bul. SEV/VSE 17, 3 sept. 1983, pp.1019-1023

5. Wasserkraftanlage Atatürk am Euphrat. Schweizer Baublatt, 2 October 1984

6. Procontrol-Decontic k Modular Binary Control System for Power Stations. Publication BBC, CH-T 100204, E/F/D

7. Procontrol P. Publication BBC, CH-T 30-01, E/F/D

8. Electronic Equipment for Synchronizing and Paralleling Three-phase Generators. Publication BBC, CH-E 3.0077.0, E/F/D

9. M.Cuénod, E.Kocaoglou. Optimal Setting of the Speed Control Parameters of a Hydro-electric Plant. Bulletin ASSPA, 1982-2

10. Static excitation equipment for hydro-generators. Publication BBC, CH-E 2.30422.0 E/F/D

11. F.Peneder, R.Bertschi. Slip Stabilization. Brown Boveri Review 1974, pp.448-454

12. F.Peneder, R.Bertschi. Optimizing the Setting of Slip Stabilization Equipment by Digital Computers and Special Computing Procedures. Brown Boveri Review 1978, pp.590-597

13. F.Peneder, R.Bertschi. Slip Stabilization Equipment, Transfer Function and the Relevant Experience. Brown Boveri Review 1978, pp.724-730

14. M.Cuénod, E.Kocaoglou. Optimal Setting of the Slip Stabilization Parameters of a Hydro-electric Plant. Bulletin ASSPA 1985-2, pp.18 to 27

15. Modular Protection System "Modures" Publication BBC, CH-E6 30-01/E/F/D

16. SOREL Système oscilloperturbographique. Publication Thomson-CSF, France

17. Automation and Control System BECOS 10. Publication BBC, ED 80651/E/D

18. Data System INDACTIC 61. Publication BBC, HEDT, 80580/E/D

19. B.Bachmann, H.Cvetko, M.Schurtenberger. Micro-computer Application in Large Hydro-electric Power Plants. Paper IFAC 86, Istanbul

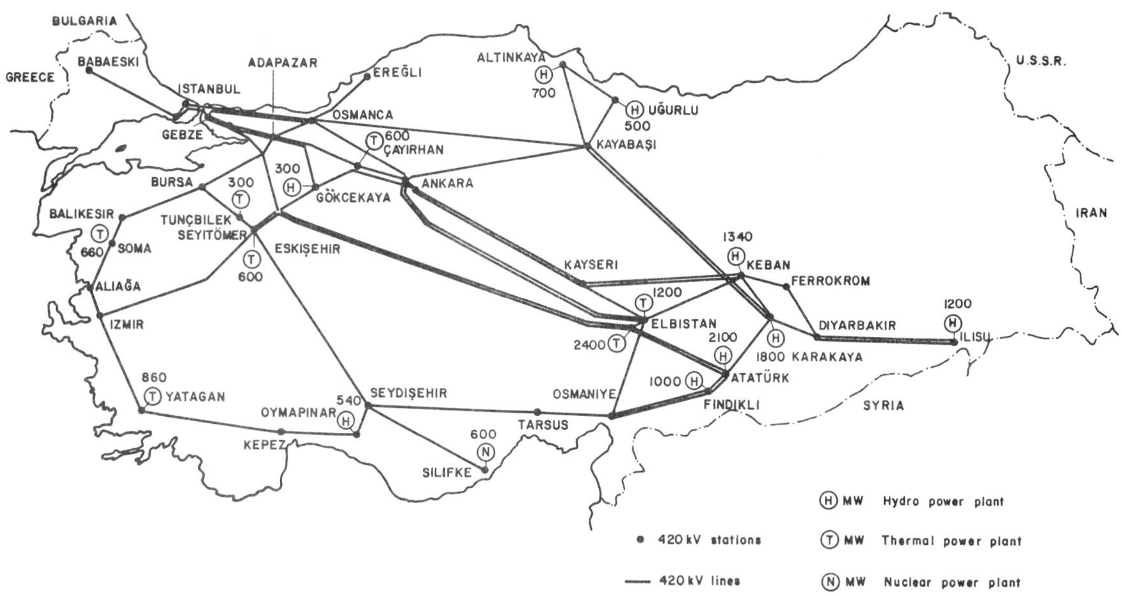


Fig. 1 Development of the 420 kV Turkish Grid (planning stage in 1990)

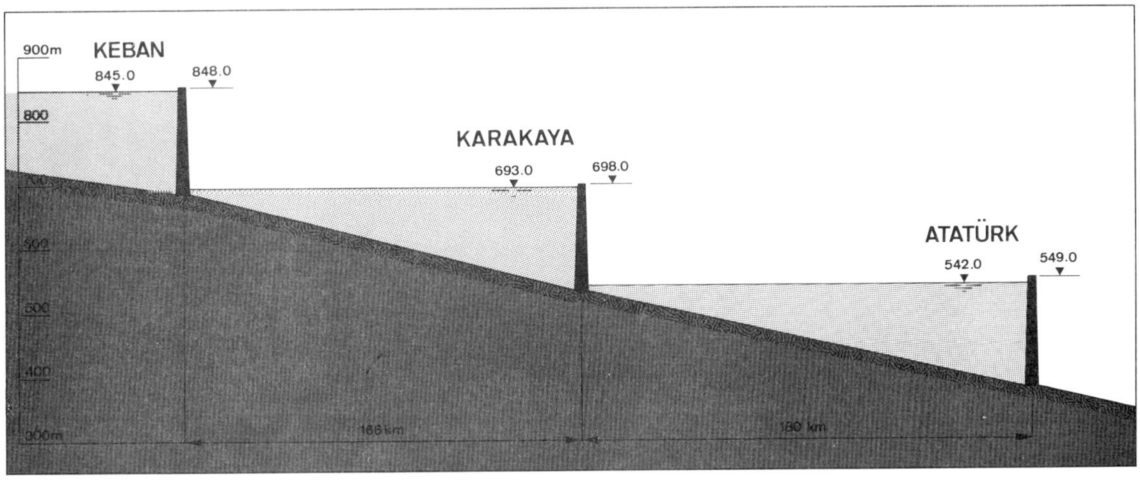


Fig. 2 Development of Lower Firat Longitudinal Section

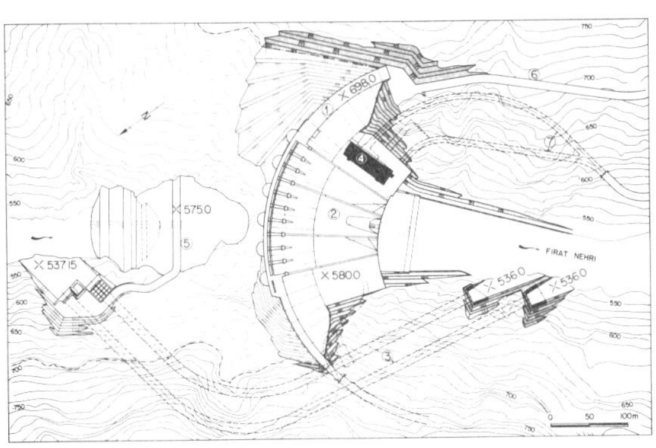

1. Dam
2. Spillway and Powerhouse
3. Diversion and bottom outlet
4. Control building
5. Upstream cofferdam
6. Access road
7. Access galleries

Fig. 3 General Layout of the Karakaya Plant

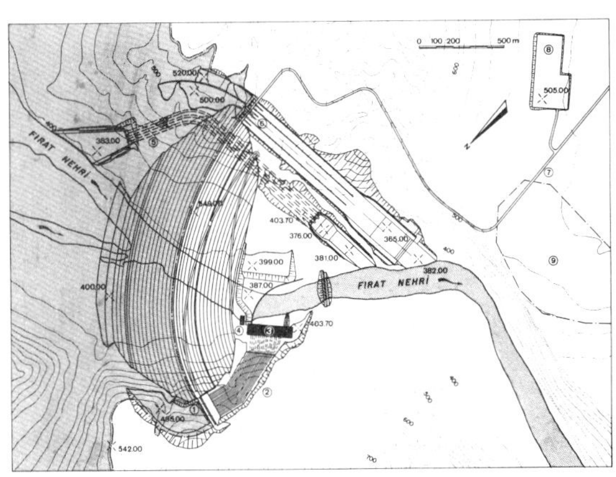


1. Intake structure
2. Penstocks
3. Powerhouse
4. Control building
5. Diversion tunnels
6. Spillway
7. Access road
8. Switchyard
9. Permanent village

Fig. 4 General Layout of the Atatürk Plant

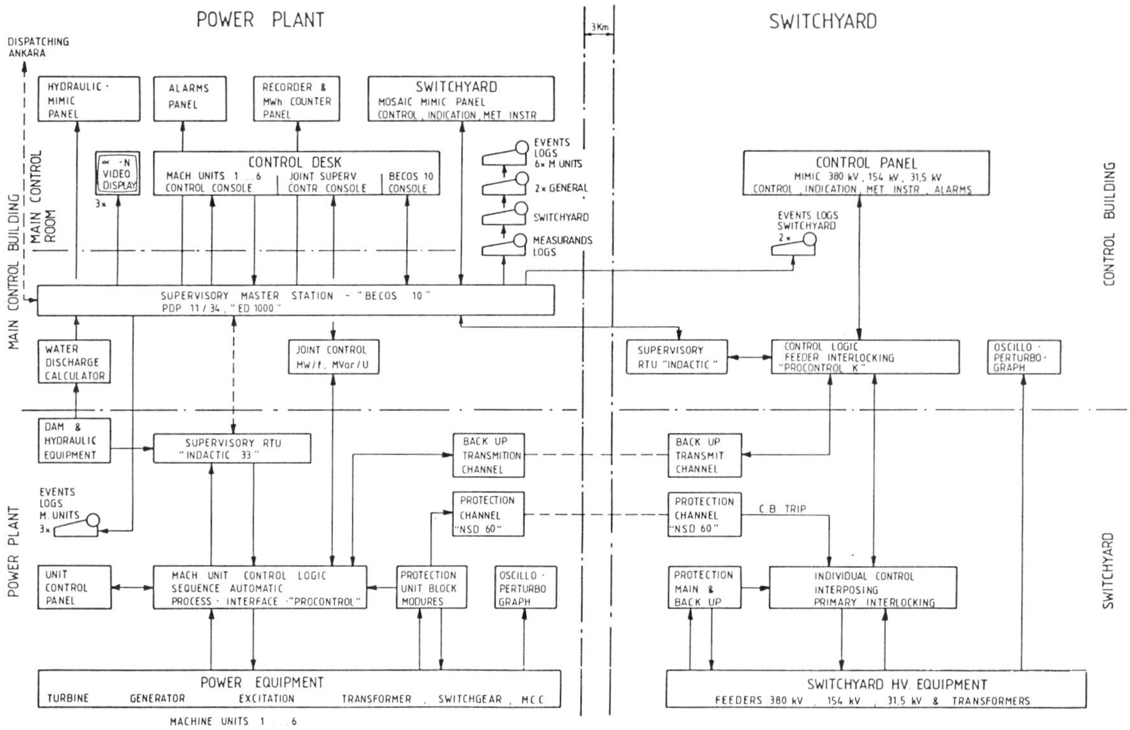


Fig. 5 General Configuration of the Control and Monitoring System of the Karakaya and Atatürk Plants

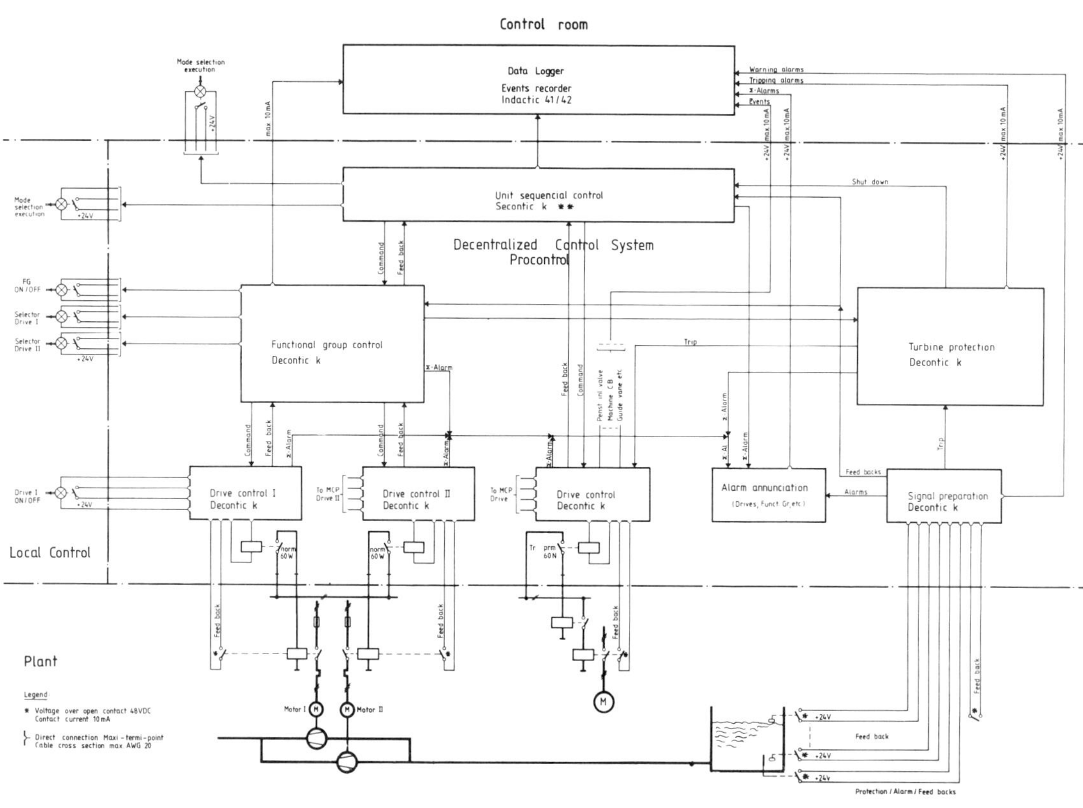


** Karakaya: SECONTIC K, Atatürk: PROCONTROL P

Fig. 6 Block Diagram of the Process Sequence Controller

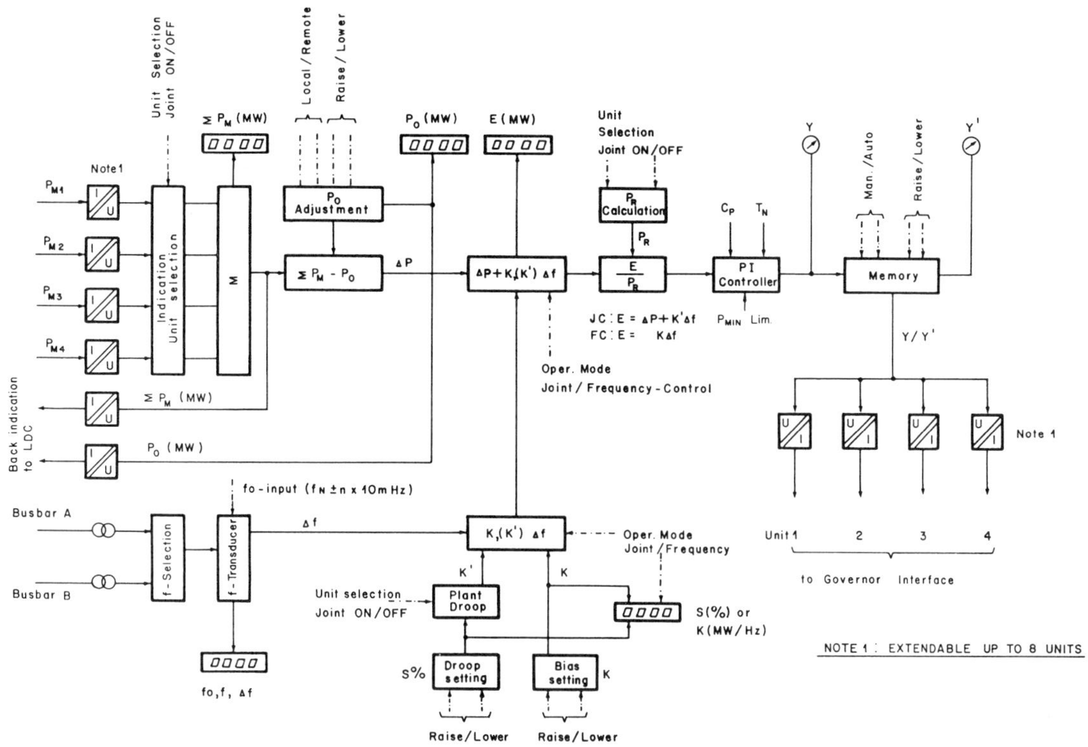


Fig. 9 Functional Diagram of the Joint Control

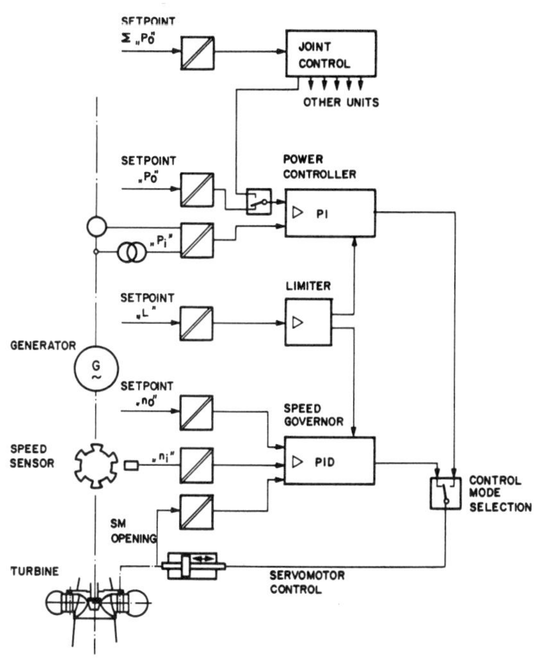


Fig. 7 Functional Diagram of the Speed Control

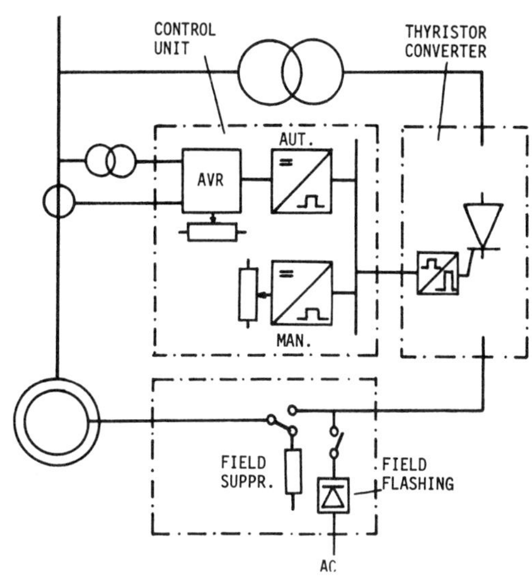


Fig. 8 Functional Diagram of the Voltage Control

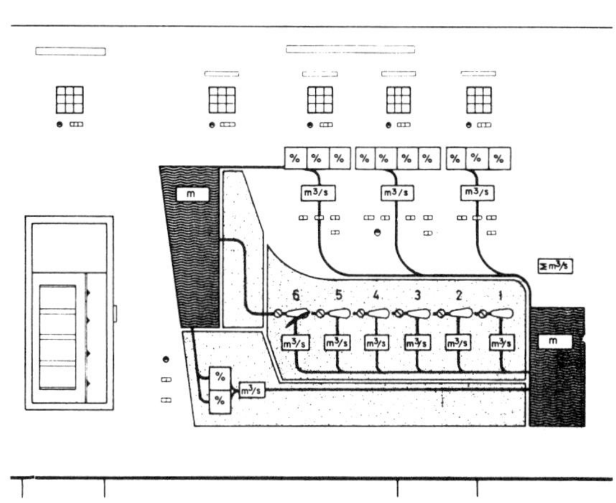

Fig. 10 Hydraulic Board

# THE APPLICATION OF INTEL 8080 MICROCOMPUTER SYSTEM TO MEASUREMENT AND CONTROL OF ENERGY AND POWER OF STEAM

R. Zarzycki

*Research and Development Center of Industrial Automation, Wiatraczna St. 15,04, 364 Warsaw, Poland*

Abstract. The new intelligent measuring system, based on INTEL 8080 microprocessor, is announced. The system acts as a flow computer for saturated or superheated steam in industrial plants. Main functions and features of the system are described briefly. Its application to steam production supervision, steam cost accounting and boiler control are discussed.

Keywords. Computer control; data acquisition; heat systems; steam plants; boilers.

## INTRODUCTION

Superheated or saturated steam is one of the main energy carrier in industry /at least approx. 2,68 MJ of energy in kilogram /. Apart of driving turbines it is used broadly as a heating member in food and chemical industries. As a result of the recent energy crisis a great attention is paid to steam savings and to an accurate steam cost metering. Wells /1983/ gave a good example of big savings achieved by steam boiler optimization, header pressure control and, mainly, by boiler management.

In order to measure steam mass flow rate by means of an orifice plate one should calculate square root of the orifice difference pressure and should compensate for steam specific volume changes resulting from steam pressure and temperature changes. In order to measure steam energy flow rate /thermal power/ one should know the value of steam enthalpy and compensate for its changes too. These measuring problems were discussed by Crowell /1982/ and Gervase-Williams /1984/but their solutions were not fully introduced yet into the present market of flow computers.

Microprocessor chips allow to build intelligent measuring systems capable to perform complex functions and to obtain a high accuracy of measurements. An example of such system is described in this paper.

## μMEC-80 STEAM FLOW COMPUTER

The μMEC-80 Steam Thermal Power and Energy Flow Computer was developed recently in the Research and Development Center of Industrial Automation in Warsaw. Its basic configuration and all the connections were presented on Fig. 1. The Flow Computer is equipped with a precalibrated segment of steam conduit with a built - in orifice and a thermometr as well as attached to it measuring transmitters of steam pressure, orifice difference pressure and steam temperature. For steam flow measurements over a wide turndown ratio two difference pressure transmitters of different ranges may be installed.

Three standard current signals from the above mentioned transmitters are delivered to the Flow Computer, which may be shortly described as a specialized microcomputer based on INTEL 8080 microprocessor and equipped with 6 kbyte EPROM memory, 1 kbyte RAM-CMOS memory and 7-level priority interrupt system. The microcomputer's architecture is a standard one and requires no comments. Then in the following only these features of the Flow Computer are described which are important to the user or interesting from a technical point of view.

Input analog signals are switched through CMOS multiplexer to a dual-slope integrating analog-digital converter. The converter contains an opto-isolation circuits; all other inputs and outputs of the Flow Computer are equipped with the same isolation. Digitized input readouts are automatically checked and adjusted, throuqh software, against two internal standards: zero readout and full-scale readout, thus eliminating zero and sensitivity drift errors. The following input data processing consists of "live zero" correction, digital low-pass filtering and converting into engineering units.

All the data in the Flow Computer are processed by means of a floating-point atithmetic package /1 byte for an exponent, 2 bytes for a fraction/. Firstly a steam mass flow rate is computed according to the equation

$$\dot{M} = C\alpha\varepsilon D^2 \sqrt{\frac{DP}{V\,/P,T/}} \qquad /1/$$

where:
- C - basic flow constant
- $\alpha$ - Reynolds' Number factor
- $\varepsilon$ - expansion factor
- D - orifice diameter
- DP - orifice difference pressure
- V - steam specific volume
- P - steam pressure
- T - steam temperature

In order to compute Eq. /1/ the value of steam specific volume V should be calculated. It is done by interpolating steam tables data which are kept in EPROM memory of the Flow Computer. An algorithm of the interpolation here is as follows:

a. analyse the steam pressure readout and find the nearest pressure data in the tables,
b. repeat it for the steam temperature readout,
c. having found four points of the tables find which three of them are the closest to the point described by the actual steam pressure and steam temperature,
d. put an interpolating plane on these three closest points /No 1,2,3/ and solve the planes' equation

$$\begin{vmatrix} P - P_1 & T - T_1 & V - V_1 \\ P_2 - P_1 & T_2 - T_1 & V_2 - V_1 \\ P_3 - P_2 & T_3 - T_2 & V_3 - V_2 \end{vmatrix} = 0 \qquad /2/$$

in order to get steam specific volume.

Some illustration of the interpolation method may be seen on Fig. 2.a.

Secondly a steam heat flow rate is computed according to the expression

$$\dot{E} = \dot{M}\ H\,/P,T/ \qquad /3/$$

where H is steam enthalpy. The same interpolation method, as the one used for steam specific volume, is used now to get the value of steam enthalpy from the second part of steam tables. It should be noted that when using interpolation of tables instead of using equation of state the Flow Computer may be easily modified to other applications, e.g. gas flow measurements. The EPROM memory volume for keeping the steam tables was reduced by arranging the common value of exponent for groups of data.

The steam mass flow rate according to Eq. /1/ undergoes a rectangular integration

$$M = \sum_{\tau=0}^{\tau} \dot{M}\ \Delta\tau \qquad /4/$$

where $\tau$ is time, $\Delta\tau$ is an integration step /here equal to 10s/ giving the total mass of steam. The same computation is applied to the steam heat flow rate giving the total thermal energy, which has been carried by steam in period of time $\tau$, according to the expression

$$E = \sum_{\tau=0}^{\tau} \dot{E}\, \Delta\tau \qquad /5/$$

The computed values of total mass and total energy, as well as a content of a hours counter CP, are kept in RAM-CMOS memory with a battery back-up.

All the values computed according to Eq. /1/, /3/, /4/ and /5/ as well as steam pressure, temperature and enthalpy are displayed digitally in the Flow Computer and are printed on demand, or in specified periods of time, by means of a local printer. Steam pressure, temperature and flow rate /mass or heat/ are checked every 1s against digitally presetted lower and upper limits and suitable alarm signals are generated if necessary.

A special attention was given to alarming an operator if a possibility of wetness of steam exists. As it may be seen on Fig. 2.b an actual steam temperature is compared to the saturation temperature, corresponding to the actual steam pressure. This original feature allows a boiler operator to prevent steam from wetness and to keep fuel consumption as low as possible at the same time.

The $\mu$MEC-80 Flow Computer provides for one standard output current signal which may be used in a control system as it was discussed in the next chapter. This signal may be generated as proportional to the steam mass flow rate or heat flow rate.

The Flow Computer is also equipped with V24 /CCITT vol. VIII/ serial interface, so all the measured and calculated information may be transmitted to a remote printer, to a supervisory computer or to a local area network controller-interface.

Important data and the microcomputer action are protected against powerfails by means of a sophisticated power-down and restart circuitry. Periodical self-checking of the microcomputer is included in the Flow Computer software.

The software system structure was presented on Fig. 3. The main program concerning flow calculations according to Eq. /1/ and /3/ is starting at POWER-ON interrupt signal and is performing endlessly. Its interpolation subroutine computes equation /2/ for the steam specific volume and enthalpy. Several interrupt subroutines perform other services, total mass computation/Eq. /4/ and total energy computation /Eq. /5// included. This software is acting without any operating system and was built and loaded in EPROMs modularly. It means that a next module of software will be entered and performed when the previously executed module confirms the presence of this EPROM chip in its socket. This technique allows greater flexibility.

The $\mu$MEC-80 Flow Computer was imbedded in a heavy water - tight and dust - proof metal case and may be installed in severe heat plant enviroment. It requires not any of complicated manipulations. Its power consumption was minimized in order to achieve high reliability.

## APPLICATION

The most obvious application of the $\mu$MEC-80 Flow Computer was already presented on Fig. 1. The instrument installed at an output of a steam boiler serves for supervision of steam production in this boiler. A boiler operator can minimize fuel consumption maintaining steam temperature at the prescribed excess in relation to the saturation temperature. The first such implementation of the $\mu$MEC-80 Flow Computer was done in a boiler house of Building Elements Factory in Warsaw in 1986. Similar supervision of steam production of the whole steam plant may be carried by means of a multipoint version of the Flow Computer. Here up to four steam flow rates may be measured in different input or output sections of a main steam collector conduit. As it is well known dividing of the whole steam load between steam boilers should be done economically, taking into account individual steam production efficiency and steam production cost of every boiler. Therefore this load allocation should be based on very accurate measurements of steam flow rates. Moreover, multipoint steam flow measurements in a boiler house allow real-time computing of steam flow balance. Such a balance is an ideal tool for detecting possible leaks and losses of steam, detecting too high steam consumption for the boiler house itself and, generally, identifying bigger inefficiences.

The most numerous applications of the μMEC-80 Flow Computer are expected in the area of steam cost monitoring and accounting. The high cost of every energy source forces the users to measure the steam heat flow rate and the total heat delivered. It is worthwile to do these measurements as accurate as possible than to pay for unknown energy delivery.

Among other possible applications of the μMEC-80 Flow Computer are real-time calculations:

- of an efficiency of a steam heat exchanger /the two-point version is needed here/,
- of a global thermal efficiency of a furnace-boiler set /the modified extended version is needed/.

Last but not least applications of the Flow Computer in boiler control systems will be discussed now. The Flow Computer is provided with one output standard current signal and therefore may work as an additional improvement in such complex and sophisticated boiler control systems as the one reported by Someyama and others /1981/ for boiler start-up and run. The main disturbance in a steam boiler steady-state work is caused by a rapid changes in steam demand. Real-time accurate measurement of a true steam mass flow rate or a steam heat flow rate is a base for any predictive /feedforward/ control system. Moreover, in order to optimize the work of a furnace-boiler set one should take into account, apart of the steam heat flow rate, a fuel thermal value , a condensed water flow and an oxygen con tent in fumes. All the above mentioned values may be precessed by an intelligent unit in order to calculate energy balance and to control the fuel consumption in an optimal way. Those additional algorithms may be embedded into the multipoint version of the Flow Computer. The suitable control system was proposed on Fig. 4.

## CONCLUSIONS

The instrument described in the paper may be treated as the example of a small, autonomous, dedicated and ruggedized industrial station. Intelligent steam flow measurements allow greater accuracy of measurements, better steam production control and more economic steam distribution management. There is a high probability prediction that in the near future many steam boilers will be equipped with steam flow computers which will work in a distributed control and management systems.

## ACKNOWLEDGMENTS

The author thanks his collegues from the research centers OBRAP and BISTYP in Warsaw for the valuable discussions.

## REFERENCES

Wells, R.D. /1983/. Computer cuts steam cost by $ 1.3 million. Istrumentation Technology., 30, 43-46.

Crowell, W.H /1982/. Analog and hybrid device measures precise heat content of steam. Control Enginering. 29, 118-121.

Gervase-Williams, K. /1984/. Steam meters put in perspective. Control and Instrumentation., 16, 37-39.

Someyama, T., I. Kano and T. Kawazae. /1981/. An automatic boiler control system employing microprocessors. Proceedings of the 8 th IFAC WORLD Congress., Kyoto, 3053 - 3058

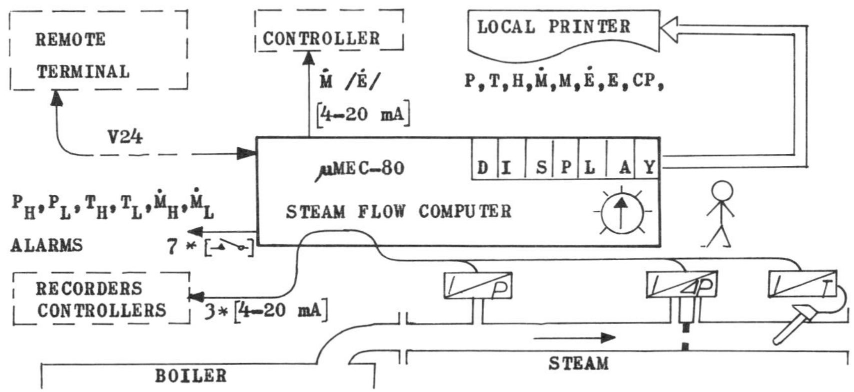


Fig.1. µMEC–80 Flow Computer connections

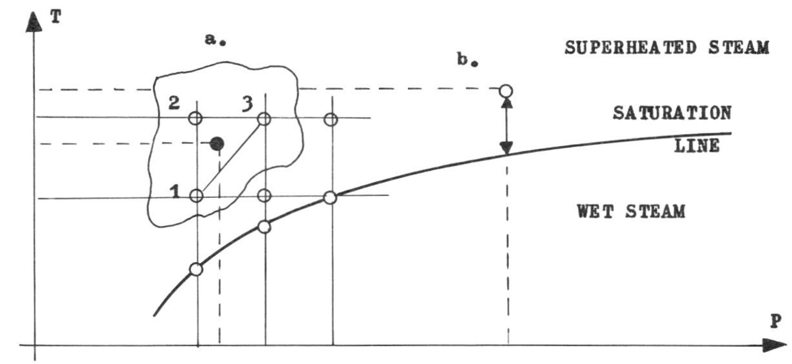


Fig.2. a. Interpolation method b. "Wet steam" signalization

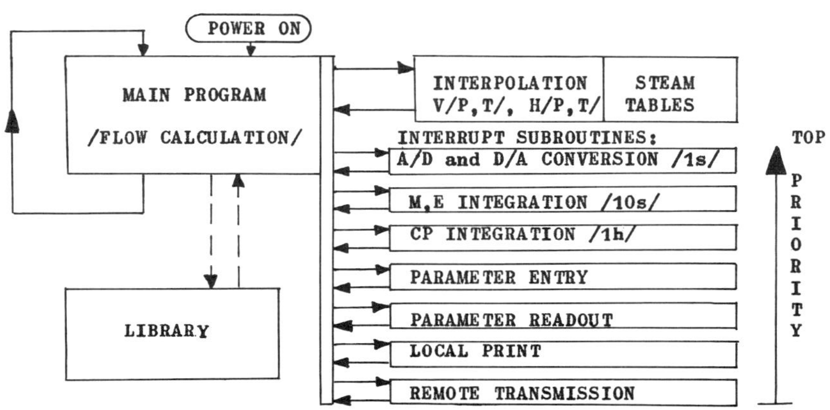


Fig. 3. µMEC–80 software system

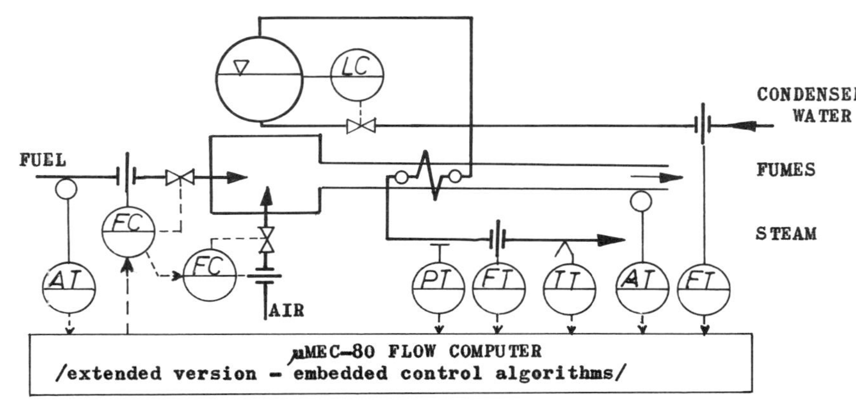


Fig. 4. An intelligent control system for a furnace-boiler set.

# SPEED CONTROL OF KAPLAN TURBINE BY USING MICROPROCESSOR

**B. T. Akgün*, A. E. Harmanci**, M. K. Sarioğlu** and M. Güleç*****

**Department of Electronic and Telecommunication Engineering, University of Yildiz, Istanbul, Turkey*
***Department of Control and Computer Engineering, Technical University of Istanbul, Istanbul, Turkey*
****Department of Mechanical Engineering, Technical University of Istanbul, Istanbul, Turkey*

Abstract. This paper deals with the speed control of *Kaplan* turbine by means of digital control technique and implementation of a controller for this purpose.The controller is a microprocessor-based discrete-time system,and implements P,I,D control algorithms.Derivative control is used to guarantee the stability of the system together with noise eliminating filter.Integral control is used to eliminate some threshold which the mechanical system has.This controller implements also a prescribed frequency-load characteristic of the system.In this work,the precision which is obtained by the microprocessor speed control of hydraulic turbine-generator is better than 0.5 %.The step response of the system can be adjusted by changing the parameters of controller.

Keywords. Direct digital control,hydraulic turbines,speed control.

## INTRODUCTION

In operation of hydroelectric systems there are two important control elements which increase the efficiency of operation and the stability of the system. These control elements are the speed regulator and the voltage regulator.In this paper,the speed regulator will be taken under consideration.

The mechanical speed regulator of *fly-ball* type has been used for many years. Later on upon development of electronic components,electronic-analog type speed regulators have been used.Nowadays as a result of wide use of microprocessors, speed regulators containing microprocessor are being extensively applied in the speed control.

There have been many papers which discuss advantages of using microprocessor in the speed control application of hydroelectric plants (Teichmann,1980 ,Elfstadius, 1983 ,Glattelder and Steinbach,1984) .

Moreover the implemention of speed control by microprocessor have the flexbility and the superiority in the followings:

i) The possibility of obtaining the controlled variable at the output of the alternator.

ii) The adjustment of the speed droop in accordance with a prescribed function.

iii) To obtain easly an adjustable control sensitivity.

iv)To be able to apply complex control algorithms which can not be implemented by analog controllers.

## CONTROLLED SYSTEM (PLANT)

The controlled system is a *Kaplan* turbine which is present in Mechanical Engineering Department of Technical University of İstanbul.The schematic diagram of turbine and control system under consideration is shown in Fig.1.

The *head* of the turbine is obtained by a pump driven by DC motor supplied from a *Ward-Leonard* set.The water obtained from the pump is applied to the turbine by the *penstock*.The water at the output of the turbine is pumped back to water-tank.There are three possibilities in applying the control forces to the speed control of the turbine:

i) By adjusting the *wicket gates*.

ii) By adjusting the *runner-blades*.

iii) By adjusting both of them.

In our work only the wicket gates have been controlled.In the model under study a *water-brake* is used instead of the alternator for loading the turbine.Since there is no alternator in the model,to obtain the controlled variable (speed) , a small alternator has been mounted on the shaft of the turbine.The output of

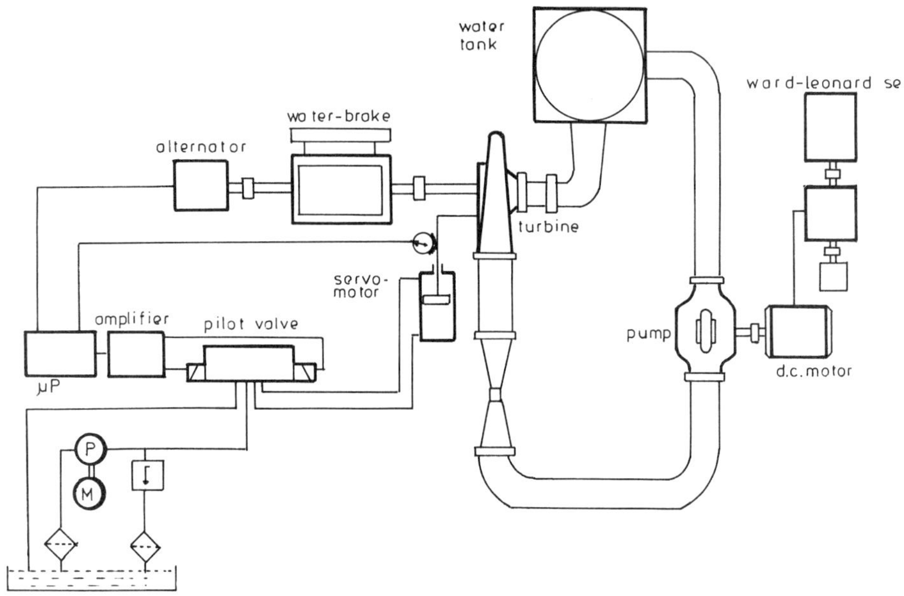


Fig.1 Kaplan turbine and control system.

the alternator is fed back through controller to control the speed of the turbine by means of wicket gates. A second feedback obtained from the position of the wicket gates is also used to realize a prescribed frequency-load characteristic.

The electronic part of the controller consists of a microprocessor and an amplifier supplied by this microprocessor. Electronic part of the controller generates an actuating signal by using two electrical signals applied to its inputs. The actuating signal is applied to pilot valve which generates a proportional hydraulic control signal. This signal controls the position of a servomotor and as a result servomotor applies a proportional position control to the wicket gates.

The controllability of the hydroelectric system shown in Fig.1 is poor. This is because of the facts that, due the lack of alternator, the inertia of the system is small and the water-brake can not give rise a stable counter-torque as in the practical case the alternator can. Moreover the water inside the water-brake has a random motion which produces high frequency oscillations.

## CONTROLLER WITH MICROPROCESSOR

The microprocessor part of the controller is a MPF-1P (MPF-1P user manual) system containing Z-80 microprocessor unit. The program written in machine language occupies 2 Kbyte and buried in ROM. Furthermore this system contains also 4 Kbyte RAM.

In the following the algorithm which

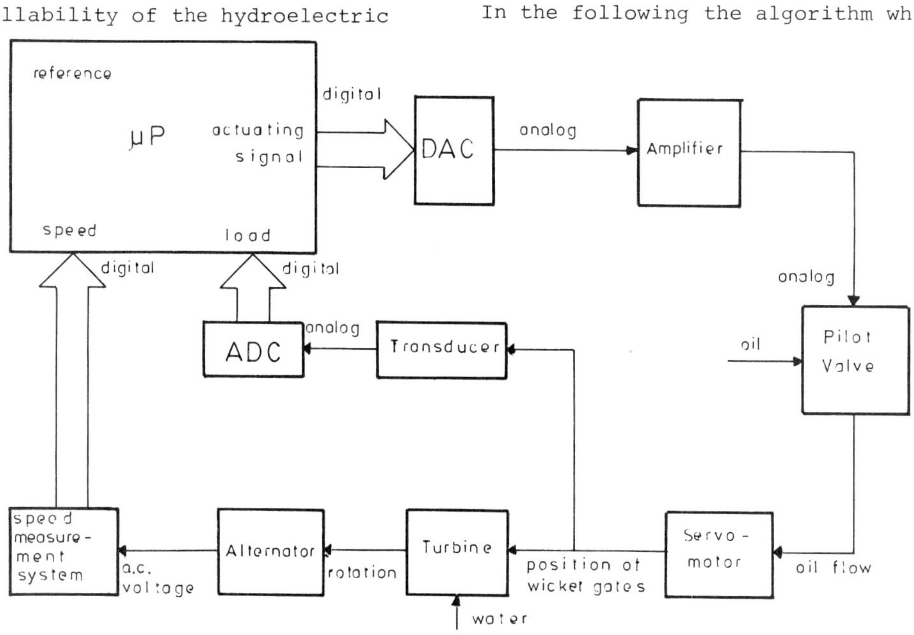


Fig.2 The diagram of controller with microprocessor.

forms the basis of the program will be explained. One cycle of the program which controls the entire system takes 5 msec with 4 MHz clock frequency.

Digital signals obtained from the continuous speed and continuous load measurements are applied to the input of the microprocessor (Fig.2) .Of these, the load feedback is used for the determination of the reference of the speed signal (speed droop) .The speed feedback may be thought to be main feedback. The microprocessor operates to produce PD actuating signal according to its written algorithm. This discrete signal is converted to an analog signal and applied to an analog amplifier which supplies the pilot valve. The pilot valve produces hyraulic signal which is proportional to actuating signal. The output of the pilot valve is applied to the hydraulic servomotor which gives control commands for positioning the wicket gates.

The speed of the turbine is measured by a small alternator. In the following sections the operation of each component of the controller with microprocessor will be explained.

## MEASUREMENT OF THE SPEED

In the system under consideration which is shown in Fig.3 the speed is measured by determining the period of the alternator output voltage. This approach makes smaller error involving in the data processing and its duration.

The output voltage of alternator is converted into TTL level throug a half-rectifier and pulse shaping circuitry. The rising edge of the pulses are used only for triggering counter circuit. In this way the small variations associated with the threshold is minimized. The reference clock is triggered by rising edges of the pulses obtained from the alternator. After this triggering process the following steps must be realized:

i) Stopping the counting.

ii) Loading the counted value into the latch.

iii) Interrupting signal to microprocessor.

iv) Resetting the counter.

v) Starting the counting.

The microprocessor can read the measured period from the latch at any instant of time. The actual period is obtained by adding the constant duration of the overhead operations (2 μsec) to the counted period.

The microprocessor controller operates when the speed of the turbine is below or above the reference speed. For the speed of the turbine smaller than a prescribed value, the counter sets the overflow flag. When minimum or maximum speed is reached in the control algorithm *exception processing routine* runs. The reference period is taken as 20 msec (50 Hz) .Suppose that the counted number (in 1 MHz standart clock ) is $m$. The period $Tp$ in microseconds is given by,

$$Tp = m+2 \qquad (1)$$

Due to the irregularity in geometry of the poles of the alternator, some correction has to be made in this formula (1). The following corrected formula is obtained,

$$Tp = (T_1+T_2+..+T_n) / n \qquad (2)$$

where $n$ is the number of the poles.

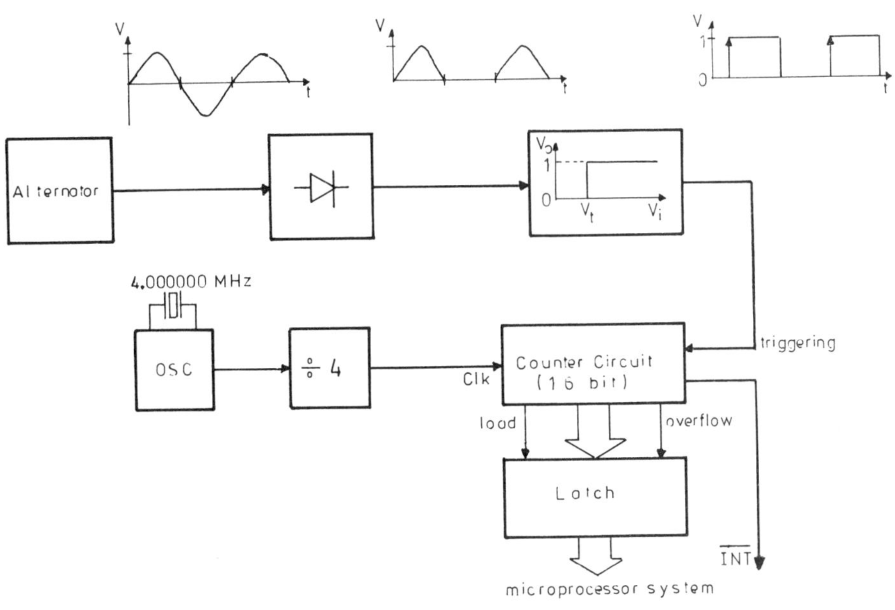


Fig.3 The schematic diagram of speed measurement system.

The method of measurement explained above is similar to *M/T* method ( Ohmae T. et al. ,1982) .The only disadvantage of M/T method is the increase of its detection time.In the method which has been developed in this work this disadvantage has been eliminated by taking the average for *n* period consisting of the last period plus *n-1* periods.Thus the detection time has not been increased.

## CONTROL ALGORITHM

The standart feedback control system is shown in Fig.4 .In this control system microprocessor implements PD control algorithm and at the same time it generates the error *e* which is the difference between *r* and *h*.The I control is achieved by means of hydraulic servomotor. For continuous PD controller,the input and output relation is given below,

$$u = K.(e + Tv.\dot{e}) \qquad (3)$$

where *e* is the input,*K* is the proportional gain and *Tv* is the derivative constant.

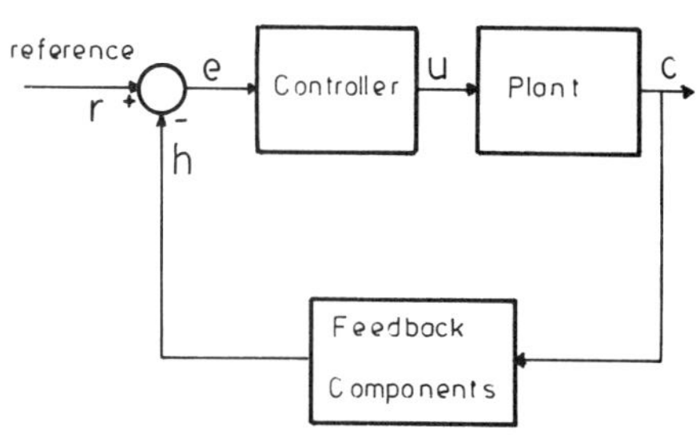


Fig.4 The standart continuous close-loop controller configuration.

For the controller operating in discrete-time and implementing PD control,the input and output relation is,

$$u[nT] = Kp.e[nT]+Kd.(e[nT]-e[(n-1)T]) \qquad (4)$$

$$Kd = Kp.Tv/T$$

where *T* is the sampling time,*Kp* is the proportional gain,*Kd* is the derivative constant.

### P Control

The operating characteristic of an alternator for parallel operation is given as frequency versus load characteristic *(speed droop)* .The desired frequency-load characteristic of the alternator is loaded in the memory of the microprocessor and it will be called *reference-characteristic* .In discrete-time proportional control signal is determined by,

$$u_p[nT] = Kp.(r[nT]-h[nT]) \qquad (5)$$

$$r[nT] = f(load[nT])$$

### D Control

To decrease the error resulting from the data processing and to eliminate the effect of the varition of the reference derivate,the derivate control is used on the feedback path.Moreover,since the reference is obtained also by using the load feedback signal,noise which may exit in the load has been eliminated.In this way the derivative control is applied only to the speed.The derivative is obtained digitally by taking the difference between two consecutive sampling values. To get rid off the undesired operation due to the measurement error the derivative control is transformed to *averaging derivative algorithm* (Föllinger,1978) . By using the above analysis the derivative algorithm achieved in microprocessor is given below,

$$u_d[nT] = Kd.1/6.\{h[nT]+3.h[(n-1)T] -3.h[(n-2)T]-h[(n-3)T]\} \qquad (6)$$

To decrease the effect of noise,each sampling value is compared with the previous one,since the inertia of turbine system is large,the diffence between the values of two consecutive sampled data can not be large.If this difference is larger than a prescribed value which means that error occurs in sampling,the previous sampling value is used.

## CONTROL OF THE HYDRAULIC PILOT VALVE

The pilot valve generates a proportional control signal by means of the hydraulic operation.As it is shown in Fig.5 the solenoids are supplied from an analog PID controller through *Voltage to Current* converters.The PID controller has been used to eliminate the hysteresis effect of the hydraulic pilot valve which has been observed in practise.

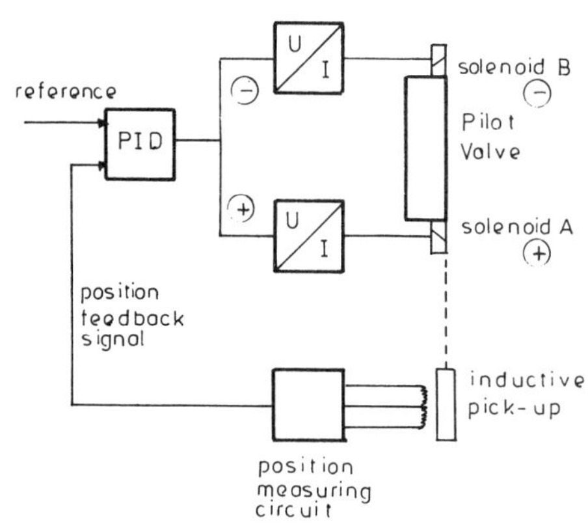


Fig.5 The control of hydraulic pilot valve

The output range of PID controller is *-10V,+10V* .The solenoid operates one each time,according to sign of the output voltage of the PID controller,and produces a proportional position control signal.

Although the input voltage and output position transfer relation of pilot valve has been linearized by using PID control, there is another practical difficulty to overcome,in this control loop,that is input voltage-output oil flow transfer characteristic of the valve.This voltage and oil flow characteristic has a thresh-

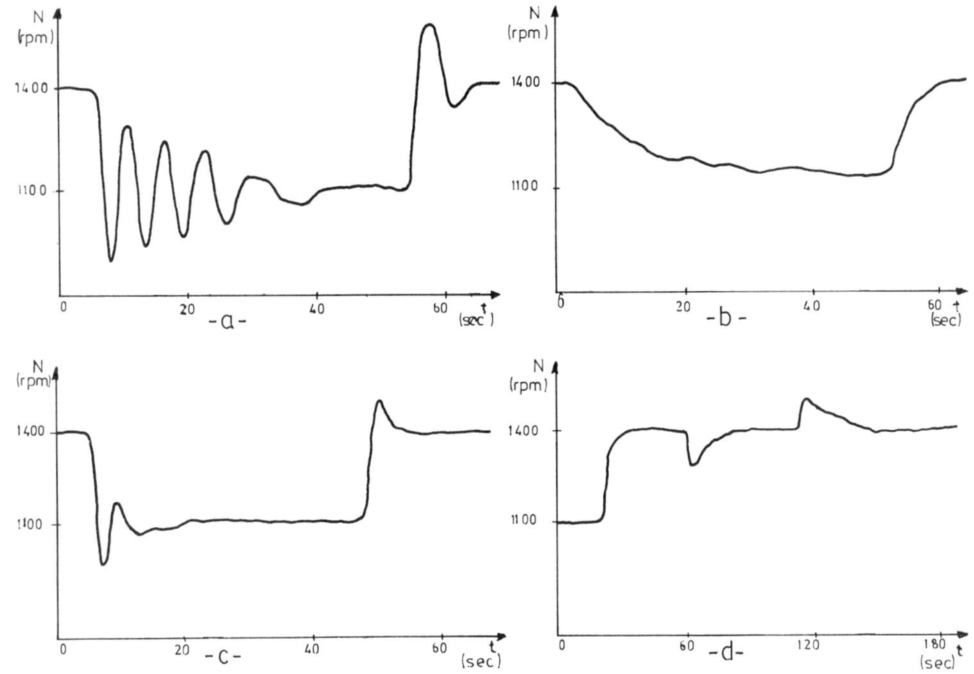


Fig.6 Experimental results.

old.This threshold is the result of the fact that after applying the input voltage to valve up to a certain value it does not give to any motion and hence no oil flow.The threshold must be decreased to certain small prescribed value.This threshold is controlled by using a special integral control algorithm in the microprocessor.

## CONTROL OF HYDRAULIC SERVOMOTOR

The gains of hydraulic servomotor are not equal for both directions of motion of the piston.This is due to internal structure of cylinder and piston arm.The value of the gains in forward and backward direction have been made equal in the algorithm used.

## EXPERIMENTAL RESULTS

The overall system behaviours have been illustrated by the response curves given in Fig.6 .

In Fig.6.a a step input response of complete system with proportional control is shown.Since orginally the forward and backward gains of hydraulic servomotor are not equal,the step input responses of the overall system are not symmetrical.

In Fig.6.b the responses are shown by applying different control gains obtained in the algorithm used for microprocessor part of controller.These input responses have been completely changed.

In Fig.6.c. the change in the step input response of the system is shown when a derivative control is added to the original algorithm (for Fig.6.a) .

In Fig.6.d the overall behaviour of the system is shown as a result of the algorithm used with appreciate parameter values in the controller.As it is observed in the figure,first a step reference input applied to system and the response is recorded.No oscillation is detected.Then the rest of the curve represents the responses of the system as the maximum load applied and removed an a step manners.

## CONCLUSIONS

In this work all the non-linearities (hysteresis,thresholds,and gain differences) stemming from the pilot valve and hydraulic servomotor have been eliminated.Moreover to secure the stability and to increase the accuracy of the control,an adjustable threshold has been produced and a special integral control has been used in the control algorithm.

## REFERENCES

Elfstadius, K.H. (1983).Microcomputers for Hydro Plant Control.Water Power and Dam Construction,November,15-17

Föllinger, O. (1978). Regelungstechnik. Elitera-Verlang,Berlin.Chap.7,pp.204-207.

Glattfelder, A.H.,and J.Steinbach (1984). Decentralized Hydropower Control.Water Power and Dam Construction,July,16-20.

Micro-professor,MPF-1P User's Manual (1983). Multitech Industrial Corporation.

Ohmae, T., T.Matsuda, K.Kamiyama, and M.Tachikawa (1982).A Microprocessor-controlled High-Accuracy Wide-Range Speed Regulator for Motor Drives. IEEE Transactions on Industrial Electronics,Vol-IE-29,No.3,207-211.

Teichmann, H.T. (1980). The Role of Automotion in Modernizing Hydro Plants. Water Power and Dam Construction, September,27-29.

# REAL TIME CONTROL OF THE POWER PROVIDED BY TWO AEROGENERATORS FROM A SINGLE SPECIALIZED PROCESSOR

**C. Kraif, S. Billo, P. Rouan and L. Protin**

*Laboratoire de Traitement en Temps Réel des Énergies Renouvelables, Ecole Nationale Supérieure Universitaire de Technologie (E.N.S.U.T.), Université de Dakar, BP 5085, Dakar-Fann, Senegal*

Abstract. The purpose of this paper is to describe a power adaptive research automatic device that will make it possible to optimize the power provided by two aerogenerators of the generating station from a single specialized processor. As this unit is to be set up in the Sahel, the self sufficiency of our energy system has been developed by means of a push-pull power supply. The input voltage is supplied by a 24 V buffer battery. This device, made up of two Europe format standard cards adapted to G64 bus is designed to work within a multiprocessor system. The power card consists of two DC to DC converters and allows permanent adaptation of the load of its aerogenerator, whatever the wind fluctuations may be.
The numerical control card allows the generating by software of the optimal duty ratio on the static converter. Such a device makes it possible to adapt several strategies of control by software. We, thus, propose and extremal power research algorithm which enables the optimization of both aerogenerators taking the wind speed into account.

Keywords. Multigenerators, multireceiver energy unit ; aerogenerator ; power optimization ; numerical control ; power converters ; galvanic insulation ; optimal control algorithm.

## INTRODUCTION

Given the considerable importance of renewable energies in developing countries (40%) and particularly in rural areas (50% to 90%) /1/ our laboratory in DAKAR ENSUT has set up as a short-term objective to develop a multigenerator, multireceiver energy unit.
As this central generating station can be set up in the Sahel, our priority objectives have been, on the one hand, the reliability of the central ensured by a fully numerical control which reduces hardware to the minimum in favor of software and on the other hand the self sufficiency of our energy system has been developed by means of a switch power supply.

The unit is composed of two aerogenerators of small and medium power (100 W and 1 KW) whose interest consists in providing energy for windspeeds as low as 2 m/s.
Before we work out some management appropriate for the coupling of these two elements, it is necessary to optimize the working of the generators. Our aim is the constant research of the maximum of energy delivered at any time by the aerogenerators, whatever the wind speed fluctuations may be.

In this regard, various methods of optimization have been suggested, the source of energy being either photovoltaic-generator or an aerogenera tor /2/, /3/.
For some years now, our laboratory has shown interest for this problem and in sept 1984 we built the numerical control for a 1 KW aerogenerator, starting from the SDK 85 INTEL KIT /4/.

The object of this paper is to describe a power adaptive research automatic device that will make it possible to optimize, in real time, the power provided by two aerogenerators of the generating station from a single specialized processor.

Such an outline has improved the reliability of our unit. Once described the device set up in our laboratory we show the power characteristics diagrams of the two aerogenerators and from them we infer the equation of the transfer optimal characteristics.

We then propose a new extremal power research algorithm taking into account the windspeed so as to improve the dynamic response of the system.

## SYSTEM DESCRIPTION

The experimental connecting up "fig. 1" includes a power card and a numerical control card with galvanic insulation so as to avoid interactions between power commutations and control. The self sufficiency of our energy system has been developed by a switch power supply provided by a 24 V buffer battery, which is the only power supply of the multigenerator multireceiver energy unit.

### Power card

Adaptation by means of static converter. The aerogenerator provides a three-phase A.C voltage which varies in time and whose power characteristic P = f(U), once rectified, shows a maximum for a given Uopt value of voltage on charge terminals "Fig. 2a". If we connect directly aerogenerator to load, the working point Q will most of the time be far from the optimal working point ; it is therefore necessary to insert a converter which modifies the apparent load impedance according to power fluctuations.

In this way, the load and converter characteristic passes through the optimal working point Qopt of the aerogenerator for a given windspeed.

The converter used here is a DC to DC converter which allows the checking of the aerogenerator power flow to a load R.L "Fig. 2b".

Choice of the structure of converters. Among the static converters existing today and considering the nominal voltage of 220 V given by the 1 KW aerogenerator alternator, we chose a buck-converter. Likewise, we chose a boost-converter for the 100 W aerogenerator which releases a 20 V voltage that is not sufficient to connect it to a 24 V buffer battery.

The use of these two different structures requires the use, in each case, of input and output filters adapted to these structures.

Description and realization of converters. The buck-converter has been made up in bipolar technology because it is easy to use and comprises a switching aid network. This converter has been built around an UAA 4002 integrated circuit which, by itself, performs the functions of basic control and active protection of the power transistor BUX 24.
The input interface of the integrated circuit provides the necessary adaptation between the output signal on the control card and the logical processor of this circuit. A relatively low frequency of 3KHz has been chosen. It is the result of a compromise between the drop in voltage due to commutations which varies according to the frequency and size of the passive elements of the buck-converter that increases as the frequency decreases.

The boost-converter has been made up in MOS technology well adapted to low working voltage. It is composed of a VN 64 Power MOSFET and its protection an control circuit in CMOS Technology.

The working frequency of this converter is also of 3 KHz. Each converter is numerically controlled by software, according to a law which allows the optimization of the working out of each aerogenerator.

Numerical control card.
The numerical control card allows the real time generating of the optimal duty ratio of each converter so that each aerogenerator can deliver its maximum power to its respective load at any time, whatever the windspeed.

To develop the software, we have replaced the processor card by a GOUPIL III Calculator. Its synoptic diagram is given in "Fig. 1".

Data acquisition. For each aerogenerator the information given to the microprocessor is the following :

- Voltage $U_A$ and current $I_A$ delivered by the aerogenerator
- The rotating speed $w_A$ of the aerogenerator
- The windspeed $V_A$.

A differential connecting up allows us to pick up and calibrate the analog data $U_A$.
A current-voltage transducer delivers an image voltage of current $I_A$, which is then shaped and calibrated.
In order to know the speed of the aerogenerator, the frequency of the alternator voltage is measured.
Before measuring, a circuit composed of a step down transformer, a low-pass filter, a voltage limiter and a Schmitt trigger allows to convert the analog speed information into a TTL digital information.

Such information is then sent into a specialized timer circuit (MC 6840) programmed as a frequency detector.

This device has the advantage of allowing the insulation of the information. The information $U_A$, $I_A$ or $V_A$ is then converted into an eight bits word, by a data acquisition and conversion system including an eight channel-multiplexer, a sample-hold and a successive-approximation A/D converter (25µs).

After galvanic insulation, the binary information (voltage, current or windspeed) is sent to the microprocessor by means of the parallel interface adapter PIA 6821 (Motorola).
The amplified and insulated control signals provided by this circuit allow, on the one hand the selection of different measure channels and on the other hand the sampling and conversion control.

Galvanic insulation. For a good working of our device, we have insulated the control card as well as the analog signals coming from various sensors of our numerical control which is thus protected from any prohibitive voltage or interference that may occur.

As for supplies, the problem has been solved by using a switch power supply with several windings developed in the laboratory "Fig. 1".

When we examine the optocoupler characteristic, we notice that it is particularly well adapted to the transmission of logical signals. We have used fast optocouplers in order to insure the galvanic insulation of binary information.

Production of the converter control signal. The control of each converter is realized by software, by two cascade connected timers from a specialized triple timer-circuit MC 6840.
The first one has been astable programmed to generate the working frequency of the converter (3KHz).

The second one has been monostable programmed, so that it is possible to vary the variable duty ratio Rc. We use the following formula :

$$(N + 1)2T = 1/f$$

in which N represents the hexadecimal word to be fed into the timer and T the period of the pilot clock (1µs).
The program which allows the generating of this variable duty ratio is realized in assembler language. The address decoding of the various circuits of the control card is ensured by a HARRIS 256 x 4 Bits generic PROM.

Finally, it is necessary to realize the galvanic insulation and to shape the output signal of each timer before getting into the basic control circuit of the converter.

Power Self-Sufficiency of the System
A 24 V buffer battery represents the input power of the switch power supply of our system. The latter is of push-pull type, whose power switches are voltage-driven Power MOSFET so that the control needs few power.
To ensure the stability of the control circuits supply voltage, regulators have been added to the switch power supply.

Device calibration
To calibrate the data acquisition and conversion system we have replaced the aerogenerator and rectifier by a continuous power supply 300V/10A.

The duty ratio of converter has been chosen as equal to one. Having proceeded to the gain adjustment of each channel, we have drawn the transfer characteristics of these channels.

Data acquisition is automatically realized by an assembler program then a statistical processing of these measures, through a basic program lead to the calibration graph of each channel :

Voltage channels : $U_{A1} = 4{,}985\ V_{IN\ A/D} + 0{,}1496$

$U_{A2} = 29{,}94\ V_{IN\ A/D} + 1{,}2126$

Current channels : $I_{A1} = 0{,}5982 V_{IN\ A/D} + 0{,}0179$

$I_{A2} = 0{,}399\ V_{IN\ A/D} + 0{,}071$

Wind channel : $V_{A1} = V_{A2} = 0{,}1495 V_{IN\ A/D} + 0{,}0057$

where $V_{IN\ A/D}$ is the input voltage of A/D converter.
We have calibrated the anemometer and determined its transfer characteristic by means of a control and calibration set :

$$V_{(m/s)} = 12{,}3762\ V_{A\ (V)} + 0{,}693$$

## STATEMENT OF POWER CHARACTERISTICS

Experimental method and combined software. After program initialization, we make the duty ratio Rc of converter vary from 0 to 1 by successive steps of 0,05 "Fig. 3".

For each value of Rc, we proceed to eight series of successive measures including information $U_A$, $I_A$, $V_A$, $R_c$ and $w_A$.

We have inserted in the loop some time delay that takes into account the aerogenerator inertia, that is to say its eletromechanical time constants $\tau$elm. For the 1 KW aerogenerator $\tau$elm = 5,9 s ; J = 0.51 Kgm²/rd /4/, we chose an approximate delay of 3 $\tau \# 20$s.
According to relation (1) we conclude as a first approximation a 5 s delay for the 100 W aerogenerator. We intend to verify its electromechanical time constants later

$$\frac{\text{Jaero 1 KW}}{\text{Jaero 100W}} \sim \left(\frac{D}{d}\right)^2 \qquad (1)$$

D = 2d = 5 m.

All real time measures are stored in memory by an assembler program. A basic program is used to compute the mean values then to sort them out according to an increasing windspeed from 2 to 10 m/s ($\pm$ 5%).
We thus get the power characteristics of each aerogenerator "Fig. 4 and 5".

The anemometer has been fixed to the 1 KW aerogenerator mast and placed under the turbulence zone created by the propeller.

Equations of each optimal transfer characteristic. By studying the graphic network $P = f(Rc) V_A = cte$ of an aerogenerator we notice that each maximum of power corresponds to a given optimal duty ratio.

We have drawn the $V_A$ windspeed V.S optimal duty cycle Rcopt curves.
As a first approximation, we may assimilate these curves to straight lines of equations :

AERO 1KW : Rcopt(t) = 0.049 Va(t) + 0.44
AERO 100W : Rcopt(t)= 0.03 Va(t) + 0.44

For best precision we have used a basic program of curve parabolic adjustment to define the equation of the transfer optimal characteristic of each aerogenerator.

AERO 1 KW : $Rcopt(t) = -0.0135 Va^2(t) + 0.2236 Va(t) + 0.0157$

$3 \leq Va \leq 10$ m/s

AERO 100 W : $Rcopt(t) = -0.0195 Va^2(t) + 0.2141 Va(t) + 0.0421$

$2 \leq Va \leq 7$ m/s

When Va < Vamin, the aerogenerators hardly produce any energy.

## PROPOSAL OF A NEW ALGORITHM

The previous research work /5/ made it possible to establish that the method called "proportional method" which gives the best results consists of having the converter cyclic ratio vary in proportion to the aerogenerator power variations according to the following law :

$$R(t + Te) = R(t) \pm K|\Delta P|$$

For the 1 KW aerogenerator, k = 1/2 ; optimal inertia time = 125 ms.

This method, to begin with, requires a phase of transition for power maximum research and is particularly efficient only for comparatively stable windspeeds. To palliate this disadvantage, we propose a new algorithm for extremal power research which takes into account the windspeed in dynamic working conditions "Fig. 6a".

After carrying out the windspeed acquisition, the processor, starting from the transfer optimal characteristic of each generator, computes the corresponding cyclic ratio and sends it to the converter using MC 6840 timer.

We use the dead time resulting from the inertia time of the 1 KW aerogenerator, during which the processor does not work, to optimize the 100 W aerogenerator working.
Due to load fluctuations and equipment obsolescence, the former algorithm does not allow stable optimal working in time ; we then used an extremal power research algorithm shown in "Fig. 6b" and which is identical for both aerogenerators.
We develop a positive disturbance of the cyclic ratio and calculate the corresponding $\Delta P$ power variation. According to the sign of $\Delta P$, we increase or decrease the cyclic ratio of a quantity $\Delta R = K \Delta P$, and the power maximum is reached when $|\Delta P| \leq \varepsilon$ "Fig. 6b".

The first tests of this algorithm were carried out with the help of combined basic and assembler software on GOUPIL III computer. The results achieved show the practicability of the device which converge towards the optimal working point whatever the external conditions of load and wind are. A systematic study concerning the algorithm optimization, from the convergence time point of view, is currently developed.

## CONCLUSION

The device described in this paper makes it possible, starting from a single processor, to carry out the entirely numerical control of two aerogenerators of 100 W and 1 KW, belonging to a multireceiver multigenerator central power station.

The practical development shows the basic contribution of numerical techniques and particularly of microprocessors on equipment simplification and versatility by adapting software.

The result is increased reliability necessary for possible use of the generating power station in the Sahel area. Moreover, we have developed the energy self-sufficiency of the set.

We propose, for the setting up of the device, a new algorithm of extremal power research which takes into account the fluctuations of windspeed or load and the obsolescence of the equipment. The first results show the feasibility of the device and the convergence towards the optimal working point of each aerogenerator.

We are planning, in the future, to optimize in real time the algorithm of extremal power research by developing it entirely in assembler language using a C. compiler.
We also want to study the speed of the device convergence to the extremal point, in response, on the one hand, to a step of windspeed by simulation, and on the other hand to real fluctuations.

## References

/1/ P. MATARASSO
"Projets énergétiques et projet de développement".
III° séminaire sur l'énergie solaire. Trieste (du 6 au 17 Sept. 1982).

/2/ C. BOISVINEAU
"Optimisation du fonctionnement d'un générateur photovoltaïque" : Asservissement extrêmal de la puissance fournie à une charge en courant continu.
Thèse de Docteur 3° cycle INP Grenoble (Mars 1981).

/3/ G. CHAMPENOIS
"Optimisation du fonctionnement d'une éolienne DARRIEUS : Commande adaptative par asservissement extrêmal numérique".
Thèse de Docteur de 3° cycle INP Grenoble (Mars 1984).

/4/ C. KRAIF
"Optimisation de la puissance fournie par un aérogénérateur commandé par microprocesseur.
Thèse de Docteur Ingénieur U.S.T.L. Montpellier (Septembre 1984).

/5/ C. KRAIF - C. MASSELOT
"Power optimization for an aerogenerator controlled by means of a microcomputer".
Congrès EPE sur l'Electronique de puissance Bruxelles (du 16 au 18 octobre 1985).

# A CENTRALIZED AGRICULTURAL IRRIGATION SYSTEM

A. F. Çalikoğlu, G. Tanyeri and O. Arkut

*Aselsan Military Electronics Industries Inc., Ankara, Turkey*

Abstract. Efficient use of water and energy resources is an important point in agricultural Irrigation Systems. This paper describes a centralized control system aimed at efficient and programmed use of 100 deep water well pumps under the supervision of a central station. Each pump unit has a microprocessor based controller which identifies its users with a special identity card. The controller sends status data to the central station and receives time schedule of users via UHF radio equipment. The system will stop excessive use of water thus reducing costs and increasing yield.

Keywords. Agriculture, Centralized Plant Control, Computer Control Energy Control, Microprocessors.

## INTRODUCTION

The Geotechnical Operations and Underground Waters Department of the State Water Works drills 500 deeep water wells each year. Nearly all of these wells have electric motor driven pumps. The total number exceeds 6500 with the motor power ratings ranging from 7.5 to 190 kW

Being distributed over a wide area, the pump stations cannot be operated efficiently.

Currently the pump stations are operated manually by farmers under the supervision of irrigation technicians and the users are charged according to the size of their land and the type of cultivation.

In order to overcome the operational difficulties of this method a computer controlled, centralized irrigation system was seeked, by which the following goals will be reached:

1- Excessive use of water will be avoided hence the yield increased
2- Waste of electrical energy will be reduced
3- A justly method of billing will be introduced
4- Immediate notification of fault conditions will be carried
5- The required number of service and support personnel will be minimized
6- The management of the irrigation system will be much easier
7- A very simple user interface should be established
8- The activities of the well pumps must be observed from a central cite
9- There must be a method of identification of the users by the system.

The plain of Harran in the south east region of Turkey was selected for the pilot application of this method. Of the 385 pump stations, 100 are to be fitted with Pump Control Units (PCU) and will be operated under the control and supervision of a Central Station (CS).

Last year, around 2.5 billion TL worth of electrical energy was consumed in the 385 pump stations, a substantial amount of which is believed to be due to over irrigation.

## OPERATIONAL ASPECTS OF THE IRRIGATION SYSTEM

The irrigation need of a land depends on the type of cultivation and rain conditions heavily. Excessive irrigation causes deterioration of soil characteristics, decreasing yield. In order to achive good yields, the irrigation policy must be determined by skilled technicians and applied accurately.

At the start of the irrigation season, the farmers must hand in their land size and the type of cultivation according to which a master irrigation schedule will be prepared. In some cases, the type of cultivation must be arranged according to pump capacity and other farmers' needs.

The master irrigation appointment schedule may be modified according to varying meteorological conditions later.

The observation of the operational status of the pump stations and notification of failures at a central station is also of prime importance in order to have a full grip on system behaviour.

The geographical layout of the wells should be within 10 km. radius from the Central Station enabling convenient RF communication means.

## DESCRIPTION OF THE CENTRALIZED SYSTEM

The system consists of a powerful microcomputer based CS and intelligent PCUs. (See Fig. 1)

The CS is located at the Irrigation Control Center whereas the PCUs are located in the cabins accomodating the motor, the control panel and the compansation panel.

As seen in Fig.2 there is an access hole leaved for the farmers to insert their identity cards.

Each farmer is given an identity card which is recognized by the PCU. The booking of appointments of the pump units are entered by the operator at the CS console and then transmitted to the related PCU. VHF radio and modems are used for communication of the PCUs with the CS. A printer output of the allocation is distributed to the farmers. Using his identity card, each farmer has complete control of the pump station during the allocated time. Additionally, the status of the well can be observed, tests performed and remote start/stop operation of the PCU performed from the CS. During unallocated time intervals, control of the pump is disabled. The CS can control up to 120 PCUs and each PCU can be operated by 30 users.

As soon as a user inserts his identity card, the PCU checks his appointment, starts the motor, and notifies the CS with a mesage including user ID. This information is both stored on a disk file and also displayed on the screen. The motor is stopped by the next insertion of the identification card or elapsing of the allocated time for the user. The stop condition and the elapsed time are transmitted to the CS by the PCU where it is logged for future reference and billing purposes.

Due to the distributed processing nature of the system, the RF link between the CS and the PCUs is non critical and uncrowded most of the time.

If sufficient margins are provided, temporary blocking of the communication channel doesn't effect the irrigation policy.

Fig. 1. The Centralized Irrigation Control System

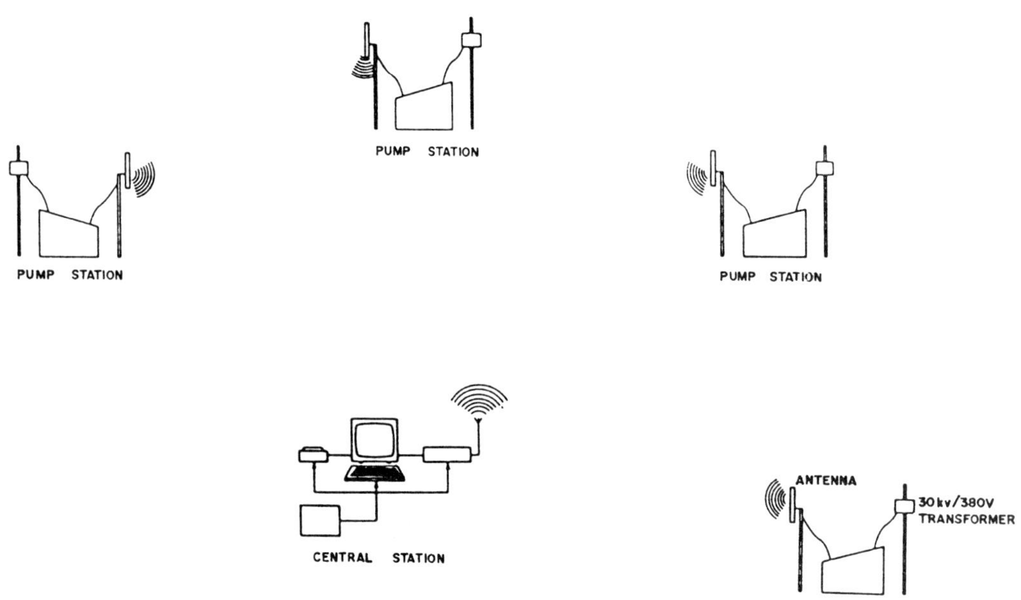

Fig. 2. The Pump Cabin

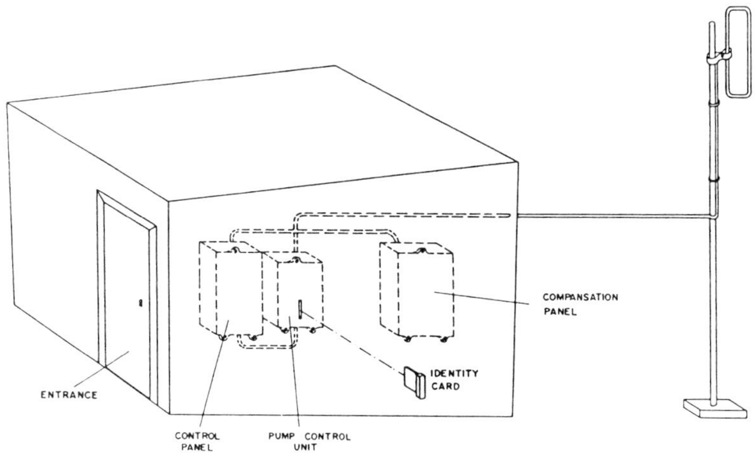


THE PUMP CONTROL UNIT

The PCU is a microprocessor based remote terminal designed to control the utilization of well pumps, record the running time for each user and communicate events with the CS.

The PCU consists of three basic modules, namely, Identification Module (IM), Control Module (CM), and RF Communication Module (RFCM) (See Fiq. 3)

The IM is used to read the identity card of the users. Due to dusty and hot operating environment, commercially available card readers were not found suitable and a rugged, magnetically coded in house module was developed.

The CM is a microcomputer with six ECB-Bus boards, all in CMOS technology.

The Single Board Computer is CMOS 8085 microprocessor based and its RS232C serial port is used to interface with the RFCM.

The System Controller Board has several functions, watch dog timer for system checking, RS232C to connect to a service terminal and dip switches to set PCU address.

The Relay Board controls pump motor contactors.

The Optoisolated Input Board reads the position of eight status switches indicating contactor status, intrusion alarm, oil level, water level, water flow, motor temperature and alarm inputs. An abnormal status causes the PCU to send an alarm message to CS and stop the motor.

The PCU software is developed around a Real Time Multitasking Executive Kernel, which facilitates a terminal task and a debug task burried into final system to ease field service. Tasks are written in Pascal and assembly language.

Important parameters such as the appointment schedule, running time and user codes are stored in battery backed up RAM to protect these data in case of power failures.

The RFCM has a simplex two channel 2W VHF radio and a 1200 baud FSK Modem Board. An RS232C port is used to interface this module to the CM.

Although designed specifically for this project, the moduler construction of the PCU enables easy adaptability to various applications via simple software and hardware module configuration changes.

Fig. 3. The Pump Control Unit

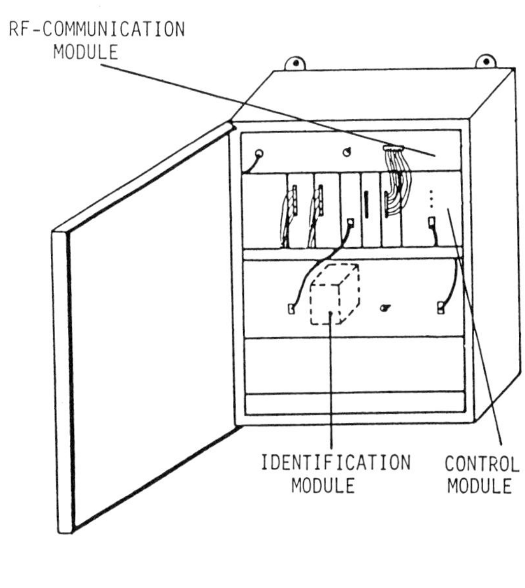

Fig. 4. The Central Station

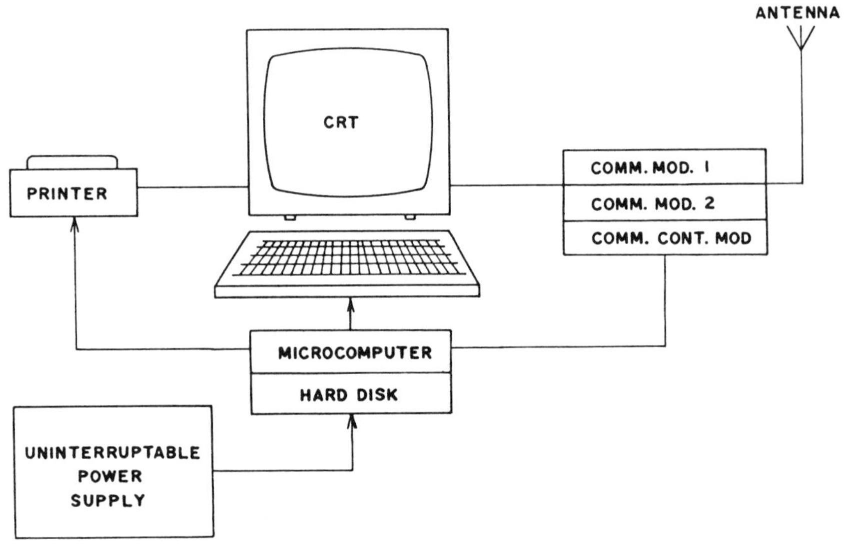


## THE CENTRAL STATION

Tne CS is constructed around a 68000 based microcomputer with a 20 MB hard disk unit which has an additional 730KB 3.5" floppy disk used for back up purposes, an 80 column dot matrix printer, a 14" color monitor and two RFCMs controlled by the Communication Control Module (CCM)(See Fig. 4).

The computer has an on board RAM of 1MB, HB-IB and RS232C video and keyboard interface boards.

The RFCMs are the same as those in the PCUs. Having two such modules makes it possible for the CS to maintain data communication on two channels simultaneously.

The CCM is an intelligent interface between the computer and the RFCMs.The purpose in including such a module is to reduce the communications overhead of the CS.

The CCM is based on a Single Board Computer, and a System Controller Board both identical to the CM boards, and a two channel Serial Interface Board used to communicate with the RFCMs. Battery backed up Real Time Clock of System Controller Board is used by CS to keep time and date.

## CS SOFTWARE

The CS System Software is a BASIC language operating system providing a structured BASIC language with powerful interrupt handling and graphics capabilities.

The application software includes a simple DBMS to store, update and retrieve customer information.

The user interface of the system is realized with soft keys working on an interrupt basis marking use of the interrupt handling capabilities of the operating system.

The CCM software is constructed over a multitasking kernel and communication and message handling routines, almost the same as in the PCUs. The CCM software is completed by tasks maintaining successful data communication with the CS computer and diagnostic routines for self-testing of the CCM.

## CS FUNCTIONS

The functions of the CS can be summarized as follows :

Real time system monitoring. Any status change in a PCU is sensed by the CS via a PCU message, displayed both on the graphic screen and printed. The necessary action is carried on either automatically or under operator control, based on the type of status change. The status changes are those detected by the opto-isolated I/O board on the PCU.

Pump operation recording. The CS records and updates pump working interval for each farmer and as totals.

System Structure Manipulation. The CS provides facilities for the generation and expansion of the irrigation system included in application software.

Customer appointment arrangement and pump operation monitoring. The CS software enables the operator to arrange customer appointments for each PCU, and on request, can display customer operations on PCUs. This is just an informative monitoring that enables control of water consumption.

The operator can also make use of the system's back up facilities to have copies of PCU files under software control avoiding loss of data due to disk errors. This makes system regeneration possible in case of mass storage failure.

The system-test property of the CS enables the operator to test a single PCU, a group of PCUs or all the PCUs in the system. The CS also has self test facilities.

## CONCLUSION

With the advence of microcomputer technology, it is common practice to produce fairly complex systems with very simple user interfaces. Such systems have added benefits of remote test and diagnostics as well as high reliability.

Application of microprocessor technology to an agricultural irrigation system is leading to substantial benefits both in yield and energy consumption as well as system monitoring and billing.

The distributed processing nature of the system facilitates a fast response time and minimum load on the RF communication link.

Currently the prototypes are under evaluation and the full operation of the system is scheduled to 1987.

# MULTIPROCESSOR CONTROLLED ROBOTS

**D. Naunin, K. Cevik and H.-C. Reuss**

*Institut für Elektronik, Technische Universität Berlin, Einsteinufer 19, D-1000 Berlin 10, German Democratic Republic*

Abstract. This paper deals with the control of industrial robots. A completely digital control system for a six-axis-robot has been developed in the Institute of Electronics at the Technical University of Berlin. This modular control system consists of several units, each with its own intelligence. The axes of the robot are moved by dc servomotors, which are driven by transistor-H-bridges. Each axis is controlled by an 8 bit microcomputer, which measures the state values (position, speed and current) of the dc motor. The microprocessor then gets a new command value from the master computer, and calculates the set variable in form of a PWM signal. The implemented state feedback control algorithm is also discussed. The motor, power converter and microcontroller are regarded as one unit called the "intelligent drive". The master computer is responsible for calculation and coordination of the motion sequences of the robot (PTP,CP). Direct user control and on-line programming have been realized.

Keywords. Robots; multiprocessing systems; digital control; microprocessor; motor control; power converter; dc motor.

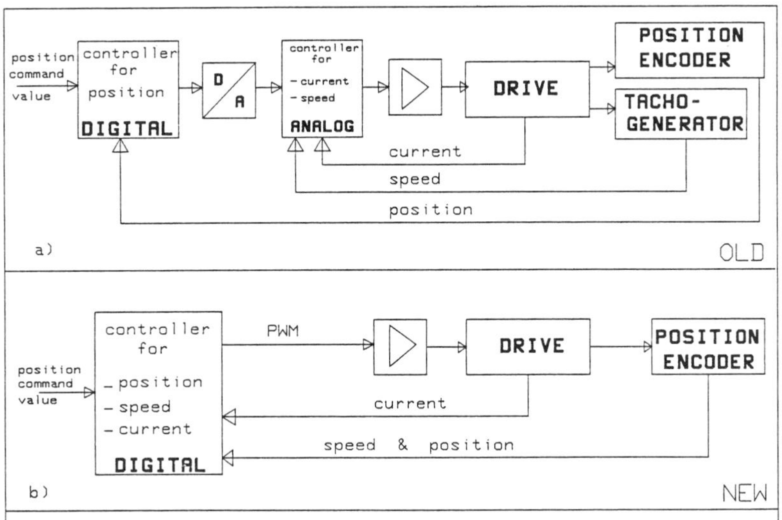


Fig. 1 Control Structures

## INTRODUCTION

Most industrial robots use dc servomotors because of the ease of speed control. Robotic drives need a position control with inner speed and current control loops. Speed and current are normally controlled by analog circuits, while the position controller is integrated in the main digital control computer of the robot (Fig.1a). This structure includes some disadvantages such as complicated analog circuits and interfaces, inflexible control algorithms, separate detectors for position and speed, and great demands on the main control computer. For these reasons a new concept for a robot controller has been developed in the Institute of Electronics at the Technical University of Berlin (Fig.1b).

The entire robot control system is structured in several hierarchical levels.

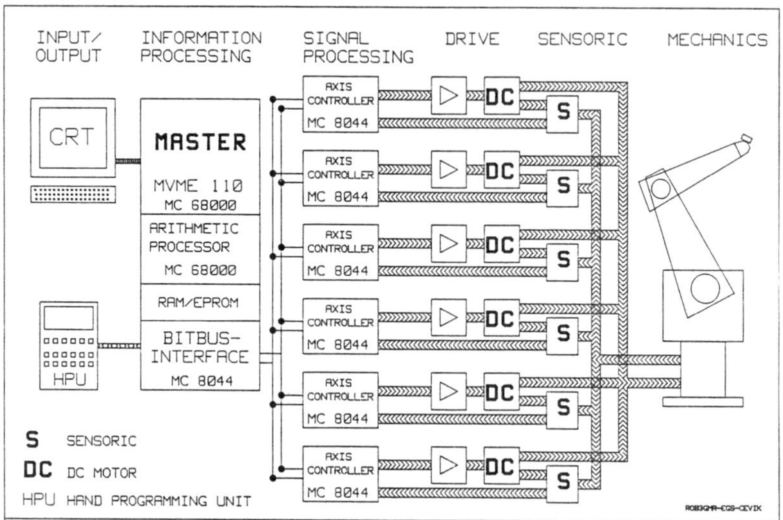


Fig. 2 Multiprocessor System

Each level should be as "intelligent" as possible in order to reduce the work load of the next upper level. This structure makes it possible to integrate the high future demands on the sensors. Each axis of the robot is controlled by its own 8 bit computer (INTEL 8044). All axis-controllers are connected to the master by a high speed serial bus (Bitbus, 2.4 Mbps, HDLC/SDLC protocol), in order to get their instructions and controlling inputs. The axis-controller detects all state values of the dc machine (current, speed, position) and calculates the set value according to a programmed control algorithm. This means that there are no longer analog components in the control loop. Moreover, the axis-controller surveys all sensors of the drive (Fig.2).

The man-machine-interface is realized in two ways: the hand programming unit (HPU) and the system console (CRT). The hand programming unit is suitable for the teach-in of real moving sequences and the system console allows numerical programming with on-line execution. The master computer includes two parallel working 16 bit microprocessors (MC 68000), one managing the entire system, the other calculating the coordinate transformation and interpolation. The bitbus interface, as primary station, controls the overall network and issues commands and reference input signals to the secondary (slave) stations.

## STATE FEEDBACK CONTROL

In order to control the position of the robot drives, the slave computers are programmed as controllers for the state feedback control of a dc motor. Therefore, a mathematical control law - which is a complex function of the states of the system involving state feed back control - is implemented on the program memory of the axis controller. In order to design a controller using feedback, it is necessary to set up the mathematical model of the drive. The parameters are determined of the step response and manufacturer's information.

Neglecting the nonlinearities, a state space model of the plant may be written in the form

$$\begin{aligned} \dot{\underline{x}}(t) &= \underline{A}.\underline{x}(t) + \underline{b}.u(t) \\ y(t) &= \underline{c}'.\underline{x}(t) \end{aligned} \qquad (1)$$

The state vector x(t) has the components armature current, motor speed and position. The system matrix $\underline{A}$ and the input vector $\underline{b}$ contain the parameters of the plant.

The transfer function can be obtained from the matrix equation (1). For the synthesis of the digital control loop with state feedback control the time-discrete state space model of the plant must be determined. The sampling time T is of great importance here. The sampling period must be as short as possible to avoid distortions of the informations about the plant. However the control algorithm needs a finite calculating time. If the calculating time $T_c$ (sampling, calculation and output) is much smaller than the sampling period T, the time-discrete state space model becomes much easier. In order to achieve a quasi-continuous control, a sampling time of 1/6 to 1/10 of the dominating time constant of the plant is proposed. For position control the mechanical time constant of 14.6 ms dominates. Therefore in this research a sampling time of T=2 ms has been chosen.

With the help of the continuous state space model and the sampling time, the time-discrete state space model can be calculated as shown in Eq.(2).

$$\begin{aligned} \underline{x}_{k+1} &= \underline{\Phi}(T).\underline{x}_k + \underline{H}(T).\underline{u}_k \\ y_k &= \underline{c}'.\underline{x}_k \end{aligned} \qquad (2)$$

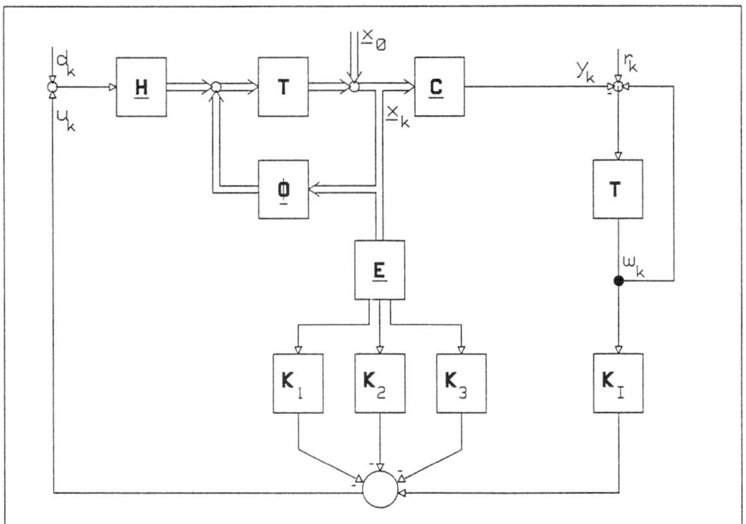


Fig. 3 Closed Control Loop

where

$$\underline{\Phi}(T) = \underline{\Phi}(t=T) = \sum_{k=0}^{\infty} \frac{T^k}{k!} \underline{A}^k$$

$$\underline{H}(T) = \int_0^T \underline{\Phi}(t).\underline{b}.dt = \sum_{k=0}^{\infty} \frac{T^{k+1}}{k+1} \frac{\underline{A}^k \underline{b}}{k!}$$

The time-discrete state space model of the plant describes the values of the state vector $\underline{x}(t)$ at the sampling moments $t=kT$ $(k=0,1,2,\ldots)$, presuming that the set variable $u(t)$ is constant during the sampling interval. An expansion into a series, with the help of a Pascal program, developes the matrices of the continuous system $\underline{A}(t)$ and $\underline{b}(t)$ into the discrete system matrix $\underline{\Phi}(T)$ and the input matrix $\underline{H}(T)$. Then the time-discrete transient function can be obtained in terms of the z-transform:

$$G(z) = \underline{c}'.(z\underline{E} - \underline{\Phi}(T))^{-1}.\underline{H}(T) \qquad (3)$$

The time-discrete transfer function is used to design a state feed back controller including an integrator (Fig.3). This is achieved with the help of the pole assignment method, by which the closed loop system poles are placed in prespecified positions in order to have complete control over the system dynamics. The integrator has the following advantages:

- the steady state error disappears completely
- the controller reactes to small permanent disturbances with great torques
- the integrator smoothes the motions on a continuous path

The closed control loop (Fig.3) can be described by Eq.(2) and Eq.(4)

$$\begin{aligned} w_{k+1} &= w_k + r_k + y_k \\ u_{k+1} &= -\underline{K}'.\underline{x}_k + K_I.w_k \end{aligned} \qquad (4)$$

With the assumption that the output of the integrator is a further state variable, an expanded state space model can be written as in Eq.(5):

$$\tilde{\underline{x}}_{k+1} = \begin{bmatrix} \underline{x}_{k+1} \\ w_{k+1} \end{bmatrix} = \tilde{\underline{\Phi}}(T)\, \tilde{\underline{x}}_k + \begin{bmatrix} 0 \\ 0 \\ 0 \\ 1 \end{bmatrix} r_k \qquad (5)$$

with
$$\tilde{\underline{\Phi}}(T) = \begin{bmatrix} \underline{\Phi}(T) - \underline{H}(T).\underline{K}' & K_I.\underline{H}(T) \\ -\underline{c}' & 1 \end{bmatrix}$$

Since the stability of the entire system is influenced only by the eigen values of the system matrix $\underline{\Phi}$, the zeros of the characteristic polynomial are of great importance:

$$\tilde{\Delta}(z) := \det[z\underline{E} - \underline{\Phi}(T)] \qquad (6)$$

Assuming that the plant is completely controllable, the poles of the command transfer function can be preset randomly. Therefore the gain factors $k_1, k_2, k_3$ and $k_I$ can be determined with the help of the controllability matrix or the first standard form. For the pole assignment the following points must be regarded:

- for a stable control loop all poles must be placed within the unit circle on the z-plane
- the amplitude of the set variable increases with greater distances between the poles of the plant and the poles of the controller
- the control loop becomes faster if the poles are placed nearer to the origin

If the poles of the command or disturbance transfer functions are placed in a sector of +/− 30 to 60 degrees apart from the real axis inside the unit circle, a good dynamical behaviour of the control loop is realized. The fastest controller is given when all poles are placed in the origin (dead beat controller).

## INTELLIGENT DRIVE

The goal in developing the hardware is to bring the microcontroller and converter into a small compact form, i.e. the axis-controller (Fig.4) consists of only two VLSI-chips (single chip computer and timer). Currently, these chips together

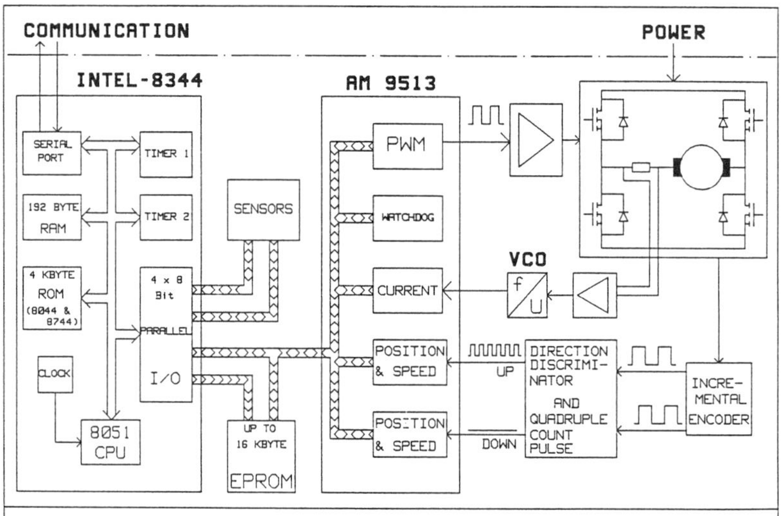


Fig. 4 Axis Controller

with a few smaller integrated chips (VCO, PLA, drivers) are soldered on one half of a board (Europe format). The other half is reserved for the power converter. In later versions the size of the board will be reduced to a small package mounted directly on the motor enclosure. This "intelligent motor" needs only a power source and a serial bus connection to the master. This will mean the reduction of the number of cables in a robot.

## HARDWARE

The 8044 microcontroller contains a powerful CPU with on-chip peripherals such as the SIU (serial interface unit), two 16 bit timer/counters, 5 source interrupt capability, 192 bytes on-chip RAM and four 8 bit parallel ports. The SIU is designed to perform serial communication with little or no CPU involvement. The SIU supports data rates up to 2.4 Mbps and the HDLC/SDLC protocol.
One timer is programmed to be the system clock (T=2 ms). An interrupt starts the sampling period which includes sampling of the state values and calculation and output of the set value (ca.375 $\mu s$). The rest of the time is used for less critical tasks such as programming the SIU and handling the I/O ports. The I/O ports control the limit switches, the emergency button, the power converter, the motor brake and the releasing of the PWM signal. Parameters and measured values of the control system are stored in the on-chip RAM. Therefore the system and control parameters can be changed during operation and the implementation of an adaptive control system is possible. State value detection is made by the allround system timing controller STC 9513 (AMD) with five programmable 16 bit timer/counters:

- counter 1 : watch-dog timer
- counter 2 : PWM-generator
- counter 3 : forward-counter for position/speed
- counter 4 : backward-counter for position/speed
- counter 5 : A/D-converter for current measurement

Since the set variable of the motor is a PWM signal, there is a ripple on the armature current. For this reason an integrating A/D converter is used to average over one period of the PWM signal. The A/D converter (Fig.5) is realized with the help of a voltage controlled oscillator (VCO). The armature current creates a proportional voltage across a

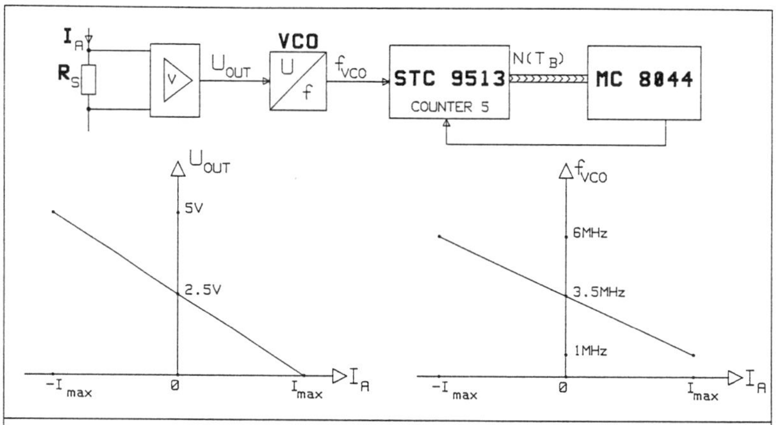


Fig. 5 Current Measurement

shunt. This voltage is amplified and level-shifted by an inverting operational amplifier and given as input to the VCO. Timer 5 counts the pulses (1.5 to 6 MHz) during constant time intervals. The resolution of the current measurement depends on two factors:

- the length of the measurement time
- the maximum counting frequency of the system timing controller

With a counting frequency of 6 MHz and a measurement time of 125 $\mu s$ a resolution of about 9 bits is obtained.

The software extends current measurement by two further features:

- permanent current monitoring
- drift compensation

On each motor axis an incremental encoder is mounted. In the incremental encoder an optical chopper disk provides two square wave signals (2500 pulses per revolution) with a phase difference of 90 degrees. A programmable logical array (PLA) detects the direction and provides a pulse for each received edge. Depending on the direction the pulses are given to counter 3 or 4. Therefore there are 10,000 counted pulses per revolution. By adding the contents of the counting registers the new position is calculated. The pulses received during one sampling interval give the speed. Since the state feedback controller needs instantaneous values, the measured average values of the speed are approximated by shortening the sampling interval. The tests that were run show that the tachogenerator can be replaced by a high resolution incremental encoder.

The power converter for the dc motor consists of four SIPMOS-FETs with very fast inverting diodes built in a four quadrant H-bridge. The bridge is driven by counter 2 providing the PWM signal (23.4 kHz). At the maximum counting frequency of the system timing controller (6 MHz) the modulation range is divided into 256 steps, giving a resolution of 8 bits.

The study of the entire measurement system shows that the counting frequency of the used timers is of great importance for its accuracy and resolution.

## SOFTWARE

The 8044 has four register banks, each with 8 registers. These banks are used to realize a multi-level, multi-tasking real time operating system. The software modules (Fig.6) are organised in four hierarchical levels corresponding to their priority. A maximum of eight functions belong to each level and work with one register bank. The "kernel" calls all needed functions corresponding to their priority. The system clock has the highest priority in the system (sampling time). When timer 0 (one of the timers in the 8044) releases an interrupt, first timer 0 is started again with a variable standing in a RAM cell (changeable sampling time). Then it is checked to see which level was interrupted. The necessary variables are stored and the register bank is switched over. At this moment the functions belonging to level 3 are started: measurement of the state variables, control algorithm and output of the set variable. Afterwards, the interrupted program is continued and the other functions are executed, i.e. communication with the master, interrogation of the limit switches, drift compensation.

There is a set of commands which can be sent from the master and which can be executed by the slave independently:

- go to reference point
- switch on controller
- switch off controller
- drift measurement
- reset
- read actual position value
- go x increments
- read error
- read RAM
- write RAM

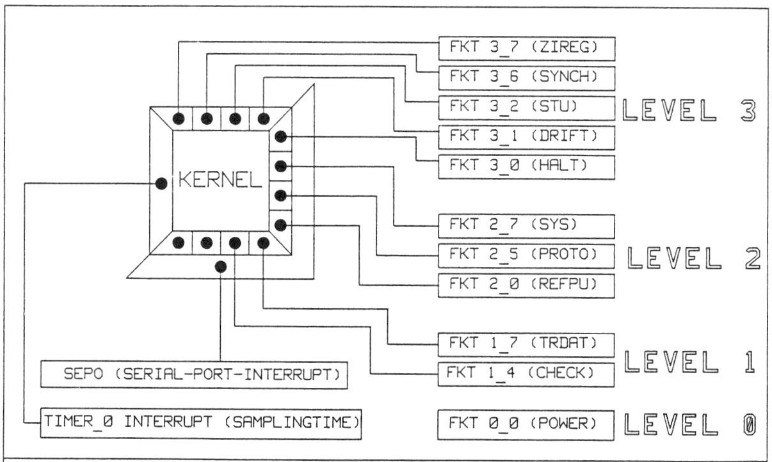


Fig. 6 Software Structure

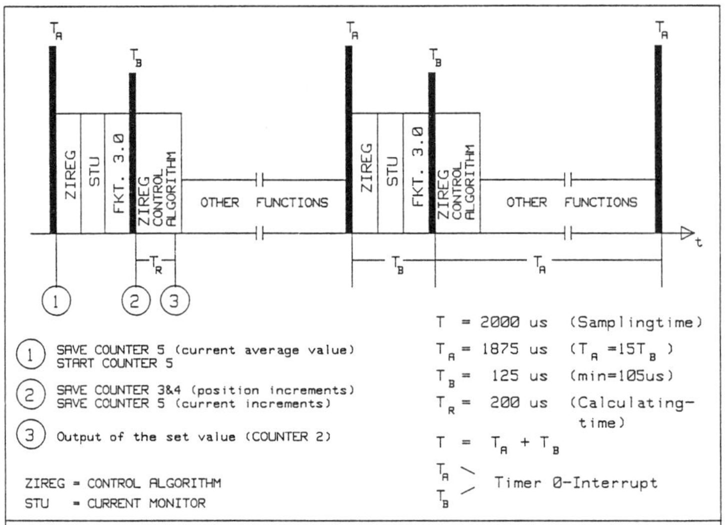


Fig. 7 Software Timing

This program system is easy to change and to adapt to new requirements.

## RESULTS

The state feed back control using the microcomputer control circuit described has created good results in the dynamic response to the command and disturbance variables. Fig.8 shows the step response to 1024 increments of axis 3 (upper arm) of the robot. The dead weight of the robot arm in the horizontal position is a permanent disturbance superimposed by the step response. The small negative set variable at beginning, i.e. t=0, provides the holding current, while not influencing the characteristic shape of the state values. Because of the integrator, the reaction of Due to the integrating component in the control algorithm the output variable is delayed by two sampling intervalls. For the set variable there is a reserve of ca. 30 per cent after the step. The actual position reaches its command value without overshoot after 70 ms.

## CONCLUSION

A digital control circuit applying only a few integrated chips has been developed to control the dc servomotors of a robot. Control circuit and motor form an intelligent drive which is fed by use of only two power wires and three control wires. The control algorithm employs the state feedback control method for the three state variables (current, speed and position) and shows a good dynamic response. No tachometer is needed.

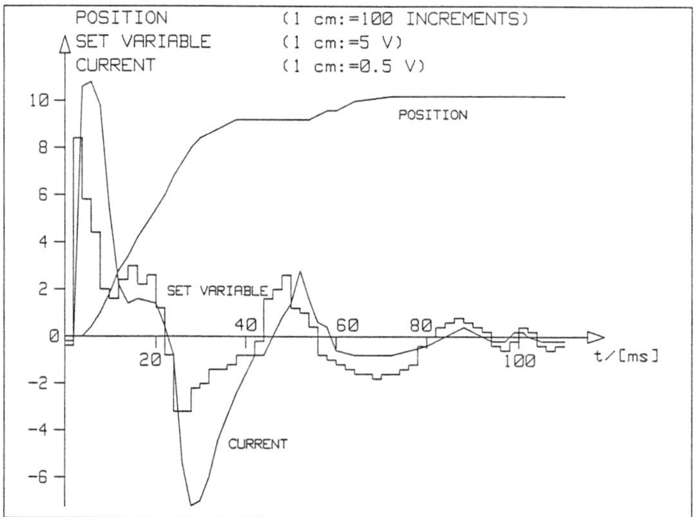


Fig. 8 Transient Function of Axis 3

## REFERENCES


Föllinger,O. (1974). Lineare Abtastsysteme. R.Oldenburg Verlag, München, Wien.

Naunin,D., K.Cevik and E.Baum (1984). Vollständig Digitale Regelung von Roboterantrieben. Elektronik. Franzis-Verlag München, Heft 10.

Hartmann,I. (1982). Digitale Regelkreise. TUB-Dokumentation Weiterbildung Heft 7, Berlin.

Naunin,D., K.Cevik and H.C.Reuss (1986). Intelligente Digitale Regelung von Roboterantrieben ohne konventionelle Wandler. VDI/VDE-Bericht Nr.598, Düsseldorf.

# MULTI-MICROCOMPUTER ROBOT CONTROL SYSTEM USING A SPECIAL ROBOT-ARITHMETIC-PROCESSOR FOR ADVANCED APPLICATIONS OF ROBOT MANIPULATORS

**K. Rathgeber and M. Schnell**

*Section Control Systems Theory, Institute of Control Engineering, Technical University of Darmstadt, Schlossgraben 1, D-6100 Darmstadt, Federal Republic of Germany*

Abstract. The performance of complex and advanced applications of industrial robots requires large computational power. The control system presented in this paper is designed as a multi-microcomputer system. The main purpose of this system is to create a pleasant environment for present and further studies in robotic research.
The Advanced Robot Control System (ARC) uses multiple microcomputers based on iAPX 8086 processors. A local bus of each processor allows additional memory and input/output expansions for increased throughput. Nevertheless, the rising calculation effort for the kinematic and dynamic relations shows the insufficient computing power of standard microprocessors.
Therefore a fast floating-point processor has been developed. This Robot Arithmetic Processor (RAP) is built up with bipolar bit-slice components, special arithmetic modules and hardware control mechanisms for on- and off-line diagnostic functions.

Keywords. Special purpose computers; Microprocessors; Hierarchical Systems; Multiprocessing Systems; Microprogramming; Diagnostic on Chip; Computer Architecture; Robots; Closed Loop Systems; Real-Time Control; Cartesian Robot Control.

## INTRODUCTION

The performance of complex and advanced applications for robot-manipulators, e.g. handling or assembly tasks with external sensor-feedback, requires large amount of computing power. The control-strategies for these applications are time-critical tasks. A multi-microcomputer system is very suitable for robotics control because it can provide the computing power required. The single components of such a control-system can fullfil different time-critical tasks. A real-time system can be designed by coordinating the single subsystems.
There are two opposite requirements in constructing such a multi-microcomputer control-system.

- The primary goal is to provide the computing power and speed for real-time calculations of cartesian and joint-based control loops available.
- On the other hand there should be a pleasant user environment for programming and coordinating the several processors of the control system. The system should offer the feasibility for users to create and test software and to analyse the results of the implemented control-strategies on the multiprocessor system or on a host-machine.

The system we designed is called ARC (Advanced Robot Control). ARC is a multi-microcomputer system with one master-unit and several slave processors.
Within this system we have implemented a special purpose processor, the Robot-Arithmetic-Processor (RAP) to meet the computational requirements of present and future robotic control applications.
The development of the Robot-Arithmetic-Processor aims at very fast processing of kinematic and dynamic transformations for robot-manipulators.
The algorithms implemented on RAP are kinematic independent to provide the control of different n-degrees-of-freedom manipulators.
There are two main goals of this paper:

- to discuss the architecture of the ARC-System including the Robot-Arithmetic-Processor
  and
- to present some principles for advanced applications with simultaneous position and force-control running on the Advanced-Robot-Control-System.

## MULTIMICROPROCESSOR CONTROL ARCHITECTURE

The required performance for a robot control system can be obtained with a multi-microcomputer system. Figure 1 presents the system structure using 4 identical single board computers based on Intel iAPX 8086 processors. The 8086 boards run at 8 MHz and each board includes 64 Kbytes of ROM, 16 Kbytes local RAM, 16 Kbyte dualport RAM, interrupt controller and one serial and parallel interface. A 8087 floating-point co-processor can be added to enhance the performance of the system. The global multiprocessor bus - an AMS-M-Bus, compatible with Multibus I - connects all processors and the global memory for fast interprocessor communication. The dualport memory permits together with the local onbord memory a high degree of parallel processing. The local 16-bit private bus systems are used for I/O-modules, intelligent controllers, local memory expansions and for fast interfaces to local processor-subsystems. The system is expandible with additional processor boards at the global bus, but a limit results from the maximum transfer rate on the bus (10 Mbytes/second).
To create a pleasant and functional software environment, it is useful to have one master processor avoiding system overhead on the other processors, for example synchronization and multi-

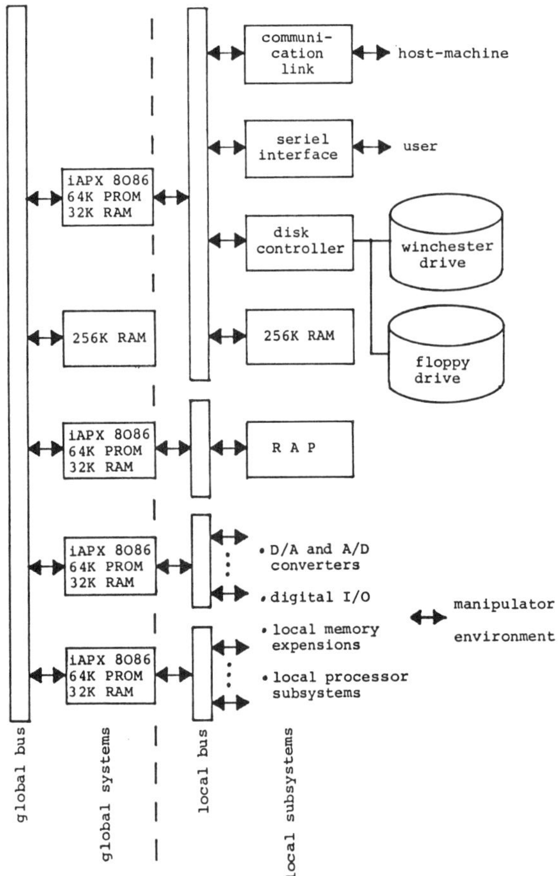


Fig. 1 Multi-microcomputer architecture

processor-supervision. These functions together with the high level tasks are implemented on the master unit using an Intel RMX multitasking operating system managing a winchester- and a floppy-disk, terminals and a communication link to a host-computer, at the moment a LSI 11/34 machine. The connection to the robotarm and the environment is realized with interface-components attached at the local bus system of two slave-processors. According to the requirements, the interfaces can be flexibly adapted to different robot tasks.
One processor of the system is qualified for arithmetic intensive tasks using a special processor described in the next paragraph.

## RAP - ROBOT ARITHMETIC PROCESSOR

It is necessary for the real-time implementation of control tasks to compare different hardware-solutions with respect to the required computation performance for robot tasks. The most common realization would be an application with standard 16- or 32-bit microcomputers with hardware-arithmetic-support. The great advantage of these systems is the quantity of available hardware- and software-components and a good development support. Nevertheless, the arithmetic computation performance is today not high enough for robotic algorithms, although it is possible to use mathematic co-processors for the floating-point arithmetic (e.g. time for one 32-bit floating-point multiplication with Intel 80287 ≈ 11,9 µs, Motorala 68881 ≈ 5,5 µs, National 32081 ≈ 4,8 µs). Looking at the computation-speed, an application with signal-processors could provide good results, but they have not yet on-chip hardware support for 32-bit floating-point arithmetic. Transputers would have the required performance, but they are at the moment too expensive and difficult in handling. Nowadays, the realization is only possible with a mini-computer (too expensive) or with special hardware designed for a specific problem.
We decided us for a micro-programmed solution, because this leads to a flexible system structure. Modifications are not complicated, so it is very practicable for different test implementations. We designed an independent microprocessor subsystem using bipolar bit-slice components and special floating-point building-blocks.
Figure 2 shows the block diagram of the Robot Arithmetic Processor, RAP.

### Floating-Point Path

The main modules of RAP are two VLSI-circuits with a one-bus building-block architecture, a floating-point ALU and multiplier /1/. They can handle both single (32-bit) and double (64-bit) precision formats and operations, as well as full 32-bit two's complement integers. The floating-point operations and data format is conform to the requirements of the IEEE standard 754, version 10.0. Basic instructions - add, subtract, multiply, divide, conversion to and from 32-bit two's complement integers, absolut value, compare - are executed onchip. Transcendental functions are calculated by a microprogram with a table lookup for fast processing. The devices can represent infinity, NaN (Not a Number), denormalized and zero operands and are capable to operate with all rounding modes. The treatment of exceptions, such as divide-by-zero, overflow,

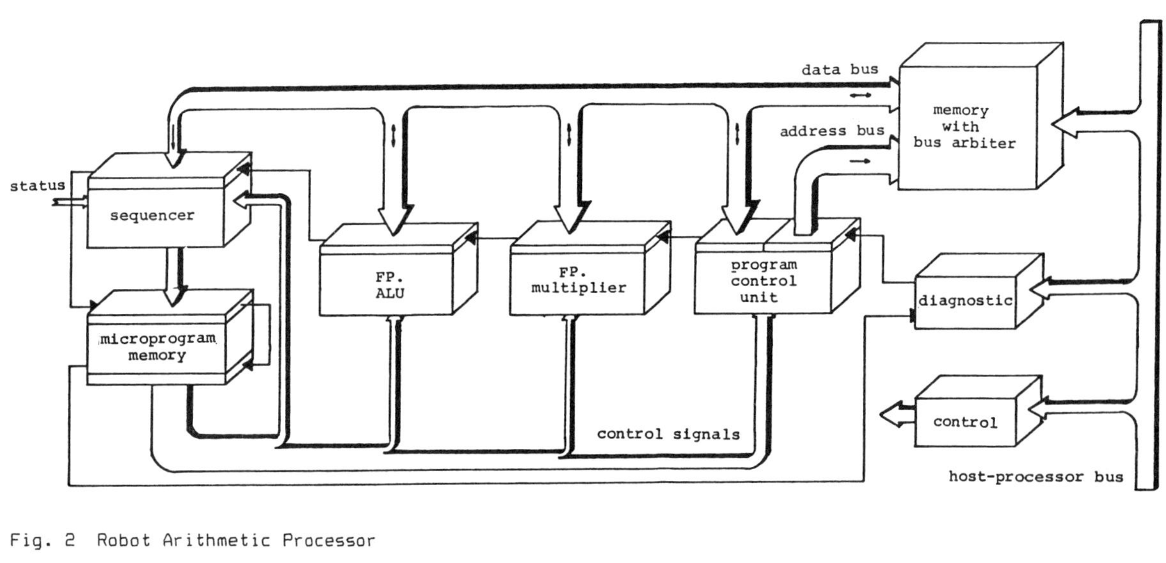


Fig. 2 Robot Arithmetic Processor

underflow, invalid and inexact operations are fully implemented. A very important feature for real-time control applications is a fast mode of operation which removes the time penalty of underflow exception handling by substitution of zero for denormalized numbers.
As the units use dedicated circuit arrays to perform the required functions, the arithmetic processing is faster than designs which rely solely on sequential, clocked logic. For example, the array flowthrough time is under 180 ns for single precision floating-point multiplication and addition. Data input and output transfer may occur at a rate of one transfer per clock cycle. One or both operands can be stored internally which allows repeated operations with a constant or multiple functions performed on two operands. Although the circuits had been designed for optimal flowthrough time, they have also the capability of pipeline-operations, because the next operands can be loaded into the input-registers during operation in the floating point array. Together with the possibility of parallel computation in the ALU and in the multiplier, vector- and matrix-arithmetic can be implemented very efficiently.

Control path
A control unit is required to build an independent processor-subsystem which can supply the arithmetic-chips with data at a high transfer rate utilizing the high performance of the floating point units. We have realized an universal micro-programmed control-unit to get a great flexibility for the implementation of a wide range of algorithms for robot control systems.
A micro-programmed machine uses a coherent sequence of micro-instructions to execute the machine instructions. Without any overlapping instructions, the throughput of the system would be to small. To improve the performance of the control unit, a pipeline technique is used with three stages of pipeline registers. We use the most efficient structure, an address-instruction-data based microprogram control architecture /2/, although this technique is the most difficult solution for the software development. The three required steps (determine microprogram address, fetch and execute micro-instruction) for a complete microcycle are overlapped by this technique which results in a shorter microcycle. This pipeline architecture offers significant speed improvements except in the case of conditional jumps where the pipeline must be cleared before execution.
In order to access instructions and data in an orderly manner, a Program Control Unit (PCU) is used to provide the most efficient mechanism for program control. The PCU is built up with four 4-bit ALU-slices. This 16-bit ALU has the capability of using its internal registers as the program counter, stack pointer and internal temporary registers for address calculations avoiding the inefficient directly produced memory addresses from the microword. This of course provides considerable flexibility in the architecture and also allows a much greater repertoire of instructions to be executed.
The 8086 host processor has direct access to the memory of RAP in a 8 Kbyte page addressing mode, because the address range of RAP is $32 \times 2^{16}$ bit = 256 Kbyte which is a quarter of the address space of the 8086 processor. This is not a restriction of the data transfer rate, since it is not necessary to move more than 8 Kbyte of data in real-time.
RAP can interrupt the host processor for communication and synchronization and - in the other direction - the host processor can set a signal in the status register of RAP.
RAP is attached to the local bus of the 8086 - a 16-bit SMP-M bus - via one interface board which can be replaced to use RAP in connection with other bussystems, e.g. Multibus I or II, VME bus, Q-bus.

Diagnostic Path
We have implemented hardware control mechanisms in RAP to utilize the following functions:

- on- and off-line diagnostics
- microprogram download
- debugging during software-development

Diagnostic is used to verify the functionality of a component or of the whole system and to identify the location which failed. A processor like RAP is difficult to test comprehensively because the combinatorial logic is very complex and includes sequential elements. Due to the complexity the logic should be partitioned into several smaller logic blocks. The problem of observation and control of sequential logic is difficult, because the output is a function of both the inputs and the state stored in a register. The design of RAP avoids the lack of access to the state registers by breaking the feedback path in the sequential logic block. The state can be controlled by injecting test data and output can be observed by reading the registers. Now the system is easy to test because the sequential system is converted into a pseudo-combinatorial logic.
A very feasible realization is the Diagnostic-On-Chip (DOC) technique, which works without a considerable increasing hardware effort /3/. DOC uses a buried shift register, called the shadow register, to load and unload diagnostic data. It allows diagnostic testing in the background - transparent to the normal system operation - and on-line diagnostics. The system needs not to be taken into off-line mode during diagnostic which is beneficial for maintenance purpose. For proper work all registers in the control path, e. g. pipeline and instruction register, must have these diagnostic capability. All DOC-elements are connected by a serial ring-line for minimal expenditure of hardware, controlled by the 8086 host-processor (see Figure 2).
Another advantage of this technique is the capability of using RAM- instead of PROM-elements for the microprogram memory. This feature is called Writeable Control Store (WCS) and increases the flexibility of RAP with regard to the development of different robot control algorithms. The working method of RAP can be adapted according to the requirements of the algorithms. Additionally the basic functions can be changed during operation by loading new parts of microcode.
The hardware of RAP together with the host-processor allows an easy software development without necessity of an expensive bit-slice processor development system. The program development utilities are:

- Assembler
  We use a relocatable macro meta assembler running on an IBM-PC which differs from an ordinary microprocessor assembler in that way, that the user can define the instruction set and hardware configurations. This development step can be divided further in defining and assembling of microprogram, generation of machine instructuion addresses out of the microprogram, assembly of machine program and formatting of micro- and machine program.
- Download
  The DOC-technique together with the Writabe Control Store allows downloading of the formatted microprogram, whereas machine-code can be loaded directly into the memory of RAP.
- Test and Debug
  The user can run the programs and debug them

with additional hardware facilities of RAP and with a software control monitor implemented on the 8086 host processor. The off-line test is supported by start/stop and single-step functions. Hardware breakpoints are also useful for the real-time test of the system.
Using these utilities, a wide range of algorithms for robot tasks can be implemented.

Computation Performance
The resulting computation performance is listed in Table 1 for 32- and 64-bit floating-point operations. Together with the number of mathematic operations of the control algorithms it is possible to estimate the attainable sampling periods of the control loops. The table shows two different calculation methods, a pipelined operation and an operation with memory-to-memory transfer which corresponds to the maximal execution time without using the parallel datapaths and pipeline facilities. Further reduction of the execution time can be obtained through parallel use of the two floating-point units. This method is very suitable for vector- and matrix-operations. An example of a vector-vector multiplication is described in the appendix. The computation performance is between 3 and 5,5 MFLOP (330 - 180 ns per operation) depending on the vector dimensions. If the whole algorithm is implemented on the microprogram level, the result would be the shortest sampling period, but the size of the microprogram is limited to addressable memory.

TABLE 1 RAP: EXECUTION TIMES, SINGLE FUNCTION

| operation | | computation of memory-to-memory transfer | computation of pipelined operation |
|---|---|---|---|
| multiply | 32 bit | 720 ns | 450 ns |
| | 64 bit | 1170 ns | 900 ns |
| add substract compare | 32 bit | 720 ns | 450 ns |
| | 64 bit | 990 ns | 720 ns |
| divide | 32 bit | 2250 ns | 1980 ns |
| | 64 bit | 4230 ns | 3960 ns |

For real-time applications it must be taken into account that data transfer between RAP and the host processor requires in proportion to the RAP arithmetic a large time. Nevertheless, the ratio between arithmetic and transfer operations of RAP is less than the ratio of standard microprocessors. Therefore the estimation of the computation time cannot be restricted to the number of mathematic operations. The worst-case execution time - memory-to-memory transfer - is a first evaluation if parts of the algorithms are microprogrammed and data transfer is avoided as far as possible.

## APPLICATIONS

Over the last few years several applications in robotics have been investigated. The achievement of robot-assembly and handling tasks requires not only the use of additional sensor-systems, called external sensors in oppsite to the internal sensors like joint-position sensors, but also complex control-strategies and algorithms. Examples for external sensor systems are wrist-mounted force/torque-sensors or proximity-sensors. The control strategies, like simultaneous motion and force control in contour-finding and -tracking tasks or the performance of the peg-in-hole problem require large amounts of computationl power to realize good performance of the applications.

Control Hierarchy
The performance of robot tasks can be stressed out in a hierarchical control structure, which is presented in figure 3. The robot tasks can be devided into three levels:
- Configuration level
  The specification of the desired trajectories in the configuration space is based on the application dependent formulation of the robot task.
- Cartesian control level
  Depending on the desired trajectories and the specified sensor-configuration the cartesian control algorithms and the cartesian sensor-data (e.g. force/torque signals) are processed on this level.
- Joint control level
  The joint torques are calculated as functions of the desired trajectories in joint coordinate space and the sensor feedback from joint-based sensor systems like joint-position sensors.

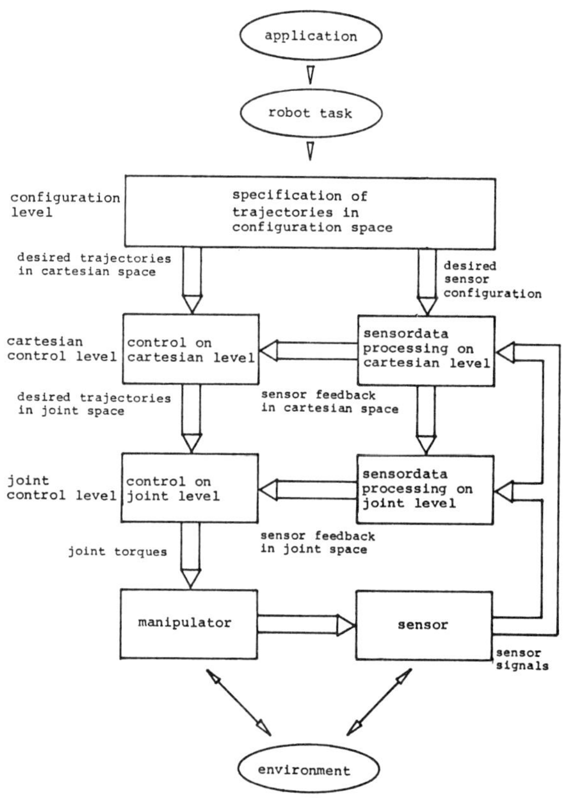


Fig. 3 Control hierarchy

General Requirements
From this hierarchical point of view we have worked out some general requirements concerning robot applications:
- Servo rates
  The realization of a high task-performance requires short servo cycles on cartesian level. In spite of the complexity of the control schemes there are two posibilities to meet the requirements: the use of a control system with very large computation power or the split of the system's servo rate. Most control strategies allow the splitting of the servo rate into two levels. At the high level the calculation of the control algorithms and the servo equa-

tions takes place. The lower level computes the configuration dependent parameters at a slower servo rate.
- Sensor data processing
  The utilization of external, in most cases complex sensor-systems necessitates additional computional power on the control system.
- Storage capability
  In order to evaluate and compare control structures in a quantitative way the system must provide the capability to store large amounts of data in real-time.
- Evaluation of experimental results
  The fast evaluation of test runs is enabled by storage and display facilities on the control system or on a host machine.
- Universal algorithms
  The design of a robot control system as a testbed for robotics applications requires the implementation of configuration independent universal algorithms to make quite sure that the control strategies are well guided for adaptation to different tasks, changing sensor configurations and distict kinematics of robot manipulators.
- Programming language
  The program development process is supported by using appropriate programming languages:
  Highlevel languages, like PL/M or FORTRAN on task configuration level and machine dependent languages (assembler- and microprogramming) on control and sevo level.

Referring to these requirements the control system must provide the following facilities and attributes:
- Modular system architecture
  The robot control system must be divided into modular subsystems to extract the parallelisms in the control algorithms. Several time critical tasks can be processed on such a modular system architecture.
- Large computational power
  The computation of the control loop equations and the sensor data processing in real-time tasks requires a multiprocessor system to provide the large amounts of computing power.
- Large amount of memory
  The system must have the facility to store large amounts of data in real-time. For later analysis data such as joint positions or sensor signals (e.g. measured contact forces) must be stored in every servo cycle.
- Intelligent subsystems
  The complex sensor data preprocessing is achieved by intelligent subsystems. These subsystems calculate the sensory feedback for the control algorithms. This concept reduces data transfer to other processors of the control system.

ARC-System for Robot Control

We have taken these requirements into account when designing the Advanced Robot Control System as a testbed for our further studies in robotics research. The distribution of the computational burden on the system's processor units is shown in figure 4.

- The master processor with the multitasking operating-system processes all high-level tasks and manages the global state memory.
- The time-critical numerically-intensive tasks - in particular kinematic and dynamic calculations - are executed by one 8086-processor with attached RAP. Data transfer to other processors reduces to a minimal value.
- The third processor uses two 8088-processor subsystems connected via dual-port memory at the local bus. The flexibility and the performance increases through sensor data preprocessing and local gripper control.

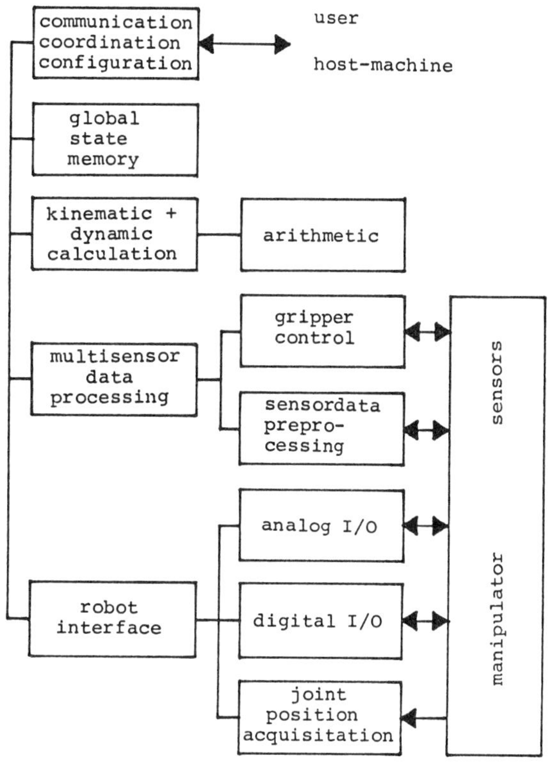


Fig. 4 Robot Control System

- The servo loops of the manipulator are connected via local interfaces to the fourth 8086-processor using analog and binary I/O-modules and a joint position acquisition system. This processor executes all manipulator specific tasks (e.g. robot supervision, joint control) except the kinematic and dynamic calculations.

## REAL-TIME IMPLEMENTATION - AN EXAMPLE

The ARC-System is designed as a testbed for real-time experiments in manipulator position and sensor control. As an example for implementation of robot control schemes the cartesian closed loop control with prescribed dynamics /4/ is discussed.

The control strategy is based on the idea to maintain a prescribed motion along desired cartesian trajectories and to get well defined responses in each direction of the task coordinate system. The manipulator being controlled with this strategy is a MANUTEC r3 with a 6-link-arm.

The real-time control algorithms must achieve the following steps:
- Measurement and preprocessing of the joint position and of the signals from a wrist mounted force/torque sensor.
- Computation of the end-effector position (forward solution of the kinematic problem), calculation of the cartesian velocity and acceleration based on the determination of the Jacobian matrix and of the time derivative of the Jacobian.
- Calculation of the desired joint velocity based on the cartesian control loop.
- Computation of the inverse dynamics of the system and calculation of the compensation-torque.

The computational burden of the control scheme is spread across the processor units of the ARC-system. The number of mathematic operations can be used to estimate the computation performance required in real-time control of the MANUTEC r3 arm.

The total number of operations for the calculation of the kinematic transformations, the inverse dynamics and the cartesian controller is approximately 3500 floating-point operations. The maximum servo-rate of the system with prescribed dynamics would be about 4 msec. To reach even a higher performance within this system, the sampling period is divided into two different cycles:

- Control level
  On the high-rate control level the system realizes the cartesian controller, computes the desired joint velocity and calculates the compensation torque resulting from the inverse dynamics of the manipulator.
- Parameter level
  On the parameter level, which works on lower rate, the control system refreshes the vectors and matrices of the inverse system.

Taking the splitted servo rates into account the system reaches a control level cycle rate of 2,5 msec and a parameter level rate of 5 msec. These cycle times are quiet feasable both for tasks with high speed requirements and for sensor-guided handling and assembly tasks in robot applications. The implementation of the proposed control scheme on the Advanced Robot Control System is under progress.

## CONCLUSION

This paper has presented the structure of the Advanced Robot Control System (ARC). Within this system we have integrated a special purpose Robot Arithmetic Processor (RAP). The ARC-System is designed as a testbed for studies in robotic research.

The architecture of the control system meets the requirements arising from hierarchical control schemes of advanced robot applications. The computation performance of the system is sufficient for real-time control of industrial robots, using universal and kinematic independent algorithms. The control system offers the feasibility to create and test control schemes and to analyse the results on a host-machine. Applications like the cartesian closed loop control with prescribed dynamics can be achieved in real-time.

A later version of RAP with reduced hardware effort could be a processor for industrial control systems. This RAP has to be programmed only with the specific robot parameters. The kinematic and dynamic relations would be computed for a n-degrees-of-freedom manipulator in a 'black-box' system.

## REFERENCES


/1/ B.Sackett, G.White (1985).
Solution For a High Performance Floating Point Co-processor Design.
Application Note Weitek Corporation, 15 March 1985

/2/ J.Mick, J.Brick (1980).
Bit-Slice Microprocessor Design.
McGraw-Hill, 1980

/3/ J.Birkner, V.Coli, F.Lee.
Shadow Register Architecture Simplies Digital Diagnosis.
Monolithic Memories Application Note AN-123

/4/ E.Ersü, K.Rathgeber, M.Schnell, W.Neddermeyer (1985)
A Robot Arithmetic Processor Concept for Cartesian Closed-loop Control with Prescibed Dynamics.
Proc. of the 1st IFAC Symposium on Robot Control, November, 6-8, 1985, Barcelona


## APPENDIX

An example of a vector-vector multiplication in a 32-bit floating-point format illustrates the working method of the Robot Arithmetic Processor (see figure 5).

$$d = d_n = \underline{a} \cdot \underline{b} = \sum_{i=1}^{n} a_i \cdot b_i$$

with $d_i = d_{i-1} + c_i$ , $d_0 = 0$ , $c_i = a_i \cdot b_i$

For fast processing, the microprogram uses the pipeline facilities and parallel computation in the floating-point ALU and multiplier. For example, three parts are executed during step 10:

- load vector-element $b_4$ from the main memory into one input register of the multiplier (memory addressed through a register of the Program Control Unit; multiplication will be carried out in step 14 and 15)
- multiply $a_3$ and $b_3$, first step
- add $c_1$ and $c_2$, second step

The status signals of the floating-point-chips can be tested by the sequencer for exeption-handling.

For a multiplication of 2 vectors with the dimension n, 4n+4 steps are required (1 step = 90 ns). Computation performance P of a computer is expressed through the number of floating-point operations per seconds (FLOP). The number of operations is 2n-1 for this example , the performance of RAP is:

$$P(n) = \frac{2n-1}{4n-3} \cdot \frac{100}{9} \text{ MFLOP}$$

An estimation is possible using the two boundaries:

$P_{max} = P_{(n\to\infty)} = 5{,}5$ MFLOP, $P_{min} = P_{(2)} = 3$ MFLOP

The result for a matrix-vector multiplication (matrix-dimension n x m) is:

$$P_{(n,m)} = \frac{2nm - m}{4nm - 2m + 5} \cdot \frac{100}{9} \text{ MFLOP}$$

$P_{max} = 5{,}5$ MFLOP, $P_{min} = 3{,}9$ MFLOP

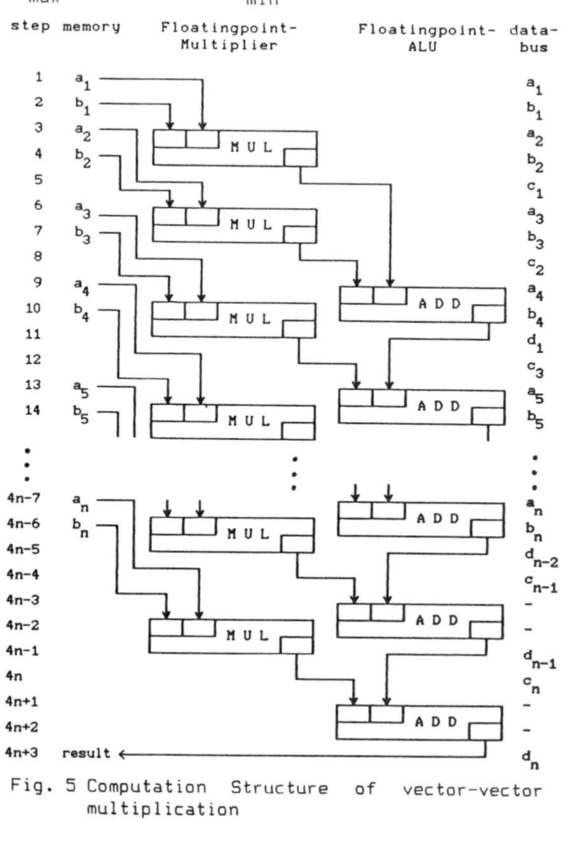


Fig. 5 Computation Structure of vector-vector multiplication

# EXTERNAL CONTROL AND GRAPHICS SIMULATION OF A TRACKING ROBOT

M. B. Zaremba and W. J. Bock

*Université du Québec à Hull, Hull, Canada*

Abstract. In advanced robotic systems operating in an unstructured environment, hierarchical, real-time control is needed to handle large volumes of sensory data and to execute the required operation. The paper concentrates on methods for visual guidance of tracking robots. The first part of the paper discusses optical sensors for robot vision and an industrial implementation of a robot guidance system using a photogrammetric 3-D vision sensor. Next, the paper describes a graphics system for the simulation of robot motion in an unstructured environment. The purpose of the system is to facilitate the design and analysis of higher lever tracking algorithms that will make obstacle avoidance possible and minimize the possibility of losing the target. The system takes the robot dynamics into account and simulates the use of various vision sensors.

Keywords. Robotics; sensors; vision systems; target tracking; graphical simulation; hierarchical control; microprocessor control.

## INTRODUCTION

The evolution of robot control has led from simple point-to-point control, through supervisory continuous path control, up to multi-robot, flexible systems using AI techniques. Two areas can be distinguished as essential for further expansion of robotics in complex manufacturing systems: sensory capabilities and effective control techniques applied on the level of task and action planning.

The use of sensed information in determining proper motion strategies leads to greatly enhanced robot performance capabilities, particularly for robots working in dynamically changing environments. Although artificial vision plays the main role, all available information will be employed to make robot technology as effective as possible, including information obtained from acceleration sensors, acoustic sensors, magnetic sensors, and force sensors. The initial part of the paper discusses a 3-D vision sensor designed at the National Research Council of Canada and optical range sensors used in machine vision applications.

The sensory information constitutes the perception part of the robotic system. The role of the control part is to connect perception to action. The control tasks are executed on different levels (Fig. 1).

The *Task Planning* level supervises the operations to be performed by the robotic system. The *Action Planning* level plans a path to a specified goal, using sensory information that provides a model of the work environment. The *Path Control* requires a model that reflects the state of the environment as the plan is being executed.

The problems of task and action planning are of particular importance, and subject to intensive research in the area of autonomous vehicle guidance. (Crowley, 1985; Keirsey, 1985). Proper mission task control and vehicle navigation should result in getting a vehicle from one place to another in a reasonable time without any collisions with the obstacles. Similar problems also occur in flexible manufacturing systems. (Freund, 1984; Shin, 1985).

The three above levels are executed externally for commercially available robots, whereas *Robot Position Control* is usually executed by the robot controller unit, which generates signals to joint servos.

Figure 1 identifies a *Tracking Control* block. In the case of robots tracking an object, when the robot positions are determined by the trajectory of the target, the control aspects of the design are stressed rather than path planning. The main requirements focus on real-time response and stable, reliable operation. The tracking problem is often encountered in industrial practice, particularly in the domain of visual guidance of arc-welding robots (Baudot, 1983; Masaki, 1981). An analysis of the control system for a tracking robot equipped with a 3-D vision sensor is given later in the paper.

The operation of the tracking robot in an unstructured environment renders it necessary to acquire information on the environment and subsequently elaborate obstacle avoidance strategies while not losing the target. Artificial intelligence techniques have recently been introduced to control robot motion in a dynamically changing environment. The final part of the paper presents an outline of a graphics system for the simulation of a tracking robot in such an environment. The system, currently under development at the University of Québec in Hull, also makes it possible to simulate range sensors and the 3-D vision sensor described earlier in the paper.

## OPTICAL SENSORS FOR ROBOT VISION.

Primary information to be sought with machine vision systems is always spatial information, including position, orientation and shape measurements. Research effort in this area is mainly focused on one problem: how to obtain more useful information

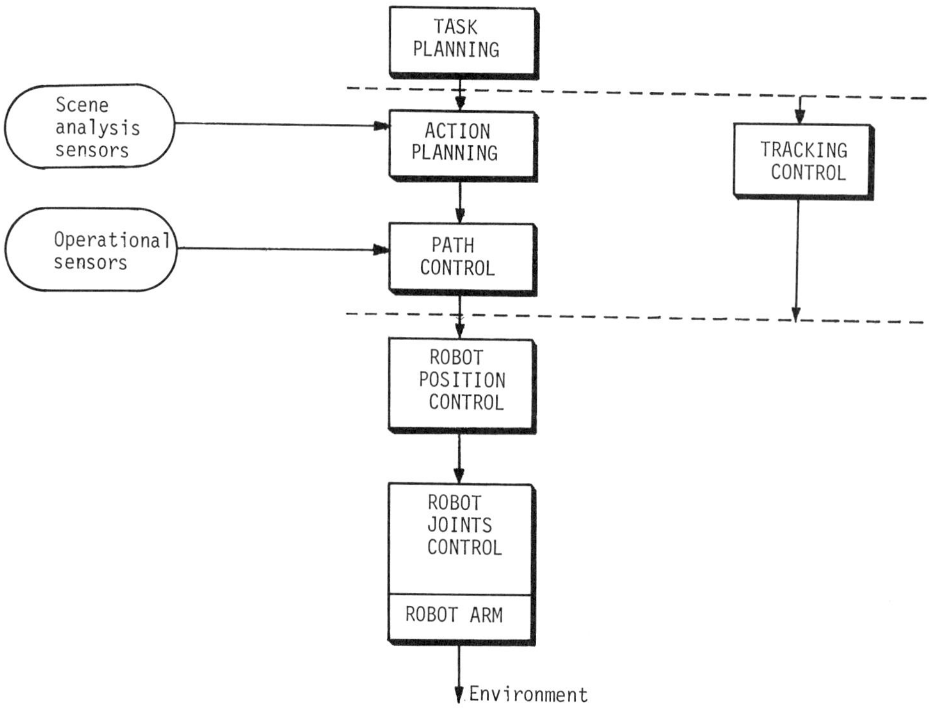


Fig. 1. Hierarchical structure of robot control.

in less time. This means simply how to overcome two major limiting factors situated at the level of physical detecting and at that of post-detection processing. Therefore, new physical sensors are being invented to reduce the measuring time and new measuring techniques that reduce post-processing (e.g. optical computing) are gradually being introduced.

Range measurement appears to be the most important to acquire in each machine vision system. Basically, four generic range measurement techniques are known and described in the literature (Strand, 1985).

The most common and the largest class uses a triangulation concept based on geometric measurements. Most triangulation techniques are active. There are passive methods that use ambient illumination, but they are rarely used because they require a great deal of post-detection data processing. The simplest active method consists of projecting a single point onto the object and subsequently imaging this point on a detector. This point is then scanned across the object to provide the range data over the object field. An example of such a method, using scanned laser beam and a single-element sensing detector, is described in (Kanade, 1984). The idea can be easily extended by projecting a line and using a 2D detector or even further by projecting a grid pattern that codes each position.

A second class of range-measuring techniques is based on the time-of-flight measurement. A signal carrier (acoustic or optical waves of known velocity) is sent from the measuring system, reflected off the object and detected by the observer after the measured period of time. Due to timing limitations, however, range accuracy of the sensors is low (about 2 cm). The method requires fast detectors with high SNR (signal-to-noise ratio), because the pulses of light are short and therefore carry very little energy.

Interferometric techniques are based on the comparison of the phase difference between the measuring beam sent to the object and reflected back, and the reference beam. Phase difference between the beams results in interference pattern and final measurement variable is light intensity.

Commercial interferometric systems are available for numerous applications and cover the distance scale from nanometers to kilometers with extreme accuracy (1 part in 10).

The fourth group consists of diffraction methods. If a spot on an object is coherently illuminated, the scattered light forms a speckle pattern on a screen. The correlation length of this pattern depends on the distance. These methods provide high data acquisition speed with MHz data rates achievable.

When considering the application of any particular method, it is important to see how efficient the measurement process is. This can be described in terms of the useful bits of range data obtained per bit of measured data; if one has to take many measurements to obtain a single bit of range data, the method would obviously be inefficient and slow. To develop efficient implementations, careful analysis and computer simulation of the real measurement conditions are usually needed to keep the number of measurements and the execution time to a minimum.

A 3-D vision sensor based on photogrammetry technology was developed at the National Research Council of Canada (Pinkney, 1978). The sensor provides precise information on both position and orientation of a special target with reference to the optical centre of the camera. The sensor is therefore particularly well suited to situations where the information required relates to the location rather than to the shape of an object. The operation of the sensor is fast enough for real-time robot control (30 - 60 Hz).

CONTROL SYSTEM OF A TRACKING ROBOT.

The vast majority of vision systems applied for robot guidance have been based on image processing. From the point of view of real-time control, image interpretation is in many cases computationally expensive and may introduce considerable noise into the control system.

This paragraph discusses an application of the NRCC vision sensor mentioned above for the control of a tracking robot. A dotted target is mounted in a fixed position with reference to the tracked object. Given the offsets, the location of the object with respect to the robot end effector can be determined and kept at the desired value by the control system.

The diagram of the control system (Zaremba, 1986a) is shown in Fig. 2.

The task of the robot is to unload a part from a swinging carrier and deposit it in a container. The various phases of the operation include target selection and acquisition, subsequent part stationkeeping and possible target reacquisition, through to the final materials handling sequence.

A set of targeted mechanical parts is located on a carrier suspended on an overhead conveyor. The carrier swings freely in a 6 degrees of freedom motion. Consequently, the robot is required to track and stationkeep on the selected target before an attempt to remove or load a part is made.

The robot used in the development is an industrial Cincinnati Milacron T3 robot, equipped with a camera mounted on its wrist. The gripper action is monitored through a force/torque sensor. The robot controller and the external computer are connected by serial DEPC (Direct External Path Control) communication link.

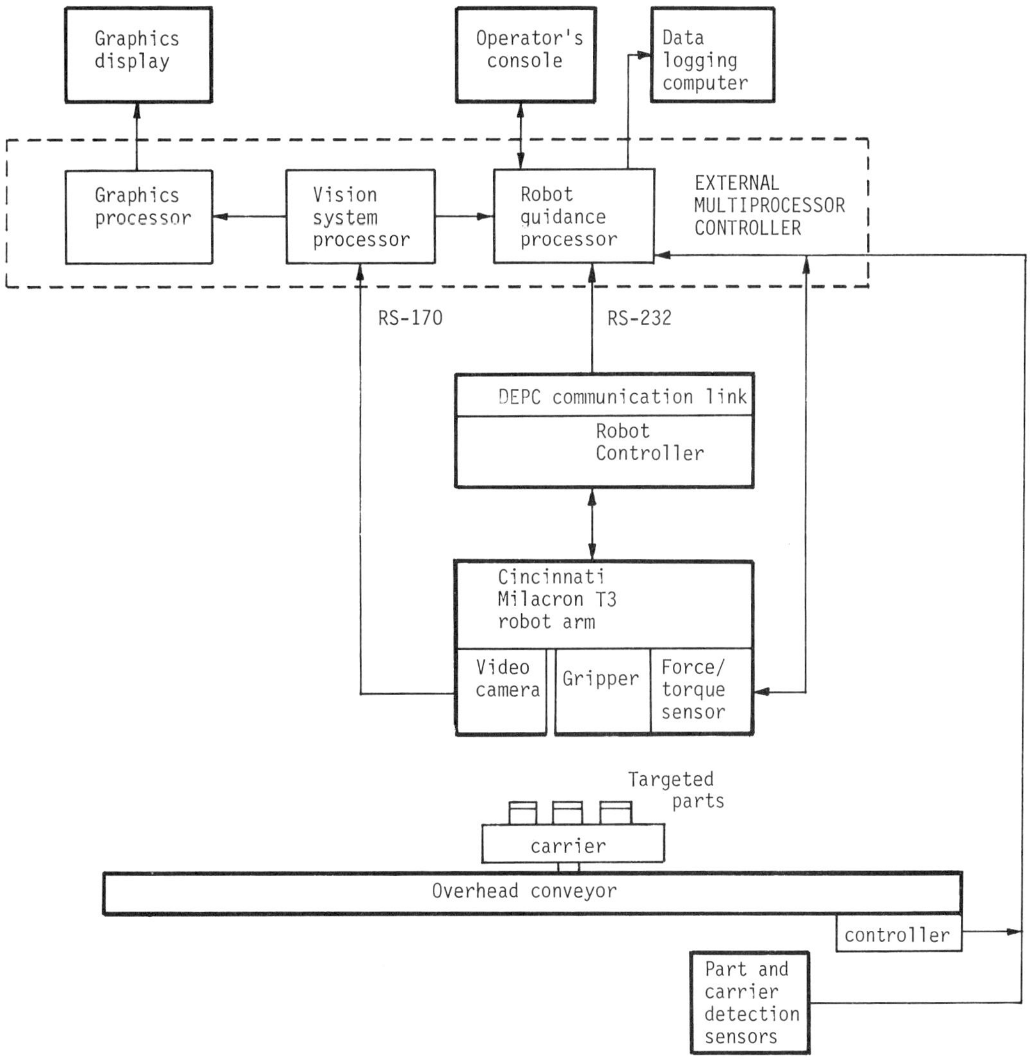


Fig. 2. Robot guidance system.

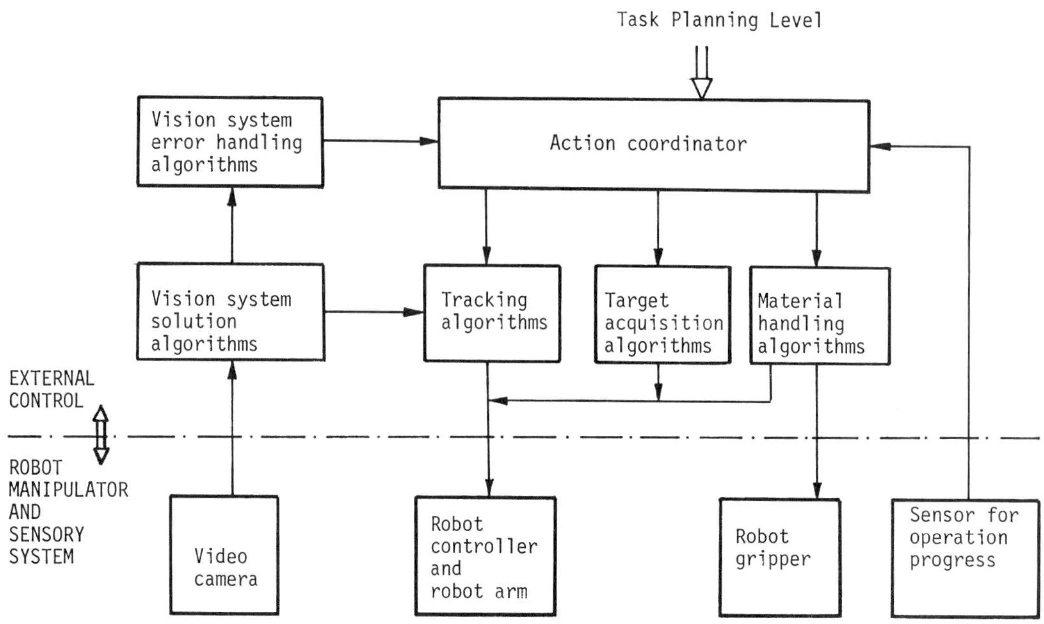


Fig. 3. Control software organization.

The external control is executed by three Intel 8086 processors and a mathematical co-processor 8087. The first processor serves to obtain the vision system solution. The second is dedicated to graphics display. The control functions are handled by the third processor. During the initial development, an additional data logging computer was used to store and analyze the data.

The structure of the control software is shown in Fig. 3.

The position and orientation of the target with respect to the camera is obtained directly from the vision system. The solution is in the form of 6 coordinates: 3 linear - x, y, z, and 3 angular - pitch ($\theta$), yaw ($\Psi$), roll ($\phi$). The vision system is therefore able to immediately define the transformation matrix S between the target and the camera coordinate systems:

$$S = \begin{bmatrix} \cos\Psi\cdot\cos\phi+\sin\Psi\cdot\sin\theta\cdot\sin\phi & \sin\Psi\cdot\sin\theta\cdot\cos\phi - \cos\Psi\cdot\sin\phi & \cos\theta\cdot\sin\Psi & x \\ \cos\theta\cdot\sin\phi & \cos\theta\cdot\cos\phi & -\sin\theta & y \\ \sin\theta\cdot\cos\Psi\cdot\sin\phi-\sin\Psi\cdot\cos\phi & \sin\Psi\cdot\sin\phi+\cos\Psi\sin\theta\cdot\cos\phi & \cos\theta\cdot\cos\Psi & z \\ 0 & 0 & 0 & 1 \end{bmatrix} \quad (1)$$

The tracking algorithm calculates the predicted position and orientation of the robot arm. The prediction consists of smoothing and extrapolating the desired values calculated up to the latest sensory solution. Therefore, for the vision sampling instants $t_v$, the transformation matrix describing predicted position and orientation of the end effector (tool) with respect to the robot world coordinate system is:

$$R_D\,(t_v + t_{pred}) = R_D(t_v) + dR_D(t_{pred}) \quad (2)$$

The actual desired robot position and orientation $R_D(t_v)$ is derived from the kinematic chain as in Fig. 4, and calculated according to the formula:

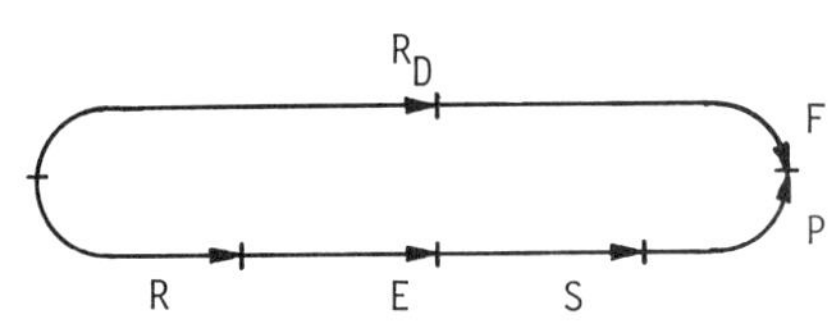


Fig. 4. Kinematic chain for the target tracking system.

$$R_D\,(t_v) = R(t_v)\cdot E\cdot S(t_v)\cdot P\cdot F^{-1} \quad (3)$$

where all the elements are homogenous 4-by-4 transformation matrices:
E - from the vision sensor frame to the end effector frame,
P - from the tracked part frame to the target frame,
F - from the tracked object frame to the end effector frame.

Tracking, although the most important, is only one of a few operations supervised by the Action Coordinator. The coordinator, based on the sensory information and on operator commands, selects the phase of the robot cycle to be executed. The major situations are (Zaremba, 1986b):
- Part selection,
- Target acquisition,
- Normal tracking with the target within the camera field of view (FOV),
- Target reacquisition in the situations when the target is partially outside the FOV,
- Approaching and grasping the mechanical part,
- Materials handling sequence,
- Termination of the cycle.

The robot guidance system has been implemented showing the advantages of the photogrammetric vision system, the hierarchical structure of the system controller, and the prediction-based tracking.

Excluding the target acquisition phase and tracking phase, the system operates in a structured environment. There are no obstacles intersecting the robot path. A further step to enhance the system capabilities is to extend its operation to an environment which is not fully structured.

## GRAPHICS SIMULATION.

The design and verification of the algorithms for the control of a tracking robot in an unstructured environment with obstacles requires the application of simulation tools. A program for graphics simulation of robot motion in such an environment is currently under development at the University of Québec in Hull. The simulation is performed on a Jupiter 7+ graphics terminal connected via DMA link with a VAX-11/750 minicomputer under ULTRIX-32 operating system.

The purpose of the program is to analyze algorithms on the task and action planning level, taking into account the inertia of the robot. The robot dynamics are accounted for by the generation of a "forbidden volume" (Ramirez, 1984) around the robot arm. The dimensions of the volume are the functions of the velocity and inertia of robot links.

The program organization is shown in Fig. 5.

The main module is the trajectory planner, activated when an intersection between the forbidden volume and an obstacle is detected. The planner works out a strategy that will allow the robot arm to avoid the obstacles and reacquire the target if the target is lost from the camera FOV during the process.

The module makes use of the information in the environment data base, the models of the variable parts of the environment obtained from the sensory data, and the knowledge base. The knowledge base can contain the priority rules for actions to handle dangerous situations, dynamic parameters of the robot, collision avoidance algorithms, the tolerances of the environment model allowed, etc.

## CONCLUSIONS.

Sensory capabilities as well as effective control strategies and techniques determine the applicability of robotic systems for industrial processes involving tracking tasks in an environment that is only partly structured.

The analysis and implementation of the sensory-based robot guidance development presented in the paper demonstrated that the NRCC photogrammetric vision system combined with the hierarchical structure of the system controller and prediction-based tracking provide real-time, industrially applicable solution for a flexible system with a certain degree of intelligence.

Operation in an environment that is fully unstructured imposes more severe requirements on the higher level control algorithms. In order to provide a tool for the design and analysis of such algorithms, a graphics simulation system is being developed at the University of Québec in Hull.

## ACKNOWLEDGEMENTS.

The work reported in this paper was supported by the University of Québec in Hull, grant SIR-52, and by the National Research Council of Canada, PILP program office.

## REFERENCES.

Baudot, W., and others (1983). Visually guided arc-welding robot with self-training features. Proc. Robots 7 Conference.

Brady, M., and Paul, R. (Ed.) (1984). Robotics Research: The First International Symposium. The MIT Press, Cambridge, Mass.

Crawley, J.L. (1985). Navigation for an intelligent mobile robot. IEEE J. Robotics and Automation, RA-1(1), 31-41.

Freund, E., and Hoyer, H. (1984). Collision avoidance for industrial robots with arbitrary motion. J. Robotic Systems, 1(4), 317-329.

Kanade, T., and Asada, H. (1981). In B.R. Altschuler (Ed.), 3D Machine Perception, SPIE, vol. 283.

Keirsey, D., and others (1985). Autonomous vehicle control using Al techniques. IEEE Trans. Software Engineering, SE-11 (9), 986-992.

Masaki, I., and others (1981). Arc welding robot with vision. Proc. 11th ISIR.

Paul, R. (1981). Robot Manipulators: Mathematics, Programming, and Control. The MIT Press, Cambridge, Mass.

Pinkney, H.F.L. (1978). Theory and development of an on-line 30 Hz video photogrammetry system for real-time three-dimensional control. Proc. ISP Symp. Photogrammetry for Industry.

Ramirez, C.A. (1984). Robotic intelligent safety system. Proc. Robots 8 Conference, 19, 50-62.

Shin, K.G., Malin, S.T. (1985). A structured framework for the control of industrial manipulators. IEEE Trans. System, Man, cybern., SMC-15 (1), 78-90.

Strand, T.C. (1985). Optical three-dimensional sensing for machine vision. Optical Engineering, 24(1), 33-43.

Zaremba, M.B., Poissant, M.A., and Hageniers, O.L. (1986 a). Robot target tracking system. Proc. British Robot Association -9 Conference.

Zaremba, M.B., Poissant, M.A., and Hageniers, O.L. (1986 b). Sensory information for supervisory control of a tracking robot. Proc. RAI/IPAR '86 Conference.

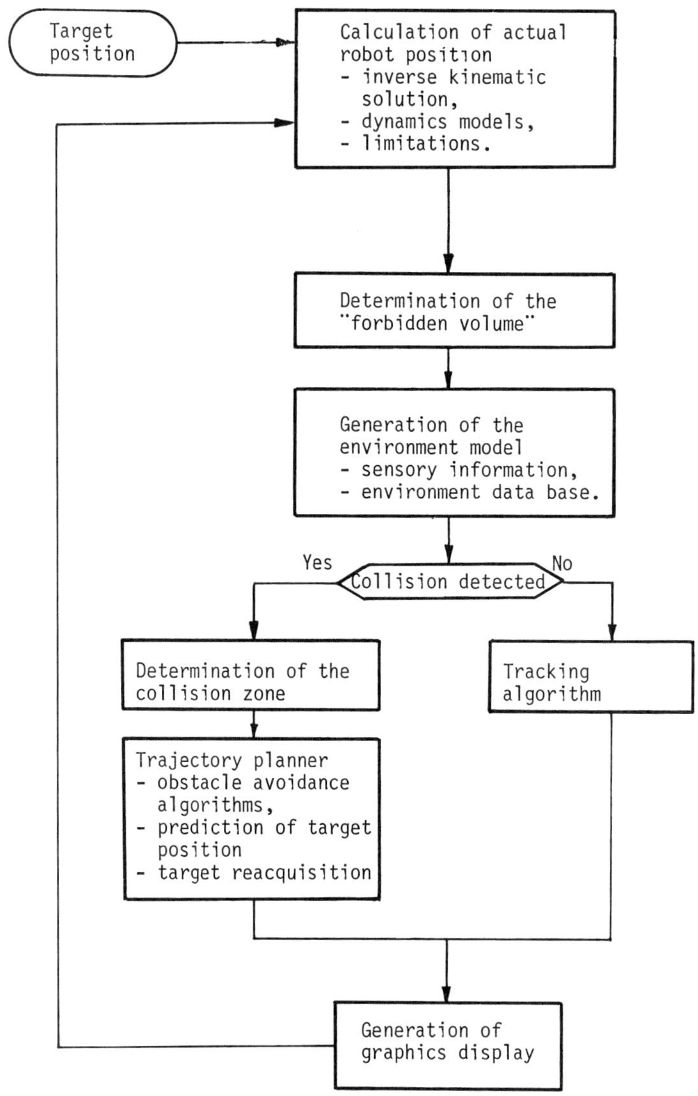


Fig. 5. Organization of the program for graphics simulation of robot motion.

# OPTICAL PROXIMITY SENSOR SYSTEMS FOR INTELLIGENT ROBOT HANDS

**E. Ersü* and G. Kegel****

**Isra Systemtechnik GmbH, Schöfferstr. 15, Darmstadt, Federal Republic of Germany*
***Control System Theory Laboratory, Institute of Control Engineering, Technical University Darmstadt, Darmstadt, Federal Republic of Germany*

Abstract. Optical proximity sensors are low-level vision sensors delivering low- and mid-range multidimensional information about the robots end-effector environment. Due to the fast signal processing they can be easily integrated in real-time robot control tasks. The paper presents the basic mechanical, hardware and software design principles of such a sensor, which uses distance measurement via optical triangulation as the basic method. For special robot tasks, special mechanical and hardware arrangements of the basic sensor type are needed. Two examples are shown for demonstration purposes. Possible applications are simple distance sensor devices, two-dimensional orientation sensors and optical robot teach-in units. Accuracy and efficiency of the sensor system are documented by using the sensor for recognizing holes and following arbitrary unknown contours.

Keywords. Position measurement; computer peripheral equipement; optical distance sensors; robot teach-in.

## INTRODUCTION

Progress and further development in robot assembly tasks will be inclose relationship with new sensor technologies. Sensory information enables an assembly robot to perform systematic movements referenced to arbitrary operational objects.
Due to the fast processing of the sensory systems information, in comparision with image processing systems, optical proximity sensors will become a very important part of the technological development of low-cost sensors. The variety of optical proximity sensors leads from simple reflex-couplers to line-sensors similar to image processing systems as shown in [1].
The paper presents the sensory design principles using position sensitive semiconductor devices and the measurement methods utilizing these devices e. g. triangulation. The major problem of optical distance measurement is the dependence of the object-reflected light on the surface and orientation of the object, and the decrease of light intensity with the square of the object's distance. Experiments to control the light intensity by adapting the output power of the light emitting diode, shown in the publications of [3] and [4], enlarged the maximum range successfully. Using diffuse reflected light of a very narrow focused light beam delivers the measurement from influence on object's surface or orientation.
Appropriate microcontrollers make a fast signal processing possible ( > 1 kHz ).

## STRUCTURE OF BASIC OPTICAL PROXIMITY SENSOR DESIGN.

The goals of robot-sensor development are on the one hand cost-effectiveness and flexibility and on the other hand robustness against mechanical and environmental influences. For this reason the proximity sensor principle published here uses a focussed infrared lightbeam at a wavelength of 950 nm coming from an infrared light emitting diode and reflected at the remote object.
Two main reasons make the use of laser devices impossible. On the one hand a laser doesn't accomplish the required dimensions, and on the other hand a laser beam endangers human eye.
Because of the roughness of the surfaces of most of the assembly-object surfaces the narrow lightbeam is diffusely reflected. A part of the reflected light falls on an optical lense and is focussed as a spot on the surface of a position sensitive semiconductor element. The edge currents of the spot-position-detector are used to evaluate the maximum of the spot position.

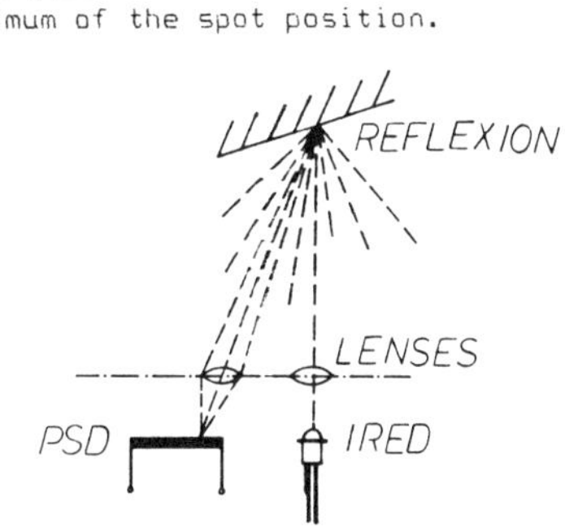


FIG. 1. Incidence of rays

The light-spot generates a photo-current which is devided into two position detector edge-currents with values inversely proportional to the distance between spot-position and detector edge.

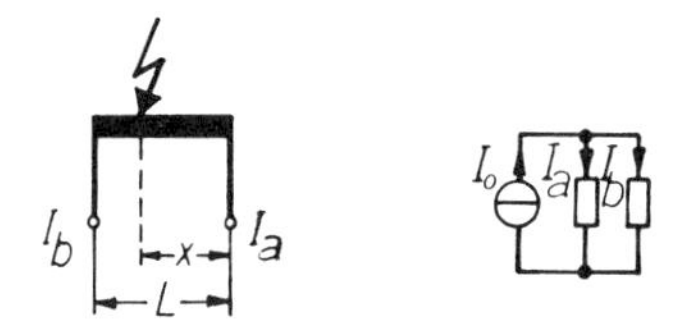

FIG. 2. Edge-currents

The following equations can be obtained by geometrical considerations:

$$(1) \quad I_a = I_o \cdot \frac{R_o \cdot \frac{L-x}{L}}{R_o} = I_o \cdot \frac{L-x}{L}$$

$$(2) \quad I_b = I_o \cdot \frac{R_o \cdot \frac{x}{L}}{R_o} = I_o \cdot \frac{x}{L}$$

$$(3) \quad \frac{I_a}{I_b} = \frac{I_o \cdot \frac{L-x}{L}}{I_o \cdot \frac{x}{L}} = \frac{L-x}{x}$$

$$(4) \quad x = \frac{L}{(1+\frac{I_a}{I_b})} = L \cdot \frac{I_b}{I_a + I_b}$$

The result of the division of the two edge-currents $I_a$ and $I_b$ is leading to the light spot position x and doesn't depend on the light intensity $I_o$. Therefore the measurement procedure becomes independent from the object's surface, color and orientation, as long as the intensity of the reflected light is high enough. However, experiments have shown that the theoretical independence from object's orientation and light intensity will only be satisfied if the object's surface is a real diffuse reflector. In [2] the application of a number of infrared diodes around the position-sensitive-detector improves the measurement accuracy for which one pays, however, by a reduced distance range and a higher amount of hardware.
Because of the jamming of environmental illumination changes, the analog edge-current signal processing is very complicated too. For this reason four general steps are necessary. At first the emitting-power of the infrared diode has to be controlled, to reduce the decrease of light intensity for larger distances. The analog PI-controller applied here varies from the method of digital approximation control as shown in [2], beeing more simple. To distinguish the useful edge-currents from enviro-mental interferences the emitted light has to be modulated and the edge-currents have to pass through an analog bandfilter. Finally an optical infrared-filter is put in front of the collector lense, similar to those published in [1] and [3].

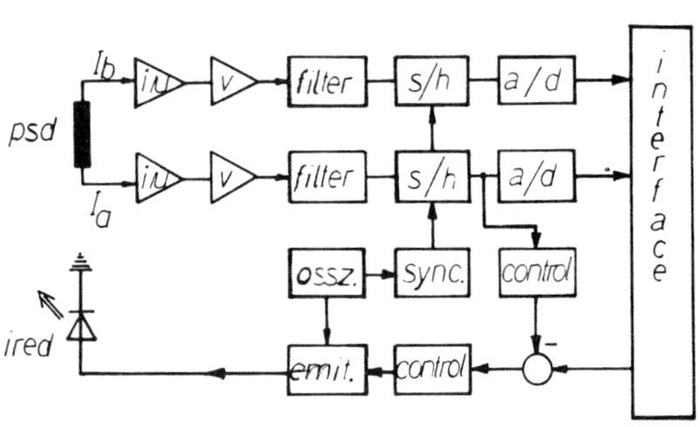


FIG. 3. Blockdiagram

The preprocessed analog signals are sampled and converted to digital values. The microcomputer is now able to perform the following tasks.
The edge-current values pass through a digital low-pass filter according to the software realization of equation (4).

$$(4) \quad x = L \cdot \frac{I_b}{I_a + I_b}$$

Depending on the geometrical assumptions, the result x of the division is assigned to the desired distance d. In the simplest case this assignment is expressed by the equation

$$(5) \quad d = const \cdot \frac{1}{x} ,$$

but in most sensor applications varying in geometrical and optical dimensions, the assigment

$$(7) \quad d = f(x)$$

is better performed in a look-up-table. Therefore this solution was chosen finally. The distance d is checked on plausibility. The maximum velocity between robot and object leads to the maximum change of the distance value in one cycle. The microcomputer has to ensure that all values are between the minimum and maximum distances fixed by the geometrical assumptions. Modulation and sampling of analog signals fix the period to be

$$(8) \quad t_{anl} = \frac{1}{f_{mod}} .$$

A modulation frequency $f_{mod}$ of 10 khz leads to an analog sampling period of 100 microseconds. In this time the microcomputer has to do digital-filtering, plausibility checks and distance assignment. Even with the time cycle of modern robot con trol systems a new sensory value has to be generated once every millisecond only. Therefore up to twenty distance values can pass through a digital low-pass filter or can be averaged by special algorithms.

## HARDWARESTRUCTURE

The basic requirements for the integration of sensory devices into robot control systems is a flexible digital interface with standard hardware components. Standardized low- and high-speed serial and parallel interfaces must be provided when the sensory signal processing is done by sensory processors and not by robot control systems. Such a concept is developed in the frame ARC ( Advanced Robot Control ) shown in [8], where each sensor processsor is interfaced via dual-port memory chips to the robot control. This concept of one sensor-processor ( 8088 ) can integrate various sensors of similar type depending on the complexity of the corresponding algorithms. An alternative, which is now under realization, is to use one microcontroller ( e.g. 8051 ) for each sensor, where a serial interfacing ( low- or high-speed ) will be used. A characteristic feature of a "smart-sensor-system" is the "stand-alone" structure of the implemented higher level operations. A stand-alone microcomputer smart-sensor-system with RS 232 serial interface with this items was developed at the THD system theory robot lab.

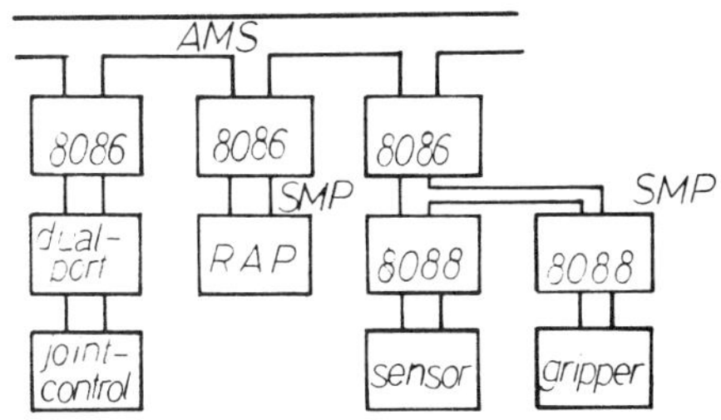


FIG 4. Hardwarecomponents ARC

The various types of hardware structures shown here will solve most of the sensor-system configuration problems by transfering the algorithms from one system to another.

## ALGORITHMS

The result of the analog-digital-converting is a 16 bit word from which only the upper ten bits are used. In most applications also the upper nine or eight bits are sufficient. These integer values are converted into the standard 32 bit floating point format and pass through a digital low pass filter to eliminate high-frequency noises.

$$(9)\quad I_{j_{out}}(k) = -a_1 I_{j_{out}}(k-1) + b_0 I_{j_{in}}(k).$$

The following computation of the spot-position on the position-sensitive line is done with equation (4).

$$(4)\quad x = L\,\frac{I_b}{I_a + I_b}$$

by assigning a 32 bit floating point value to the variable x. An integer operation converts this floating point number into an integer value to address the look-up-table input $x_{INT}$ of equation (7).

$$(7)\quad d_{INT} = f(x_{INT}).$$

If necessary, the successive values $d_{INT}(s)$ will pass through a second low-pass digital filter

$$(10)\quad d_{OUT}(k) = -a_1 d_{OUT}(k-1) - \ldots - a_n d_{OUT}(k-n) + b_0 d_{IN}(k)\ .$$

If the parameters $-a_1 \ldots -a_n$ and $b_0$ have the value 1/n+1, the equation is equivalent to an averaging algorithm. Adapting these parameters to the special application leads to a sensor information free of disturbing noises and unreasonable values. Unlike the methods published in [3] and [4] the higher level signal-processing is here completely digital and this gives the user the opportunity to vary the filter-algorithms and extend the signal-processing with plausibility-checks.

## VARIATION OF THE BASIC SENSOR DESIGN

In all sensor applications the basic principle described in chapter one is the major part of the various sensor types. The proximity sensor based on triangulation produces a distance information nearly independent from the object's orientation or surface. The construction is similar to the publication [3] or [4].

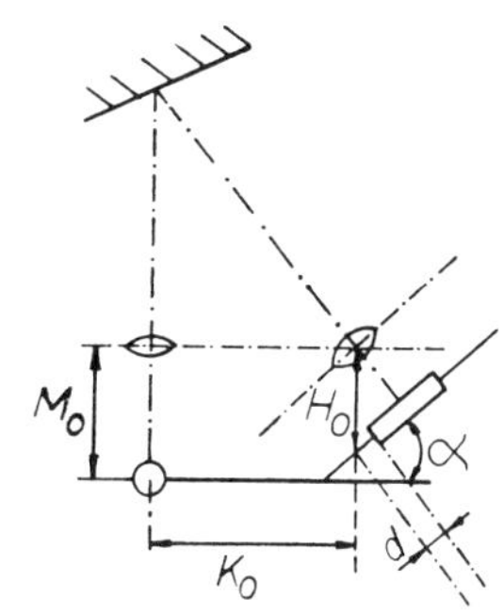


Fig. 5. Distance sensor

The geometrical sensor set-up leads to the equation :

$$(11)\quad d = H_0 K_0 \cdot \frac{1}{\cos\alpha} \cdot \left[\frac{1}{x - b} - \frac{\sin\alpha}{H_0}\right].$$

In this case $\alpha$ is the triangulation-angle. In the simplest version $\alpha$ and b are zero,

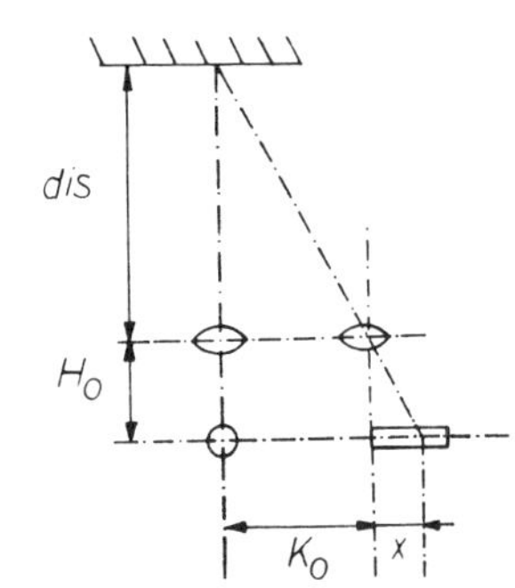


Fig. 6. Standard distance sensor.

which leads to the equation

$$(12)\quad d = \frac{H_0 K_0}{x}.$$

A type of this distance sensor was built up in the THD robot lab and leads to a measuring range of 50mm - 300mm with an accuracy of 0.5% at a distance of 100 mm. This application can be used in various robot tasks like following a moving object along the optical axis of the distance sensor. The distance between a sensor mounted on the robot-endeffector and an object is measured and used as the actual value in the distance contol-loop of the robot control system.

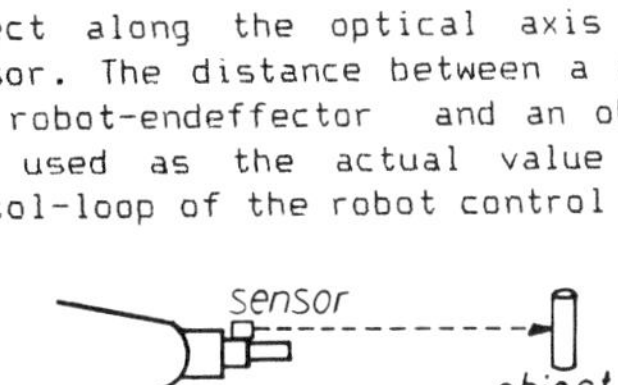


Fig. 7. Following moving objects

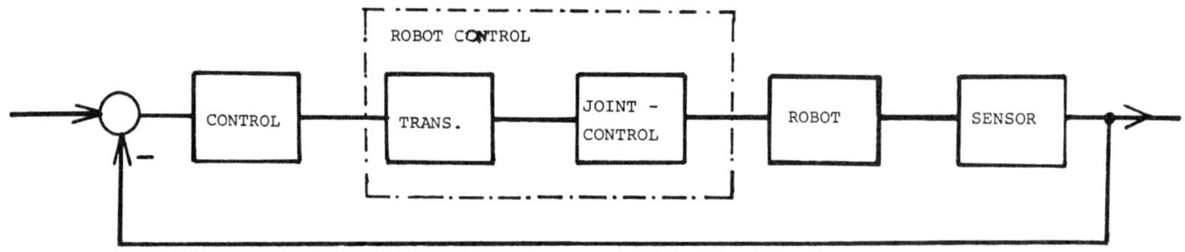


Fig. 8. Distance-control-loop.

By using two or more infrared emitters, the sensor will be capable of finding the orientation of an object.

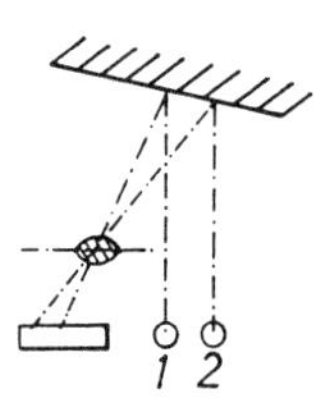


Fig. 9. Measuring object's orientation.

Determining the orientation of an object can be useful to prevent gripping of a wrongly oriented object, thereby loosing or damaging it. An arbitrarly oriented object lying on a working platform is gripped parallel when using an orientation sensor.

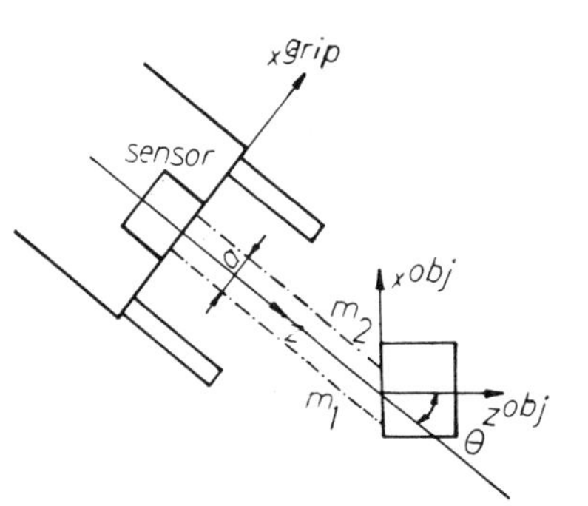


Fig. 10. Object orientation relative to gripper orientation.

The two emitter-rays, $m_1$ and $m_2$, give the orientation angle $\theta^{OBJ}$ and the average distance $z^{OBJ}$.

$$(13)\quad z^{OBJ} = z^{GRIP} + \frac{m_1 + m_2}{2}$$

$$(14)\quad x^{OBJ} = x^{GRIP}$$

$$(15)\quad y^{OBJ} = y^{GRIP}.$$

Therefore, the first step is a coordinate transformation from the coordinate frame $\underline{X}^{GRIP}$ to the frame $\underline{X}^{OBJ}$, which is equivalent to shifting along the z-axes by $\frac{m_1 + m_2}{2}$. The second step is a rotation about the new axis $Y^{GRIP} = Y^{OBJ}$ through an angle

$$(16)\quad \theta^{OBJ} = \arctan \frac{m_1 - m_2}{a}.$$

Gripping an object means then rotating the gripper in Z-X coordinate plane about the $Y^{OBJ}$-axes through the orientation angle $\theta^{OBJ}$ and then moving along the Z-axis with the distance $z^{OBJ}$.

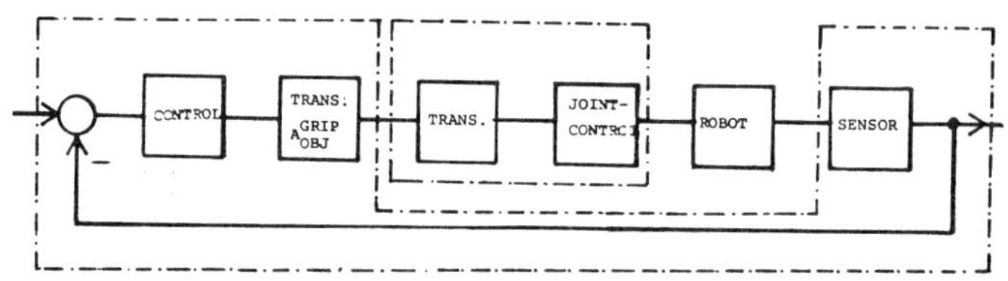


Fig. 11. Control loop for orientation control

A more complex extension of this method is to use four infrared diodes around a two-dimensional position sensitive detector. This gives the robot control the opportunity to measure the object's orientation in two dimensions. The two-dimensional position sensitive detector has the same features as the one-dimensional sensor. Because of this the corresponding hardware is the same as described in chapter one.

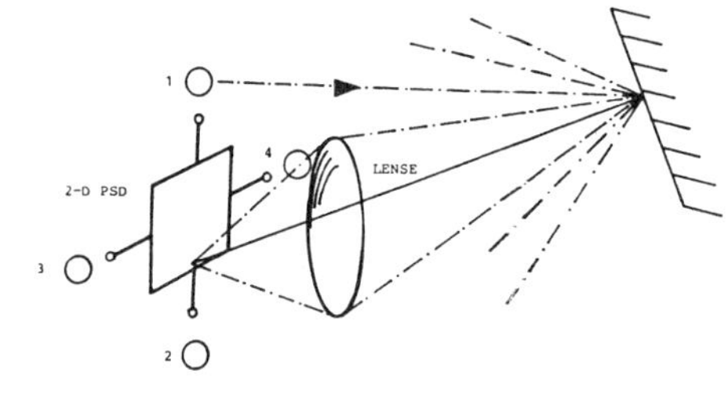


Fig. 12. Two-dimensional orientation measuring

The geometrical set-up leads to two angles $\theta^{OBJ}$ and $\psi^{OBJ}$ calculated out from the four emitted rays $m_1 \dots m_4$.

$$(17)\quad \theta^{OBJ} = \arctan \frac{m_1 - m_2}{a}$$

$$(18)\quad \psi^{OBJ} = \arctan \frac{m_3 - m_4}{a}.$$

Gripping the object means now rotating the gripper in the X-Y-plane through $\theta^{OBJ}$ and in the Y-X-plane through $\psi^{OBJ}$ about the coordinate $\underline{X}^{OBJ}$.
The last variation of the basic sensor principle leads to an optical teach-in device. The basic idea is letting the robot follow the motion of an optical spot describing the path to be programmed.

The robot follows the visible and infrared light-beam combination of an equipment in the users hand. The infrared light is scattered by a special optical lense and therefore is equivalent to the scattered reflection of the light beam in the distance sensor device. A part of this light finds its way through an optical infrared filter and an optical lense, focusing the light as a small spot on a two-dimensional position sensitive detector. The spot-position, calculated from the four edge-currents of the detector, is then used by the tracking algorithm in the robot control system.

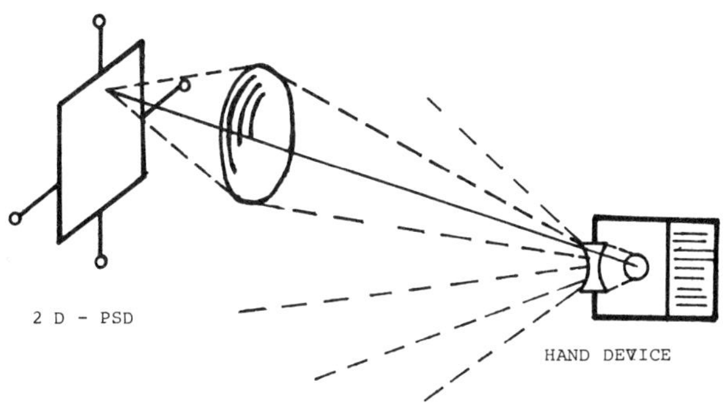


Fig. 13. Optical teach-in

The algorithm which assigns a robot motion to the spot position is very simple.

(19) IF $|I_{j+} - I_{j-}| > I_{IMOV}$ THEN

IF $I_{J+} > I_{J-}$ THEN MOVE IN DIRECTION j
ELSE MOVE DIRECTION -j

ELSE DO NOT MOVE

## SENSORAPPLICATIONS

To demonstrate the advantages of the sensor-system, four applications are described here.
At first a single-distance sensor is used to follow an unknown arbitrary contour. The distance $z^{contour}$ between the sensor-guided robot and the reflection-point of the contour is held on a nominal value of 120 mm by moving the robot in Z-direction vertical to the movement along the X-axis.

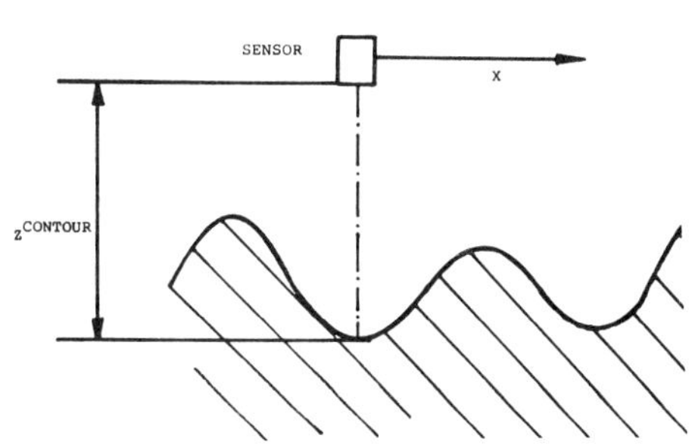


Fig. 14. Following an unknown contour

The $z^{contour}$ sensor control-loop is similar to the control-loop of following moving objects. Because of the integrating behaviour of the process, especially of the internal joint controllers, only standard P- or PD-controllers can be used. The PD-controller is droped out because the differentiating behaviour is disadvantageous on the presence of noise. A standard P-controller calculates the difference between the desired value $z^{contour}_{nom}$ and the actual value $z^{contour}$. The following diagrams document the actual distance value $z^{contour}$ along the contour with various P-values K of the contour-following control-loop.

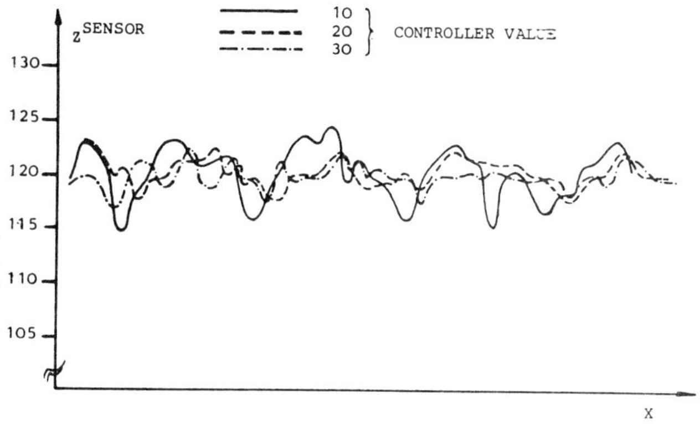


Fig. 15. Contour following diagram

A second tested application is using the single-distance sensor for hole recognition in a plane environment. The robot performs a linear movement along the X-axis and the distance value in Z-direction is registered.

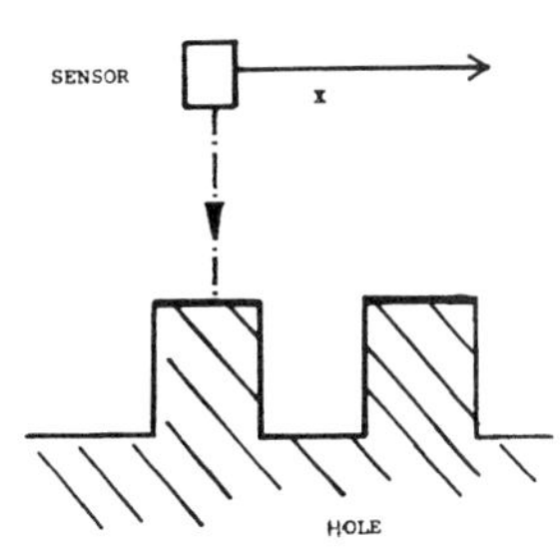


FIG 16. Hole recognition

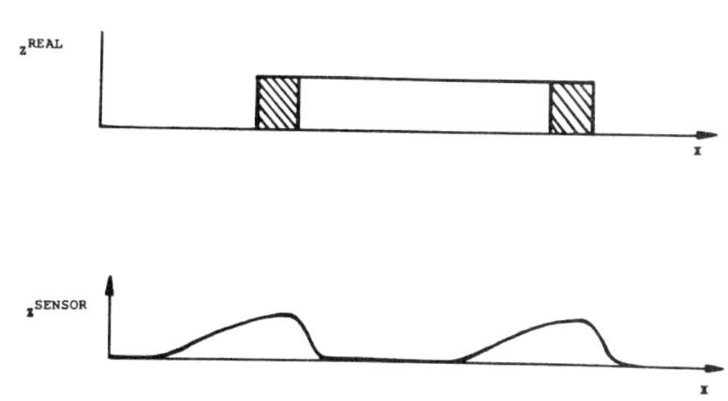


FIG 17. Sensor value

A third possible application uses a combination of three single-distance devices mounted in a u-formed endeffector to follow the sharp edge of an assembly plate.

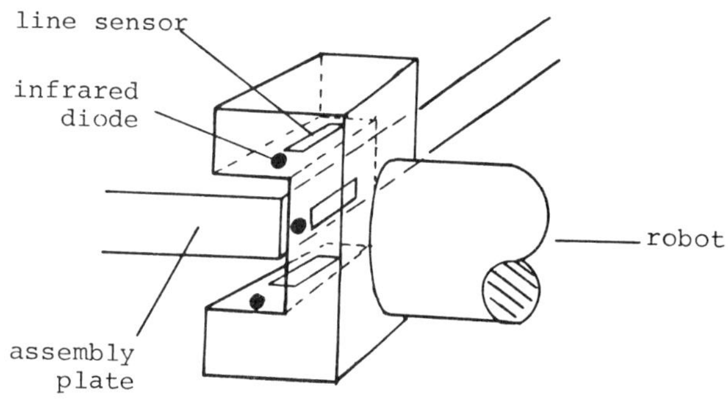


FIG 18. Edge-following endeffector

A last very promising application is a three-dimensional sensor-cube to measure the dynamic behaviour of the robot endeffector-tip. The sensor cube is provided with three orthogonal mounted two-dimensional position sensitive areas. This complex sensor device is able to measure small dynamic missalignments of the robot endeffector-tip after reaching a desired teached position, by using six space fixed infrared diodes as cartesian position reference.

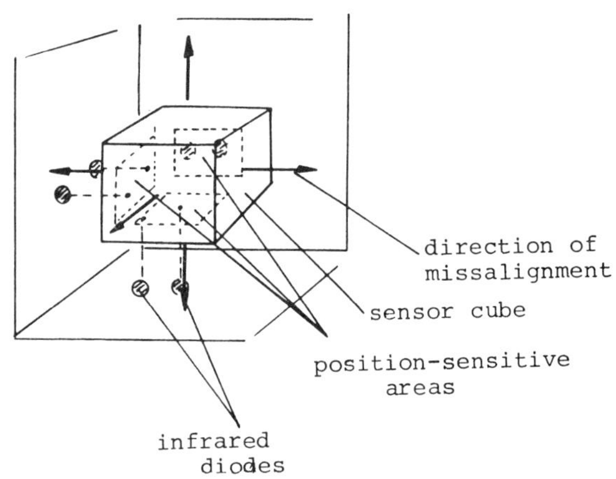


FIG 19. Dynamics measurenment

## CONCLUSIONS

Basic design principles and some of the possible extensions and advantages of optical proximity sensor systems are discussed. As a low-level vision system they can be integrated in various robot tasks very effectively. Due to the fast signal processing they can be integrated in real-time control loops utilizing low- and mid-range envi-romental information with respect to the endeffector. Further developments are expected in five major areas.

On the one hand the sensor equipments will become smaller in size and the analog signalprocessing will be integrated in the sensor head. The optical lenses and filters will be replaced by special microlenses. On the other hand new sensor combinations will be designed for special applications. Furthermore the sensor devices will be integrated into multisensor grippers together with reflex-coupling matrices, gripping-force sensors and slip sensors to detect the behaviour of a gripped object. Moreover the accuracy of the sensor system will be improved to values better than the repeatability accuracy of the mechanical robot system. Finally hierarchies of often redundant multi-sensor information structures will be a major requirement for complex robot assembly systems. However, all robot assembly requirements couldn't be fulfilled by one universal sensor. The various applications will still require a number of different solutions. A small number of the possible sensor applications is presented here.

## REFERENCES

[1] KEMPKENS, K. AND BECKER, N.
REGELUNG VON ROBOTERBEWEGUNGEN AUFGRUND OPTISCHER ABSTANDSMESSUNG.
VDI/VDE GESELLSCHAFT MESS- UND REGELUNGSTECHNIK
STEUERUNG UND REGELUNG VON INDUSTRIEROBOTERN
12./13.5.1986 LANGEN/FRG

[2] KANADE, T. AND SOMMER, T.
AN OPTICAL PROXIMITY SENSOR FOR MEASURING SURFACE AND ORIENTATION FOR ROBOT MANIPULATION.
ROBOTICS RESEARCH 1984

[3] HIRZINGER, G. AND DIETRICH, J.
SENSORSYSTEM FüR ROBOTER.
VDI AUSSPRACHETAG
SIGNALVERARBEITUNG IN MULTISENSORSYSTEMEN
18./19.2.1986 KARLSRUHE/FRG.

[4] DIETRICH, J.
OPTISCH-ELEKTRONISCHER ENTFERNUNGSMESSER.
PATENT P 3502634.0 ANGEMELDET

[5] KONSTANTAS, A.
ERWEITERUNGEN AN DER SIGNALVERARBEITUNG EINES OPTISCHEN DISTANZSENSORS.
DIPLOMARBEIT 1986
TECHNISCHE HOCHSCHULE DARMSTADT FRG

[6] BARZ, R.
OPTISCH ELEKTRONISCHER SENSOR ZUR MESSUNG VON OBJEKT-ORIENTIERUNGEN IN ROBOTERMONTAGE AUFGABEN.
DIPLOMARBEIT 1986
TECHNISCHE HOCHSCHULE DARMSTADT FRG

[7] KEGEL, G.
AUFBAU VON OPTISCHEN NäHERUNGSSENSOREN FüR SENSORGESTüTZTE ROBOTERMONTAGE AUFGABEN.
STUDIENARBEIT 1985
TECHNISCHE HOCHSCHULE DARMSTADT FRG

[8] E.ERSü, K.RATHGEBER, M.SCHNELL, W.NEDDERMEYER
A ROBOT ARITHMETIC PROCESSOR CONCEPT FOR CARTESIAN CLOSED-LOOP CONTROL WITH PRESCRIBED DYNAMICS.
PROC. OF THE 1st IFAC SYMPOSIUM ON ROBOT CONTROL, NOVEMBER, 6-8, 1985, BARCELONA

# DEVELOPMENT OF A CONTROL SYSTEM FOR A FILM PROCESSING DEVICE AND TECHNOLOGY TRANSFER TO AN INDUSTRIAL COMPANY

**M. Ardac, B. Schürmann and H. Rake**

*Institut für Regelungstechnik, RWTH Aachen, Federal Republic of Germany*

Abstract.
In cooperation with a small industrial company, the Institute designed a microcomputer based control unit for the automation of a film processing device. Special emphasis was laid on the detailed description of the concept and the operator interface. The concept supports all developing processes known and is extendible. The process steps are described by functional diagrams which aid an easy programming by students. A programmable testing device is utilized to inspect the product automatically. Due to the strict definition of both partners' tasks the cooperation between partners from distinct disciplines was accomplished successfully.

Keywords. Process control, film processing, computer applications, operator communications, automatic testing, technology transfer

## INTRODUCTION

One of the most outstanding phenomena in the recent years is the rapid rise in the use of microelectronic technology in several areas of everyday life, particularly in the automation of technical devices. This widespread application of microelectronics has lead a photolab equipment company to take direction in establishing a cooperative work with the Institute for automatic control of the Technical University Aachen.

The aim of the company in this cooperative work was twofold.

- In specific the aim was the automation of a film processing device using microelectronics.
- In the long run, the company intended to integrate microelectronics into all its activities using this experience as a basis for educating its personnel and getting aquainted with this new technology.

Therefore a more general aim was to become self-sufficient in using microelectronic technology and to manage independent production.

The basic idea in this cooperative work was the automation of positive and negative film developing; a central process in photography (Fig. 1).

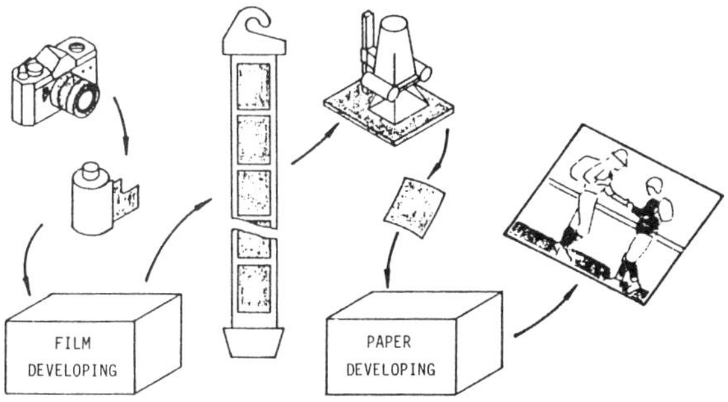


Fig. 1. Developing processes in photography

All the steps in the development process are already well defined. Any desired changes can be produced during the exposure of positives or negatives or by changing the paper used. Therefore if any changes are desired at the output it is not necessary that they are done during the development process.

Tasks necessary for the automation of this film processing device are

- to develop a concept for the control equipment of the device
- to develop a model for the automated form of the device
- to design and develop the hardware and software for a microcomputer based control unit
- to instruct the employees of the company in handling, servicing and adapting the hardware and software on their own
- to adapt the control unit to suit for mass production
- to introduce a method for automated testing in mass production.

During the early proceedings of the cooperative work, the tasks to be taken over by either the institute or the company were defined.

The design of a model for the automated form of the device was undertaken by the photolab equipment company, due to their extensive experience on the development of similar devices. A second task they were to accomplish was the adaptation of the control unit to mass production.

Other tasks were performed by the institute under the condition that they had to be realizable within the financial power of a small industrial company.

The experiences showed the effectiveness of the university in introduction and utilization of new technology in industrial companies. This fact is especially true for small companies because they can benefit from technological developments taking place in the universities and get acquainted with the know-how without having to pay large amounts of money for establishing their own development department.

## SYSTEM ANALYSIS AND DESIGN

The basic idea of the concept is built upon the mode of operation of an already existing device which is used in developing and fixing film and paper by a rotational method (Fig. 2).

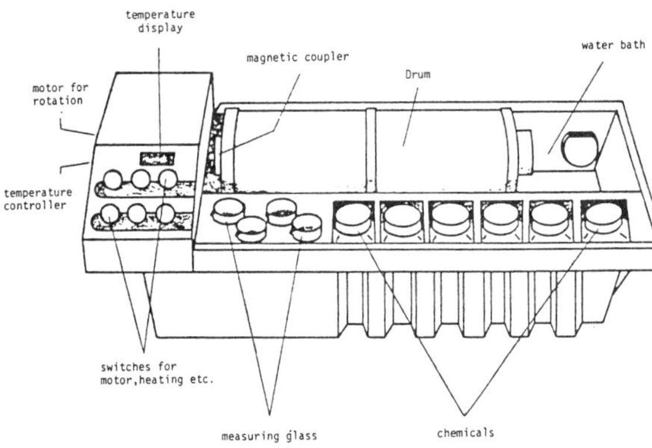


Fig. 2. The film developing device for film and paper (before automation)

The operator initiates the process by placing the material to be developed in the drum and filling in a certain amount of a chemical. The drum is then placed in the developing device and is left to rotate there. During the rotation period the temperature of the drum and its contents are kept at a specified degree, by heating or cooling the water bath surrounding the drum. After a specific period of time the operator empties the drum and fills in a new chemical. These steps of rotating and filling are repeated for several times using different chemicals depending on the nature of the development process.

The operator is responsible for the exact execution and especially the exact timing of these steps. All these steps which are observed and carried out manually were intended to be automized in the new developing device.

The above mentioned requirements, aims and important parameters such as efficiency, cost limit etc., were to be determined through an agreement prior to the cooperative work. The interface of the unit to the operator and to the process was also determined at an early stage of the cooperation. The intention of these early decisions was to provide freedom to both partners in completing their tasks independent from each other.

The factors which require a more detailed emphasis in order to develop the concept further can be seen on Fig. 3. These factors, namely, target developing processes, technical details of the device and operator communication interface were used to define the requierements on the control unit.

All of the conventional film developing processes had to be explained by a simple scheme, so that every process would be easily programmable with a minimum modification and that future processes

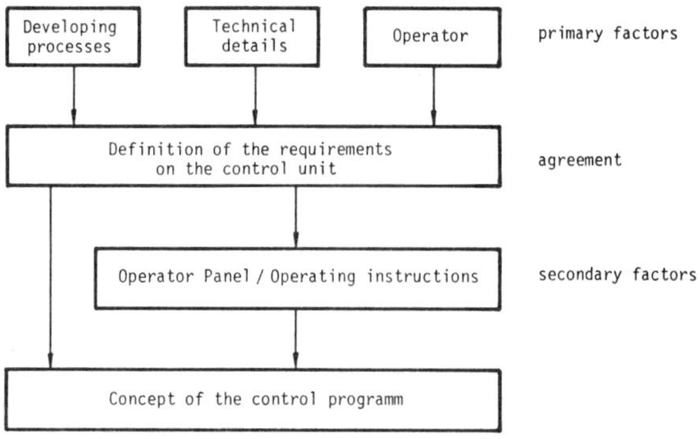


Fig. 3. The factors influencing the automation concept

would be easily adapted into this scheme. Therefore another advantage of this method lies in the availability of the device for future use. Such a simple scheme which fulfilled the desired criterion could be found for developing processes of black and white or colour pictures and films (Fig. 4).

In this scheme all the processes consisted of a combination of a few types of process steps, such as rinsing, chemical treatment etc. For each process these steps are combined in a unique sequence and only a few parameters for each step (e.g. time) need to be given in order to operate the device.

| Process step \ Developing processes | chemical usage | colour A | B | C | D | E | F | G | H | I | K | Black & White B/W |
|---|---|---|---|---|---|---|---|---|---|---|---|---|
| 1 Rinsing | | | | | X | X | | | | | | |
| 2 Developer | C | X | X | X | X | X | X | X | X | X | X | X |
| 3 Rinsing | | | | | | X | X | X | | X | | X |
| 4 Stop bath | C | X | | X | X | X | | X | | | X | |
| 5 Rinsing | | | | X | X | X | | X | | | X | |
| 6 Reversal bath | C | | | | | | | | | X | | |
| 7 Re-exposure | | | | | X | X | | X | | | | |
| 8 Developer | C | | | | X | X | | X | | X | | |
| 9 Intermediate rinsing | C | | | | | | | | | X | | |
| 10 Rinsing | | | | | X | X | | X | | | | |
| 11 Bleach bath | C | X | X | X | X | X | X | X | X | X | X | |
| 12 Rinsing | | | | X | | X | X | X | X | | | |
| 13 Fixer bath | C | | | X | | X | X | X | X | X | | X |
| 14 Rinsing | | X | X | X | X | X | X | X | X | X | X | X |
| 15 Stabilizer | C | | | | | X | | | X | ↓ | | |
| manual treatment (N: Netting, S: Stabilizer) | | | N | N | N | | | N | | S | N | |

Fig. 4. A simple scheme adaptable for all developing processes

Timing charts and functional diagrams were used to describe events taking place in each step.

The criteria used for determining the technical details of the device were that all the automatic functions would be realizable as valid and long-lasting. It was very soon recognized that the control signals to be utilized were nearly totally independent of the way those functions were realized. This made the description and definition of the tasks for the control signals possible at an early phase of the concept formation. For example, to control the filling phase, signals for the pumps and valves are generated. The state of the device is identified by testing suitable sensory signals. Due to the early definition of the control signals, a simulation model for the device was realized which was used to develop and test the control unit. Thus experimentation was possible without having to wait. The completion of the actual device and the time required for development of the final device was reduced considerably.

The design of the operator communication interface is one of the most important tasks in the automation of a technical device. The most essential feature of the design is that it has to avoid the operator from feeling at a loss over the device. Furthermore the operating panel had to fulfill conflicting demands. On the one hand it had to offer manifold operating possibilities to provide extensive use, on the other hand it had to be clearly arranged and easy to use.

In the final form (Fig. 5) the operator may choose to run the device using only a single botton for predetermined processes, or may input new developing processes by combining the previously mentioned steps, modifying their parameters, storing them and subsequently running them automatically.

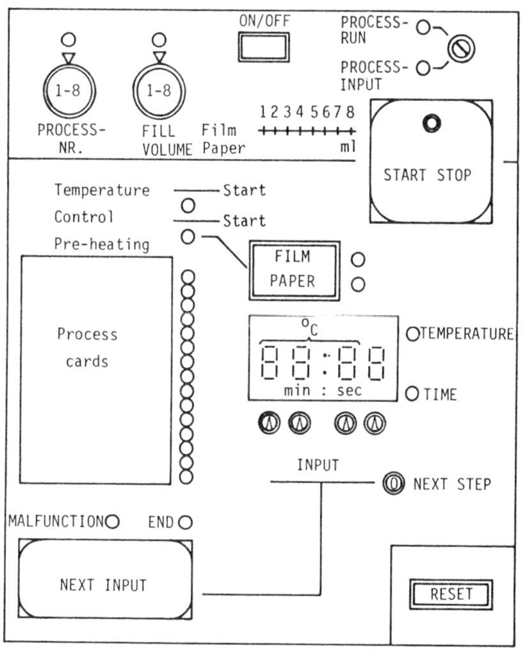


Fig. 5. Concept of the operator panel of the control unit

A detailed consideration of the structure of control and its interfaces at the very start of such a project was experienced to be extremely profitable. The discussions were mainly based on the requirements on the automatic control and the extend to which they would be actualized, without too much emphasis on secondary questions such as programming techniques and language, hardware etc. If the aims, requirements and interfaces, are well defined through such discussions, it is possible for partners from differing disciplines to work effectively on the production of seperate parts of the same device which would then be assembled with ease.

It is advisable to prepare an operator instructions manual on the basis of the outlined requirements which would permit the future operators to evaluate the concept of the device effectively and figure out any possible shortcoming beforehand. Early detection of shortcomings through such a process enables one to eliminate or correct them without causing excessive expenses. All the activities extending from the structure and definition of data to the preparation of flow charts and structograms are accomplished only after this phase of decisions.

METHODS FOR THE DESCRIPTION OF THE CONTROL STEPS

As was mentioned earlier, all the developing processes already known and most probably those that will come up in the future can be described by a sequence of a few types of process steps. This section deals with the methods used for describing each of these process steps.

Timing charts and functional diagrams are the aids used in describing the flow of each process step.

These representation means are generally known to engineers who are familiar with conventional control techniques. Thus the new technique was described through known methods which reduced the number of new subjects to be mastered and simplified the introduction to this new technique.

Timing charts give a wide general view on the timing condition of the control events to be accomplished. In Figure 6, a timing chart of the process step of chemical treatment is given as an example.

A subprogram arranges all the necessary operations during this step. Only the actual time for the chemical tratment (indicated by a double line in Fig. 6) needs to be given as an input. All the other timing conditions are either fixed or can be calculated. Because the graphical representation of these timing charts provide a quick comprehension of the relations between control signals (indicated by arrows in Fig. 6), the timing conditions can be readily observed.

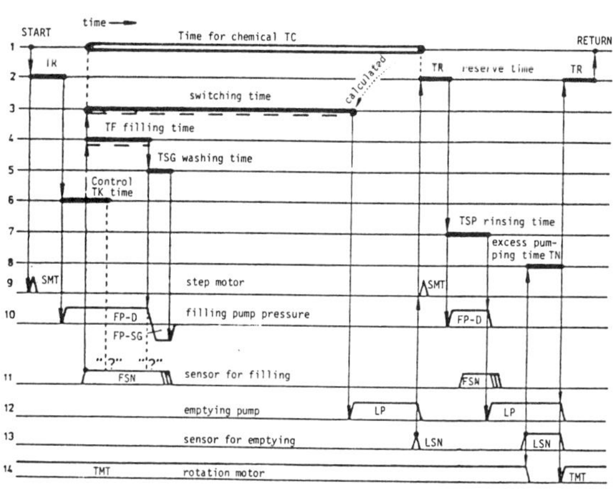


Fig. 6. Timing chart for chemical treatment step

Though a functional diagram is much less comprehensive from this regard, it is very powerful in indicating all the switching conditions definitely and precisely. Any necessary precautions to be taken in an exceptional case, such as a disorder or malfunctioning, are also indicatable by a functional diagram, whereas in a timing chart, these indications are vague and complex.

Using functional diagrams, it is possible to represent each subprogram by flowcharts and Nassi-Schneiderman diagrams (structograms). Thus a complete description or documentation of the control program can be obtained without any additional effort.

Devoting extensive effort in describing these process steps at the very start was observed to be very beneficial. It permits one to focus on the major problems from the start, instead of having to spare time on minor matters such as the form of programming. It results in more reliable flowcharts which facilitates the testing of software and reduces the duration of programming considerably.

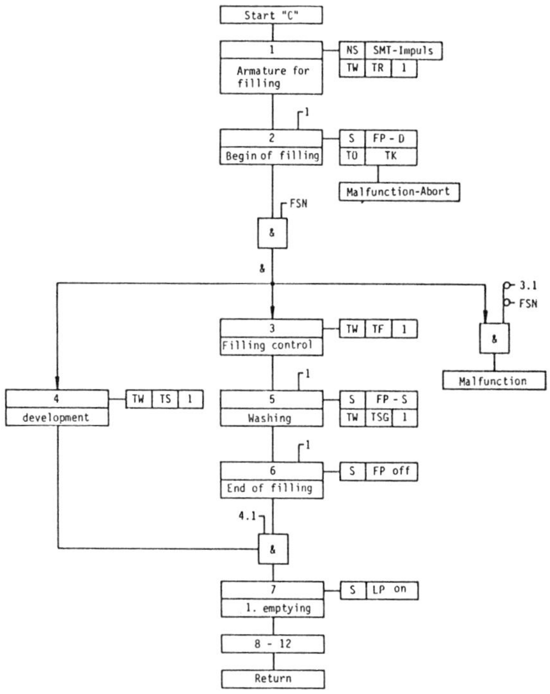


Fig. 7. First part of the functional diagram for chemical treatment step

## CONCEPT OF PRODUCTION INSPECTION

The production of such a high technological device requires a detailed inspection of its functionability. This process can be done within reasonable financial limits, if efficient and powerful test equipments and methods are utilized. Expenses resulting from the introduction of microcomputer technology are essentially due to the cost of testings in production and service.

Two steps were specified for the testing of the control unit of the device.

The first step made a reliable detection of the disorders within the unit possible without demanding extensive knowledge from the personnel working on this particular task. During this step all the function blocks, down to the peripheral hardware drivers, were to be automatically tested in the shortest time possible. Any disordered unit was to be removed from the production to proceed on to the second step. In the second step, the unit was tested by higher qualified personnel equipped with additional test devices to locate the disorder in the microcomputer unit. A full automatic location of the disorder is extremely complex and even impossible in most cases. Therefore, the capability and experience of the personnel is also highly important at this stage for locating and repairing the disorder within a reasonable period of time.

For actualizing the distinct functions required in those two steps, a programmable testing device with reasonable cost was purchased (Fig. 8). The device consists of several conventional testing tools required for the inspection process. It can be connected in the place of CPU of the unit under test, by a processor specific interface. This device can be programmed for the automatic testing of the functional blocks by using a language similar to BASIC. The device also contains many software functions generally necessary for trying out the hardware of microcomputers such as BUS, RAM, ROM test macros, which can be used for developing customer specific programs.

A test program, which runs almost automatically was developed for the first step of the inspection. The personnel in charge of inspection has to accomplish only a few tasks such as

- connecting the unit to be inspected
- starting the testing program
- observing the display of the testing device
- keeping to the instructions of the device
- ruling out the unit if necessary.

Several distinct programs were available for use as testing tools for the processes in the second step. In preparing these tools, the difficulties which come up with the analysis of the disorders in microcomputers were taken into consideration, thus providing a very effective base in inspection.

A diagram showing the means by which the disorders were located was prepared for the purposes of education of the personnel, as well as for daily use.

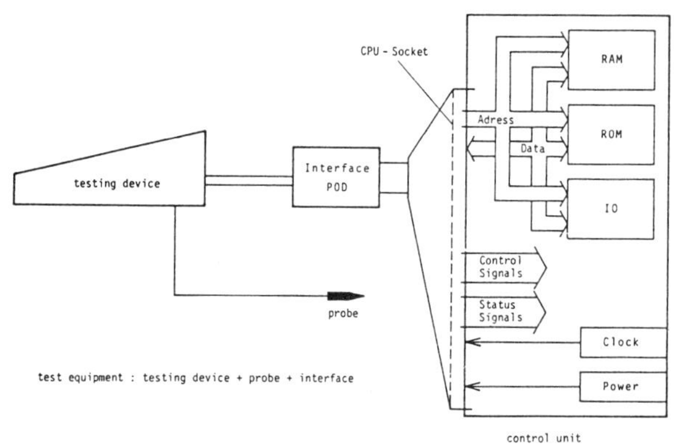


Fig. 8. Universal test equipment for microcomputer applications

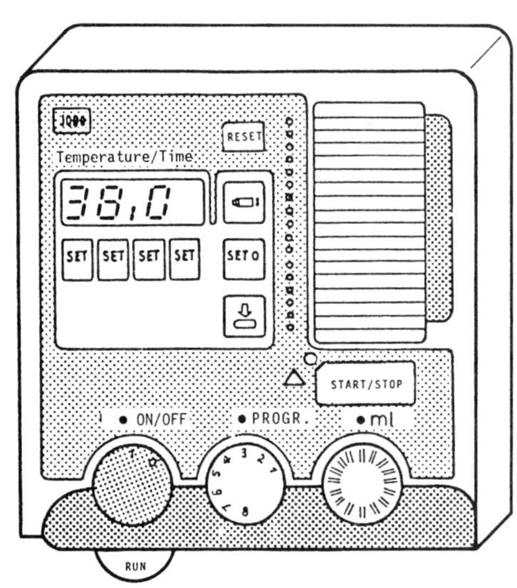


Fig. 9. Final form of the control unit

TECHNOLOGY TRANSFER

As shown in Fig.10 the cooperational work begins with the conferring of a contract. The very first activity of the institute was to prepare a catalogue of its own duties and the specifications of the automatic device. This catalogue describes all the aims and requirements of the cooperation and the several steps of the realization. Thus it is also used as the basis of the working plan. The institute also presented a concept of the manual for operating instructions for the planned device. After this step it was possible for both partners to work in cooperation but almost independently. Being its tasks strictly defined the institute could also employ students for most of the practical work such as hardware design and programming. The students could get experiences for their future activitites in the industry and benefit from the modern equipment purchased for this project. The trainings practised by the members of the institute for the employees of the company represented a valuable experience and a good completion of the theoretical work in the university.

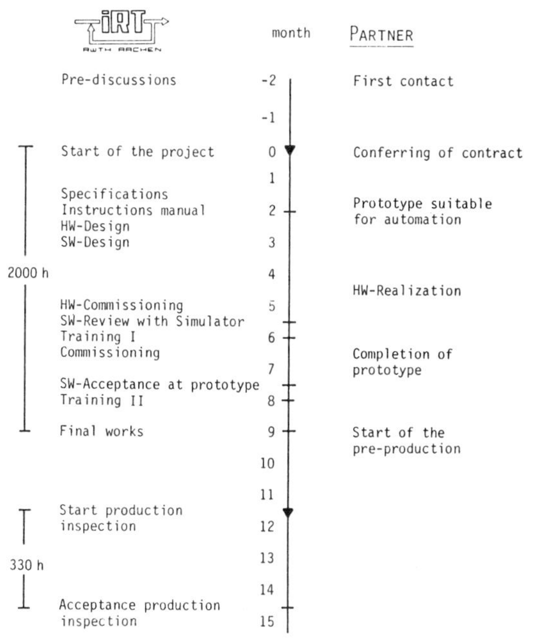


Fig. 10. Time flow of the cooperational work

# SUBSTANTIAL IMPROVEMENTS OF THE PERFORMANCE OF AXIAL WINDERS BY ADAPTIVE STATE SPACE CONTROL

**W. Höger and D. Schröder**

*Lehrstuhl für Elektrische Antriebstechnik, Technische Universität München, Arcisstr. 21, D-8000 München 2, Federal Republic of Germany*

ABSTRACT

The stationary and dynamic behaviour of axial winders will be described. Axial winders are normally used in multimachine-drive-systems. The single drives of this system are coupled by an elastic web of paper, rubber or steel, which should be processed. Starting from the results of the analysis of the stationary and dynamic characteristics new concepts with adaptive controllers on a state space approach will be developed. The results, obtained from this new approach will be compared to the results of conventional approaches. Furthermore practical experiments show the advantages of the new concepts.
Off-line identification of the system parameters can be realized with the computer-control-system, so putting into operation is very much simplified.

KEYWORDS

Axial Winder, Web Force, Continuous Processing, Control, Adaptive Control, State Space Control, Observers

INTRODUCTION

In many industrial plants with continuous processing of running webs - for example in the paper -, the textile- or the metal-industries - the material has to pass seve-ral processing stations. All sections of the continuous process are coupled by the

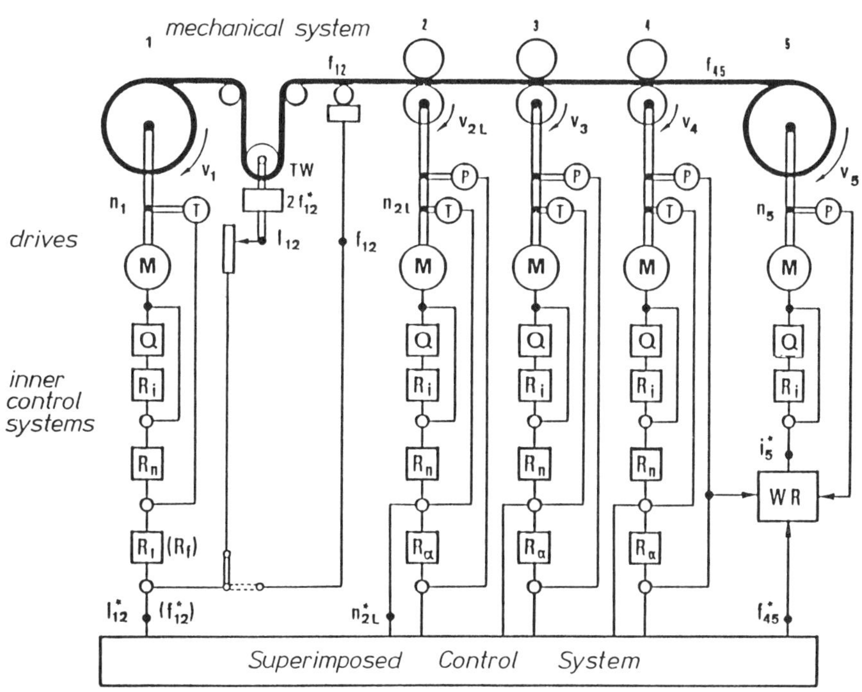


Fig. 1 Structure of a System with Continuous processing and multimachine drives with speed control

material. At the beginning and at the end of the plant there are frequently winders installed. The structure of a continuous processing system is shown schematically in fig. 1.

In the system - shown in fig. 1 - the web will be processed in different stations. In these stations, called nip sections, there are driven and undriven rollers for the transport and the processing of the material. In the nip sections transport the cylinders transfer forces depending on the technological process. The cylinders are controlled in torque, speed or angle by electrical drives. A superimposed guidance system controls the whole process and therefore the speed relations of the different drives are depending on the technological requirements. The winders at the beginning or the end of the system are a store of the material. But from the analytical point of view the winders are a special kind of nip section, where the process is beginning or ending, so that the winders are coupled with the drive system only from one side. From the technological point of view there is the demand that we should have a constant flow of material at the first processing station and furthermore that the web tension is constant too. During the nipwinding process the tension of the different layers of the material should be controlled in such a way that no stress up to plastic deformation of the material is reached /7/.

From these requirements - constant flow of the material and controlled tension in a narrow range - follows, that the force and stretch in the material must be kept within a narrow margin in the stationary and the dynamic state. This is necessary for the unwinding as well as for the upwinding process /4/.

A simple solution to keep the force constant is to use a dancing roller. Another solution is a sensor for the tension in the material and control the force in the material directly /5/. But the disadvantage of the sensor are the random disturbances of the output, so that the signal must be smoothed and delayed.

The most frequently used method is a feedforward control of the force and the tension of the material. This solution is characterized by a calculation of the torque of the electrical motor driving the winder and which is variable with the diameter of the roll.

## STATE OF THE ART OF THE WINDER CONTROL

The state of the art for the control of winders is the feedforward control of the torque of the machine. With the knowledge of the flux in the machine this can be realized by the control of the armature current (fig. 2). The basis for this feedforward control of the tension or stretch in the material are the relations between the electrical characteristics of the drive and the mechanical characteristics of the processed material and the roll /5/.

These relations are used in a normalized form.

speed of the material : $v = n_M \cdot r/r_o$ (1)

force in the material : $f = m_f \cdot k_w / r$ (2)

torque in the air gap of the machine : $m_M = i_A \cdot \varphi$ (3)

friction-torque : $m_R = f(n, i_E \ldots)$ (4)

torque at the roll : $m_f = m_M - m_R - m_b$ (5)

acceleration-torque : $m_b = T_{\Theta o} \dfrac{d(n \cdot \vartheta)}{dt}$ (6)

inertia : $\vartheta = 1 + k_\vartheta \cdot (r^4 - r_{min}^4)$ (7)

controller : $f = \dfrac{(i_A \cdot \varphi - m_R - m_b) \cdot k_w}{r}$ (8)

Starting from these fundamental equations and some more equations for the stationary state a feed forward control can be realized. This conventional feed forward control calculates the reference values for the armature current and the flux starting from a desired reference value of the force f* in the material and from the given acceleration set point dv*/dt of the whole system with some real values and the parameters of the system. This calculation is realized in analoguous form nowadays normally in the block WR in fig. 2. In the block RR the radius r of the roll is calculated. This can be done by the evaluation of the ratio of the pulse frequency $f_{pk}$ of the speed of the cylinder k next the winder and the pulse frequency $f_{pw}$ of the roll of the winder itself.

The control of the armature current is done by a variable limitation $i_A^*$ of the output signal of the speed controller of the winder. The speed controller of the winder is overdriven because an additional reference value $\Delta n^*$ is added to the general reference value n* of the speed for the whole system. Therefore the speed controller is active only during the start of the processing or if the material is teared. During normal operation the speed controller has no in-

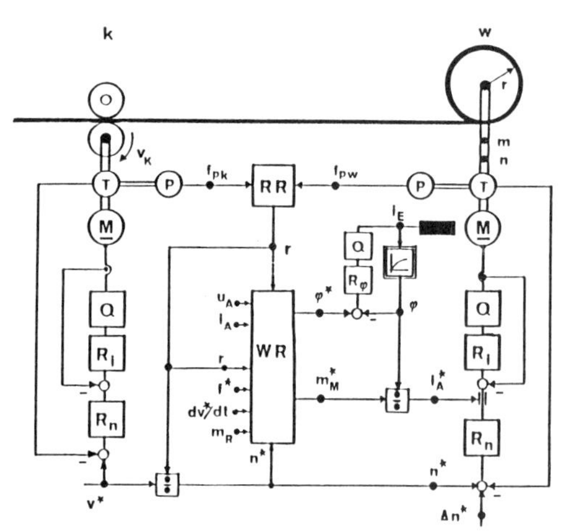


Fig. 2 Schematic feedforwarded control for axial winders

fluence in the stationary and dynamic state of operation.

The realization of the control with the two signal processing blocks WR and RR is complicated and not very precise if an analogue solution is used. It should be kept in mind that the equations are valid only if small disturbances occur and the system is in a state near stationary operation.

Furthermore the limitations of this model become clearer if we consider parasitic attributes of the system like friction, iron losses, the nonlinearity of the excitation etc.. Therefore we decided to use digital control with a multi microprocessor system. The advantages of digital control are higher accuracy and better reproduction. These are familiar aspects nowadays. A much more important aspect is that new approaches for the control can be examined with the same hardware and these new solutions can be much more detailed and complex than analogue solutions. Another advantages are the implementation of off line identification algorithms to recognize the parameters of the systems.

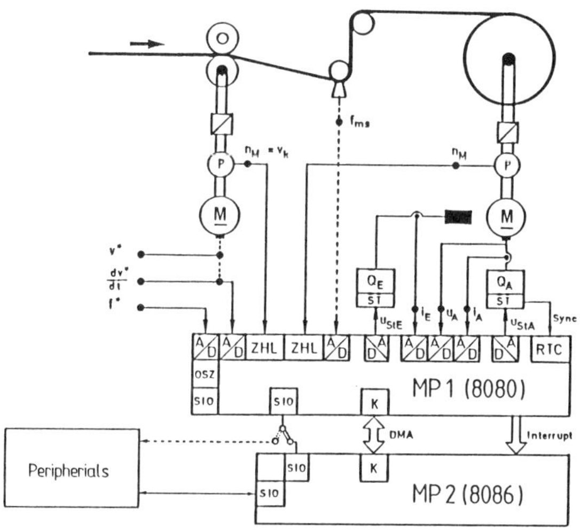


Fig. 3 Schema of μP-control system for axial winders

PRINCIPLE CHARACTERISTICS OF THE ELECTROMECHANICAL SYSTEM

The dynamic behaviour of the electromechanical system is determined mainly by the spring-mass-system of the elastic material which should be processed and the inertia of the roll and the electrical machine (fig. 4). Another resonance is produced from an elastic coupling of the roll and electrical machine itself, but this resonant system should be neglected at the beginning of the investigation. Both resonant systems are damped weakly so that undesired oscillations of the web tension can occur. The resonance of the dominant mechanical system is particularly distinct if the frequency of this resonance is clearly deeper than the cross over frequency of the speed controlled system of the drive next to the winder. We assume the speed of the drive next to the winder impressed and we assume furthermore the mechanical system between the electrical motor and the roll

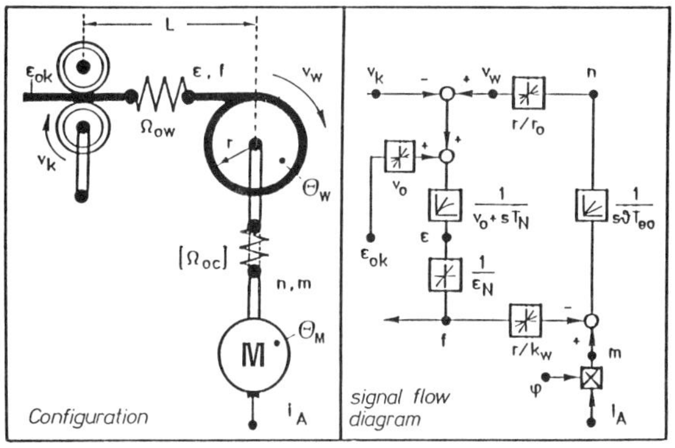


Fig. 4 Principle mechanical construction and simplified signal flow diagram

as rigid and without any backlash. Under these assumptions the electro-mechanical system can be considered as a second order system. The signal flowgraph is shown in fig. 4.

If such a resonant system is disturbed very undesired oscillations occur. The characteristic frequency $\Omega_{ow}$ depends on the length, the cross section and the modulus of elasticity of the material and furthermore on the inertia of the winder which is variable with the radius of the roll.

$$\Omega_{ow} = \sqrt{\frac{E S_o R^2}{L(\Theta_w + \Theta_M)}} = \frac{1}{\sqrt{T_{\Theta o} r_o T_N N k_w}} \cdot \frac{r}{\sqrt{\vartheta}} \qquad (9)$$

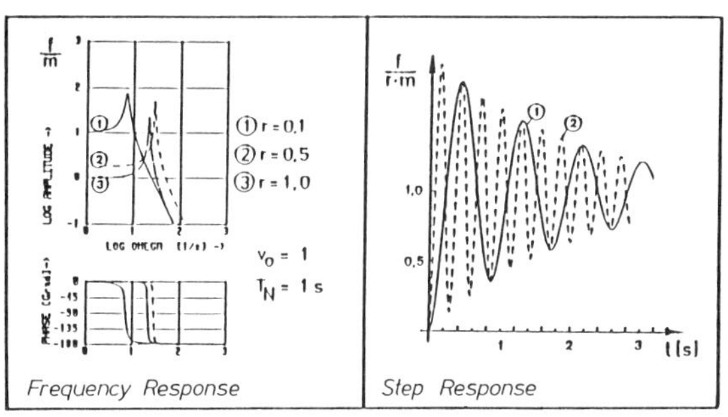


Fig. 5 Frequency response and step response of the force f at different radii

This response is not influenced by the parameters of the conventional feedforward control shown in the upper chapter because only reference values and linearized quasistationary measured values are used. One important objective of this investigation is the implementation of a feedback control strategy. This strategy should use modern control concepts like an adaptive state space approach. But as a first logical step control strategies with a sensor for the web force should be investigated.

CONTROL STRATEGIES WITH A WEB FORCESENSOR

Force control is possible if the force can be measured. A simple solution is shown in fig. 6. The reference value f* has a feedforward path to control the armature current directly. The force controller $R_f$ is in the feedback path and is correcting the errors. So this solution is similar to

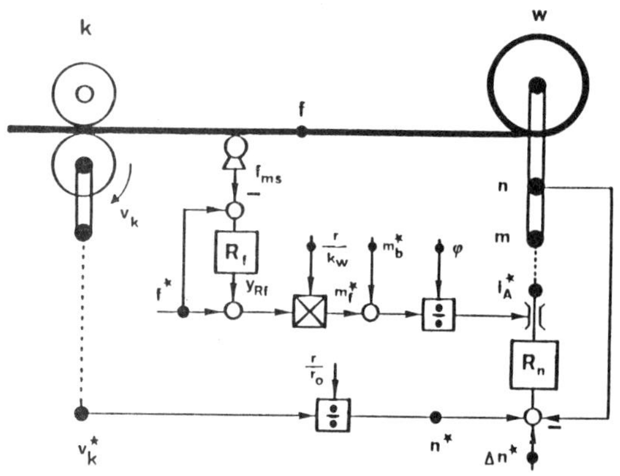


Fig. 6 Winder control with web force sensor and subordinate armature current control

the solution shown in the upper chapter. The difference is the additional feedback path with the force controller only. This solution has been used often in industrial plants. The feedforward path is important for the dynamic of the response, the feedback path is responsible for the accuracy. But the solution has a decisive drawback. The resonant system - consisting of the elastic material and winder mainly - is included in the closed control loop. So stability problems arise and bad dynamic performance is the result.

Much better results can be obtained if a speed control loop is subordinated to the force control loop (fig. 7). In this solution the reference value n* of the speed control again has a feedforward path from the general speed reference value v*. The force control loop only adds a correcting signal to the subordinated speed control loop. The main advantage of this solution is that the speed and the web force are decoupled in the dynamic state, because the resonant mechanical system is split into two nonresonant subsystems. But there is one problem again. In cascaded control loops the optimization starts in the inner loop and the other loops should be optimized without changing the parameters of the inner loops again. This well-known step by the step optimization of the different control loops can be realized in our solution only if the cross over frequency of the speed control loop is definitely higher than the characteristic frequency $\Omega_{ow}$ of the mechanical system.

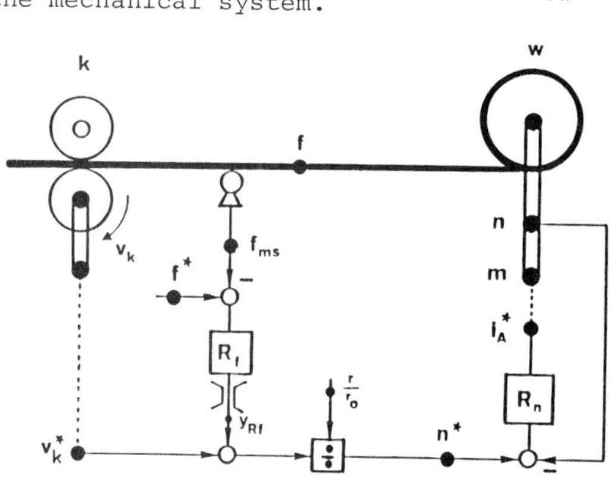


Fig. 7 Web force control with subordinate speed control

## ADAPTIVE STATE SPACE CONTROL WITH OBSERVERS

If state space approach is considered then all states of the system must be used to get the desired result. These states are torque, speed and force. Theoretically pole assignment is possible if for example proportional feedback is assumed and the plant is linear and time invariant.

In the practical situation of the winders, which are driven from dc machines and fed from converters with natural commutation normally, there are limitations in the field and armature current dynamics($T_{ersi}$). These limitations can't be changed by pole assignment /6/. Furthermore there are nonlinearities like the excitation characteristic or the friction and backlash. In addition there are time-variant parameters like the radius r of the roll, the inertia $\vartheta$, and the speed of the web $v_o$.

It is yet possible to give the analytical functions for the feedback parameters, which provide an optimal damping characteristic of the system:

torque $$k_m = -\frac{v_o \cdot T_{ersi}}{T_N} \tag{10}$$

speed difference $$k_v = \frac{T_{\Theta o} \cdot r_o}{2T_{ersi}} \cdot \left(\frac{\vartheta}{r}\right) - r \cdot \frac{T_{ersi}}{\varepsilon_N T_N k_w} \tag{11}$$

force $$k_f = \frac{T_{\Theta o} \cdot r_o}{8T^2_{ersi}} \cdot \left(\frac{\vartheta}{r}\right) \cdot \varepsilon_N T_N - \left(\frac{r}{k_w}\right) \tag{12}$$

As shown in the upper equations the optimized parameters of the control system depend on the dynamic limitations of the actuator, which are taken into consideration in analytical functions. These analytical functions can be calculated in a micro-computer system so that the controller parameters are adapted automatically.

There is one remaining problem, the sensing of the web force in the material. Another interesting signal is the peripheral speed difference between the driven cylinder next to the winder and the roll of the winder. As mentioned above the direct sensing of the force is delayed and is often not available. Therefore observers will be used. The observer shown in fig. 8 is a model of the plant. The input signals are the armature current $i_A$ and the speed of the two drivers $n_m$ and $v_k$. Furthermore we need as

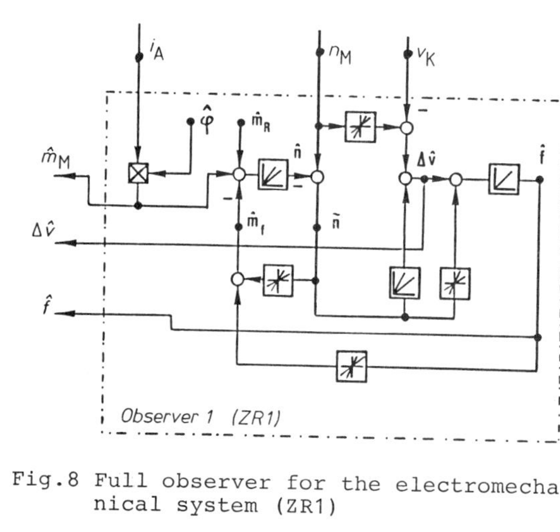


Fig.8 Full observer for the electromechanical system (ZR1)

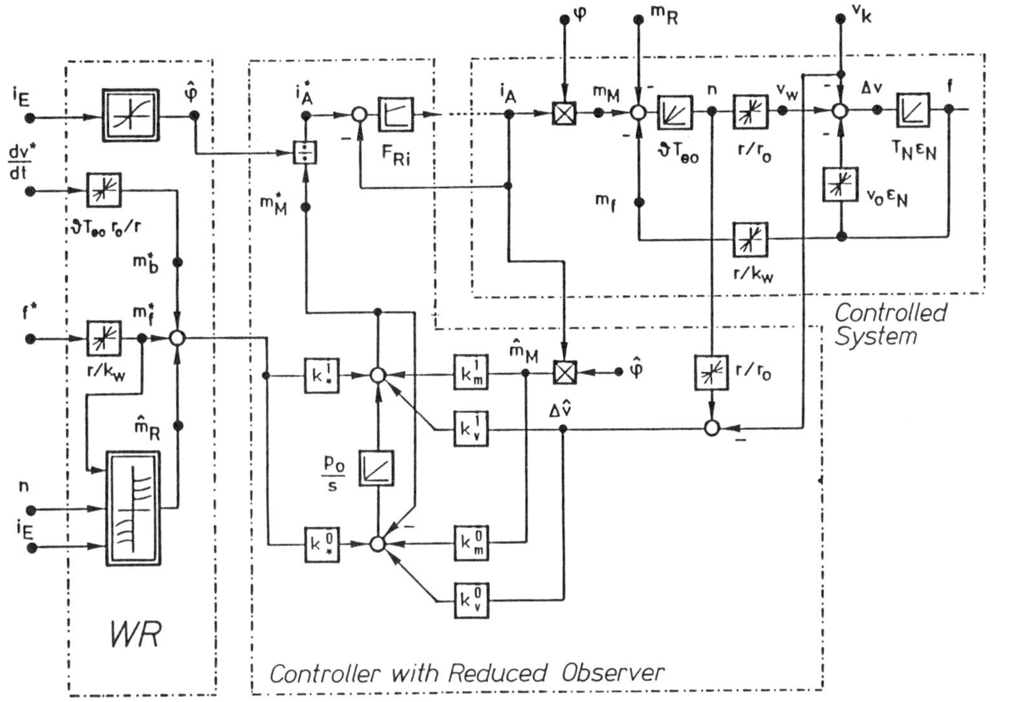


Fig. 9 Signal flow graph of the system with reduced implicit observer

input variables the estimated flux $\hat{\varphi}$ in the machine and the friction $m_R$ in the system at all operating points.

The observer also has to be adapted from analytical functions, because some parameters $(r, \vartheta)$ vary by a factor of 10 and even more during the winding process

$$h_1 = 3\,\omega_o\,\vartheta T_{\Theta o} \tag{13}$$

$$h_2 = 4\,\omega_o^2\,\varepsilon_N T_N\,k_w \vartheta T_{\Theta o}/r \tag{14}$$

$$h_3 = 2\,\omega_o^3\,\varepsilon_N T_N\,k_w \vartheta T_{\Theta o}/r \tag{15}$$

A more simple solution is shown in fig. 9 where a reduced observer is used. The observer is first order and implicit. The results are obtained with a design strategy developed in /3/. The advantages of this design strategy are the simple realization on a microcomputer and the robustness against simplifications.

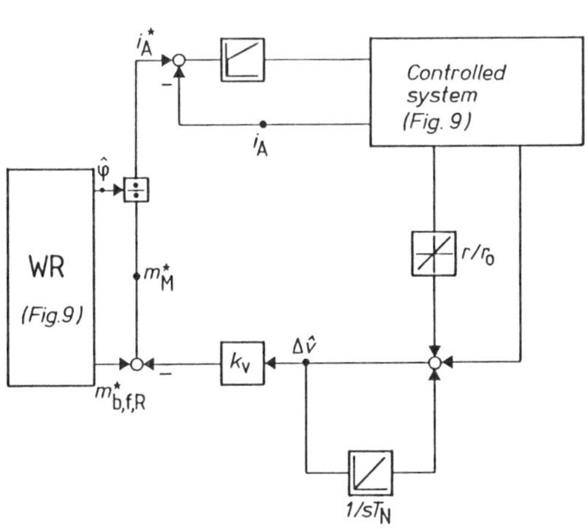


Fig. 10 Simplified control strategy (ZR3)

An important simplification e.g. was the neglection of the resonant coupling between the electrical machine and the roller.

One disadvantage of both systems is the parametrization depending on the web properties $(\varepsilon_N)$. This disadvantage can be eliminated by a further reduction, shown in fig. 10. This solution (ZR3) is similar to the conventional feedforward control with only one additional feedback path $(k_v)$.

An advantage of ZR3 is the sole application of well-known plant parameters and signals which can be measured easily. A drawback of this solution is the loss of one order of freedom for the pole assignment. Nevertheless of the simple structure of the whole control system the damping of the resonant system can be improved considerably

## EXPERIMENTAL RESULTS

The following results are obtained when during the optimization the charcteristic polynomials corresponding to the "Doppelverhältnisse" are used /2/.

In the following figures practical results (left) are compared with results gained from simulation (right, -.-.-).

In addition results are shown if the conventional feedforward control (right, ----) is applied. During the investigations we assume a radius r = 0,1 of the roll, which is an unfavorable point of operation. Fig. 11 shows the step response of f/f* if control systems with ZR2 and ZR3 are used. Both results show a good damping of the system, practical results and simulation are identical. The solution with ZR2 has a better dynamic response compared with the response of solution ZR3. Nevertheless the solution with ZR3 gives much better results than the conventional feedforward control method.

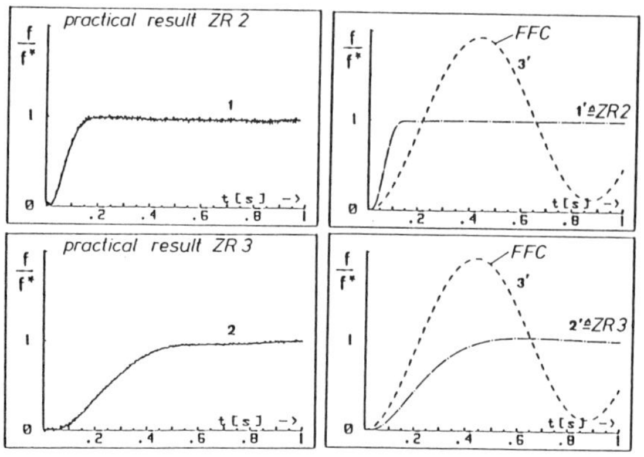


Fig. 11 Step response of the web force

The results, obtained with a full observer (ZR1) are identical to ZR2.

A very important disturbing variable for the force in the material is the speed of the drive next to the winder. In fig. 12 the reference value of the drive next to the winder was varied 3% in a step. The lefthanded figure shows the insufficient response with the conventional feedforward control used up to now and the righthanded figure shows the response with the solution ZR1. The results with ZR2 and ZR3 are quite similar to ZR1.

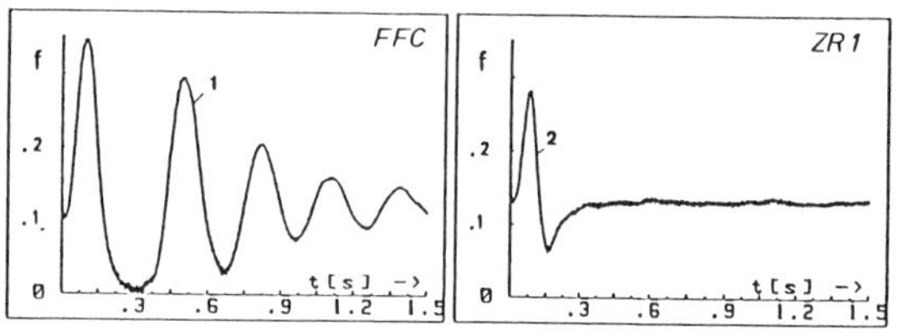


Fig. 12 Setting of the web force after a step in the speed reference

Furthermore a very intense resonant process exists if the roll of the winder is not circular and the speed of the roll is

$$N_W = \Omega_{OW}/2\pi \qquad (16)$$

In fig. 13 are the results without (A) and with (B) an adaptive state space controller.

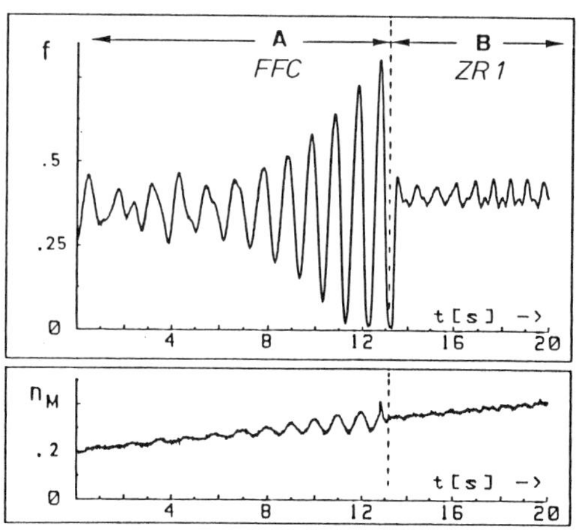


Fig. 13 Web force and winder speed with a noncircular roller

In the experiment the roll had a very small deviation from circularity. The speed of the winder was accelerated and with increasing speed we came up to $N_W$. In the experimental phase A we used the conventional feedforward control and switched to state space control in phase B. The results show clearly the advantages of the new solution.

## CONCLUSION

An axial winder is a resonant mechanical system with a wide range of parameter variations. The state of the art "feedforward control" does not influence the resonant behaviour and the conventional single loop web force control even leads to a deterioration of performance.

It has been shown that we can improve the dynamic response considerably by means of state space control strategies.

The controller and observer parameters are given through analytical equations, so an extreme range of process parameter variations can be tolerated.

Both identification of process variables and adaptation of the control system, and calculation of the control algorithm can be performed by a microcomputer system.

## REFERENCES


/1/ Brandenburg, G.: Prozeßmodelle für durchlaufende elastische Bahnen in kontinuierlichen Fertigungsanlagen. Tagung: Prozeßmodelle VDI/VDE GMR. VDI-Bericht Nr. 276, Düsseldorf 1977

/2/ Brandenburg, G.; Zäh, M.: Das erweiterte Dämpfungsoptimum. Ein analytisches Optimierungsverfahren nach Doppelverhältnissen für Regelstrecke mit Zählerpolynom. Veröffentlichung voraussichtlich in Automatisierungstechnik

/3/ Hippe, P.; Wurmthaler, C.: Ein einfaches Verfahren zum Entwurf von Zustandsreglern mit Störgrößenkompensation. Reg.technik 29, 1981, S.91-96,S.131-136

/4/ Kessler, G.: Electrical drive systems in the prp-industries. 3rd intern.IFAC conf.on instrumentation and automation in the paper, rubber and plastics industries, Brüssel 1976

/5/ Kessler, G.; Brandenburg, G.; Schlosser, W. und Wolfermann, W.: Struktur und Regelung bei Systemen mit durchlaufenden elastischen Bahnen und Mehrmotorenantrieben. Regelungstechnik 32, 1984, S.251-266

/6/ Schröder, D.: Analysis and synthesis of automatic control systems with controlled converters. 5. IFAC-Congress, Paris 1972, Session 22.1

/7/ Wolfermann, W.: Bahnzugkraft und innere Spannungen beim Wickeln von elastischen Stoffbahnen. Regelungstechnik 26 1978, S.105-140

# FAULT-TOLERANT MICROPROCESSOR-BASED CONTROL SYSTEM FOR MULTIPLE LIFT INSTALLATIONS

**J. Duato**

*Computadores y Automática, Department of Ingeniería de Sistemas, Universidad Politécnica de Valencia, P.O. Box 22012, 46071 Valencia, Spain*

Abstract. A fault-tolerant control system for multiple lift installations is proposed. The system has a modular structure. It is composed by a system controller and as many motor controllers as lifts. The motor controller supports transient failure recovery through software routines and watchdog timers. Each lift may be assigned to two motor controllers, thus taking advantage of the existence of similar components. The system controller has a redundant architecture and, among others things, governs the system reconfiguration in the case of a failure. The additional cost of true redundancy is less than 1% of the overall installation cost. Also, the main causes of failure are analysed as well as the suitability of the proposed architecture to recovery them.

Keywords. System failure and recovery; lifts; computer architecture; controllers; microprocessors; motor control.

## INTRODUCTION

There are many industrial applications requiring large mass displacement with accurate final position control, but without speed control. This is the case of cranes, lifts, etc. Typically, an on-off control is applied in this kind of applications, with some mechanisms in order to increase the final position accuracy. Now, the research carried out in the field of power semiconductors allows the use of induction motors controlled by microprocessors, getting an accurate final position.

An application of this kind is the lift motor control. In this case, the acceleration must be controlled during the speeding-up and slowing-down parts of the displacement. Also, the final position of the lift must be controlled. However, the displacement at nominal speed does not need to be controlled because of the inherent properties of the induction motors.

In some previous papers (Albertos, 1983, 1984; Duato, 1985), the hardware and software of the basic controller have been described. The motor controller is modelled by a finite automaton with several internal states. In each state, a different control strategy is carried out, also generating the speed reference.

Among the main goals of the system to be designed are high system security and reliability. Although some improvement in system reliability can be achieved through software techniques, only true redundancy allows the reconfiguration of the system in the case of failure. Unfortunately, that generally implies an increment of the system cost not justified if only a control cabinet is concerned. However, if a battery of lifts must be controlled, a redundant system controller can be developped without a significative additional cost.

In this paper, a fault-tolerant microprocessor-based control system for multiple lift installations is described. This system is modular and uses an improved version of the controller proposed by Albertos (1983) as the main building block. Most of the troubles detected in the industrial application of our first development are corrected and they influence the issues of the fault-tolerant design proposed.

First, a brief description of the motor controller is given. Secondly, the system controller as well as the possibilities of reconfiguration are analysed. Finally, the main causes of failure and the system cost are studied.

## MOTOR CONTROLLER

This subsystem receives commands from a system controller and executes them, controlling the motor. A block diagram of the motor controller is shown in figure 1.

A three phase induction motor is used, thus providing reduced maintenance cost and increased reliability. The power circuit speeds-up the motor by stator voltage control, allowing to slow-down it by static phase inversion.

The inclussion of current sensors in all the branches of the circuit allows to check the power circuit as well as to block it in case of a shortcircuit. Also, the current sensors allow a more reliable phase inversion.

The speed measurement is obtained through an incremental encoder with dual pulse train and rotation sense detection. This configuration supplies two redundant values of the speed, wich will be compared between them and against a speed estimation performed by software, for increased reliability.

The heart of the controller is a microprocessor-based circuit with a classical architecture. This circuit includes several counters that generate interrupts, obtain the speed measurements and trigger the thyristors (semipacks) of the power circuit. It also includes ports for interprocessor communication. The figure 2 shows a block diagram of the circuit.

A multimode control has been implemented. The behaviour of the system is modelled by a finite automaton. The state transition is produced by internal conditions or external commands. Figure 3 shows the normal states and the transitions between them.

The controller implements the digital control algorithm adequate to each state, also generating the speed reference as shown in figure 4. A great number of fault conditions are analysed in each state, producing the transition to one of the emergency states if an anomalous situation is detected.

Moreover, some techniques for fault-tolerant systems (Siewiorek, 1982) have been included in the controller. First, a watchdog timer verifies periodically that the program is running correctly. This is achieved through timer interrupts. Also, a simple circuit generates an interrupt if an instruction is fetched from an invalid address. This interrupt allows to implement a fault recovery routine, wich uses redundant variables and checks them for coherency, in order to determine the state previous to the fault.

If the system includes a system controller, as will be described in the next section, then the motor controller is periodically requested by the main controller to send its state. If the motor controller does not respond, then the main controller reconfigures the system, as will be seen later.

Thus, the motor controller constitutes a fault-tolerant subsystem that has few redundant circuits (the incremental encoder and the current sensors). These circuits as well as the software routines are oriented to detect the failures rather than to correct them, because there is no possibility of local reconfiguration (with the exception of the encoder).

The described configuration supports one of the more frequent kinds of failure: the loss of control by the microprocessor due to electromagnetic noise. Other kinds of failure would require alternative elements, such as power circuitry, etc., thus increasing the system cost.

However, if a battery of lifts must be controlled, there will be similar elements in the system, thus yielding a source of possibilities to reconfigure the overall system. The motor controller presented in this section has been designed as a basic module to allow the realization of complex systems with several lifts as well as simpler ones, with only a lift and without an intelligent system controller. This can be achieved without any change.

## SYSTEM CONTROLLER

This section describes the architecture of the system controller. The overall system has been designed with a modular hierarchical architecture. For a given installation there is a system controller and as many motor controllers as lifts. First, the architecture of the overall system is presented and then the internal structure of the system controller is analysed.

Figure 5 shows a block diagram of the system. The communication between each motor controller and the system controller is carried out by means of two unidirectional ports. There is a handshake protocol to allow the transfer of data at a maximum rate of 100 Kbytes/s, limited by the execution time of interrupt routines. Really, the data transfer rate required is much lower, allowing to send the information twice, for redundancy purpose. Thus, there is an input-output port pair for each motor controller. This configuration has been prefered over the common bus, because the former has a greater redundancy, thus allowing to isolate the faulting motor controllers.

Each motor controller may be connected to two different lift motors by means of two contactors: one as the principal motor and another as the secondary one. Each incremental encoder supplies two pulse trains, both connected to two motor controllers by means of multiplexers. Each multiplexer selects the encoder wich is the source of the pulse train pair to be input to the corresponding motor controller. Both, the multiplexers and the contactors are governed by the system controller, allowing to use each motor controller with any of two motors.

The configuration proposed allows to reconfigure the system in case of a permanent failure in the encoder and / or the motor controller. First, if a pulse train fails, the lift can arrive to the destination stage simply by taking into account only the second pulse train. This is carried out internally in the motor controller.

A more complicated situation arises when the motor controller itself fails. In this case, the system controller detects the situation because the former does not send its state or reports the failure. Then the system controller isolates the damaged motor controller and slows-down the corresponding lift by means of an electromechanical break. Once the lift stopped, the alternative motor controller is connected to the motor and encoder, when the lift assigned as principal is stopped under normal operation. This allows the lift to arrive to the destination stage. The switch of a motor controller to another is not allowed dynamically because the new controller does not know the state of the previous one. Moreover, the first one may be running at that moment.

As can be seen, the system controller is

the heart of the system and must be fault-tolerant because it is responsible of the detection of failures in the rest of the system and of its reconfiguration.

The architecture of the system controller is shown in figure 6. It is based on the dual synchronous configuration (Siewiorek, 1982) using two 8088 microprocessors. They operate synchronously, both fetching instructions and reading data from memory, but only one of them writes the results into the memory. The values appeared on the data buses are compared at the rising edge of Read and Write signals. If that values are not equal, an interrupt is generated to both microprocessors.

Each interrupt routine tests the corresponding microprocessor, then proceeds with a fault-recovery procedure. If a microprocessor is encountered to be damaged, then the controller is reconfigured, the faulting processor becoming the master one and inhibiting the comparison of the data buses. This allows a degraded control wich leads the lifts to the desired stages and then blocks the system, waiting for maintenance.

The configuration described uses a single clock generator, because this circuit rarely results damaged or affected by electromagnetic noise. Instead, the RAM memory has a SECDED (single error correction double error detection) circuit that interrupts the master microprocessor in case of a failure. The interrupt routine associated with a single error serves statistical purposes, blocking the system if a repeated failure is encountered. If a double error is detected, a fault recovery procedure is started.

## CAUSES OF FAILURE

As mentioned above, an important goal of the system is to achieve a high reliability. In this case, that means the possibility of failure detection and system reconfiguration so that each lift arrives to the destination stage in a reasonable time.

The main cause of failure, as already mentioned, is the electromagnetic noise produced by the switching components (semipacks, contactors, etc.). This noise, normally coupled through the power supply, may produce two kinds of failures: transient failures, such as the loss of control by the microprocessor, the storage of an erroneous data into the memory or the sampling of an erroneous measurement, and permanent ones, such as the trigger of a wrong thyristor, leading to a short-circuit.

Other situations, such as excesive heating of the circuits, great variations on the main power supply, etc., also may result in a permanent damage of some circuits.

The transient failures generally start the corresponding fault recovery routine, in order to get the state previous to the failure. In some cases, such as the sampling of an erroneous measurement, the failure is detected comparing two redundant measurements. On the other hand, the permanent ones result in a re-configuration of the system, allowing the lifts to arrive to their respective destination stages and then blocking the system. The single element in the system wich does not allow the reconfiguration is the motor, because each lift has only one. Fortunately, the motor is one of the more reliable elements in the system. Nevertheless, it is convenient to calculate the power of the motors with a security margin.

Another kind of event that is not a failure, but impedes a lift to arrive to the destination stage, is the main power supply failure. In this case, the system can do nothing to recovery the fault. The solution to this problem is very expensive, because it requires accumulators and power inverters. At present, we are analysing the possibility to profit the lift inertia to lead it to the stage next to the position where the main power failure happens. In this case, only a little accumulator is required to supply the system controller and the electromechanical breaks.

## SYSTEM COST

In this section, the cost versus reliability trade-offs are briefly analysed. If a battery of lifts must be controlled, a single microprocessor can act as system controller and motor controller, sampling the speed of all the lifts and triggering all semipacks. This solution is cheaper than the proposed one, but the following considerations are of concern:

a) The cost of the microprocessors and associated circuitry represents less than 1% of the overall installation cost.

b) The maintenance cost is lower if the system is modular.

c) The use of a unique type of motor controller for all installation sizes reduces the production cost.

The additional cost of true redundancy (contactors and multiplexers) of a system with several non-reconfigurable motor controllers, as well as the cost to make the system controller redundant, are also very reduced, compared with the overall cost.

## CONCLUSIONS

A fault-tolerant controller for multiple lift installations has been proposed. The cost increase due to the inclusion of true redundancy is less than 1% of the overall installation cost.

Fault-recovery routines and invalid instruction address detectors has been included in all the microprocessors, thus allowing them to be tolerant to the more frequent cause of failure: the electromagnetic noise.

Each motor controller may be connected to two different motors and encoders, thus allowing the reconfiguration of the system if a motor controller fails. The communication path between the motor controllers and the system controller is also redundant.

The system controller has a dual synchronous architecture, thus yielding the possibility of internal reconfiguration. Moreover, the RAM memory has a SECDED circuit for increased reliability.

ACKNOWLEDGEMENT

The author wishes to express deep gratitude to Professor Albertos for his suggestions and comments to this paper.

REFERENCES

Albertos, P., and J. Duato (1983). Position control with power induction motors. III International IFAC Symposium on Control in Power Electronics and Electrical Drives, Lausanne, Switzerland.

Albertos, P., J. Duato, and J. M. Valiente (1984). Control con microprocesador de motores asíncronos trifásicos. II Simposio Nacional sobre Automática en la Industria, Zaragoza, Spain.

Duato, J., P. Albertos, and J. M. Valiente (1985). Controlador de posición y velocidad para motores de ascensor basado en µP. Regulación y Mando Automático, 147, 155-159.

Siewiorek, D. P., and R. S. Swarz (1982). Reliability and availability techniques. In The Theory and Practice of Reliable System Design. Digital Press, Bedford, Massachusetts. pp. 77-174.

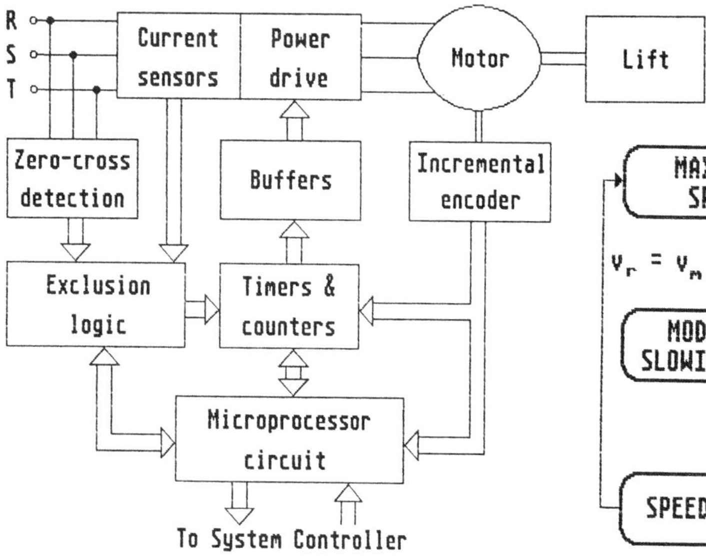


Fig. 1. Motor controller block diagram.

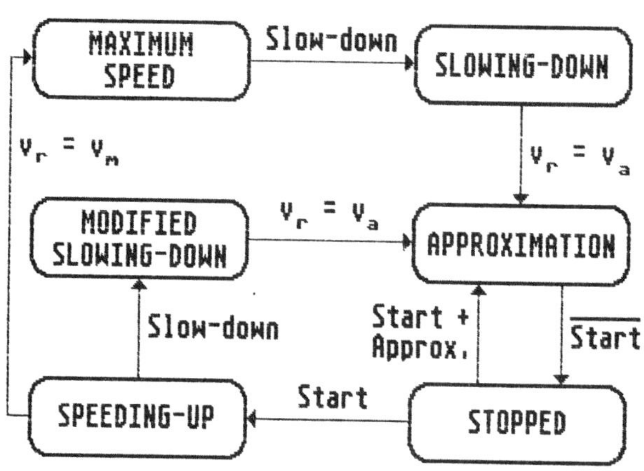


Fig. 3. Normal state diagram.

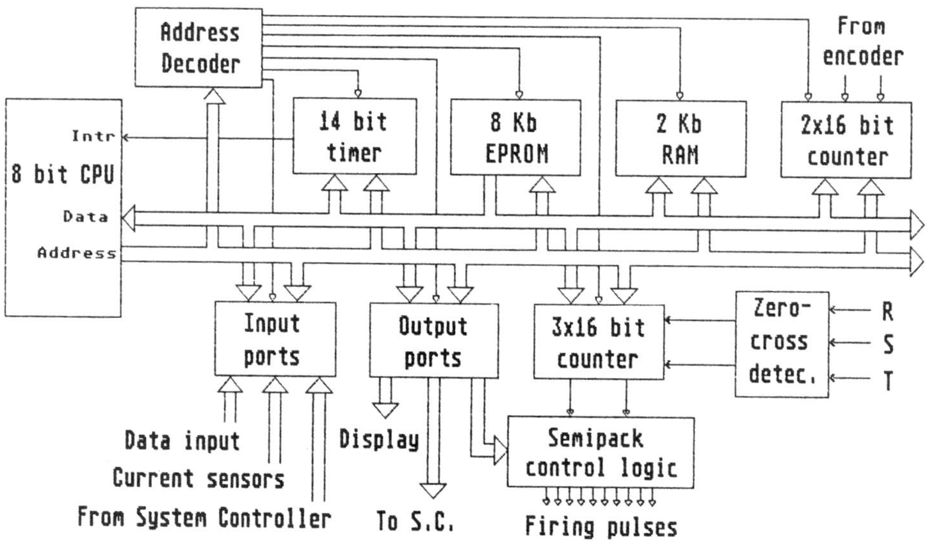


Fig. 2. Microprocessor circuit layout.

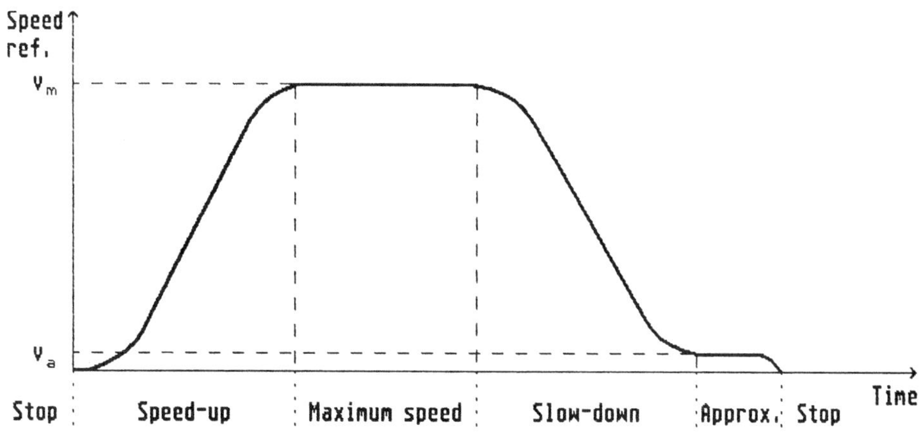


Fig. 4. Speed reference.

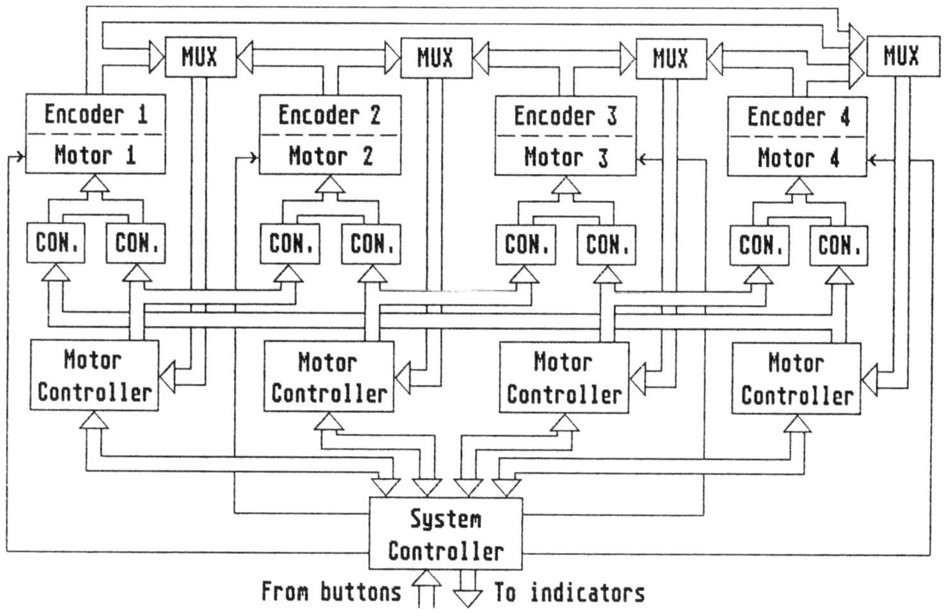


Fig. 5. System block diagram.

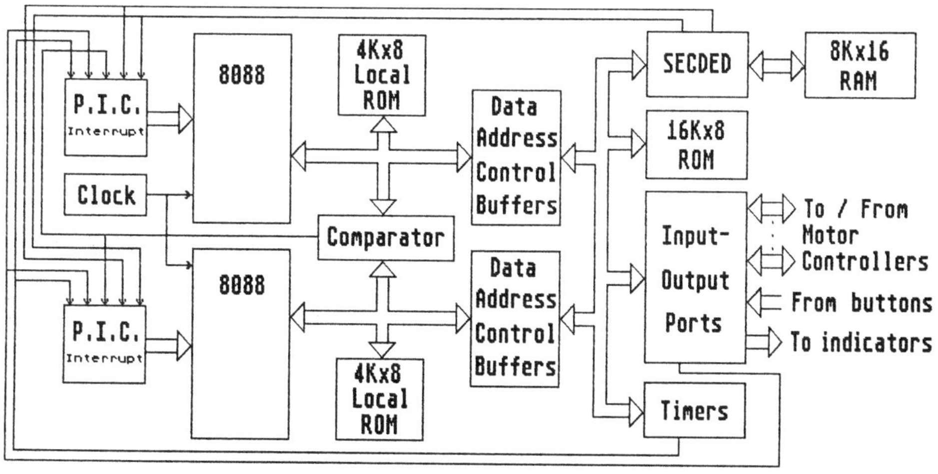


Fig. 6. System controller circuit layout.

# DISTRIBUTED CONTROL SYSTEM FOR MACHINES IN MANUFACTURING

G. Haidegger*, G. Csurgai*, L. Frittmann* and L. Revesz**

*Computer and Automation Institute of the Hungarian Academy of Sciences, H-1502 Budapest, P.O. Box 63, Hungary
**Telecommunication Engineering Factory, Hiteka, Budapest, Hungary

**ABSTRACT**

The Computer and Automation Institute, Hungarian Academy of Sciences has developed a high-speed, idustrial Programmable Logic Controller system that features distributed sequential and logic control, a user-friendly man-machine interface and a high-throughput local-area network for controlling mid-size industrial processes with real-time requirements. The presentation details the features of the system-elements, the distribution of control-information and the fast realization of the control functions. The various aspects and results of industrial application show the benefits of this new system for control application.

**KEYWORDS**

computer control, communications control applications, multiprocessing systems, manufacturing processes, local-area conrol networks

## INTRODUCTION

From the end of the 70s, industial control systems, like numerous other computer-based developments, have departed from centralized control, turning towards a broader use of distributed intelligence concept [1],[2],[3]. The continuously increasing need for new control systems and the large market-demand for post-automation of already existing production systems have given large impetus for designers to find new application techniques and develop new control systems for driving them. As recent studies have shown [4], the application-field of Programmable (Logic) Controllers in Hungary increases at an annual rate of 25%. As Fig. 1. indicates, the ratio of applications for already existing environment is representing the larger part of the market, with a factor that could be surprising at the first glance.

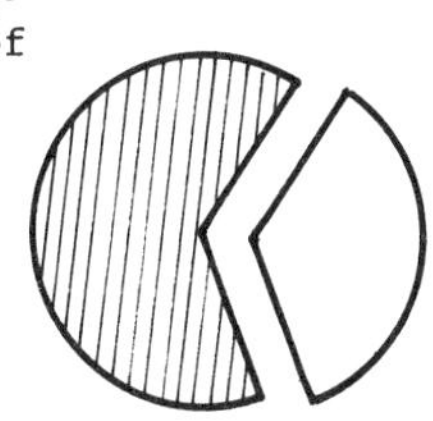


Fig. 1. Applications of PLC-s

## NEW FEATURES

A couple of years ago there were seperate cathegories for the various types of industrial electronic devices. The terms of these devices are still in wide use, though the functions and services performed by them are far from being the same as then. Cathegories such as 'process controllers', 'CNC-s','Robot-controllers', 'PLC-s', etc. achieved their rather high reputation for wide-spread industrial applicability. But as computer-science, micro-electronics and information-science have evolved, the functions and services of these devices have changed, overlapped, developed and set out new categories without changing their original name.

What are the main characteristics of this evolution in case of the 'PLC'?

- higher level logic and sequantial functions are programmable,
- analog and digital signals are equally common in most applications,
- the processing of information is implemented exclusively by digital computation,
- "smart" sensors and actuators make use of the results of "mechatronics" technology,
- instead of the use of monoprocessor architecture decentralized computing power and distributed intelligence is applied on multiprocessor architecture,
- apart from the 8-bit wide processing elements, 12-, 16- and even 32-bit wide elements are as well needed,
- the local contol functions are not enough, when larger sites and several devices are used, thus controllers must easily be integrated,
- the user must find a high-quality man-machine interface,
- the control system must bear a high level of self-diagnostics, and give a useful help at repair,

- the user programs must easily be loadable and changeable,
- the elecronics must easily be maintained and serviced.

## BASIC CONTROL UNITS

The Institute has developed a control system that meets the above mentioned requirements. It is called "CONESYS", referring to the abbreviation of COntrol NEtwork SYStem. The system is based on Basic Control Units (BCU) that can be regarded as stand-alone high-performance PLC-s [5]. Up to 64 of these BCU-s can then be tied together onto a local-area network to share control and process data (information-flow).

The internal structure of the BCU-s are modular and expandable so different configurations (implementations) may serve for different applications. The control language is user-friendly and tailored to the specific application needs. Some features of the BCU-s are as follows:

- up to 1024 I/O variables,
- 1 to 4 KBytes of user program memory,
- selection of RAM or EPROM for user-convenience,
- 2 msec/KByte execution time for logical operations,
- 64 programmable registers for timer/counter functions,
- 36 basic instructions,
- self-diagnostics,
- RS 232C interface,
- RS 422 interface for local networking.

For distributing the control and computational tasks within a BCU (ref. to Fig. 2.), several processor elements are used:

- intelligent communication processor for control of the local-area network functions,
- universal 8-bit wide MOS microprocessor,
- floating-point arithmetic slave-processor,
- BIT-processor for high-speed bit-wide datamanipulation and computation.

A set of I/O modules are available for configuring the BCU-s. Besides 16-bit digital input and 8- or 16- bit digital output modules analog single- and multi-channel modules are also used for input and output functions. Special user- and application-oriented I/O modules can serve as smart-sensor or intelligent slave-interface.

The instructions and functions of BCU-s can be grouped as follows:

- variable control instructions:
  PRESENT, MISSING, EQUAL,
- logic operations:
  AND, OR, NOT, (, ), NOP,
- program control instructions:
  BEGIN, END, GO, STOP, CONDITION, JMP, SUBROUTIN CALL,
- register instuctions:
  LET, TEST, INCREMENT, DECREMENT, ADD, SUBTRACT, COMPARE, OUTPUT, TIMER OPERATION,
- floating-point instructions (real arithmetics):
  ADD, SUBTRACT, MULTIPLY, DIVIDE, SQUARE-ROOT.

The user instructions are performed either by the fast bit-processor or are carried out by the 8-bit wide MOS microprocessor and its arithmetic slave-processor.

The operation of the fast bit-processor can easily be traced by Fig. 3. Certain functions, such as address generation, timing, etc. is not implemented separatly in the bit-processor, but rather "lended" from the functions of the general-purpose

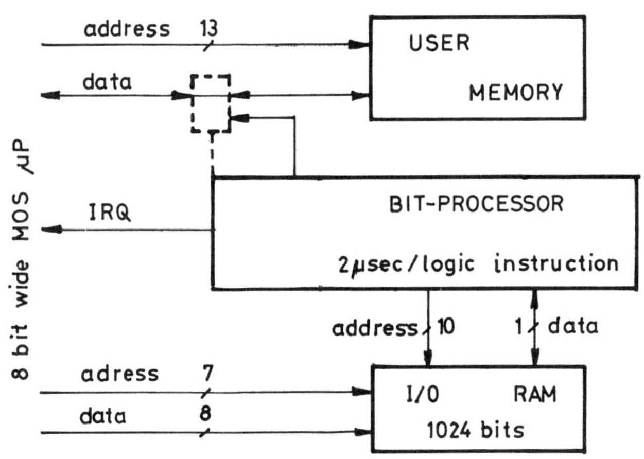


Fig. 2. Architecture of Basic Control Units (BCU-s)

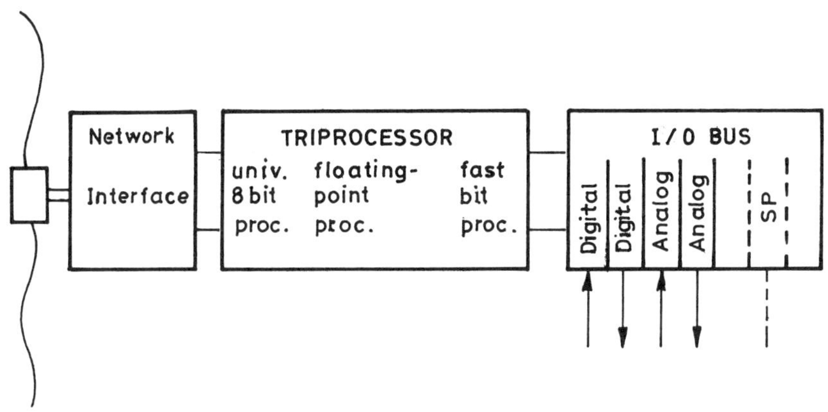


Fig. 3. Architecture of the bit-processor module

8-bit microprocessor. In this cost-effective way, the bit-processor hardware contains only three basic blocks:

- the USER memory,
- the bit-processing logic and
- the memory for storing the I/O signal values, the variables, programming registers and some system parameters.

The USER memory - that may be either RAM or EPROM - is addressed by the 8-bit processor, usually incrementally.
The bit-processor executes the instruction specified by the USER program and performs the needed operations on the variables or I/O values stored in the RAM sub-system. When the processing of bit-processor instructions comes to an end (e.g. I/O scanning, or floating-point instructions have to be performed) the module generates an interrupt-request to the 8-bit processor to take over the system control.
The I/O RAM module is a dual-ported memory that may be accessed (in both READ and WRITE operations) byte-wide from the 8-bit processor and bit-wide from the bit-processor.
Once the USER program comes to its end, the 8-bit processor transfers the OUTPUT variables to the opto-isolated OUTPUT interface modules, and copies the INPUT signals to the INPUT variables in the I/O RAM. The present limitations refere to 8 KBytes of USER memory and 1024 independent I/O variables.

## MAN-MACHINE INTERFACE UNIT

The man-machine interface unit serves as the user interface of the CONESYS system. In this sense its use is optional. As it can be seen from Fig. 4., this unit has no direct connection to physical I/O signals, but has a separated sub-system for outputing videodata and accepting user commands from the functional keyboard. The communication is served by the same type of LAN controller sub-module as in case of the BCU-s. The central processing module performs here similar type of user-programmable tasks, but consequently with remote I/O data, whose parameters have previously been transmitted. This user program can calculate various system variables based on signals belonging to unique BCU-s. This is programmed in rather the same high-level language, as in case of the BCU-s, but features additional instructions (e.g. for synchronizing processes running in various BCU-s, controlling the man-machine interface sub-system).
The user may define not only the logic content of the screen, but may also set any desired format for it. The screen is organized to display 24x60 semi-graphic characters, with programmable character-generator. In this way digital, analog, string and special code variables may be written to the screen as digits, strings, analog-proportional bars. The keyboard is implemented in form of using menu-driven "soft-key" technics for technological functions and activities.

## LOCAL-AREA CONTROL NETWORK

This network serves as communication-interface between the BCU-s and man-machine station(s). Since it has been designed for industrial use, it must conform to harsh environment-specifications. The network has high noise-immunity, it is electrically isolated from the node-processors. It features enchanced SDLC fail-safe protocol with deterministic response-time. This feature enables the network to be used for strict real-time applications. Some further characteristics of the network:

- data transfer rate: 0.3 Mbit/sec,
- network topology: bus,
- maximum distance limit: 1 Km,
- maximum number of stations: 64,
- access scheme: token-passing,
- communication media: coax cable.

## INDUSTRIAL APPLICATION

The Telecommunication Engineering Factory, HITEKA licenced the CONESYS control system for production to sell it together with its cable-manufacturing machines. When installed within a larger configuration, the machinery of the cable-manufacturing line (unvinders, accumulators, pullers, extruder(s), coolers, dryers, winders,etc. are controlled by a couple of BCU-s and a

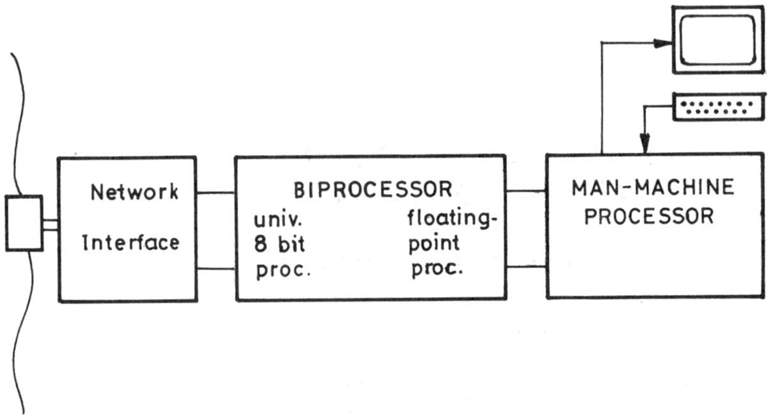


Fig. 4. Architecture of the man-machine interface unit

man-machine interface unit. The operator can control the complete production-line from the very spot, can direct the start-up and shut-down processes according to predefined requirements and has the capability to store, recall and modify previously run operating parameters as well as production data to evaluate technology parameters.
The production of CONESYS has just been started at HITEKA and a couple of cable-extruder machines will be installed by its electronic control already in this year.
Besides the cable-manufacturing industry other application fields are also aimed by the CONESYS. Reference [6] demonstrates the need for CONESYS-type devices for using them in manufacturing systems under the control of industrial cell-controllers.

## CONCLUSION

At present, when **Computer Integrated Manufacturing (CIM)** has become a reality, increasing need calls for reliable, high-performance controllers that meet the requirements of **system-integration.** The CONESYS aims for the applications of these manufacturing technologies.

## ACKNOWLEDGEMENTS

This research and development work was initiated by a group of assosiates lead by I.Rakoczy and S.S.Nagy from the Institute, while considerable help was given in the form of user-requirements from the factory HITEKA. The first industrial applications could not have been achieved without the assistance of the HITEKA team, lead by Laszlo Vesztergom.

## REFERENCES

[1] Hatvany,J.- Nemes,L.: Intelligent Manufacturing Systems, a Tentative Forcast, Preprints of the 7t IFAC World Congress, vol. 2. pp.895-899, Helsinki, 1978

[2] Hatvany,J.-Yoshikawa,H.-Rathmill,K.: CAM: An International Comparison, National Academy Press, Washington D.C. 1981

[3] Hanlon,P.D.-Weston,R.H.: Use of Local Area Networks within Manufacturing Systems, Microprocessors and microsystems, vol.6. Oct. 1982

[4] MECHATRONINFO Conference, Budapest, Sept.2-4.1985

[5] Frittmann,L.-A.Csicseri-K.Koralevics An uptodate microprocessor-based PLC: EMG 8862, Budapest CAI 28.Jan.1983

[6] Haidegger,G.-A.Krizsan-S.S.Nagy Manufacturing Systems Based on Distributed Intelligence Cell Concept Integration of CAD/CAM, IFIP WG. Conf. North Holland, 1984

# SELF-TUNING CROSS PROFILE CONTROL FOR A PAPER MACHINE

**A. Gräser* and W. Neddermeyer****

*Lippke GmbH and Co., Wilhelm-Leuschner-Str. 12, D-5450 Neuwied 1, Federal Republic of Germany*
***Section Control Systems Theory, Institute of Control Engineering, Technical University of Darmstadt, Schlossgraben 1, D-6100 Darmstadt, Federal Republic of Germany*

Abstract. The paper discusses a self tuning control algorithm for a multivariable plant applied to a grammage cross profile control of a paper machine. The plant model represents the steady-state couplings as well as the dynamics behaviour of the system. The control algorithm handles special situations including the case where the number of measurements is not equal to the number of actuators and the case where special actuators are locked.

The whole identification algorithm with a special filter and an estimation of the required measurement accuracy for a succesful identification of the steady-state couplings will be discussed. The paper shows how to proof the stability of the control system including the cases of locked actuators and identification errors.

Finally, the whole control system is verified by in house simulations and at a real plant. The obtained control results are discussed.

Keywords. Computer Control, Direct Digital Control, Paper Industry, Paper Cross Profile Control, Multivariable Control Systems, Decoupling, Stability

## I. INTRODUCTION

Paper and extruded plastics are produced continuously on fast running machines with web width up to 10 m.

The quality requirements in paper and extrusion coating productions are very high, especially for low variance in cross direction. The main quality sheet parameters of interest are

- grammage
- moisture content
- caliper.

The quality parameters across the web can be influenced by several actuators as

- slice positioners at the head box (grammage)
- steam showers (moisture)
- infrared heaters (moisture)
- air showers (caliper, smoothness)
- inductive heating coils (caliper, smoothness)

The actuators consist of several equal parts, which are controlled independently and influence areas of different size of the web.

For cross profile grammage control the slice opening of the head box is changed. Due to the physical interaction between actuator, web and the flow in the head box, a single actuator element influence broad areas of the sheet. This results in strong couplings between the single actuator responses. From the control point of view these interactions result in an multivariable plant description with strong couplings between the single control loops. The grammage control is the most demanding control loop compared with moisture and caliper ones. Therefore we will focus our interest to this quality control loop. The derived algorithm is also useful for other kinds of actuators and controlled variables (e.g. moisture).

The paper is organized in 8 sections. The second, third and fourth describes the plant behaviour, the mathematical representation and the applied control strategy. Section five is applied to the identification algorithm.

Section five to nine describe the complete multivariable control system, the stability analysis and compare some in house simulation results with commissioning results at a paper machine.

## II. PROCESS DESCRIPTION

The process has been discussed in some earlier paper. The work of Beecker an Bareiss [1] is the most complete early treatment of the subject from the cross-profil-control point of view.

Some later papers of Boyle [2,3] and Wilhelm [4,5] discusses the weight control in cross direction by actuators which are located at the wet end of the machine, too.

All these papers discusses some special aspects of cross profile control. This paper tries to give a solution for the complete problem including an identification procedure and a stability proof.

However, the major problem for the controller design is the strong interaction of the effects of adjacent cross-direction actuators at the wet end of the machine.

Fig. 1 shows the flow onto the wire. An increase of the flow within one section leads to changes in the grammage profile within a third of the web width.

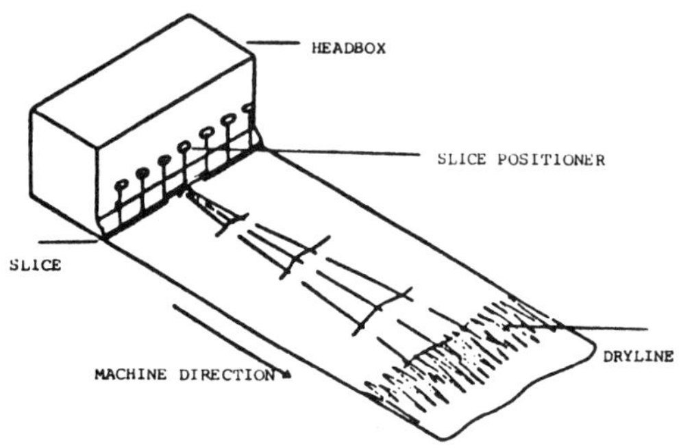


Fig. 1 Wet end of a paper machine

In order to obtain the steady-state behaviour of the systems we assume that the i-th actuator is moved. The result is a decreasing or increasing of the mass-flow at this point depending the sign of the slice positioner displacement. Due to cross flows in the head box we obtain at the dryline a typical response of the grammage profile as shown and described in some papers before (see also fig. 2).

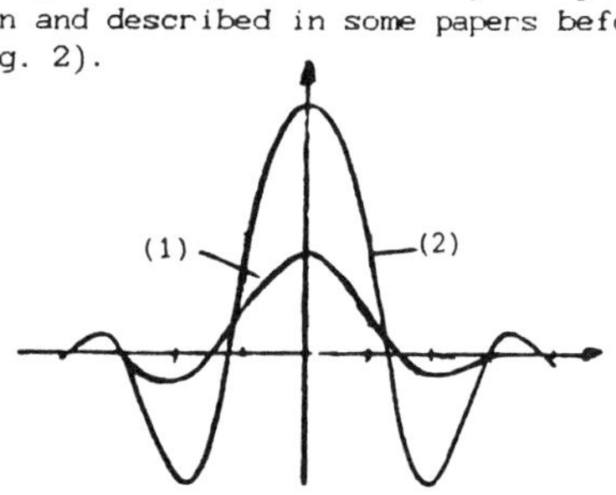


Fig. 2 Typical weight profile responses to slice displacements in one position

We call such a response "coupling curve". One can measure this effect on-line by scanning sensors mounted at several locations within the paper machine (see fig. 3).

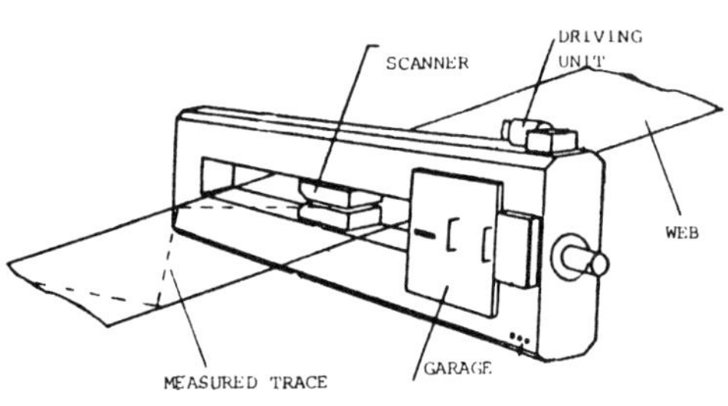


Fig. 3 Measurement unit

These units measure the web parameters (e.g. grammage, moisture, caliper) along a triangular measurement trace as shown in fig. 3. The measured values contain the information in cross-direction and in machine direction. We discuss the cross direction measurements only. These has to be calculated from the measured diagonal profil. The necessary filter algorithm is not considered here. For further information on this point see [6].

## III. PROCESS MODEL

The steady state response can be discribed by eq. 1, if the displacement of one slice position j is considered only.

$$\begin{pmatrix} y_1 \\ \vdots \\ y_i \\ \vdots \\ y_m \end{pmatrix} = \begin{pmatrix} q_{1j} \\ \vdots \\ q_{ij} \\ \vdots \\ q_{mj} \end{pmatrix} u_j \qquad \text{eq. 1}$$

with $y_i$ -The measured grammage values in cross direction.

$q_{ij}$ - The values of the coupling curve.

Considering all actuators eq. 1 expands to the matrix equation

$$\underline{y} = \underline{Q}\ \underline{u} \qquad \text{eq. 2}$$

Normally the number of measured points (m) of the web in cross direction is not equal to the number of actuator elements (n). Accordingly the coupling matrix has the order m x n. Due to the fact that the coupling curves influence a part of the web only the coupling matrix has a band-structure.

The dynamic part of the process model can be described by the lag behaviour of the actuators and a transport delay $T_t$.

These two effects are represented in eq. 3 in the frequency domain as

$$D(s) = \begin{bmatrix} \frac{1}{1+T_i s} & & 0 \\ & \ddots & \\ 0 & & \frac{1}{1+T_i s} \end{bmatrix}_{n\times n} \qquad \text{eq. 3}$$

$$T(s) = \begin{bmatrix} e^{-T_t s} & & \\ & \ddots & \\ & & e^{-T_t s} \end{bmatrix}_{m\times m}$$

with: $T_t = \frac{l}{v_{pap}}$

$v_{pap}$ - paper velocity

$l$ - distance between actuator and measuring unit

The diagonal matrix elements have generally all the same value.

$T_i = T \qquad i = 1,2 \ldots n$

Finally, the whole process model is shown in fig. 4.

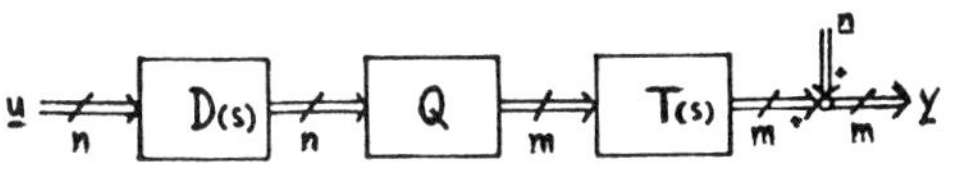


Fig. 4 Block diagram of the process model

## IV. CONTROL STRATEGY

For the control task it is most important to influence the grammage on a width of one measure point, that means one has to eliminate the coupling terms. This can be done via a decoupling matrix.

From multivariable control theory it is known, that strong couplings between control loops cause stability problems. A common solution is the introduction of a decoupling matrix which yields to decoupled single control loops. In the case of m = n the coupling matrix is a quadratic one.

To calculate the required slice position $\underline{u}$ we define eq. 4.

$$\underline{u} = R^T \underline{y}_D \qquad \text{eq. 4}$$

with $\underline{y}_D$ the required grammage profile and $R^I$ the decoupling matrix. With the goal that $\underline{y}$ should be equal to $\underline{y}_D$ ($\underline{y} = \underline{y}_D$) we obtain $R^I = Q^{-1}$. (see also fig. 5)

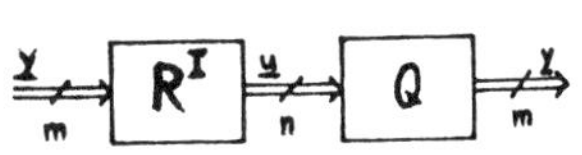


Fig. 5 Decoupling with the matrix $R^I$

In the case of m > n which is the normal one, the inverse matrix is not defined. A possible way to find still a decoupling matrix is the introduction of a performance index and the minimization of this index.

A typical performance index J for such a problem is given with eq. 5

$$J = \underline{e}^T \underline{e} \qquad \text{eq. 5}$$

with $\underline{e} = (\underline{y}_D - \underline{y})$

Minimizing the index J the resulting decoupling Matrix $R^I$ can be written as

$$R^I = (Q^T Q)^{-1} Q^T \qquad \text{eq. 6}$$

This matrix is the pseudo inverse of Q. The multivariable system is decoupled by the matrix $R^I$- and the multivariable control loop can furtheron be discussed as a system with n independent loops.

All these loops has the same time constants. This yields to identical controller parameters.

The complete control loop is described by five matrices as shown in fig. 6.

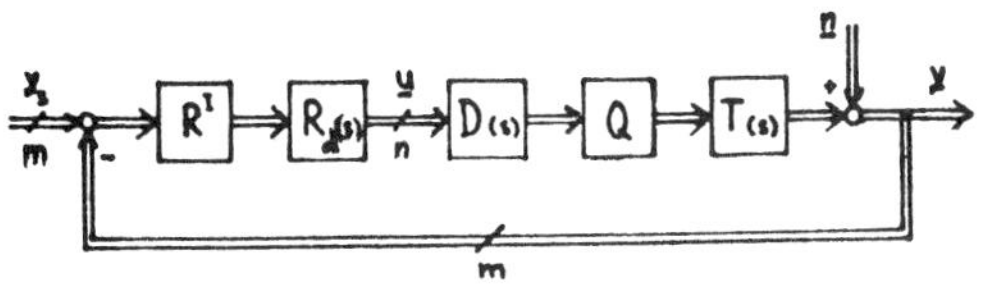


Fig. 6 Structure of the control loop

$R^I$ represents the decoupling matrix and $R_d(s)$ the dynamic part of the controller, which is a diagonal matrix with identical elements r(s).

$$R_d(s) = \begin{bmatrix} r(s) & & \\ & \ddots & \\ & & r(s) \end{bmatrix}_{n \times n} \qquad \text{eq. 7}$$

The plant model is represented by D(s), Q and T(s).

The dynamic behaviour of the cross profile with the above described strategy depends very closely on the accuracy of the decoupling matrix $R^I$. The matrix is calculated from the steady state plant model Q which is calculated from the coupling curves. The identification of the coupling curves is described in section five.

## V. IDENTIFICATION

The identification is done with a least square method. One slice position $u_j$ is changed at time k and all others remain in the old position. Equation 8 describes the response to be measured.

$$\underline{y}(k + d + e) - \underline{y}(k + d) = \underline{q}_j (u_j(k) - u_j(k - 1)) \qquad \text{eq. 8}$$

d is the transport delay

e is the response time

both are rated in sampling steps.

In order to decrease the influence of measurement disturbances the measurements can be repeated s times without slice changes.

$$\sum_{l=0}^{s} (\underline{y}(k + d + e + l) - \underline{y}(k + d - l) = \underline{q}_j (s+1)(u_j(k) - u_j(k-1)) \qquad \text{eq. 9}$$

With N changes of the slice positioner one obtains eq. 10.

$$\underline{y} = u_j \, \underline{q}_j^T \qquad \text{eq. 10}$$

$$\underline{y}^T = (\underline{y}_d(1), \underline{y}_d(2) \ldots \ldots \underline{y}_d(N))$$

$$\underline{u}_j^T = (u_{jd}(1), u_{jd}(2) \ldots u_{jd}(N))$$

with

$$\underline{y}_d(i) = \sum_{l=0}^{s} (\underline{y}(d + l + 1 + ie + (i-1)s) - \underline{y}(d - l + 1 + (i-1)e + (i-1)s))$$

$$u_{jd}(i) = \sum_{l=0}^{s} (u_j(1 + (i-1)e + (i-1)s) - u_j((i-1)e + (i-1)s))$$

For the estimation of the $\underline{q}_j$-terms eq. 11 can be used (see [7]).

$$\underline{q}_j^T = (\underline{u}_j^T \underline{u}_j)^{-1} \underline{u}_j^T y \qquad \text{eq. 11}$$

With eq. 11 an identification of a single coupling curve is carried out.

The identification algorithm calculates each point of the coupling curve indenpendent from its neighbours. Due to the physical behaviour of the flow onto the web, the coupling curve must be smooth.
This knowledge leads to a filter algorithm of the coupling curve, which is shown in fig. 7.

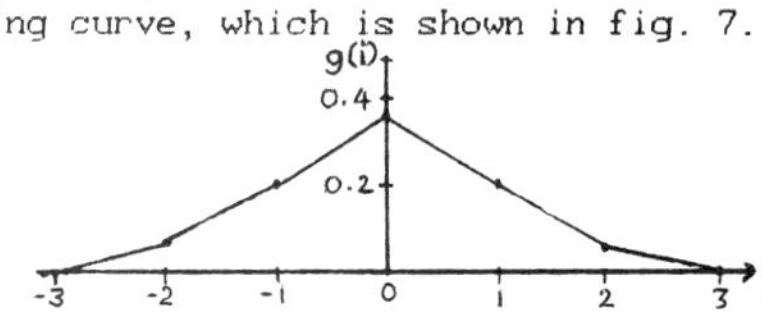


Fig. 7 Filter function

$$g(i) = \begin{cases} e^{-|i|/a} \Big/ \left[\sum_{j=-b/2}^{b/2} e^{-|j|/a}\right] & |i| \le b/2 \\ 0 & |i| > b/2 \end{cases} \qquad \text{eq. 12}$$

With the weighting factor a and the width of the filter b the filter response can be chosen.

The filtered values of the coupling curve $\underline{q}^F$ are calculated with eq. 13

$$q_l^F = \sum_{j=1}^{n} q_j \, g(j-l) \qquad l=1,2,\dots n \qquad \text{eq. 13}$$

## VI. STRUCTURE OF THE MULTIVARIABLE SELFTUNING CONTROL LOOP

The identification and the multivariable control loop strategy are combined in the whole control structure which is shown in fig. 8.

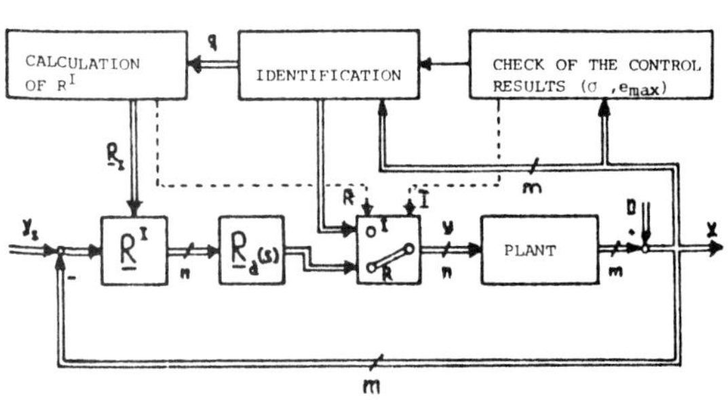


Fig. 8 Structure of the multivariable selftuning loop

The plant level in fig. 8 describes the basic control loop with the decoupling matrix $R^I$ and the controller matrix $R_d(s)$.

The supervisor level observes the control results and calculates the necessary position of the switch in fig. 8. The position of the switch indicates closed loop control or identification mode. There are several mathematic or heuristic criterions available to calculate the position of the switch. One criterion for rating the control loop result is the $\sigma$-value of the cross profile (see eq. 14)

$$\sigma = \sqrt{\frac{1}{m} \sum_{i=1}^{m} (y_i - y_{Di})^2} \qquad \text{eq. 14}$$

m - number of measured points in cross direction.

An other one is the maximum of the difference between the commanded and measured value of the cross profile

$$e_{max} = \max(|y_i - y_{Di}|) \qquad i=1\dots m \qquad \text{eq. 15}$$

A practical way is to switch the control off and start the identification if the $e_{max}$-value or the $\sigma$ value exeeds their upper limits.

The identification results are needed to calculate a new decoupling matrix. When the identification is finished this matrix will be used and the switch is set back to the basic control mode.

## VII. STABILITY ANALYSIS

The stability of the basic control loop has to be verified especially in the case of locked actuators and errors of the plant model which is used for decoupling.

In order to proof the stability of the multivariable control loop we break the loop between $R_d(s)$ and $D(s)$ in fig. 6. Then one obtains for the open loop $L(s)$

$$L(s) = R_d(s)\, R^I T(s)\, Q\, D(s) \qquad \text{eq. 16}$$

Assuming that $T(s)$ is a diagonal matrix with identical elements one can change the position of $R^I$ and $T(s)$ in eq. 16 and obtain

$$L(s) = R_d(s)\tilde{T}(s)\, R^I Q\, D(s) \qquad \text{eq. 17}$$

with $R^I = (\hat{Q}^T\hat{Q})^{-1}\hat{Q}^T$

$$L(s) = R_d(s)\, \tilde{T}(s)\, (\hat{Q}^T\hat{Q})^{-1}\hat{Q}^T Q\, D(s) \qquad \text{eq. 18}$$

with $(\hat{Q}^T\hat{Q})^{-1}\hat{Q}^T Q = S$

$$L(s) = R_d(s)\, \tilde{T}(s)\, S\, D(s) \qquad \text{eq. 19}$$

In the case of $\hat{Q} = Q$ is S equal I and the whole multivariable loop can be written as a n-single loop system since the matrices $R_d(s)$, $\tilde{T}(s)$, $D(s)$ are diagonal ones. In this case the multivariable system is stable if the n-single loops are stable.

The design of the multivariable loop is now simplified to n-single loop designs which can easily be carried out by wellknown procedures.

In such a system a locked actuator effects that one single loop is not controlled but this has no influence on the other ones. Due to the fact that the uncontrolled single plants are stable the whole multivariable system is stable, too.

The assumption $\hat{Q} = Q$ is not a serious one for a real plant. The measurement errors and disturbances lead always to an identification error $\hat{Q} \neq Q$ with the result that S is not equal I.

However, the above described facts are valid also for stability, if the matrix S is a diagonally dominant matrix. For a proof look at [8]. A stability theorem based on the concept of a set of "normalized" Gershgorinbands and a practical rule how to caculate the normalized radians are given in [9,10].
With this rule we try to answer the question: Which size of model errors are acceptable in order not to destroy stability?
This question can not be answered in general. For a specific number of actuators, measurement points and coupling curve one can get an empirical value by a lot of simulations. For the case of 64 measurement points, 18 actuators and a coupling curve which is shown in fig. 9b we did some simulations with different sizes of measurement disturbances. Fig. 9a/b shows a typical result.

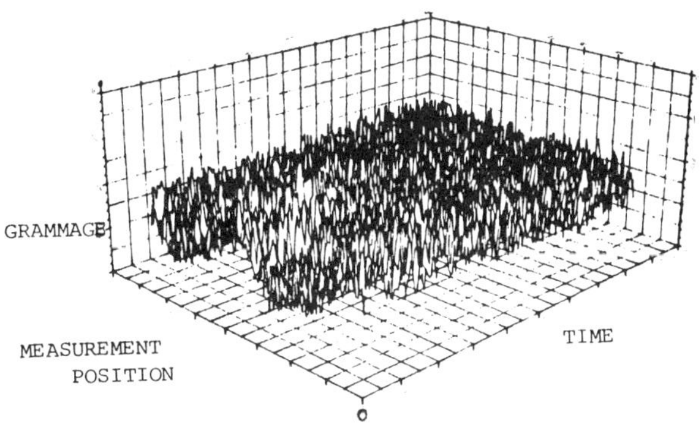


Fig. 9a Simulated cross profile

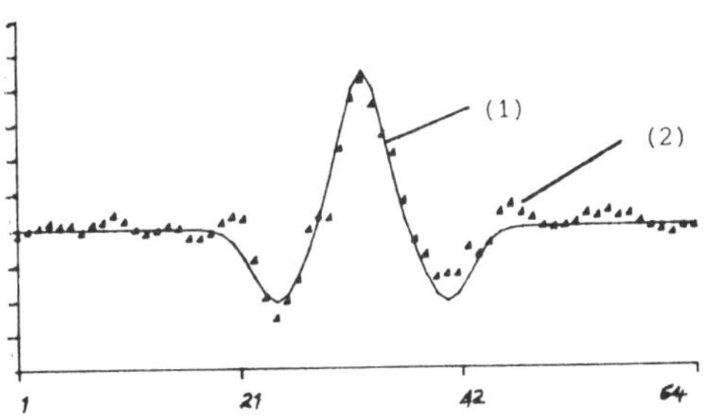


Fig. 9b Coupling curves (1) plant model
(2) identified

In fig. 9a the web grammage profile (nominal value 80 $[g/m^2]$) with a measurement disturbance of 1 $[g/m^2]$ is shown. The identification of the coupling curve is done with an actuator stimulation of 30 counts that leads to a maximum response of approximately 2.8 $[g/m^2]$. The identified coupling curve is marked by stars in fig. 9b, the real coupling curve is given by the line. With this identified coupling curve one obtains Gershgorinbands, which satisfy the stability conditions.

Due to a lot of simulations we found that for disturbances of the identification step response up to approximately 30 percent the matrix S remains diagonally dominant.

## VIII. CONTROL RESULTS

This point figures out three control situations. The first two are simulated ones. The third is measured at a real plant controlled by a Lippke 4012 system.

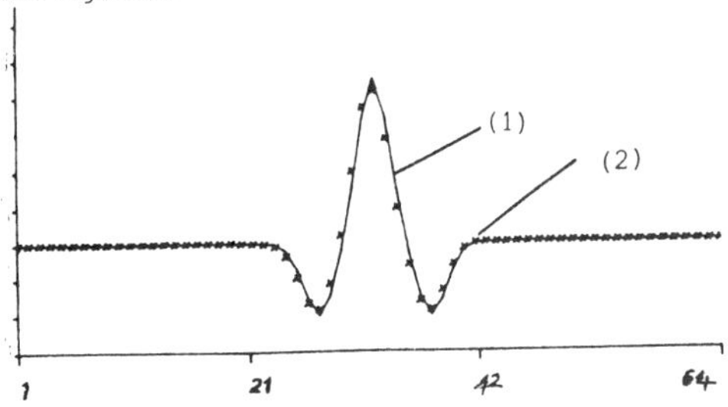


Fig. 10a Coupling curves (1) plant model
(2) identified

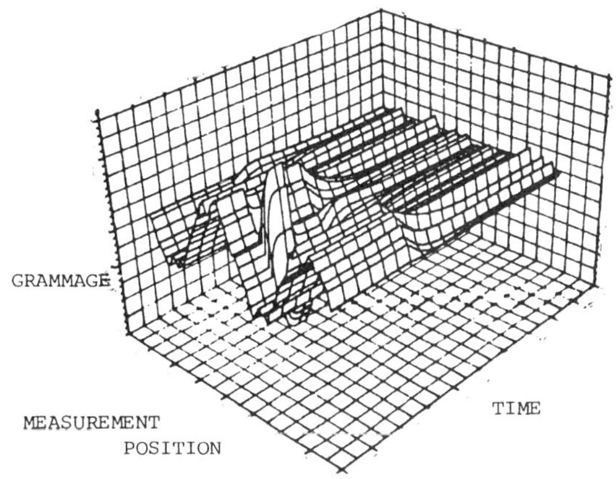


Fig. 10b Grammage cross profile

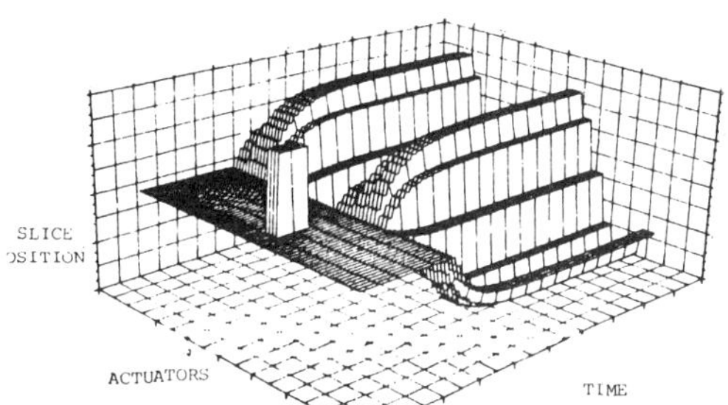


Fig. 10c Actuator positions

Fig. 10 shows the simulated coupling curve and the identified one (stars). The plant is simulated without any disturbances, the web cross profile and the actuator positions for 62 seconds are in part(b) and (c) of this figure. The σ-value calculated according eq.14 is shown in part (d) over the same time interval. The increasing part at the beginning results by the stimulation of one actuator during the identification cycle. The steady state behaviour is shown in fig. 10e by the start cross profile and the final one.

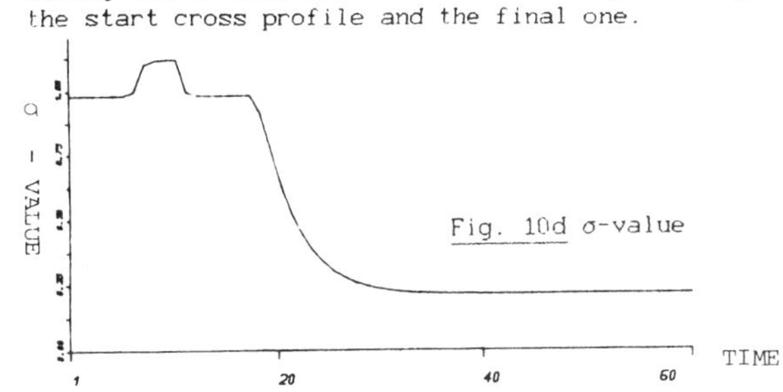


Fig. 10d σ-value

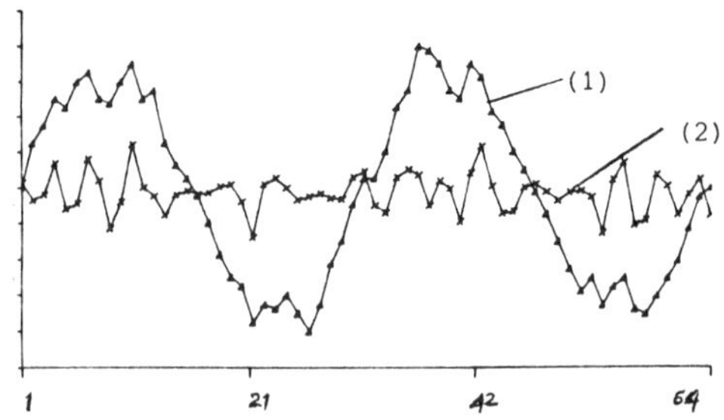


Fig. 10e cross profile (1) start profile
(2) end profile

Fig. 11 sketches the second simulation run with three locked actuators (positions: 5,6,12).

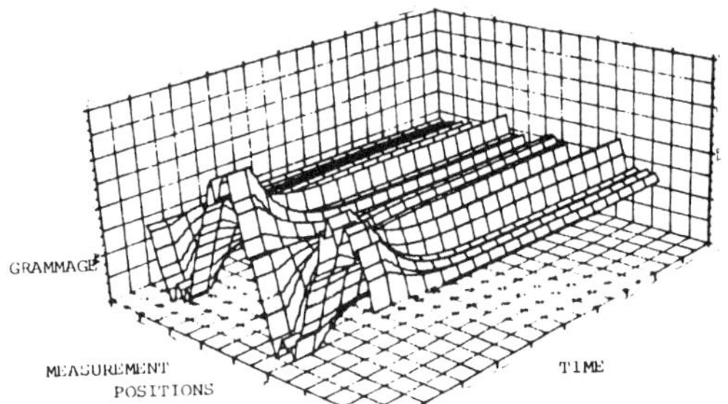


Fig. 11a Grammage cross profile

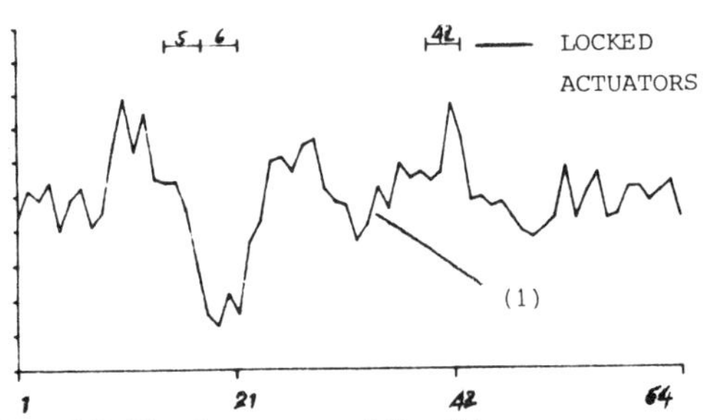


Fig. 11b Final cross profile (1)

The cross profile over the whole simulation time is in part (a) of this figure. Part (b ) shows the final profile. One can see that the final profile is not so good as in the first simulation run especially at the regions where the actuators are locked. However, the whole multivariable loop is stable.

The last fig. 12 shows the σ-value of the grammage cross profile of a paper machine which was controlled by the above described algorithm.

After an identification cycle the control loop decreased the σ-value very fast (compare fig. 12a with fig. 10d).

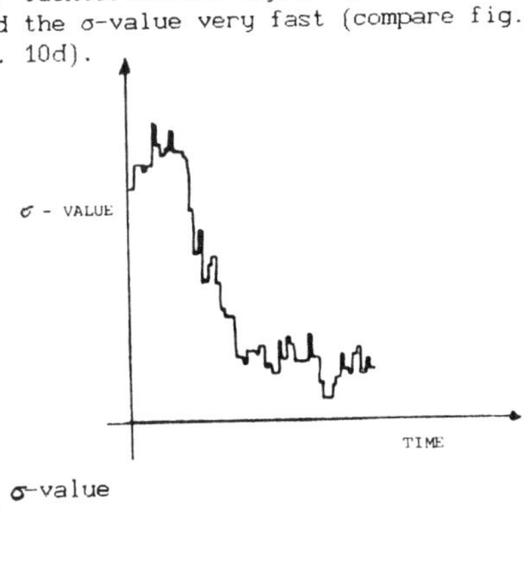


Fig. 12a σ-value

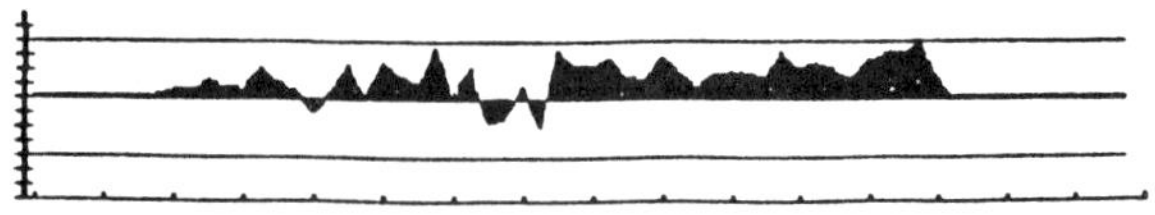

Fig. 12b Start profile

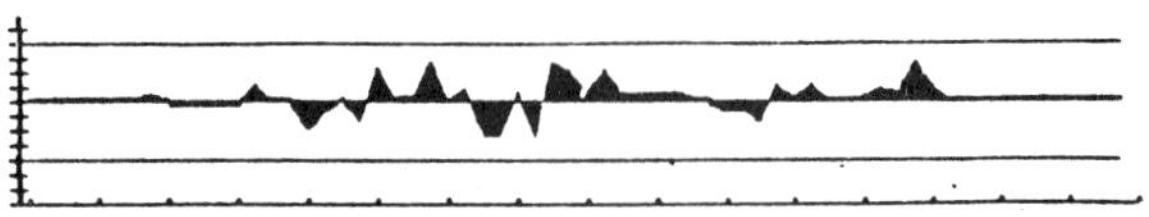

Fig. 12c End profile

The start profile and the steady state profile are given in part (b) and (c) of this figure.

## IX. CONCLUSIONS

We did this simulation and control tests with a lot of different paper qualities. The measured results were always in a good agreement with the simulated ones as shown in the example case above. This indicates, that the control loop shown in fig. 8 is an acceptable solution for the discussed control problem.

## X. REFERENCES


[1] Beecker, A.E. and Bareiss, R.A. Theory and Practive of automatic control of basis weight profiles; TAPPI, VOL.53, No-.5, 1970

[2] Boyle, T.J. Can. J. Chem. Eng. (1977)

[3] Boyle, T.J. Practical algorithms for cross-direction control; TAPPI, VOL. 61, No.1, 1978

[4] Wilhelm, R.G. Self tuning control strategies: Multi-faceted solutions to paper machine control problems ISA JOINT SYMPOSIUMS 1982, COLUMBUS, OHIO

[5] Wilhelm, R.G. Norwegian Institute of Technologie, PRP 5, Automation , Sept. 1983

[6] Richards, G.A. Cross Direction Weight Control JAPAN PULP and Paper, 1982

[7] Isermann, R. Prozeßidentifikation Springer-Verlag

[8] Mee, D.H. Specifications and criteria for multivariablecontrol system design; Joint Automatic Control Conference, Purdue University, Indiana, USA, 1976

[9] Kantor, J.C. Andres, R.P. A note on the extension of Rosenbrock's Nyquist array techniques to a larger class of transfer matrices, Int. J. Control, Vol.30, 1979

[10] Tolle, H. Mehrgroßenregelkreissynthese; Band I, Oldenbourg Verlag, 1983

# THE EXPERIMENTAL CLASSIFICATIONS OF THE LINEAR MODELS OF THE LATHE CUTTING MECHANISMS

**N. Olgac, G. Zhao and J. Guttermuth**

*Mechanical Engineering Department, University of Connecticut, Storrs, CT 06268, USA*

ABSTRACT

A preparatory stage to achieve "smart machining" is addressed, namely a forecasting model for the oscillation at the tool/workpiece interface is reported in this text. The nonlinear dynamics of the mentioned interface is suggested to be put into a very similar discretized linear format: Autoregressive Moving Average Model (ARMA). This linearization operation is normally carried on as an off-line procedure. The following questions are addressed for the ARMA models:

a) What are the ARMA forms to achieve the best representation? This question leads to a library of plausible coefficients which are envisioned to have a powerful place in "knowledge based expert system representations". A comparative study is included in the text for a broader understanding of the overall nature of the cutting tool dynamic modeling.

b) The ARMA procedure is performed over a set of data. How dependent are coefficients of the model on the length of available data points? And can the data collection period be shortened without jeopardizing the health of the respresentation? And going a step further: can the ARMA modeling be made into an on line procedure so that true meaning of predictive and adaptive control concepts hold?

Additional to the temporal concerns of ARMA coefficients a spatial dependency of the model is scrutinized. Results are tabulated considering the flexural properties of the workpiece with respect to the cutting tool dynamics.

## Introduction

An increasingly important part of research efforts in "manufacturing technology" in recent years is aimed towards smart machining. That is, the techniques in which the computerized machine tool controllers utilize sensory information in order to act autonomously to choose, alter, optimize their on-going process. In an extreme case a smart controller may even selectively choose the sensory data to operate on.

In this report the effort is concentrated around the highly dynamic behavior of the cutting tool tip vs. the workpiece interface. As many researches indicated earlier [1,3,5] metal cutting dynamics is a highly complex and parameter driven occurrence and the number of parameters involved is large. First of all the geometry of the cutting set-up is not unique. For instance, a cutting tool has a number of geometric quantitates which define its ideal or desired form (Figure 1) for the purpose of practicality these geometric identifiers like rake angle, clearance, etc. are taken as invariants for the coverage of this report. But what happens to this ideal form once the cutting starts is a major question which is not easy to answer. In addition, tool wear is a very obscure occurrence for the investigators [5,8]. The on-going geometric variation becomes a major factor in governing the tool tip/workpiece interface (in the rest of the text referred as tool/work interface). Second, the dynamic parameters are not well defined, such as the tool deformations, workpiece deflections, restoring forces, etc. Third, and most importantly the nonlinearity which is intrinsic to the machining process is an added difficulty. For instance tool/work separation is always conceivable for very short time intervals. But once interrupted, the cutting will not be in a smooth continuation, therefore causing number of nonlinearities.

The surface quality is measured by following the motion of the cutting-tool tip. In Figure 2 a typical tool tip departure from the equilibrium (nominal) cutting conditions is depicted in a 2D turning process. u and v represent the displacements of the tool from the nominal cutting position in the directions of thrust and cutting respectively. The restoring elements spring stiffnesses ($k_x$ and $k_y$) and the structural damping components ($b_x$ and $b_y$) are also shown, along with the directional mass inertias ($M_x$ and $M_y$). Because the turning surfaces are not perfectly circular or the external disturbances force the tool/workpiece interface to deviate from the ideal path. Next time around these undesirable points become excitation sources. Typical evolution of this feature leads to a condition called "chatter of a toolpiece". Chatter is very much undesirable and from a machining quality point of view should be avoided as much as possible.

Earlier works of the group [4,5] present an important finding: An approximate linear model in continuous time can be taken as a guide, incorporating a variety of chatter conditions (i.e. speed chatter, regenerative chatter, etc.). In addition one of them [4] contains an Autoregressive Moving Average (ARMA or ARIMA) modeling effort. These modeling attempts are over the discrete data values obtained by numerically solving the mentioned linear continuous model. It is shown that matching between the discretized and continuous dynamics is very strong, both over the data points and beyond.

A parallel study is reported in this text starting from the real data instead of numerically generated counterparts. A group of data of a known nonlinear system is recasted into a discrete linear format. The discrepancy between the data and the model is reviewed for both the varying observation periods and the forecasting time zones.

For the sake of simplicity and broad accumulation of prior experimental knowledge the focus is kept on turning operations. The above mentioned displacements (u and v) correspond to co-directional force components, $F_t$ = Thrust Force and $F_c$ = Cutting Force respectively (Figure 2). Considering the cutting operation taking place in the elastic deformation zone a simple linear relationship between the forces and tool displacements can be written. For the full 3-dimensional case the tool forces are:

$$\begin{bmatrix} F_t(t) \\ F_c(t) \\ F_f(t) \end{bmatrix} = \begin{bmatrix} A_{11} & A_{12} & A_{13} \\ A_{21} & A_{22} & A_{23} \\ A_{31} & A_{32} & A_{33} \end{bmatrix} \begin{bmatrix} u(t) \\ v(t) \\ w(t) \end{bmatrix} + \begin{bmatrix} B_1 \\ B_2 \\ B_3 \end{bmatrix} \qquad (1)$$

where $F_f(t)$ is the feed directional force (perpendicular to the plane of Figure 2 in positive sense, w(t) is the displacement along $F_f$, $\{A_{ij}\}_{3X3}$ is the transformation matrix from the tool displacements to the tool forces. The ease of monitoring the tool displacements compare to the forces is obvious. But since most of the machining control schemes are developed around the tool forces, the equation (1) is a necessary transformation in this study.

Consider a time series of tool forces

$$F(k) = \begin{matrix} F_t(k) \\ F_c(k) \\ F_f(k) \end{matrix}$$

which is being registered continuously by a data acquisition system as the cutting process takes place, through the conversion of equation (1). At a given time t, based on the previous data

$$\{F(i)\} \quad 0 \leq i \leq k$$

how should the controllable parameters be adjusted during the turning process (namely the depth of cut, feed rate, and the spindle speed) so that the maximum metal removal rate will be achieved (Figure 3). An important constraint here is the limitations imposed on the tool forces. It is intuitively obvious that the increasing metal removal speed causes higher forces in the 3D structure. In turn, higher tool tip movements/ oscillations emanate. The mode of operation maximizing the metal removal volume but yet still preserving the force restrictions is sought. Note that ideal nominal cutting conditions exist when no oscillations of the tool tip are present. Because the corresponding tool forces stay constant (the forces are called nominal tool forces) for this case. But in general on top of the nominal tool forces an additional dither (oscillatory expansion) is observed due to the nature of the dynamic structures. High frequency dynamics are hard to observe, comprehend and respond to by the computerized machine controllers (i.e. CNC controllers) a prediction based control operation is needed to achieve the desired "metal removal maximization".

This report explains the techniques used for the prediction of tool forces and the dependency of these techniques on data. Further discussions on "knowledge based expert systems" application to the metal turning processes and a look-up table (library) are development for different cutting conditions. Then some critical prediction (or modeling) error analysis is carried out with respect to the size of the data utilized for the representation. This effort of minimizing the number of necessary data points for a healthy prediction is addressed.

## Problem Formulation

Consider a time series of 3D tool forces as recorded by a computerized data acquisition system:

$$\{F(i)\} \quad 0 \leq i \leq k$$

where k is the running time counter and i is the dummy time elements. Both i and k are counted in discrete time increments in parallel to the speed of the data acquisition set-up. F(i) is an 3 X 1 vector, one component of which is $F_c(i)$, the cutting force, and it is the most significant among the three. This is basically due to the nature of the machining operation, which is in fact, mainly a chip shearing effort in the cutting direction. Therefore for the sake of conciseness the cutting force component is considered in place of F(i) as a representative element throughout this text.

An Autoregressive Moving Average technique is widely used to represent the time series of a nonlinear systems dynamics in the form of a discrete time, linear model [6,7,8]. The general form of the Autoregressive Moving Average (ARMA) model is:

$$F_{c,i} - \phi_1 F_{c,i-1} - \phi_2 F_{c,i-2} - \cdots - \phi_n F_{c,i-n} =$$
$$ai - \theta_1 a_{i-1} - \theta_2 a_{i-2} - \cdots - \theta_m q_{i-m} \qquad (2)$$

where $\phi_1, \ldots \phi_n$ are the n Autoregressive parameters (AR) $\theta_1, \ldots \theta_m$ are known to be the m Moving Average parameters (MA) (a misnomer instead of the "m coefficients of stochastic terms"). $a_1$, $a_2$ ... - $a_m$ are zero mean, Gaussian distributed, independent random variables, i.e.

$$E\, a_i = 0$$

$$E \sum_{j=1}^{N} (a_j \cdot a_l) = \delta(j-1)\, \sigma_a^2 \quad , \; a_j = 0 \text{ for } j \geq N+1$$

$$l = j + \text{constant}$$

$\delta(j-1)$: Kronecker Delta function

$\delta(j-1) = 1$ if $j-1 = 0$
$\delta(j-1) = 0$ elsewhere

$$a_i = \bar{F}_{ci} - F_{ci} \tag{3}$$

where $a_i$ is the difference between the actual measurements ($\bar{F}_{ci}$) and the calculated values ($F_{ci}$).

The equation (2) represents a recast of n terms of nonlinearly dependent data registrations ($\bar{F}_{ci}$) into m independent, zero mean, Gaussian noise elements ($a_i$). This form is known as ARMA (n,m) model. The residual value $a_i$ is independent of the state; $F_{c,i-n-1}, F_{c,i-n-2}, \ldots F_{c,1}$.

Not until the $n^{th}+1$ data entry is reached can equation (2) be used. All residual values prior to the $n^{th}+1$ are assigned zero.

$$a_i = 0 \quad \text{for } i = 1,2\ldots n$$

Now that a way of representing a nonlinear behavior in a linear format is in hand, how are the autoregressive and moving average parameters (i.e., $\phi i$ and $\theta j$) calculated. An accepted method for this evaluation is to minimize the sum of squares of the residual values (SSAT) throughout the sampling interval (1 to N)

$$E[\sum_{i=1}^{N} a_i^2] = SSAT \tag{4}$$

where E denotes the statistical expectation and N is the total number of data points available. Starting from the initial time i=1 and remembering that

$$a_i = 0 \quad \text{for} \quad 1 \leq i \leq n$$

$$a_{n+1} = \bar{F}_{c,n+1} - \sum_{i=1}^{n} \phi_i \bar{F}_{c,n+1-i} \tag{5}$$

$$a_{n+2} = \bar{F}_{c,n+2} - \sum_{i=1}^{n} \phi_i \bar{F}_{c,n+2-i} + \theta_1[\bar{F}_{c,n+1} - \sum_{i=1}^{n} \phi_i \bar{F}_{c,n+1-i}]$$

continuing in the equation (5) and substituting all the residual values $a_i$'s into equation (4):

$$SSAT = (\bar{F}_{c,n+1} - \sum_{i=1}^{n} \phi_i \bar{F}_{c,n+1-i})^2 + [\bar{F}_{c,n+2} = \sum_{i=1}^{n} \phi_i \bar{F}_{c,n+2-i} + \theta_1 (\bar{F}_{c,n+1} - \sum_{i=1}^{n} \phi_i \bar{F}_{c,n+1-i})]^2 + \ldots \tag{6}$$

The "SSAT" which needs to be minimized shows a highly nonlinear structure in the unknown n + m coefficients $\phi_1, \phi_2 \ldots \phi_n, \theta_1, \ldots \theta_m$.

Therefore the procedure to be followed is a nonlinear least squares technique. During this study two commercially available mathematical packages are utilized for comparison:

. ARMA by the University of Wisconsin, Professor S. M. Wu and his colleagues.
. ARIMA by IMSL group from Texas

For the majority of the study only ARMA package is considered.

Two important aspects of the ARMA representation (Eq. 2) should be noted carefully:

a) The model is recursive, i.e. once the first n points are observed, the rest can be calculated in real-time. That is, when time n + 2 is reached $F_{c,n+2}$ is possible to calculate, as far as the necessary data is concerned.

$$F_{c,n+2} = \sum_{i=1}^{n} \phi_i \bar{F}_{c,n+2-i} + \sum_{j=1}^{m} \theta_j a_{n+2-j} \tag{7}$$

Since the m-1 elements in the second summation of Eq. 7 are zeroes, the calculated value of the cutting force for time n+2 is:

$$F_{c,n+2} = \sum_{i=1}^{n} \phi_i \bar{F}_{c,n+2-i} + \theta_1 a_{n+1} \tag{8}$$

now the residual value $a_{n+2}$ which ideally is zero can be calculated as:

$$a_{n+2} = \bar{F}_{c,n+2} - F_{c,n+2} \tag{9}$$

and the progression is continued.

b) Although once the AR and MA parameters ($\phi i, \theta j$) are known the recursion is trivial, the calculations of these parameters are a rather cumbersome procedure. This is the result of successive nonlinear least square minimization process. Therefore the calculation of $\phi_i$, $\theta_j$ over the given data is generally an off-line operation (i.e. the current computing power of the micro's cannot respond fast enough to include the very last data into the calculation of the parameters). Thus predictive control is inevitable in the long term processes, when involved with metal cutting dynamics. A typical data registration along with the associated cutting parameters is shown in Figure 5. As can be observed the dynamic oscillation is about at 150 - 200 Hz. It is widely accepted [8] that for dynamic systems the most effective forms of ARMA models have the AR and MA orders differ by one (i.e. n=m+1). Therefore our search is carried out for orders of (2,1), (3,2) and (4,3). Then a comparative study is made among the three as explained below.

## Experimental Set-up and Analysis

The experimental site consists of 3 major components:

1. A Colchester/Clausing VS-12", variable speed lathe retrofitted with 3 stepping motors, which control the depth of cut, the feed rate, and the spindle speed (Figure 4).
2. The sensory unit, which measures the tool tip dynamics. This 3 D dynamometer consists of 3 LVDT's (linear variables differential transducer) for precise measurements of the tool deflections.
3. An ADAC 2000 Data Acquisition System (DAS) with enhanced I/O boards which is used for data collection and control. This unit is currently registering data at 12.8 kHz, through a 12-bit A/D converter.

The scope of the current investigation covers the use of ADAC 2000 as a data registry and processing center, not as a controller. Future plans by the investigators entail a full fledged utilization of the power provided by this system.

A set of experiments are carried out using a standard HSS tool bit with a 1 1/4" diameter Al 2024 workpiece. The characteristics of the tool tip are given in Figure 1. An exemplary data registry is shown in Figure 5. The rate of data acquisition is 3200 Hz per channel. The corresponding behaviors of ARMA and ARIMA (2,1), (3,2) and (4,3) are shown in Figure 6. The similarity among these representations is remarkably good. Table I contains the SSAT values corresponding to the three models with 1000 data points. It is observed from SSAT variations that, better modeling is performed as higher order ARMA representations are utilized. Yet the (2,1) order is preferred due to its low execution time, and it is used for the remainder of the text as the prime modeling order.

The evaluations of the AR and MA coefficients are lengthy in CPU time for large amount of data. Therefore off-line schemes are unavoidable. Figure 8 represents typical CPU times needed for ARMA package's runs over varying number of data points. On the other hand smaller data sets may jeopardize the accuracy of the fit. Figure 7 represents a set of SSAT vs. number of data points (NOD). Although larger NOD's yield smaller SSAT's, one may apt for a smaller NOD in return of a savings in computer time. This issue becomes very important when a continuous on-line up dating of ARMA coefficients is desired, in addition to achieving on-line modeling.

Next an issue of joint flexural properties of the workpiece/tool interface are studied. The workpiece is divided into a number of segments due to the computer registration limitations (Figure 9). Each segment constituted a test run, and the ARMA (2,1) coefficients are calculated for each run. The results of this study are shown in Figure 10. The variation in the coefficients from segment to segment reflect the flexural properties of the workpiece/tool interface. A further study is made to find the functional relationships between the coefficients and the position on the workpiece. A simple linear regression is performed (e.g. $\phi_1(x) = ax+b$, where x is the distance from the free end). Table II depicts the comparative results of how this spatial linear model behaves with respect to the actual data.

An interesting case occurs when the conglomerate natural frequency of the tool/workpiece couple is approached. As Figure 11 depicts an apparent property of linear systems, the "beat effect" is observed and the corresponding SSAT is very small, compare to the other experiments.

## Conclusions and Future Study

This study has proven that an extremely simplified representation of ARMA (2,1) can be utilized for a highly nonlinear system. This feature becomes very important when predictive control is required. Because the iterative steps of ARMA (2,1) are rather simplistic, the future states of the tool/workpiece interface can be forecasted sufficiently early for a preventive control action.

Most of the modeling efforts in ARMA (2,1) showed very close fittings. Even when a functional dependence of these coefficients was introduced, the results were remarkably close to the original data.

An important spatial property of ARMA coefficients is noticed during the evaluations of the experiments. This feature is due to the flexural properties of the workpiece, and certainly introduces a potential aspect for further evaluations.

## References


O. Masory, Real Time Estimation of Cutting Process Parameters in Turning, ASME Trans., Journal of Engineering for Industry, August 1984, Vol. 106.

A. G. Ulsoy, Y. Koren, F. Rasmussen, Principle Development in the Adaptive Control of Machine Tools, ASME Trans., Journal of Engineering for Industry, June 1983, Vol. 105.

Y. Koren, Computer Control of Manufacturing Systems, McGraw Hill, New York, 1983.

N. Olgac, Z. Guoquan, Time Series Applications for Predictive Model of the General Turning Mechanism, to appear, 1986.

N. Olgac, M. Devin, A Study on the Computer Modeling of the Lathe Cutting Mechanism. Int'l Journal of Modeling and Simulation, Vol. 4, No. 4, 1984.

K. F. Eman, S. M. Wu, Forecasting Control Machining Chatter. The WAM of ASME, Chicago, November 1980.

T. Y. Ahn, K. F. Eman, S. M. Wu, Cutting Dynamics Identification by Dynamic Data System, (DDS) Modeling Approach, ASME, Trans., Journal of Engineering for Industry, May 1985, Vol. 107.

S. M. Wu, S. M. Pandi, "Time Series and System Analysis, Modeling and Application", John Wiley, 1983.


TABLE I: SSAT VS. ARMA (n,m) ORDERS

| | CUT 1 | CUT 2 | CUT 3 | CUT 4 |
|---|---|---|---|---|
| ARMA(2,1) | 1203.1 | 1472.7 | 1151.0 | 1356.3 |
| ARMA(3,2) | 801.4 | 1649.3 | 740.8 | 740.4 |
| ARMA(4,3) | 767.6 | 1046.9 | 740.0 | 5511.7 |

SSAT values based on 1000 DATA points.

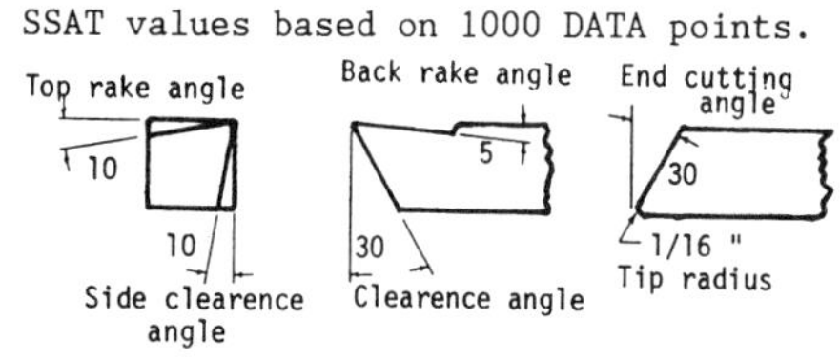


Figure 1. Cutting tool geometry.

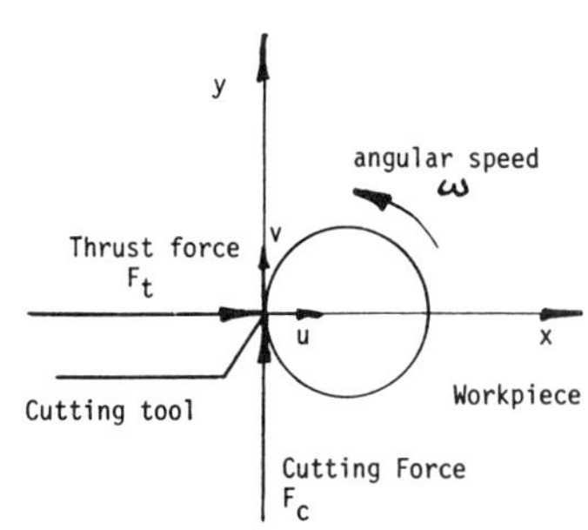


Fig.2 Cutting tool/Workpiece interface.

TABLE II: EXPERIMENTAL AND LINEAR REGRESSION COEFFICIENTS

| WORKPIECE LENGTH | $\phi_1$ | $\phi_2$ | $\theta_1$ | SSAT | $\phi_1$ | $\phi_2$ | $\theta_1$ | SSAT |
|---|---|---|---|---|---|---|---|---|
| | (EXPERIMENTAL) | | | | (LINEAR REGRESSION) | | | |
| .125 | 1.4845 | -.6747 | -.3931 | 2525.0 | 1.3850 | -.5254 | -.5853 | 1288.6 |
| 1.125 | 1.4464 | -.6901 | -.4298 | 1523.7 | 1.4145 | -.5774 | -.4978 | 1603.7 |
| 2.375 | 1.3708 | -.4453 | -.1865 | 753.4 | 1.4514 | -.6424 | -.3886 | 850.5 |
| 3.625 | 1.3184 | -.5725 | -.2163 | 539.3 | 1.4883 | -.7074 | -.2793 | 568.5 |
| 4.875 | 1.6546 | -.8881 | +.0461 | 443.5 | 1.5253 | -.7724 | -.1700 | 485.1 |

NOTE: All the SSAT's are based on 1000 data points and only a sample of coefficients are listed.

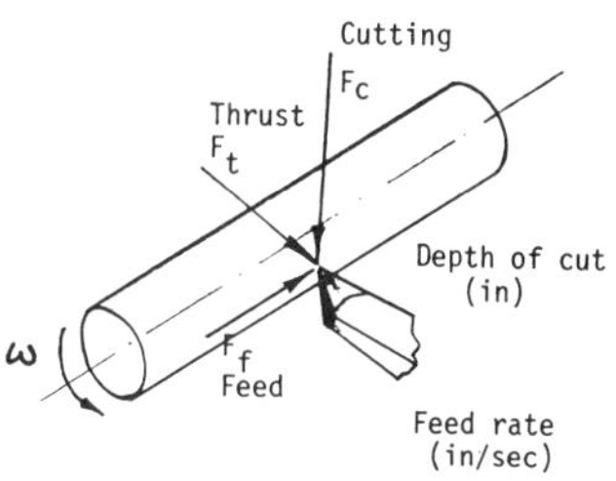


Fig.3 Three fundamental control parameters for the lathe cutting (underlined).

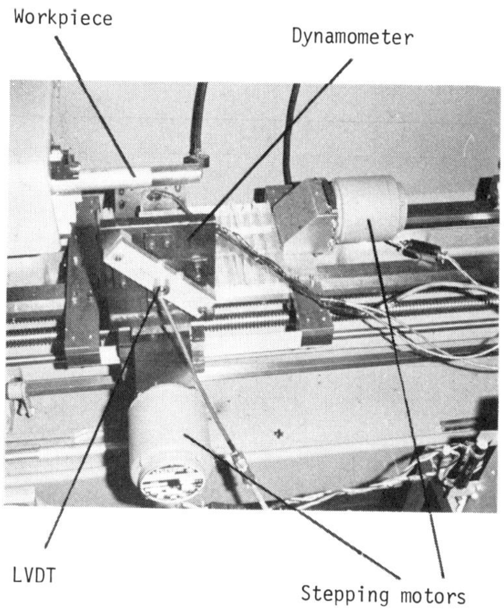


Figure 4. Experimental turning center.

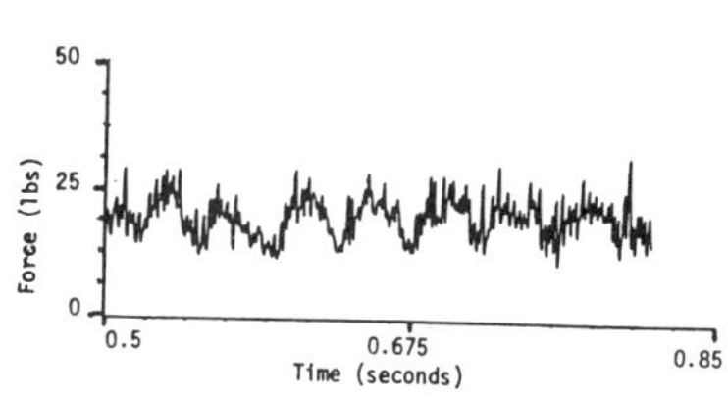


Figure 5. Experimental data registration, cutting force vs. time.

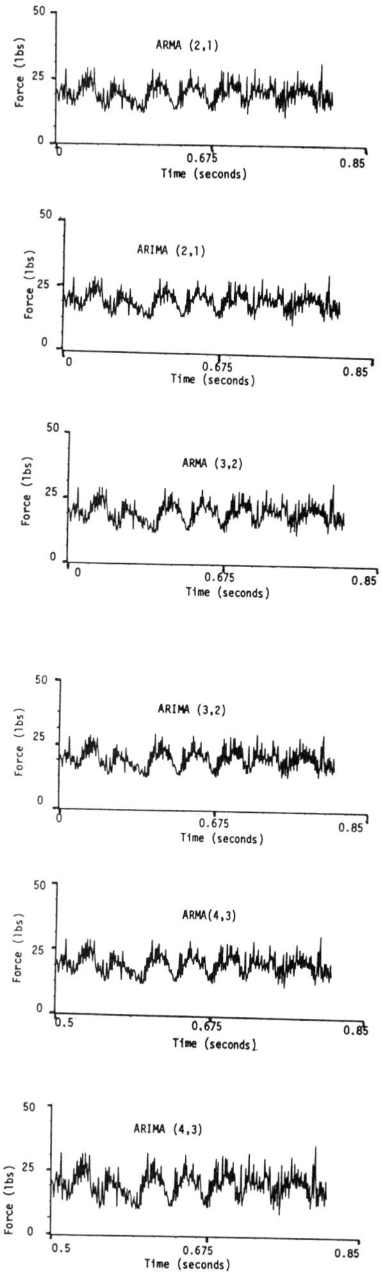


Figure 6. ARMA and ARIMA representations of cutting force data for various orders.

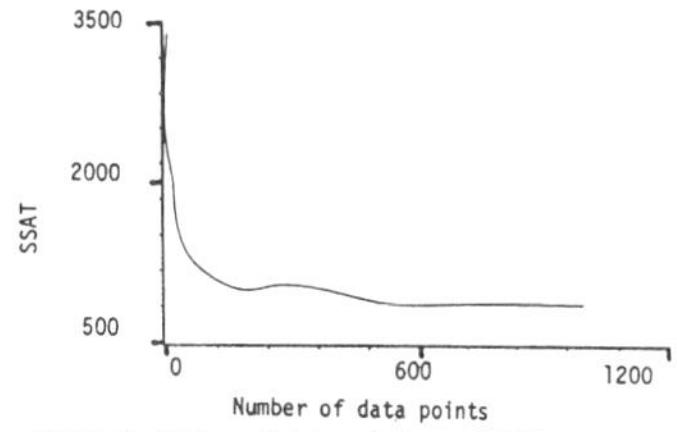


Figure 7. Number of data points vs. SSAT(sum of error squares).

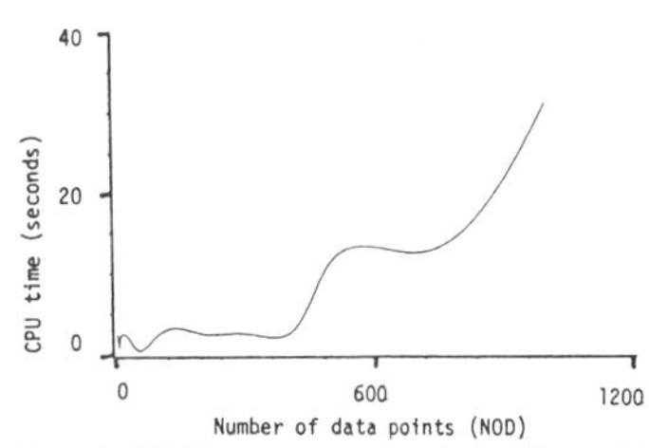


Figure 8. CPU time consumption in number of data points for the calculation of ARMA coefficients.

Figure 9. Segmental Cutting of the Workpiece

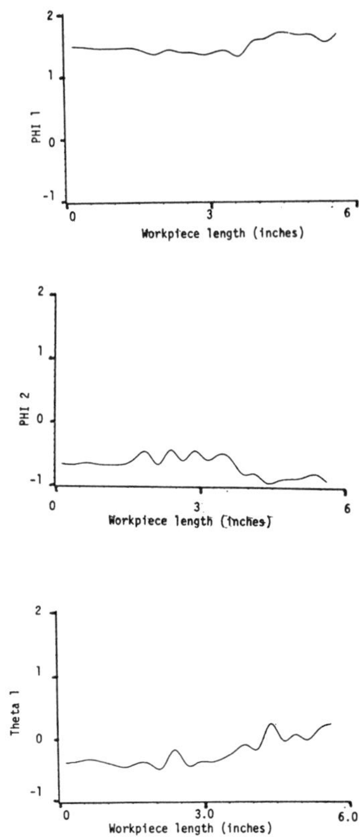


Figure 10. AR and MA coefficients over the length of the workpiece.

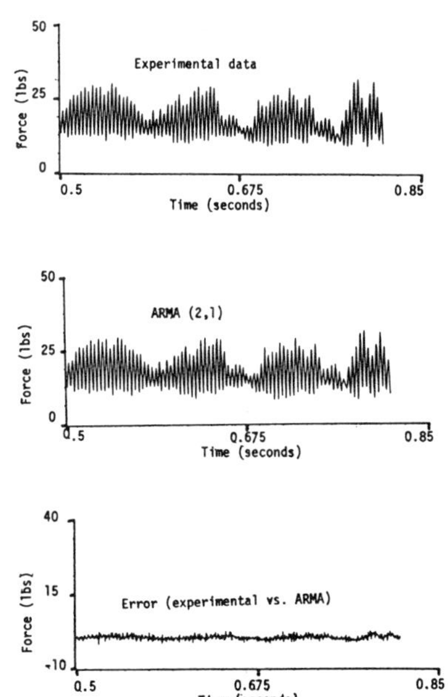


Figure 11. Experimental cutting force registrations, their ARMA (2,1) representation and the error in this modelling.

# ADAPTIVE TRACKING ALGORITHMS AND THEIR APPLICATION TO A MACHINE-TOOL

**P. Hulliger and Ch. Kohler**

*Control Laboratory, Swiss Federal Institute of Technology, CH 1015 Lausanne, Switzerland*

*Abstract : This paper describes the application of modern control concepts such as adaptive control to trajectory tracking problems encountered in an industrial environment. The control system structure and relevant algorithms, have been designed to be implemented on a microprocessor MC 68000. In typical applications such as machine tools, the main objective of the control system is to track a predetermined trajectory as closely as possible. Two fast algorithms with compensation of tracking errors are developed in this paper. They can also be applied to non-minimum phase systems. The first one is based on a PD regulator and the other one on a pole placement. The algorithms parameters are adjusted by an off line system identification procedure. Problems with conventional algorithms are shown by simulation and the effectiveness of the proposed techniques are demonstrated by investigation the problem on a modern machine-tool axis driven by an hydraulic motor.*

*Keywords : Servomechanisms, machine tool, microprocessors, PID control, pole placement.*

## 1. INTRODUCTION

Machine-tools developpement is living now in a period of great progress. On one side the evolution of performances of machine - tools allows always higher machining speeds. At the opposite always higher accuracy is requested.

This phenomenon gives serious problems to the designer of machine-tool axis control systems. In fact, modern CNC's usesat the moment simple conventional regulators like P and exceptionally PD or PID regulators. The limit of these algorithms forces the designer to introduce more and more correction factors in the control system, and this to correct, at least partially, the errors due to these algorithms. This goes against all modularity. Moreover it requests a long and expensive adaptation from the CNC for every machine-tool axis. The principal goal of this paper is to describe algorithms able to satisfy all the requirements of a modern machine tool axes control system.

## 2. TRACKING ERROR PROBLEMS IN MACHINE-TOOLS

### 2.1 Description of the problem

Machining by contourning needs action from two or more axes. Each one of them needs to follow a predefined trajectory in order to yield the wanted shape of the work piece. It is well known that using a conventional algorithm, tracking errors are always resulting. In the past the designers of machine-tool control systems were not very woryied about this problem. In fact, very low feed rate were used in contourning, as the tracking errors are proportional to the feed rate these errors remained low. Moreover in low speeds, tracking errors of first order are predominating. The resulting tracking error from two axes having each one first order tracking errors is always tangential to the final form. This fact reduces again the resulting form error. But in higher feed rate, tracking errors due to orders higher than one apear (i.e. accelerations). In all this considerations we supposed that the dynamics of all the axes are the same. This is wrong since in machine-tools no two axes used in contourning are identical, and often one axis is mounted over the other. Usually the designer reduces the dynamics of the more performant axis in order to obtain the same static closed loop gain on both axes. From this results a bad dynamic behaviour. The fact that the little axis has less inertia than the other, it will have serious consequences in the presence of perturbations.

To know better the influences from the tracking errors on the form errors we have simulated this phenomenon.

### 2.2 Simulation

A modern machine-tool axis is now extensively investigated by simulation. This machine includes a work table with low friction slides, driven by a DC motor through a ball bearing screwbolt. The table position is measured by an incremental sensor with a resolution of 1 μm. The simulation model order is five. It includes the inertia and the rigidity of the various elements, dry friction acting on them and the sensor resolution. Two machine axes are represented in a contourning operation where the trajectory to be followed is a circle whith a diameter equal to 10 mm. Fig. 2.1 shows the trajectories that each one of the two axis has to follow in order for this circle to be machined.

This operation is representative enough of the difficulties that can be encountered from the regulation point of view, the system being subject to accelerations and direction inversions (dry friction inversion). The tangential countourning speed (1 m/min) is high enough to show the difficulties of the control problem. Fig. 2.2 shows the trajectory errors from each one of the two axes driven by a PD regulator. The dynamics of the two axes are supposed the same. The real machining trajectory compared with the circle is also shown. In this case trajectory errors going to 200 μm of each axis give machining trajectory errors less than 10 μm except for the four error points resulting from the dry friction inversion. Fig. 2.3 shows the same errors but this time the dynamics of the two axes are not the same. We are considering a difference of 20 % of the static closed loop gain between both axes. These are also controlled by a PD regulator. These graphics show very well the influence of this inequality and confirm what was said above. In fig. 2.4, the adjunction of an I-term corrects the errors on each axis after starting but the errors in the resulting trajectory remain high around all the circle. Considering that in this case the dynamics of both axes are equal, the results are worse with an I-term.

## 3. ALGORITHMS DESCRIPTION

### 3.1 Introduction

The predetermined trackingtrajectory problem applied at a SISO fast system is treated. This problem implies given requirements concerning the used control law. The control of fast systems needs algorithms which requires a low volume of computation. The upper limit of the sampling period may in fact be quiet low. On the other hand if predetermined trajectories are to be followed with accuracy these algorithms need to consider the tracking errors which occur in conventional servomechanisms. This problem may be solved by superposing two controls. The first one in open loop, called "a priori control", aiming at the tracking of the predefined trajectory, the second one being a classical feedback which regulates the system around the above mentioned trajectory. This type of control requires the identification of a mathematical model in real time or not, in order to adjust the algorithms parameters. Fig. 1 shows this control system.

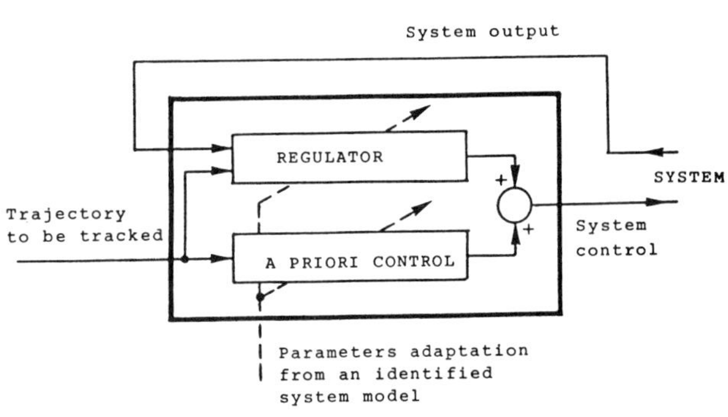


Fig. 1 : Control system structure.

On the other hand, tracking errors elimination greatly facilitates the multi-axes machine-tool synchronisation. A modular law and hierarchical structure is obtained in a natural way for the control algorithms [1].

In order to satisfy the computing requirements of the control algorithms, the problem has been formulated in polynomial form instead states variables form. This approach directly uses the system model in ARMA form and does not need a state estimator as only the system output sequence is used by the algorithms. With polynomial control laws, the problem of non-minimum phase systems occurs. This kind of systems are commun in practice [2].
A good overview of the control of such systems is given in [3], [4] and [5] . The algorithms which do not need the restrictive minimum phase system assumption require a higher computation volume and complexity. In this paper, the approach of control which cancels the zeros of the system is taken because it yields a very low computing complexity. Such control law can only be applied to minimum phase systems with well damped zeros. This kind of systems is very uncommon. For this reason, a weighting control polynomial is introduced in the law without increasing the computations. This also allows to apply these algorithms on non-minimum phase systems. If the polynomial form is well chosen, the annoying tracking errors may always be avoided.

In this paper, a polynomial objective form for the control law is firstly defined. Consequently, this objective is applied to a control law developed on the basis of a PD regulator and to another one developed from a pole placement regulator. The statical errors due to disturbances acting on the system are avoided by an integrator in the case of the PD algorithm. A disturbances estimator is introduced in the case of the pole placement algorithm.

### 3.2 Control law objective

Consider a discrete time SISO system modelled around its nominal point as :

$$A\, y_{k+d} = B\, u_k \tag{1}$$

where $u$ is the control signal and $y$ the output signal. A and B are polynomials in the backward shift operator (as are all others in this paper) $q^{-1}$. $d$ is the dead time (related to the sampling period) and $k$ represents the sampling instant. Let $e$ be the error related to $y_c$ , the reference sequence of the predetermined trajectory to be tracked as :

$$e_k = y_k - y_{c_k} \tag{2}$$

Let us define the general relation of the objective to achieve by the control law as :

$$F\, e_{k+d} + G\, u_k = 0 \tag{3}$$

where F and G are polynomials. F defines the speed distrubances elimination. G is a control weighting.

The system behaviour in closed loop related to the error is :

$$A^{*}\, e_k = B^{*}\, y_{c_k} \tag{4}$$

with, from (1) → (3) and (2) :

$$A^{*} = F B + G A \tag{5}$$

$$B^{*} = - G A \tag{6}$$

From (6), G produces tracking errors. In a conventional servomecanism, A contains $1-q^{-1}$ in factor. If it is chosen that control variations are weihgted by $G = g \cdot (1-q^{-1})$, only little tracking errors of second order will be produced.

### 3.3 General polynomial control law

A general control law which combines feedback and a priori control (feedforward) is given by

$$R\,u_k = T\,y_{ck+d} - S\,y_k \tag{7}$$

where R, S and T are polynomials.
With the expressions (7), (1) and (2), the polynomials representing the closed loop behaviour are :

$$A^* = RA + BS\,q^{-d} \tag{8}$$

$$B^* = -RA - BSq^{-d} + TB \tag{9}$$

### 3.4 Control law based on PD regulator

The structure of this control law is given by :

$$R = B + G \tag{10}$$

$$S = s_0 + s_1\,q^{-1} \tag{11}$$

By first order appoximation of an analogue PD regulator, $s_0 = g(1 + t_d/h)$ and $s_1 = -g \cdot t_d/h$, where $g$ is the regulator gain, $t_d$ the derivative time and $h$ the sampling period. $s_0$ and $s_1$ are adjusted by a conventional method to allow good closed loop dynamics.
By (10), (9) and (6), the objective (3) is satisfied by :

$$T = A + S\,q^{-d} \tag{12}$$

If B contains unstable or not enough damped zeros, G allows the weighting of the control and a good behaviour.

### 3.5 Control law based on pole placement regulator

The structure of this control law is given by :

$$R = R'B + G \tag{13}$$

By (13), (8) and (5), R' and S are found resolving the polynomial identity :

$$F = R'A + S\,q^{-d} \tag{14}$$

where F is given and represents the closed loop dynamics. This resolution requires few computations because this identity contains only independent equations. By (13), (9) and (6), the objective (3) is satisfied by

$$T = F \tag{15}$$

The remark about the G polynomial in section 3.4 remains valid in this case. The real closed loop dynamic is in fact :

$$A^* = FB + GA \tag{16}$$

If $G = g(1-q^{-1})$ is chosen reasonably, practical results have demonstrated that the dynamics given by $A^*$ are close to the dynamics set by F.

### 3.6 Compensation of the static error due to distrubances

In the case of the PD regulator, an integrator of the error given by (2) may be added to the control given by (7). This component of the control is given by :

$$u_{i_k} = u_{i_{k-1}} + s_2 \cdot e_k \tag{17}$$

where $s_2$ is the integrator parameter. Always by first order approximation of an analogue PID regulator $s_2$ is given by $s_2 = g \cdot h/t_i$ with $t_i$ the integrator constant time.

In the case of pole placement, a disturbances estimator may be used. This estimation is obtained by filtering the model error of (1) as :

$$e_{m_k} = A\,y_k - B\,u_{k-d} \tag{18}$$

$$A_e\,\hat{e}_{m_k} = A_e\,(1)\,e_{m_k} \tag{19}$$

where $e_m$ is the model error, $\hat{e}_m$ its estimation and $A_e$ a polynomial representing the dynamics estimator. A control $u_e$, fonction of this estimation, is added to the control resulting of (7) and is :

$$u_{e_k} = -R'\,(1)\,\hat{e}_{m_k} \tag{20}$$

It is important to notice that this method do not change the closed loop dynamics of the system given by $A^*$.

## 4. TESTS AND RESULTS

### 4.1 Simulation

To show how the algorithms described above improve the control of a machine-tool axis, we have controlled the simulation model from section 2.1 by a pole placement regulator with a tracking error compensator and a disturbances estimator. Fig. 2.5 shows the results of this simulation.

### 4.2 Description of the trial-installation

The proposed algorithms are implemented and tested in a real trial-installation, which is representative of one axis of a modern machine-tool; it includes a worktable sliding on low friction and prestressed guide-ways, a ball bearing screw, and an incremental high accuracy measuring device. The worktable is driven by an hydraulic motor. It turns out that the hydraulic motor is often particularly well suited for this type of applications. As it has excellent dynamical performances which are due to its small dimensions, very low inertia and high power density.

However, to exploit all its performances, sophisticated control methods must be used. This is due to the fact that it is a low damped system because of the oil compressibility. Also the oil temperature variations provoke variations in the oil viscosity. Moreover, the electrohydraulic servovalves, which are the electrical to fluid interface in hydraulic systems, modulate the hydraulic power according to an electrical signal by flow restriction. This is a non-linear phenomenon because the flow rate Q through a restriction is given by :

$$Q = k\,\sqrt{\Delta p}$$

where k characterizes the restriction and the fluid, $\Delta p$ the pressure drop through the restriction which depends on the load.

The plant is controlled by a microprocessor system based on the Motorola MC 68000.

### 4.3 Results

The described algorithms are tested on a machine tool axis trial installation. It is controlled by different algorithm configurations. The trajectory to be followed is a sequence of parabolas. The direction inversions and the dry friction inversions resulting introduce disturbances. The maximum trajectory speed is 3 m/min. The fig. 3 shows the trajectory to be followed with the respective trajectory errors and control signal obtained by using different control laws. Fig. 3.1 shows the system controlled by a conventional PD regulator. The tracking errors going to 550 [μm] are resulting. Fig. 3.2 shows a PID regulator with a tracking errors compensator. The errors remain less than 10 [μm] . Fig. 3.4 shows a pole placement regulator with tracking errors compensator and a distrubances estimator. The errors remain less than 4 [μm]. Moreover the errors peaks due to the dry friction inversion are substantially reduced.

## 5. CONCLUSION

The effects of the tracking errors on a contourned machined form have been shown. The introduction of an open loop control, called a priori control, in the algorithms allows the cancellation of this kind of errors. On the other hand, a modular and hierarchical structure is obtained for a multi-axis machine-tool control. This allows to follow the evolution in the machine-tool field with respect to the machining speed and accuracy. The high performances of the shown algorithms have been demonstrated setting them up on a modern machine tool. An excellent accuracy is obtained by these algorithms which requires a low computation complexity and volume.

## REFERENCES

[1] Hulliger, P., Longchamp, R. Adaptive tracking algorithms and their implementation on a signal processor. Proceedings $7^{th}$ IFAC conference on digital computer applications to process control, Vienna, 1985.

[2] Åström, Hagander, Sternby. Zeros of sampled systems. Automatica, 20, 1, 31-38, 1984.

[3] Åström, Wittenmark. Self-tuning controllers based on pole-zero placement. Proc. IEE, 127 D, (3), 120-130, 1980.

[4] Clarke. Self-tuning control of non-minimum phase systems. Automatica, 20, 5, 501-517, 1984.

[5] M'Saad, Ortega, Landau. Adaptive controllers for discrete-time systems with arbitrary zeros : an overview. Automatica, 21, 4, 413-423, 1985.

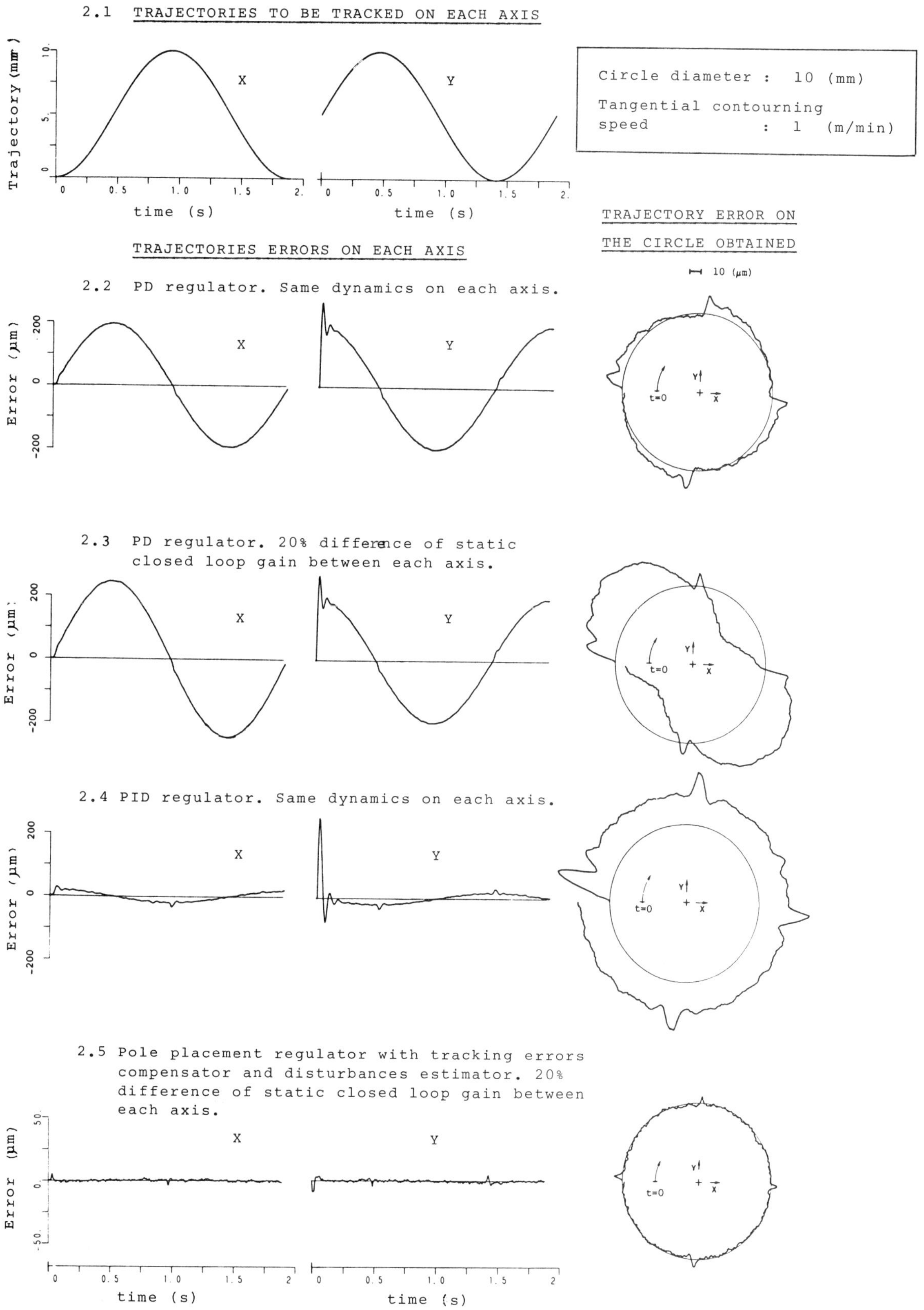


Fig. 2 : Contourning operation simulation results

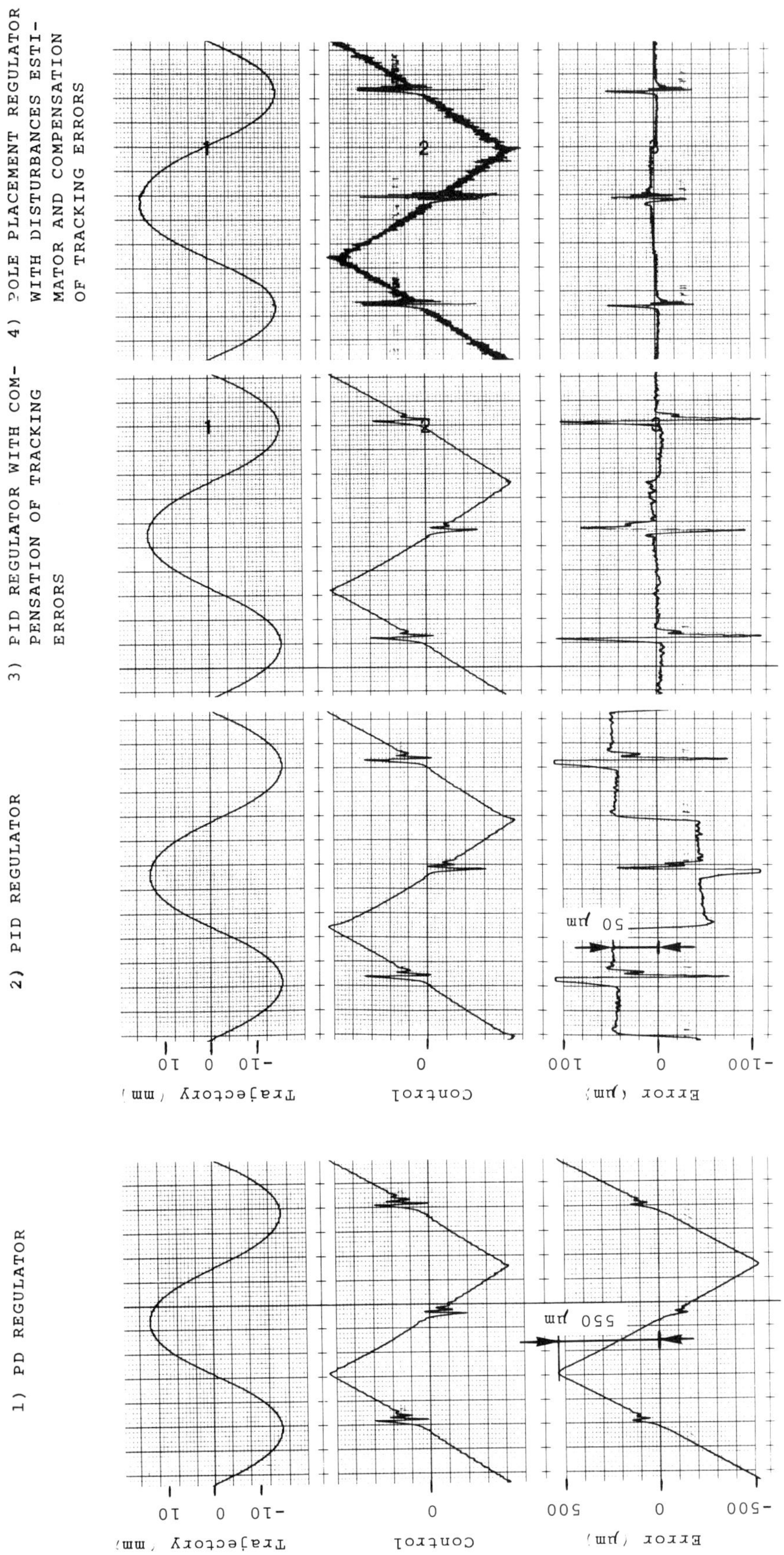


Fig. 3 : Experimental results obtained on one machine-tool axis

# MODELLING AND ANALYSIS OF A STEPPER MOTOR DRIVEN FEED-DRIVE SYSTEM OF A CNC LATHE

**B. E. Platin and B. Çelik**

*Department of Mechanical Engineering, Middle East Technical University, Ankara 06531, Turkey*

Abstract. A sixth order mathematical model of the feeddrive of a stepper motor driven CNC lathe is developed. The effects of the system parameters, such as the center distance and the speed ratio between the lead screw and the motor shaft, damping on the lead screw and on the slideways, axial stiffness of the lead screw are investigated in terms of eigenvalue loci on the complex plane. A series of guidelines are given describing how these parameters influence the dynamic performance of the machine tool.

Keywords. Computer Numerical Control (CNC); eigenvalues; machine tools; modelling; root loci; stepping motors.

## INTRODUCTION

Recent advances in electronic industry, especially in Integrated Circuit technology, have resulted in replacement of permanent control systems of Numerically Controlled machine tools by microprocessor-based controllers. This has lead to the development of Computer Numerically Controlled (CNC) machine tools of a great versatility with powerful software applications.

Since the accuracy in machining is the final outcome, the performance of a CNC machine tool depends upon the performance of its position control system. On the other hand, the performance of this control system is affected and usually limited by the dynamic characteristics of the feeddrive system of the machine tool. Undershoots and/or overshoots in positioning must be minimized to achieve an accurate machining. In machining, it is also preferred to reach high cutting speeds. Hence, the dynamic characteristics of the feeddrive system must allow accurate and also high speed operations, while not interacting with the structural dynamics of the machine tool body. All these requirements can be satisfied by a proper tuning of system parameters (Gavin, 1967), once the effects of these parameters on the system performance are determined.

The feeddrive system in a machine tool can be considered as a link between the controller and the work table of the corresponding axis, which converts electrical signals into tool positions. A typical feeddrive system consists of a driving motor and a series of mechanical parts transmitting the rotation of the motor shaft to, usually, a linear motion of a table. While DC motors and hydraulic actuators are widely used in closed position loop control systems, stepper motors with smaller step angles and higher torque capacities have been increasingly implemented in open loop position control systems, lately. Next to their accurate positioning feature under certain operating conditions, their digital nature provides a suitable match to the digital controller of the CNC control units.

There exist a good number of studies (Bell, 1973; Bianculli, 1970; Delgado, 1969; Kieburtz, 1963, Kuo, 1973; Naresh, 1974; O'Donohue, 1961; Robinson, 1969) on the dynamic analysis and mathematical modelling of stepper motors. However, in these studies, the stepper motor has been considered alone with the reflected inertia of the mechanical parts of the feeddrive system, despite the fact that most of these studies seem to be originated from machine tool applications.

In this study, a mathematical model of a stepper motor driven feeddrive system of a lathe is developed. Effects of various system parameters on the single step dynamics of the system are also presented.

## MODELLING OF FEEDDRIVE SYSTEM

The particular feeddrive arrangement taken into consideration in this work is shown in Fig. 1. The modelling of various components of this system is discussed below.

### Stepper Motor

Since the major aim of this study was to investigate the system performance for a single step operation, a simplified, linear dynamic model has been used for the stepper motor. This model, used by Çelik (1985) and Platin (1984) recently, is based on the model proposed by Kieburtz (1963) and given by following equations for a two-phase permanent magnet stepper motor

$$T = \sqrt{2}\ K_e i - K_r \theta \qquad (1)$$

$$v = Ri - K_e d\theta/dt \qquad (2)$$

where T is the torque supplied by the motor, $\theta$ is the angular deviation of the motor shaft from the equilibrium position for the step considered, i and v are the current and the voltage associated with each phase respectively, R is the electrical resistance of the phase windings, $K_r$ and $K_e$ are constants associated with the motor and named as the "restoring torque constant" and the "voltage constant", respectively.

### Lead Screw

The lead screw is a slender shaft transforming the angular motion to a translational motion. Hence, it is subjected to coupled torsional and axial vibrations. However, it has been shown by Çelik

(1985) that the dynamics involved in the torsional vibrations are much faster than the leading dynamics of the system so that the rigid body modelling is adequate to represent the torsional dynamics of the lead screw.

On the other hand, the axial dynamics of the lead screw depends on the axial supporting conditions at its ends. With the end condition shown in Fig. 1, the lead screw can be modelled by an effective mass

$$m_e = (\frac{1}{3} L_1 + L_2)m/(L_1 + L_2) \tag{3}$$

supported by an equivalent spring

$$k_e = EA/L_1 \tag{4}$$

where m is the total mass of the lead screw, A is its cross-sectional area, E is the modulus of elasticity of its material, $L_1$ and $L_2$ are the distances of the tool table to the axially supported end and the other end of the lead screw, respectively.

### Timing Belt and Pulleys

The timing belt material has a negligible mass and mainly is made of plastics reinforced with steel wires. Experiments (Bumin, 1986) have shown that it has a considerable compliance, therefore its tight side should be modelled as a linear spring.

Motor shafts are usually short enough so that their torsional compliances can be neglected. Hence the inertia of the pulley on the motor shaft can be integrated to the rotor inertia. The inertia effect of the other pulley can be taken into account by considering an increased effective inertia for the lead screw.

### Frictional Losses

Frictional losses occur primarily at the bearings of the lead screw and at the screw-nut arrangement. By the use of antifriction bearings at supports, recirculating ball-screw for the lead screw and with a good lubrication, the level of frictional losses can be reduced drastically while eliminating the dry friction, almost completely. Hence, frictions at these locations and also at the slideways can be modelled by the viscous friction model in which the friction force changes linearly with the relative velocity between two mating parts.

## PHYSICAL MODEL OF THE SYSTEM

The complete physical model of the system is shown in Fig. 2. The stepper motor is represented in a closed form. In this model, $B_m$, B and b are the viscous friction coefficients at the motor bearings, lead screw bearings and slideways, respectively. $I_m$ is the combined mass moment of inertia of motor rotor and attached pulley. Similarly, I is the combined inertia of lead screw and the attached pulley. M is the mass of the tool table. $k_k$ is the effective spring constant of the timing belt. $r_1$ and $r_2$ are the radii of pulleys on the motor shaft and lead screw, respectively.

## MATHEMATICAL MODEL OF THE SYSTEM

Using the physical model described above, a sixth order mathematical model of the feeddrive system has been obtained in standard state equations form

$$\dot{x} = Ax + Bu \tag{5}$$

where the state and the input vectors are given as

$$x^T = [\omega_m \; T_r \; F_k \; \omega_p \; F_p \; V_t] \tag{6}$$

$$u^T = [v(t) \; f(t)] \tag{7}$$

where $\omega_m$ is the angular velocity of motor shaft, $T_r$ is restoring moment of stepper motor given by $K_r\theta$, $F_k$ is the force transmitted by the timing belt, $\omega_p$ is the angular velocity of the lead screw, $F_p$ axial supporting force of the lead screw, $V_t$ is the translational velocity of the tool table, v(t) the voltage across the phase windings of the motor and f(t) is the component of the cutting force along the slideways. The elements of 6x6 system matrix A, and 6x2 input matrix are given in Appendix A.

## EFFECT OF SYSTEM PARAMETERS

In the developed mathematical model, dynamic properties of the elements of the feeddrive system are represented by constant parameters like M, $k_k$, $I_m$, $K_r$, etc. In the design stage of a feeddrive and its controller, one needs to know the effects of these parameters on the overall system performance so that their selection that suits best to the design requirements could be made. Moreover, during the actual operation of the system, values of some parameters change. For example, the effective axial stiffness and the equivalent mass of the lead screw vary with the axial position of the tool table. Hence, the sensitivity of the system performance to such parameters should also be known.

The effects of the system parameters on the system performance have been investigated by determining the variations of eigenvalue locations on the complex plane with these parameters. This method provides a useful information about the transient response characteristics of the system without actually solving the system equations. Values of system parameters used in the analysis are for a SIGMA-21-420200-F03 stepper motor and a COLCHESTER TRIUMPH lathe and they are given in Appendix B.

In the region of interest of the parameter values, the system possesses three complex conjugate eigenvaules. Hence, only the upper left quadrant of the complex plane is used to expose the variations in the eigenvalue locations.

### Center Distance Between Lead Screw and Motor Shaft

Although this parameter seems to be geometric, it has a dynamic effect. Indeed, the stiffness of the timing belt is inversely proportional with this distance. Figure 3a illustrates the eigenvalue locations for the values of timing belt stiffness ranging between 1 kN/m to 1 MN/m. As the stiffness increases, the second and third eigenvalues interchange their roles by changing their real parts more, providing a weak optimum stiffness in terms of settling time.The change in the first eigenvalue is even more drastic as it rapidly moves away from real axis keeping almost the same real part, indicating initially faster but more oscillatory responses with same settling times.

### Speed Ratio

As the speed ratio between the lead screw and the motor shaft ($r_1/r_2$) is changed between 1/4 and 8, the resulting eigenvalue loci are shown in Fig. 3b.The upper two eigenvalues are almost insensitive to this parameter. However, the changes in the first eigenvalue show a very similar trend to that for the previous parameter,but just in the opposite direction. That is, as the value of this parameter is increased, the first eigenvalue becomes closer to the real axis keeping almost the same real part, indicating initially slower but less oscillatory responses with same settling times.

Axial Stiffness of the Lead Screw

This is a parameter not only indicating the axial compliance of the lead screw but also reflecting the effect of the table position. As the tool table moves towards the driving end of the lead screw, this stiffness increases. Figure 3c shows the effect of this parameter a it is changed from 400 kN/m to 1 GN/m. As seen from this figure, the first two eigenvalues are quite insensitive to this parameter. However, the third eigenvalue moves away from the real axis keeping almost the same real part as this stiffness is increased.

Friction on the Lead Screw

The dynamic effect of the viscous friction at the lead screw bearings and at the screw-nut connection is represented by a linear rotational damper with an equivalent damping coefficient. As seen from Fig. 4 , this parameter affects chiefly the lowest eigenvalue, while its effect on the upper two eigenvalues are not that significant. As the damping coefficient is increased from 0.01 to 0.75 N.m.s/rad, only the real part of the first eigenvalue gets large in magnitude for values of damping coefficient up to around 0.5 N.m.s/rad. Hence,this parameter is very effective to stabilize the domanating low frequency behavior.

Friction on the Slideways

Figure 5 illustrates the effect of this parameter. As the value of the friction coefficient is increased from 10 N.s/m to 200 kN.s/m, real parts of all three eigenvalues become large. But, this trend is primiraly important in the third eigenvalue and less important in the first one while the second one is influenced the least amount.

## CONCLUSION

The parametric analysis on the sixth order mathematical model of the feeddrive system of a stepper motor driven CNC lathe has shown that, it is possible to determine the effects of various system parameters on the self dynamics of the system. Results have indicated that the dominating first mode dynamics can be improved by a proper selection of the center distance and the speed ratio between the lead screw and the motor shaft, as well as the damping on the lead screw. The axial stiffness of the lead screw and the damping on the slideways are shown to be more effective on the fast, high frequency third mode dynamics.

## REFERENCES

Bell, F.K. (1973). Application of mathematical modelling for a PM-step motor. Proc. 2nd Symp. Incremental Motion Control Systems and Devices, Illinois, USA, 1-9.

Bianculli, J.A. (1970). Stepper motors: Application and selection. IEEE Spectrum, 7, No.12, 25-29.

Bumin, F. (1985). Conversion of a conventional general purpose lathe to a CNC lathe. M.Sc. Thesis, Dept. of Mech. Eng., Middle East Technical University, Ankara, Turkey.

Çelik, B. (1985). Modelling and analysis of the feeddrive system of a computer numerically controlled machine. M.Sc.Thesis, Dept of Mech. Eng., Middle East Technical University, Ankara, Turkey.

Delgado, M.A. (1969). Mathematical model of a stepper motor operation as a finite positioner around a given step. IEEE Trans. Aut. Cont., AC-14, 394-397.

Gavin, J.T. (1967). The machine tool: A multiloop control problem. Control Eng., 14(Pt2), No.10, 77-83.

Kieburtz, R.B. (1963). The step motor-the next advance in control systems. IEEE Trans. Aut. Cont., AC-9, 98-104.

Kuo, B.C. (1973). Applying stepper motors in machine engineering. Proc. 1st Innovation Week Conf., 122-147.

Naresh, K.S., R.A. Elliot and C.S.R. Wong (1974). A realistic mathematical model for permanent magnet stepping motors. IEEE Trans. Ind. Electron. Cont. Instrum., IECI-21, 168-173.

O'Donohue, P.J. (1961). Transfer function for a stepper motor, Control Eng., 8(Pt2), No. 11, 103-104.

Platin, B.E. and B. Çelik. (1984). Mathematical modelling of feeddrive systems in CNC machine tools. Proc. 1st National Machine Theory.Symp., Ankara, Turkey, 193-205 (in Turkish).

Robinson, J.D. and K.C. Taft (1969). A dynamic analysis of magnetic stepping motors. IEEE Trans. Ind. Electron. Cont. Instrum., IECI-16, 111-125.

## APPENDIX A

Nonzero Elements of System and Input Matrices

$$\left.\begin{aligned}
a_{11} &= -(B_mR + K_m^2/\sqrt{2})/I_mR \\
a_{12} &= -1/I_m \\
a_{13} &= -r_1/I_m \\
a_{21} &= K_r \\
a_{31} &= r_1k_k \\
a_{34} &= -r_2k_k \\
a_{43} &= r_2/I_R \\
a_{44} &= -B/I_R \\
a_{45} &= M(h/2\pi)/I_R(M+m_e) \\
a_{46} &= -bm_e(h/2\pi)/I_R(M+m_e) \\
a_{54} &= -k_e(h/2\pi) \\
a_{56} &= k_e \\
a_{63} &= r_2m_e(h/2\pi)/I_R(M+m_e) \\
a_{64} &= -Bm_e(h/2\pi)/I_R(M+m_e) \\
a_{65} &= -I/I_R(M+m_e) \\
a_{66} &= -bI_e/I_R(M+m_e)
\end{aligned}\right\} \qquad (A.1)$$

$$\left.\begin{aligned}
b_{11} &= -K_m/I_mR \\
b_{42} &= -m_e(h/2\pi)/I_R(M+m_e) \\
b_{62} &= -I_e/I_R(M+m_e)
\end{aligned}\right\} \qquad (A.2)$$

where

$$I_e = I + m_e(h/2\pi)^2 \qquad (A.3)$$

$$I_R = I + Mm_e(h/2\pi)^2/(M+m_e) \qquad (A.4)$$

and h is the lead of the lead screw.

## APPENDIX B

Base Values of the System Parameters

$I_m = 3.53 \times 10^{-4}$ kg.m$^2$

$I = 7.79 \times 10^{-4}$ kg.m$^2$

$M = 77$ kg

$m_e = 5.45$ kg

$k_k = 1.5 \times 10^5$ N/m

$k_e = 1.59 \times 10^8$ N/m

$B_m = 2 \times 10^{-5}$ N.m.s/rad

$B = 0.03$ N.m.s/rad

$b = 10^4$ N.s/m

$h = 6.4 \times 10^{-3}$ m

$r_1 = 0.018$ m

$r_2 = 0.036$ m

$K_m = 0.44$ V.s/rad

$K_r = 200$ N.m/rad

$R = 13.8$ Ohm

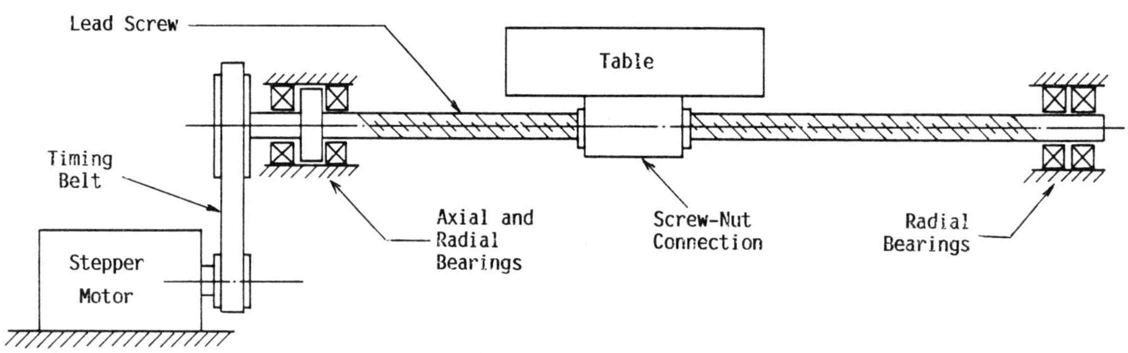


Fig. 1. Feeddrive system

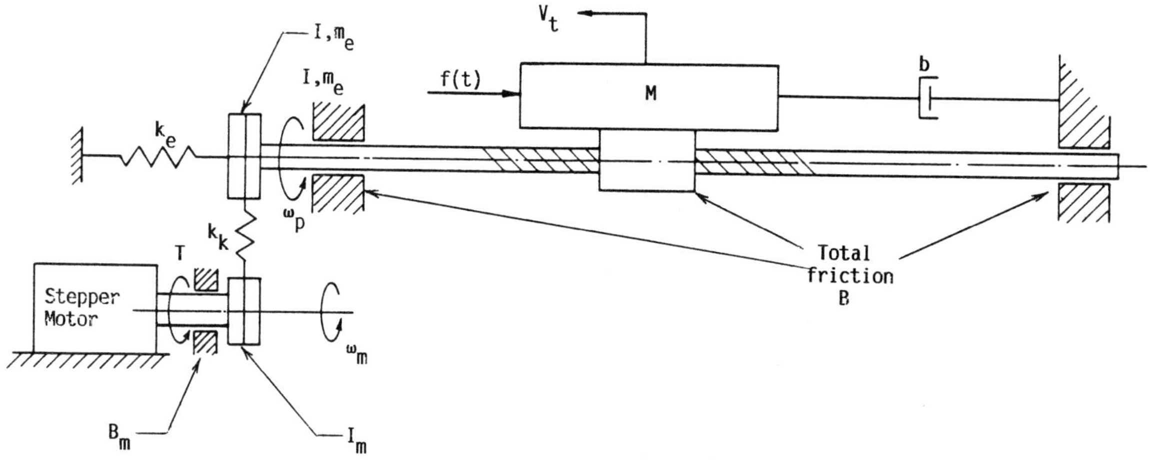


Fig. 2. Physical model of the feeddrive system

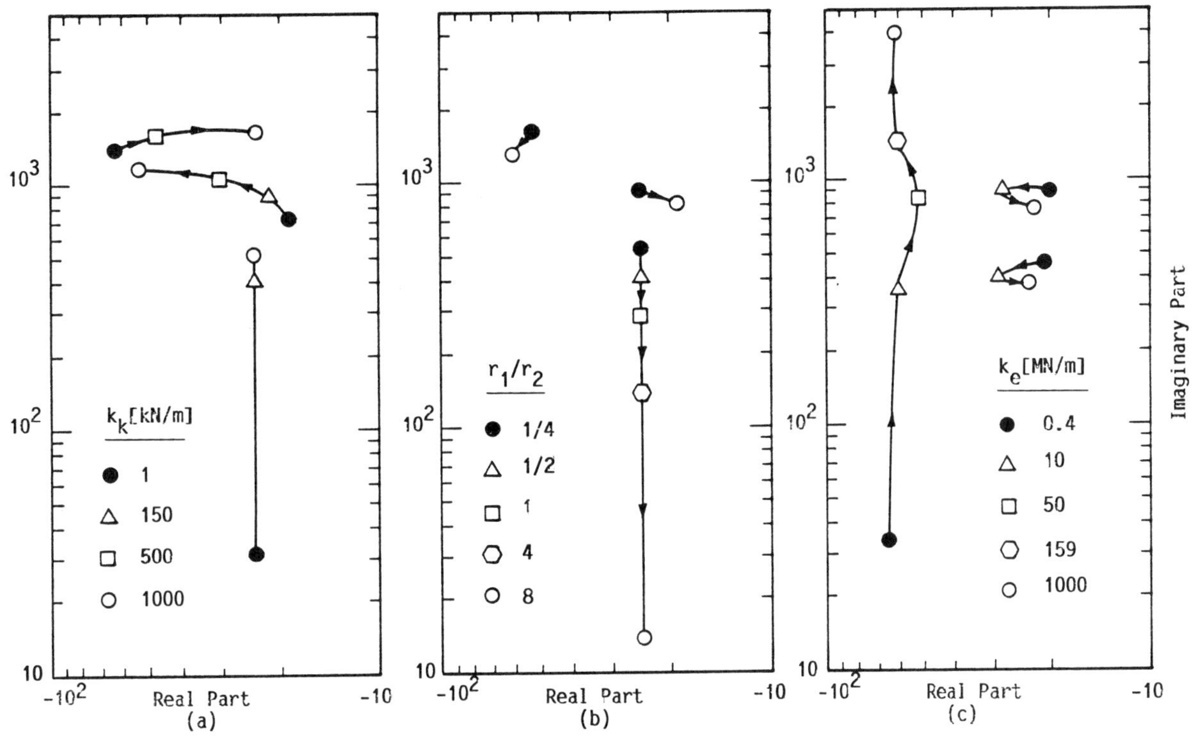


Fig. 3. Eigenvalue loci as a function of (a) Timing belt stiffness, (b) Speed ratio, and (c) Axial stiffness of lead screw

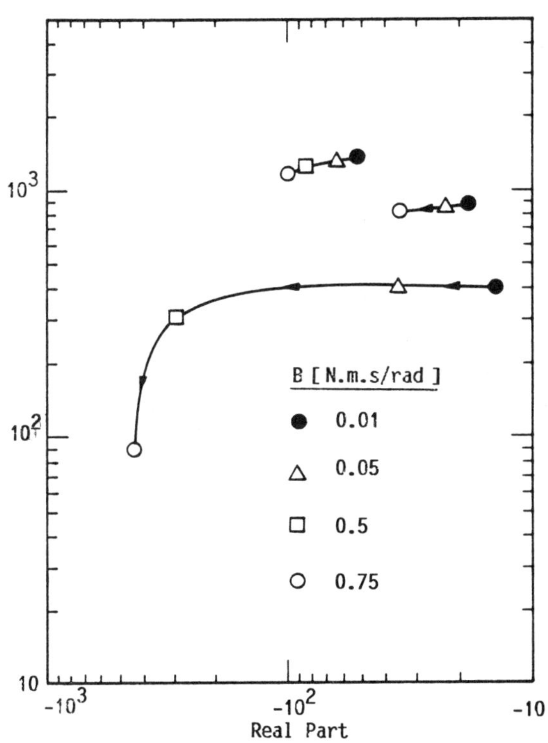


Fig. 4. Eigenvalue loci as a function of friction on the lead screw

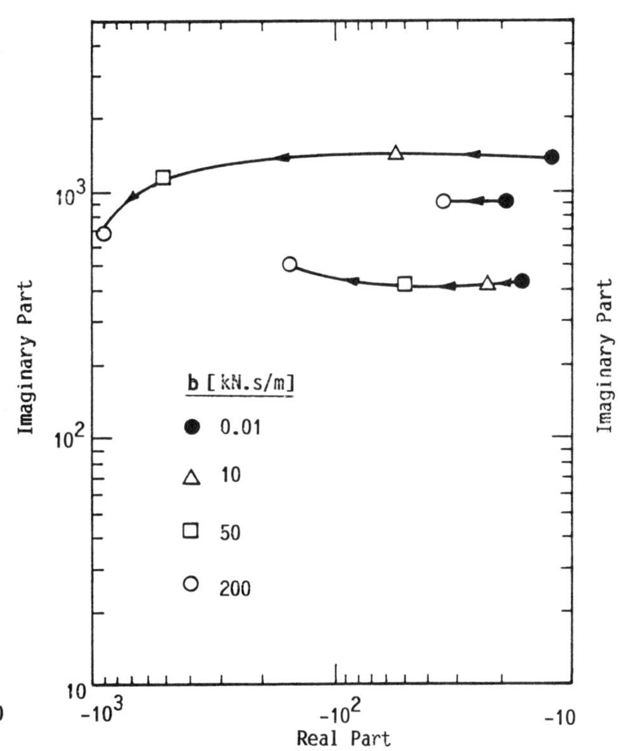


Fig. 5. Eigenvalue loci as a function of friction on the slideways

# FAST POSITION CONTROL WITH MICRO-PROCESSOR FOR MACHINE TOOLS

**U. Kunz and W. Oberschelp**

*Institut für Elektrische Energietechnik, Universität Siegen, Siegen, Federal Republic of Germany*

Abstract. Due to the development of automation in manufacturing engineering higher demands are expected regarding the accuracy and speed of machine tools and robots.
A measure for the quality is given by the contour error when working off a given contour.
This paper deals with a high dynamic position control based on a fast microcomputer. This microcomputer is coupled with a high accurate continuous contour control. Based on a model the continuous contour control generates the control function and the optimal system response of the position drives for a given contour on-line in order to suppress the deviation between the desired and actual position. The limitations of the control inputs and state variables are considered.
The real drive system of each axis will be controlled at first by the control variable acceleration resp. current and then corrected by the optimal state variables speed and position obtained from the model system. With this information about the reference state of the drive in every point of time the mechanical disturbances (load torques and backlash) can be eliminated directly.

Keywords. Optimal control, continuous contour control, position control, modelling, digital computer applications, machine tools, controllers.

## INTRODUCTION

The growing automatization in manufacturing engineering needs more and more machines with continuous contour control which move a work piece or work tool along a desired contour.
These controls are mainly applied to machine tools (milling and turning machines, electrical discharge machines etc.) and robots. The desired motion along the given contour is obtained by position drives which are controlled simultaneously by the computerized numerical control (CNC). The number of the necessary drives depends on the degrees of freedom of the movement. Electrical position drives are preferred because of its high dynamics and simple power supply. In conventional continuous contour control systems contour errors can't be avoided if the work piece or tool is moved along a curved path. The value of the contour error depends on the radius of curvature of the path and the velocity along the path (speed of machining operation). Furthermore the mechanical characteristics of the forward feed systems affect the accuracy of operation. Nonlinear friction forces and backlashes reduce the dynamics of the position controlled drives as shown for instance by Stute (1975) and Groß (1981).

In order to avoid contour errors at high speed of operation a new control method was developed by Hackstein and Kunz (1980, 1982). Further developments adapted this control to machine tools by implementation in a fast microprocessor system realized by Oberschelp and Kunz (1983, 1985). The control method is based on setting the acceleration resp. the torque of the motor of each axis with respect to the path contour and the speed of operation. The control variable is computed in consideration of the limitations of acceleration and velocity.
Since the motor torque instead of the acceleration is the control quantity unknown load torques would disturbe the motion and result contour errors. Therefore a ideal model of the drive is the reference for the real drive. The ideal state of the model is compared with the real state and state deviations which results from disturbances and parameter variations correct the control quantity of the real system in order to reduce or to eliminate these deviations. In this way the state of the real drive is controlled in a closed loop and the real system gets the behaviour of the ideal model.

The model delivers the state variables velocity and position which are the reference quantities for control loops which are superposed on the torque control (e.g. current control) of the drive. By realization in a fast microprocessor system the state controller can solve additional problems as for instance the compensation of friction torques.

## CONVENTIONAL CONTINUOUS CONTOUR CONTROL

Fig. 1 shows the structure of a conventional continuous contour control. The CNC generates on the basis of a desired contour the reference values of position $s_{1w}$ ... $s_{nw}$ for all axis drives.
The actual position values in each axis are $s_1$ ... $s_n$. Disturbances as cutting and friction forces as well as backlashes influence the control systems. A speed control in each servo drive has to eliminate these influences.
In order to obtain step responses without overshoot in the position control systems every position controller has proportional action with the gain factor $K_V$ (see Schmid 1972).
The output of the controller is the reference quantity for the speed control. During operation a deviation of the position must occur because the speed resp. the velocity is proportional to the deviation. Therefore the actual position lags behind the desired position. This lag in the position causes contour errors if the contour is curved.

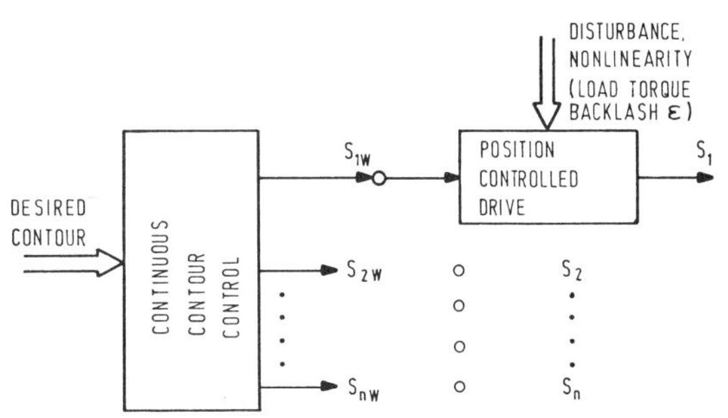


Fig. 1. conventional continuous contour control

Fig. 2a shows the basic diagram of a position control if a backlash with the value ε exists outside of the closed loop for instance if the position is measured indirectly by an encoder on the motor shaft. In this case the backlash causes not recognizable position errors.

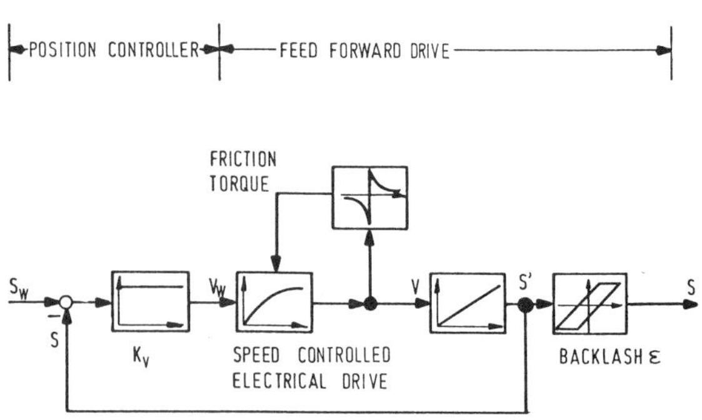


Fig. 2a. position control with indirect position measuring system

If the position is measured directly e.g. by linear measuring scale at the support the structure of the position control loop is shown in fig. 2b. In this case the influence of the backlash can be eliminated but at an expense of dynamics. The gain factor $K_V$ must be reduced for stable and satisfying working of the position control.

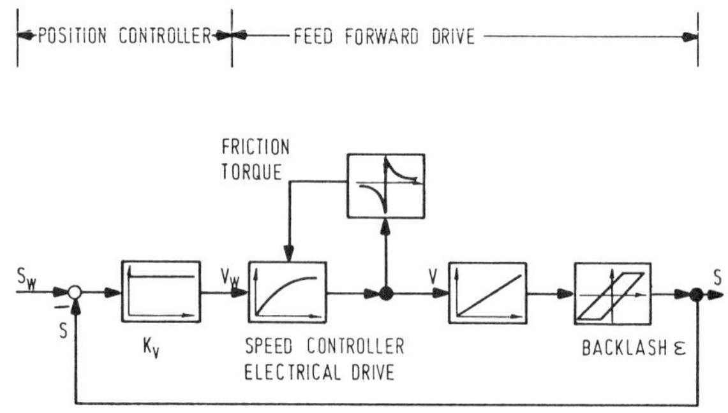


Fig. 2b. position control with direct position measuring system

If there is no backlash the dimensioning of $K_V$ is optimal for

$$K_{vopt} = 0{,}3 \ldots 0{,}4 \cdot \omega_{oA}$$

with the eigenfrequency $\omega_{oA}$ of the speed control loop of the axis drive.

Fig. 3 shows the proportional gain $K_V$ divided by the value $K_{Vopt}$ with respect to the value ε of the backlash in order to obtain stable and satisfying operation as shown by Groß (1981).

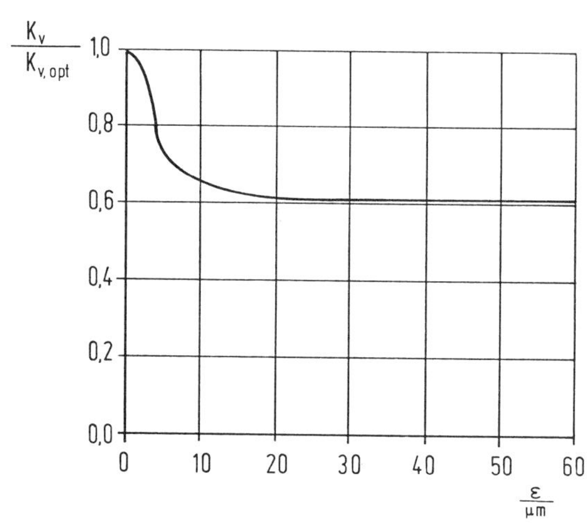


Fig. 3. Reduction of controller gain $K_V$ by backlash ε

## CONTROL METHOD FOR THE REDUCTION OF THE CONTOURING ERROR

Fig. 4 show the structure of the realized control method. In opposition to conventional continuous contour control the position drive follows acceleration resp. current references. With the information about the dynamic of the position drives a model generates the reference values $v_W$ and $s_W$ of the state. The state deviation between $v_W$, $s_W$ and v, s forms the correction quantity $\delta i$ by a state controller. $\delta i$ is added to the control value $i_w$ to minimize the contour error.

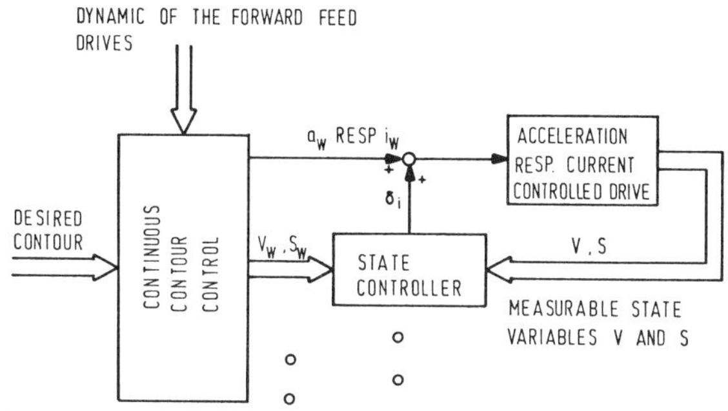


fig. 4. structure diagram of a servo drive system with model reference control

In this way a solution was developed which yields the control quantities for the position drive of each coordinate axis considering the limitations of armature current (acceleration) and speed (velocity) by an on-line computation. In order to enable on-line computation with a fast microprocessor (MC 68 000) a control was developed for such contours which consists of pieces of straight lines and arcs of circle. This is sufficient regarding usual machine tools.

## DIGITAL STATE CONTROLLER

Fig. 5 shows the structure of the digital position and speed controller. The references for the control variable acceleration $a_W$ resp. current $i_W$ and the state variables velocity $v_W$ and position $s_W$ are preset by the continuous contour control. A cascade control structure on principle is given. The deviation $s_W$ - s multiplied by the proportional sensitivity $K_V$ gives the correction signal $\delta v$ for the reference value $v_W$. The difference between $(v_W + \delta v)$ and v is the input quantity for the speed controller. For this controller a digital PI algorithm is implemented which generates the correcting quantity $\delta i$. An additional reference value $i_R$ compensates the influence of friction in the forward feed system. For that purpose the friction torque was measured off-line with respect to the velocity and the functional relation between friction torque and velocity is stored into the memory of the microcomputer.
Furthermore the capability of compensating the backlash is provided but not yet set into operation. The influence of a backlash is observed when the speed passes trough zero. For compensating the backlash in this points the drive must be accelerated with a high armature current pulse in order to pass through the width of the backlash as fast as possible.
The short-time overload capacity of servo drives up to 10 times the nominal torque gives an advantage. Since a fast microprocessor (MC 68 000) is used the cycle time for a complete computation of the control algorithm corresponding to figure 5 is lower than 250 µs.

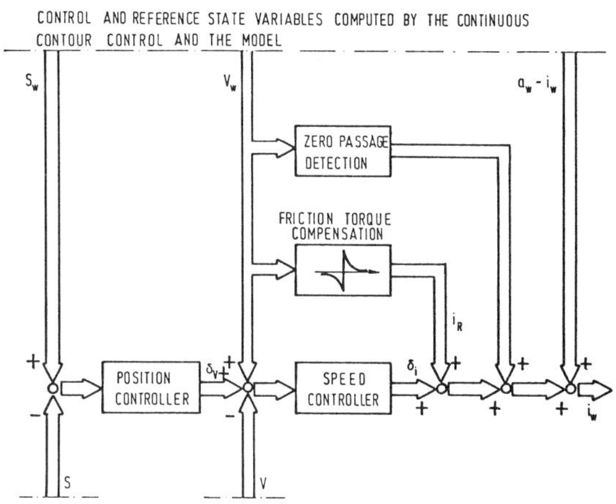


fig. 5. structure of digital speed and position controller

Fig. 6 show the flow chart of the digital control. To synchronize the continuous contour control with the state controller a interrupt handler is implemented. The INT 1 interrupt (1ms) is given by the cycle time of the continuous contour control. In this point of time new control references are delivered to the position controller (state controller).
With this requirements the state controller computes the current control input $i_W'$ in the way shown in fig. 5. The running period of time for the interrupt program is as mentioned above smaller then 250 µs. In fig. 6 this time intervall is marked by hatching (number 1). In order to increase the accuracy of the state controller a second interrupt INT 2 is generated in intervalls of 250 µs synchronously to interrupt INT 1.
In the points of time defined by INT 2 the controller gets actual input values of position and speed and generates new reference values of position and speed in the time intervalls 2, 3 and 4 marked by dotting on the assumption that the acceleration $a_W$ is constant in each intervall defined by INT 1. In this way the controller works with a cycle time of 250 µs in opposition to the continuous contour control with a cycle time of 1 ms.
The increase of the sampling frequency in the controller increases also the accuracy of the continuous contour control.

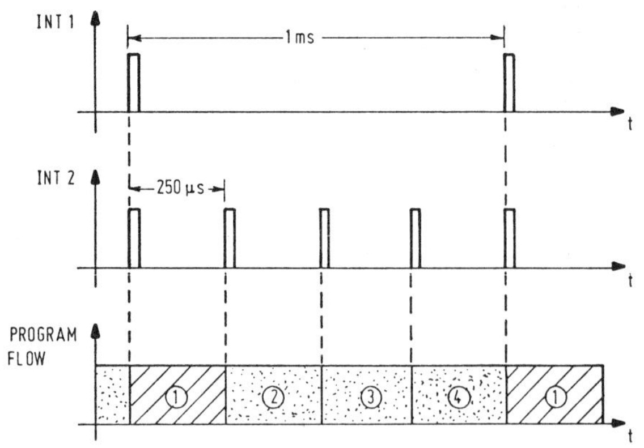


fig. 6. program flow of the digital controller

RESULTS

Some experimental results shall shown the advantages of the here described control method. The investigation were made in a test set up with two servo drives modeling a two dimensional machine tool. With this model motions in a plane defined by a cartesian x-y-coordinate system can be realized.

A simple point-to-point-motion with the coordinates (x, y) = (0,0) → (200, 200) mm → (0,0) is regarded. The desired contour is a straight line which is passed from the origin to the point (200, 200) mm and then again back to the origin.
In order to demonstrate the capabilities of the µprocessor based position controller at first a motion is regarded if the drive is controlled only by the control variable acceleration without feedback of velocity and position. The control variable is switched on the current control according to the time optimal control law. Since a friction torque in the range of 30% of the rated torque exist deviation in the velocity and position must occur. Fig. 7a shows the transient response. In the lower part of the fig. the control variable is shown. In the upper part the velocity and in the middle the deviation of the velocity with reference to the optimal velocity are presented.
As expected the velocity deviates strongly.

In the next step a disturbance compensation is added to the control. For that purpose the characteristic of the friction torque was measured off-line and stored in the computer memory.
The results of the compensation is shown in fig. 7b. Already this capability of the controller effects a great reduction of the velocity deviation although no feedback of velocity and position exists.

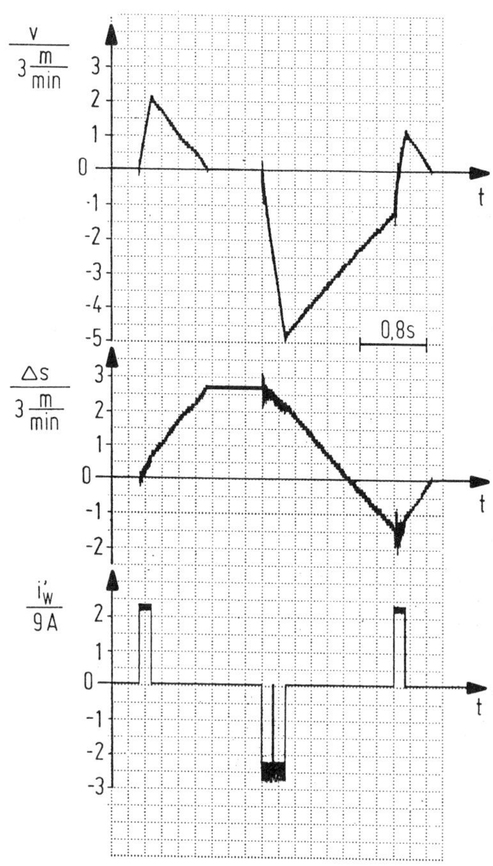


fig. 7a. transient response of current controlled drive without friction compensation

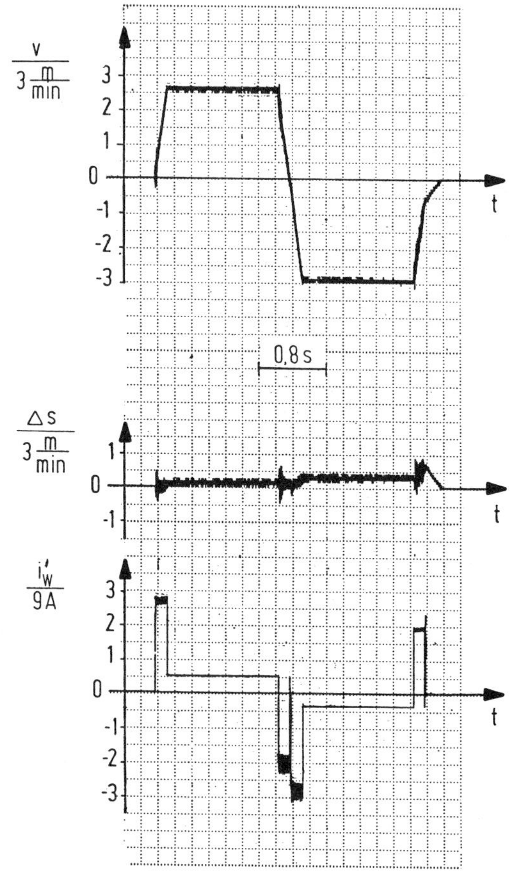


fig. 7b. transient response of current controlled drive with friction compensation

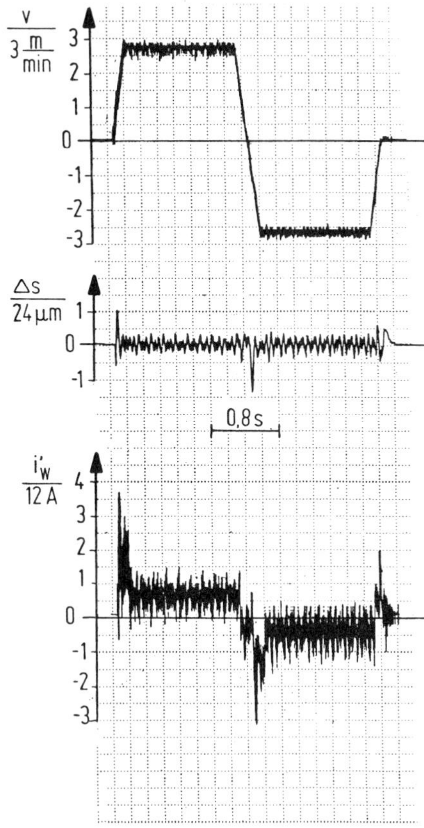


fig. 8a. transient response of the position controlled drive without friction compensation

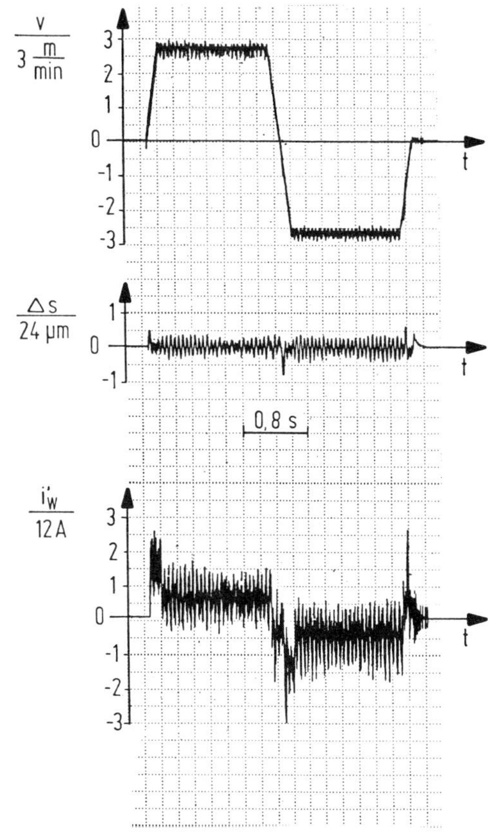


fig. 8b. transient response of the position controlled drive with friction compensation

In the next step the control loops of velocity and position are closed and the same motion as mentioned above is regarded. In the first test the compensation of the friction torque is switched off. Fig. 8a shows the corresponding transient response. In the upper part of the fig. the velocity, in the middle the deviation of position and in the lower part the armature current are drawn. The very small deviation of position is remarkable. Its value is only in the range of 6 μm at a resolution of 2 μm of the position encoder. Only in the switching points of the control variable the position deviations are greater.

Fig. 8b shows the same motion but now with compensation of the friction torques. The position deviations in the switching points are reduced now considerably.

## OUTLOOK

In the next time a two-axis milling machine will be equipped with these position controllers and the accuracy of the continuous contour control will be tested. Furthermore a compensation of the backlash will be investigated and the characteristic of the friction torque will be determined on-line in order to obtain an adaptive friction compensation.

## REFERENCES


Groß, H. (1981). Elektrische Vorschubantriebe für Werkzeugmaschinen. Siemens Aktiengesellschaft, Berlin, München.

Hackstein, D. (1980). Eine zeitoptimale Bahnsteuerung mit Digitalrechner. Dissertation, Technische Hochschule Darmstadt.

Hackstein, D. and Kunz, U. (1982). High Accurate and Fast Continuous Contour Control with Electrical Drives Third Int. MOTORCON Conference, Genf. pp. 383-391

Kunz, U. and Oberschelp, W. (1983). High Accurate Continuous Contour Control by on-line Simulation using Microprocessors. Fifth Int. MOTORCON Conference, Genf. pp. 665-675.

Kunz, U. and Oberschelp, W. (1985). Two Dimensional High Accurate Continuous Contour Control using a Multimicroprocessor System. Fifth Power Electronics Conference, Budapest. Vol. 1, pp. 141-150.

Oberschelp, W. (1985). Eine schleppfehlerfreie Bahnsteuerung mit Mikrocomputer. Dissertation, Universität Siegen.

Schmidt, D. (1972). Numerische Bahnsteuerung, Beitrag zur Informationsverarbeitung und Lageregelung. Springer-Verlag, Berlin-Heidelberg, New York.

Stute, G. (1975). Die Lageregelung an Werkzeugmaschinen. FISW-Selbstverlag, Stuttgart.

# PANTOGRAPH WITH MICROCOMPUTER

E. Adali

*Department of Control and Computer, Technical University of Istanbul, Ayazaga, Istanbul, Turkey*

**Abstract** : The combination of a personal computer and a printer provides remarkable opportunities for text editing and storing. The main idea of this project was to construct a pantograph with the similar features of a text editing computer. In order to realise a pantograph with microcomputer an X-Y table and an Amstrad CPC 6128 computer have been used. All necessary interfaces and motor drive circuits have been designed and implemented. The programming language is PASCAL.

**Keywords** : Microcomputer application, pantograph, machine tool, machine tool software.

## INTRODUCTION

As very known fact that computers provide remarkable opportunities of text editing and storing due to their editing features and their large memory capacity. Editing and updating of an old text is as important as editing of a current text. A text editor system also provides the selection of printer characters.

Editor system was the basic idea of the pantograph project. An X-Y table is used as an output device of the system. Computer provides all necessary editing and storing facilities.

The dimensions of the table are 50 * 50 cm. The driver motors of the table are stepper motors. The etching mechanism has been attached over the table. This mechanism consists of two motors; first motor is dedicated for positioning of etching tool and the second one is etching tool driver. The first motor is stepper motor and second one is a DC motor.

An interface circuit between the computer and the table has been designed and installed. The interface consists of three stepper motor drive circuits and a parallel interface between the computer and the drive circuits.

An Amstrad CPC6128 has been used as the computer of the pantograph. Computer has 128 K Byte memory and 170 K Byte diskette unit. The operating system of the computer is CPM. TURBO PASCAL-3.0 has been used as programming language of the project.

This project has been developed for ASELSAN ( Military Electronic Industry, Ankara) in their laboratory.

## OPERATING SYSTEM OF THE PANTOGRAPH

When the pantograph system is turned on, the menu of the operation is seen. Operation has three phases : Editing, rehearsaling and etching. The menu is shown in Fig. 1.

*EDIT*:

( N ) New Text
( O ) Old Text
( E ) Examination

*REHEARSAL*:

( R ) All Text
( S ) Part of Text

*ETCHING*:

( B ) From The Beginning
( P ) From Page Number

Fig.1

EDIT: In this phase user can write a new text or modify an old text.

*New Text*: When this option has been chosen,user will be prompted for the new text file name. After entering the file name user will see a new page on the screen which is given in Fig. 2.

Since the size of etched plates varies with the application, it is not possible to handle an whole plate on a computer screen. Therefore, the editing of a text is based on page logic. Each page has seven rows and forty columns. The sequence number of a page is depicted by the program. User should give the following data:

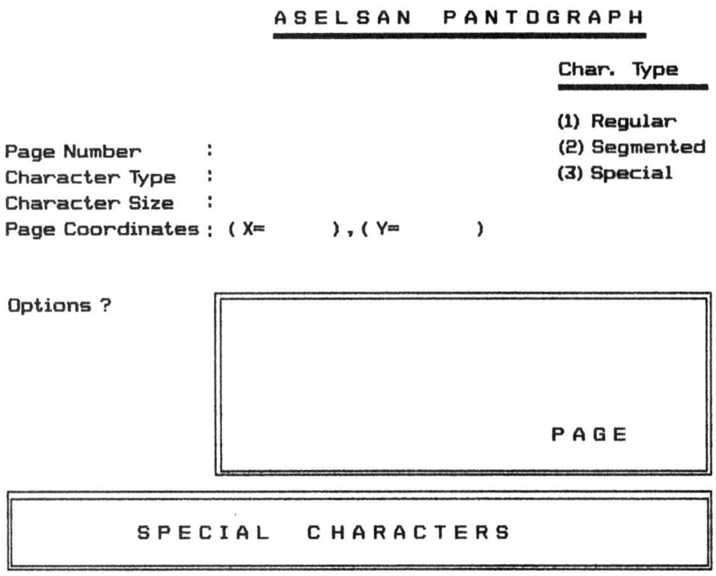


Fig. 2

*Character Set*: In the computer memory there are three type of character set. But the number of the types can be easily added latter on. The first two type consists of alphanumerical characters and the first one is in regular shape. In the second character set all alphanumerical figures are given as segmented manner. This character set gives faster etching process facility. The third character set consists of some special figures which are frequently used by the user of the pantograph. In the case of selection of special figure, user must indicate the coordinates of the left-lower corner of the figure.

*Character Size*: Only one character size is allowed in a page. The unit size is ten micron. In order to eliminate some friction error, the size value should be divided by power of 2.

*Page Coordinates*: The location of a page on plate is given by the page coordinates. The field of a page may be overlapped. By using of this method, various size of character can be etched in the same field. The unit coordinate is ten micron.

After setting the initial data cursor goes into the page. Full screen editing facility is provided in the page. In order to terminate text editing the COPY key must be used.

There are three options, after editing the page:

( N ) Next Page
( C ) Correction
( E ) End

*Next Page*: Next page is appeared on the screen. Page is blank.

*Correction*: Allows user to correct the current page.

*End*: Indicates the end of current text file and saves the data on diskette with the given file name, then returns to the menu.

*REHEARSAL*: Before attempting to etching process it is useful to see the output on computer screen. In rehersal mode the post-processor program runs faster then etching version.

*ETCHING* : Etching process is the output phase of the pantograph. The edited text is etched in this phase. The etching process is carried out by a post-processor.

## SOFTWARE

The software of the pantograph has been written in PASCAL language. But some part of the program has been written in machine language and some firmware of the computer has been used as well. The program consists of a main program and several program packages: The name and function of the packages are explained in the following paragraphs:

*Main Program*: The main program is the operating system of the pantograph program. It controls the program sequences and logic of the program. The main program also handles the menu.

*Screen*: This program generates a user friendly screen by using of written information, questions and graphics.

*Screen Input*: This program works with the Screen program. The most important part of this package is the page input feature. In the page field, inputs are given in a window. The screen input program reads all keys and seperates alphanumerical and control keys.

*Filer*: Filer works on file organization such as getting data from page,placing them into Page- Input Buffer. At the end of text editing filer saves the data on the diskette with the given file name. The same program also handles file transfer from diskette to the Page-Input Buffer and shows on the screen.

*Graphic*: A graphic program has been developed to draw lines,figures,and special symbols.This program is also used in rehearsal mode

*Character*: This program is developed to generate the necessary movement steps of each character. The logic of character generation will be described in the algorithm section.

*Etch*: In order to handle the etching process a post-processor program which is called etch has been developed. This is an interface program between editor program and the x-y table system. The etch program knows the hardware specifications of the computer and the table. Because of hardware involment some part of this program has been written in machine language.

## ALGORITHM FOR CHARACTER GENERATION

A pantograph can etch alphanumeric characters and some graphic characters just like a printer . In this project, two methods have been developed in order to etch the alphanumerical characters : 1. For regular characters and 2. For segmented characters.

*Segmented Character*: In this method, all letters and numbers are formed by lines. The angle of a line can be 0, 18, 27, 45, 63, 72 and 90 degrees. The reason for selection of these angles is to find an easy and errorless relation without calculation between the step numbers of the X and Y stepper motors. For example; in order to get 30 degrees, X motor turns two steps and Y motor turns one step at each period. Another important point of this method is the length of the segments. In this application the length of the segments are divided by power of 2. Some examples of the segmented characters are given in Fig.3.

*Regular Characters*: Although the segmented character set provides good looking letters and faster etching process, a letter quality character set is added to the pantograph software. In addition to the line etching, circle type etching has been implemented. Some examples of the regular characters are given in Fig. 4.

*Special Characters*: A special character can be formed by combining lines and circles.

## FORMAT FOR CHARACTER

In order to save memory size a technique which is similar to APT represantation has been used. In this method a character is saved in following format:

*part 1*: Etching toll is up or down
*part 2*: Plane number: 0,1,2,3
*part 3*: Shape: line or circle
*part 4*: Angle: angle is given respect to X-axis and line; or X-axis and and tangent of circle at origin.
*part 5*: Length: This value gives the length of a line or the radius of a circle. Angle: Angle of circle.

The first three parts are given in one byte and if this byte is zero, this indicates that data of a particular character has been completed.

This method allows the pantograph system to be more flexible and open ended. The number of character sets of the pantograph can be easily increased.

The post-processor,i.e. the etch program gets and interprets the data related to the character, and generates step pulses and direction signals to the X-Y table.

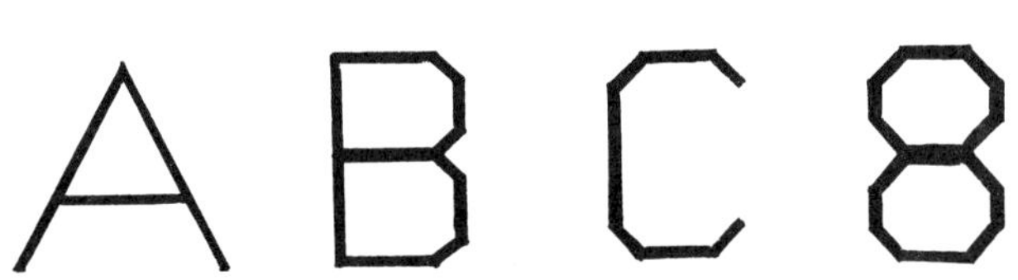


Fig. 3

Fig. 4

## HARDWARE

As far as the hardware of the pantograph is concerned the following units are appeared:

*Computer*: As computer an Amstrad CPC 6128 has been used. This computer consists of a screen, a 3 inch diskette unit and the processor unit with 128 K byte memory. There was no input and output facility of the computer itself.

*Interface*: In order to connect the table drive motors to the computer an interface block has been designed and installed to the system. The interface board consists of three parallel output ports. Each port has two bits: One for the step pulse and one for the direction. The block diagram of the interface is given in Fig. 5.

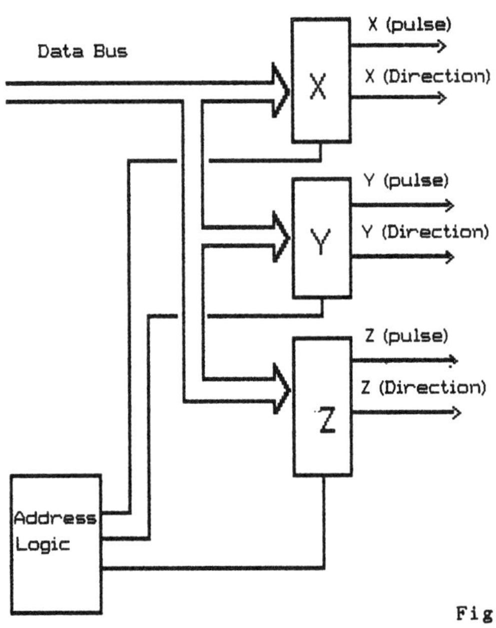


Fig.5

*Driver*: To drive the stepper motors,the electronic drive circuits have been designed and installed to the system. Since three motors are the same, the driver circuits are identical. A driver circuit has two inputs: Step input and direction input. These two inputs are TTL compatible and therefore can be connected to the interface output without any additional component. The block diagram of a driver is given in Fig.6.

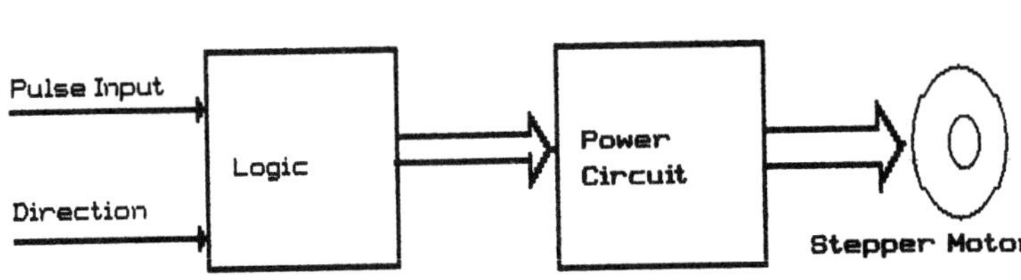


Fig.6

*Manual Inputs*: For initial setting operation, manual control inputs are required. These inputs provide the direction information and the step pulses in different frequencies for fast and slow motions. The manual input diagram is portrayed in Fig.7.

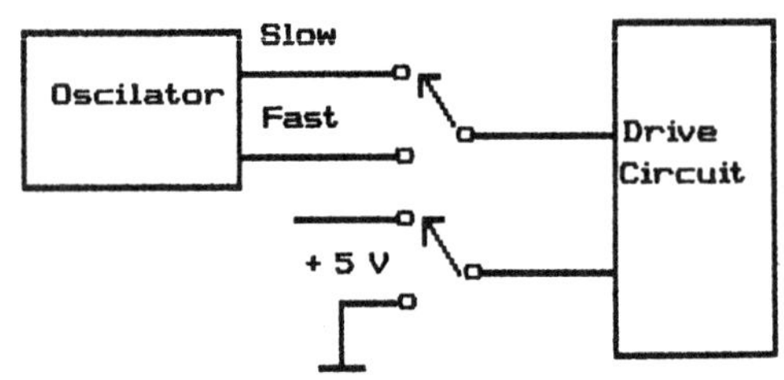


Fig. 7

*Power Supply*: The power supply unit provides all necessary DC current for stepper and DC motors.

## CONCLUSION

By this research and the application, we would like to realize a pantograph system which can work as a text editing system. In the mean time, the system has been designed in flexible and open ended manner for future developments and researches.

## ACKNOWLEDGEMENT

The author would like to thank H.Kamoy, G. Gülersoy for making this project possible in ASELSAN laboratories and many thanks to Dr. F. Tunalı for her contribution in software.

# A LANGUAGE FOR CNC

**E. Adali and F. Tunali**

*Department of Control and Computer, Technical University of Istanbul, Ayazaga, Istanbul, Turkey*

Abstract : TEZDIL is a programming language which has been designed for use with CNC tools. In this paper, TEZDIL is described as it is implemented as the software part of a project which was done for an automobile factory (TOFAS) in Turkey.

The software has two main parts: first, the programming software which includes the Editor and the Interpreter ; second, the Postprocessor software which has been developed according to the electro-mechanical properties of the tool that has been used in the research project.

First, the project is described briefly. Second, the general capabilities of TEZDIL are presented. These include the ability to interactively and user friendly edit, interpret, debug, and store user programs in an enviroment. Finally, the Postprocessor is described and its algorithms are given.

Keywords : Computer Numerical Control, Programming Languages, Editor, Interpreter, Postprocessor.

## INTRODUCTION

With the pressing needs for increasing productivity and quality-end products, industry is turning more and more toward computer-based automation. Although there are many CNC tools on the market, it is still an important interest area to work on the conversion of the conventional tools into CNC tools, especially for the developing countries like Turkey.

The language which is presented in this paper has been developed as a result of such a project which was to convert an conventional automatic copy mill into a CNC tool .

The tool which was produced in the sixties is an automatic copy mill machine and it is still used for preparation of dice and mould of automobile body parts. The maximum displacements of the tool along x,y,z axes are 550, 400 and 200 cm, respectively. On the copy mill only one DC motor has been attached as the drive motor for the three axes. The axial motions are transmitted by means of electromagnetic clutch systems. A position measuring system has already been installed by Olivetti with the Inductosyn's linear trandücers. The logic and the display units were designed by using diod-transistor logic and nixie tubes in the seventies. The general view of the tool is given in Fig. 1.

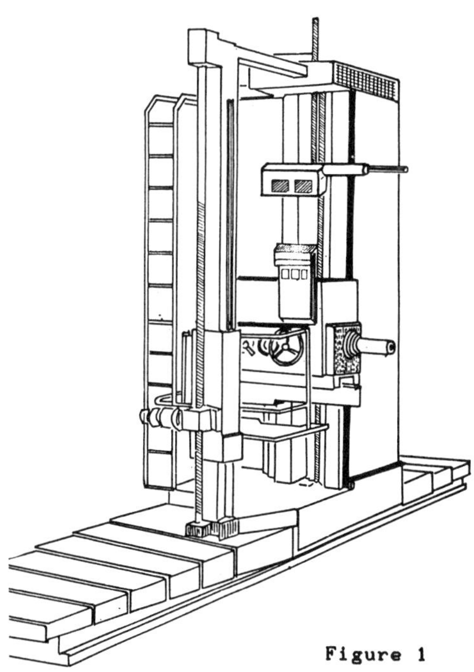

Figure 1

As the first step of the project, according to the electromechanical properties of the tool, all necessary hardware has been designed and installed [1]. Second, the software which had been used for this system has been developed and the system has been run succesfully.

Finally the software has been generalized such that it can be used as the programming language for the CNC tools. The software (TEZDIL) has two main parts : first, the programming software which includes the Editor and the Interpreter; second the Postprocessor software which has been developed according to the electro-mechanical properties of the tool that has been used in this project.

Although there are many languages which have been used for CNC tools, the researchers have been continuing to develop their languages or to design new languages which are general, highly interactive and user friendly.

TEZDIL has been also designed to be highly interactive in order to minimize the programming time and error to provide as many programming aids as possible. In addition to these properties the editor uses Turkish to eliminate the foreign language problem for the Turkish operators.

**Note:** In this paper, all computer messages which are essentially in Turkish are interpreted to English to get clearity.

## SYSTEM CAPABILITIES OF TEZDIL

As mentioned above TEZDIL has been designed to be interactive and user friendly and to provide as many programming facilities as possible. For these reasons, top level editor commands are immediately interpreted and user program steps are translated and stored in a compact internal form which can be interpreted very quickly during program execution. In the following sections, each of the capabilities which are available in TEZDIL are briefly described.

## EDITOR

At the begining of the project an operator panel has been designed and a Calculator-Based Editor has been developed. After achieving succesful results, the main idea of this editor has been transferred to a personal computer and an advanced editor has been obtained which is called as TEZDIL.

Before presenting TEZDIL , it is necessary to describe the Calculator-Based Editor briefly.

### Calculator-Based Editor

The Calculator-based Editor has been developed for the operator panel which has been attached to the system for easy use of the tool. Editor allows the user to operate manual or programmed mode .

The operator panel has 25 +2 keys for manual and programmed mode usage. In addition to the keypad there are two display sections on the panel. First one consists six seven-segment LED display elements plus one LED for sign which serves as data and command entry and also displays necessary information. The second display unit is dedicated to indicate the direction of the motion. There are six single LED one for each direction. The operator panel is given in Fig. 2.

By using this panel it is possible to operate either manual or programmed. In the manual operation can handle the tool as he did before, setting the drive speed and moving the tool in any one of the six possible direction. Operator can also stop the tool whenever he wants or move the tool up to certain destination point by giving the necessary coordinates. For example, in order to move to a given point along the x axes , panel should be used as follows:

a) SET SPEED : Press speed key ,display shows

S-----

The desired speed is given in the range of 0 to 9.

b) MOVE TO A GIVEN POINT :Press desired direction key and enter the value and then press ENTER

X

(The value)

ENTER

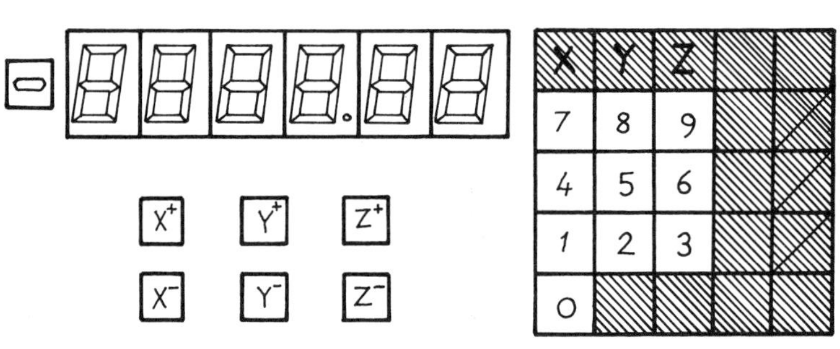


Figure 2

In manual operation mode, user might also set cutter, tolerance, and the coolant condition. Otherwise the default values are set.

The programmed mode facility allows the user straight line operation. Each step of the program consists of the following information :

PLANE : One of the three plane should be chosen; x-y,x-z or y-z.

SPEED : Operator can set the maximum speed according to the operation conditions.

TOLERANCE : The tolerance may be changed by user. The default value is 10 micron.

DESTINATION : The destination is given relative to absolute reference.

CUTTER : The information about the cutter is given. During the operation, if the cutter is different for the next step, the operation is stopped automatically and the operator is prompted to change the cutter.

IN/ON/OUT : Tracing a line may be carried out in three modes: in line, on line or out line.

COOLANT : On/off control of coolant can be chosen for each step.

According to the editor it is not necessary to give the whole information at each step. If a value is not given, the program assumes that the previous value is valid.

In order to present the programmed mode a simple example is given :

**Example :** On the x-y plane;move from (0,0) to (10.000,0), then move to (10.000,10.000), and then move to (20.000,10.000). Tolerance is 10 micron, cutter radius is 10 mm and speed is 7 for all steps. No coolant is used.

| STEP | MEANING | KEY | DISPLAY | ENTRY |
|---|---|---|---|---|
| 1 | speed | SPEED | S----- | 7 |
| | axis | X | A----- | 10000 |
| | cutter | CUT | CU---- | 10 |
| | in/on | I-O | I-O--- | 0 |
| | toler. | TOL | TOL--- | 10 |
| | cool. | COOL | COOL-- | 0 |
| 2 | axis | Y | B----- | 10000 |
| 3 | axis | X | A----- | 20000 |
| 4 | end | SPEED | S----- | 0 |

It is also possible to list and to modify the program . Modifying facilities are : append or insert new steps, delete old steps ,replace all or part of existing steps etc..

**Advanced Editor (TEZDIL)**

As seen above the Calculator-based editor is a dedicated software which can be used only with the operator panel. To allow only straightline motion was another disadvantage of the calculator-based editor.

By using Advanced Editor (TEZDIL) user programs are written by using a personal computer. All screen facilities have been used to achieve an interactive and user friendly editor.

TEZDIL contains an easy to use program editor. The program editor allows the user to create new program or to modify existing programs. Program modification facilities include the ability to append or insert new program steps, the ability to delete old steps, and the ability to replace all or part of existing steps.

Since TEZDIL works interpretatively, it is not necessary to separately edit, compile, link and load programs. It is also used to aid in program debugging since modifications to programs can be made very quickly.

TEZDIL allows circular and angular motions and it is also suitable for other kind of curves.

When the programming mode is chosen, TEZDIL displays step number and asks some questions about plane, shape, destination, tolerance etc. User should answer the questions . If there is an error, user will be prompted by computer.

As an example two steps are given :

STEP 0001

| | |
|---|---|
| PLANE ? | x-y |
| SHAPE ? | circle |
| IN/ON/OUT ? | out |
| CUTTER NO ? | 25 |
| TOLERANCE ? | 10 |
| DESTINATION (X,Y)? | 26000,11000 |
| RADIUS ? | 15000 |
| CENTER (X,Y)? | 11000,11000 |
| SPEED ? | 8,3 |
| COOLANT ? | OFF |

STEP 0002

| | |
|---|---|
| PLANE ? | x-z |
| SHAPE ? | line |
| IN/ON/OUT ? | out |
| CUTTER NO ? | 15 |
| TOLERANCE ? | 10 |
| DESTINATION (X,Z)? | 80000,38000 |
| SPEED ? | 8,7 |
| COOLANT ? | ON |

........

As seen above , TEZDIL asks just necessary questions concerning the task at each step. In step 1,since user wanted a circle, editor asked information about the radius and the center . In step 2 ,there is no question about radius and center.

If the full screen operation is chosen, append, insert, edit and create are possible. In this mode, all fields of the step are put on a line and the data from up to 20 steps is displayed onto screen. The screen should be considered as a window .It is possible to scroll backwards and forwards through the steps and to pan left and right through the fields of the steps. As mentioned above, if you define a line , radius and center fields are skipped automatically.

## INTERPRETER

As mentioned before, when the data of a step is entered, it is interpreted and stored in a compact form i.e. binary coded form, into certain memory location. By doing this each step occupies only 16 bytes. Hence, memory space usage is reduced and the postprocessor interprets each step very quickly during program execution.

In Fig.3 data saving format of each step is presented.

## POSTPROCESSOR

The postprocessor software has been developed for this particular copy mill. As known previous sections, the tool has only one drive motor and the electromagnetic clutch system. The postprocessor consists of four main parts:

MAIN : This part interprets F-1 and F-2 in Fig. 3 , and sets the speed and decides about shape . According to the shape (circle, straightline, ramp etc.) of the task a branch to other parts is performed.

The flowchart of the MAIN part is given in Fig. 4.

STRAIGHTLINE: The most important feature of the straightline program is to use a look up table between speed and break distance. By using such a table, the tool moves along a straight line as fast as possible and stops at the destination in tolerance.

The algorithm for this application is presented in Fig.5.

RAMP (or tangential line): Since there is only one motor, a tangential line (ramp) motion has been carried out as shown in Fig. 6. The flowchart is given in Fig.7.

CIRCLE : Because of the same reason, circle motion is realized stepwise as shown in Fig.8 and the algorithm for circle is given in Fig. 9.

## CONCLUSIONS

In this paper, TEZDIL,a computer numerical control tool programming language has been described. The general capabilities of TEZDIL has been emphasized. Also, it should be noted that the goal in developing TEZDIL was to obtain an highly interactive ,user friendly programming language . In many respects that goal has been achieved. However, we have knowledge that there is still much unrealized aspects in software system that has been presented in this paper.

## REFERANCE

[1] E.Adalı, F.Tunalı ,"A CNC Application With 8-bit Microcomputer "IFAC 7 th Conf. on Digital Computer Applications to Process Control, Sept. 17-20 ,1985 Vienna,Austria.

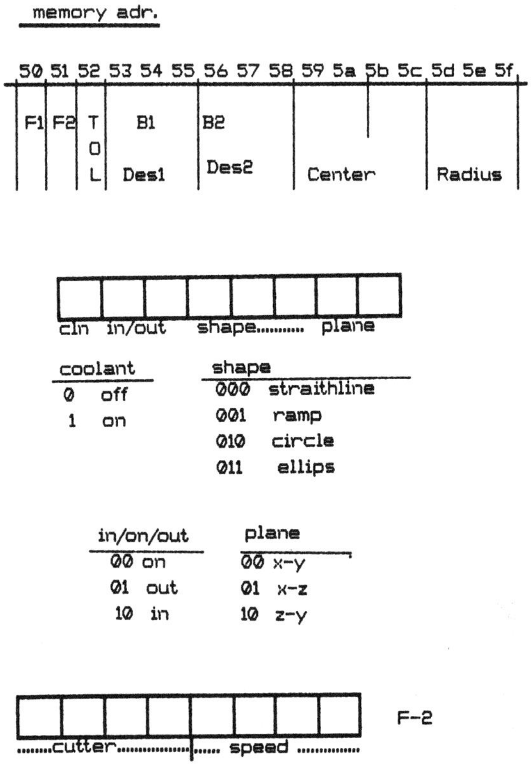


Figure 3

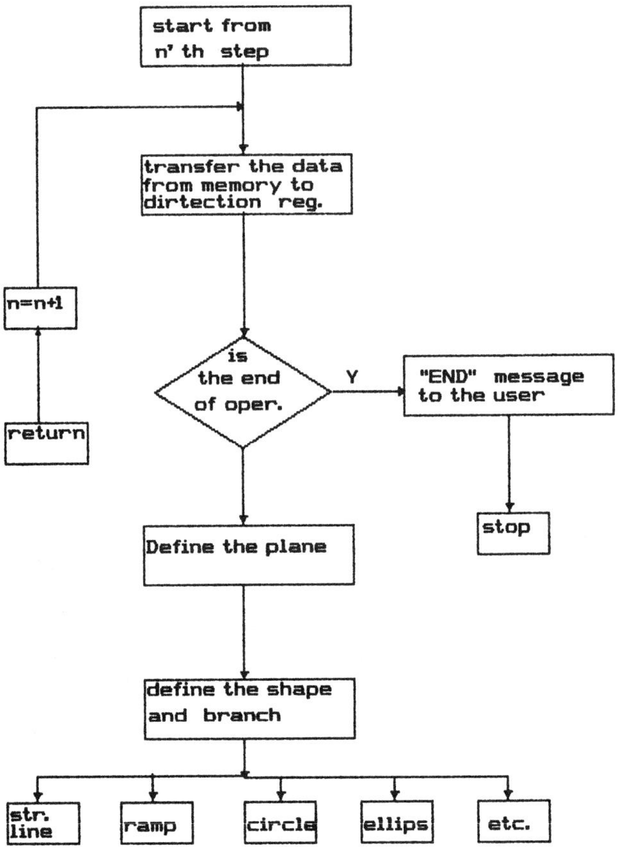


Figure 4

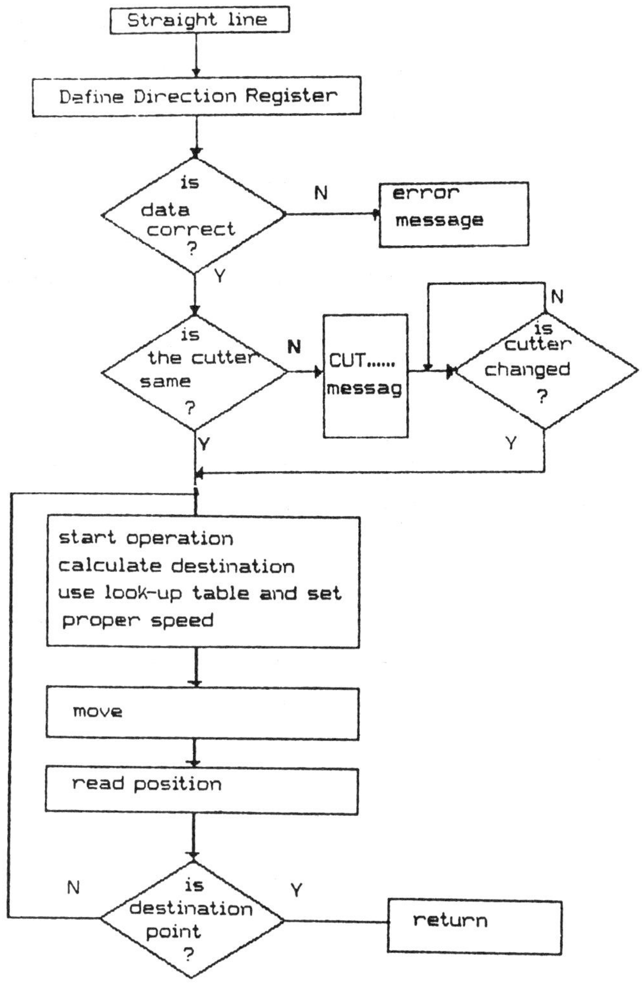
Straight line
Define Direction Register
is data correct ?
N
error message
Y
is the cutter same ?
N
CUT...... messag
is cutter changed ?
N
Y
Y
start operation
calculate destination
use look-up table and set
proper speed
move
read position
N
is destination point ?
Y
return

Fig. 5

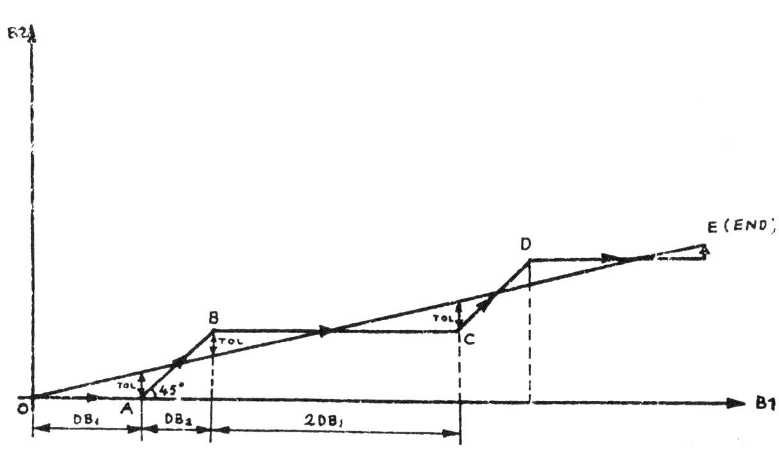
B2
E (END)
D
B
C
TOL
TOL
TOL
45°
O
A
B1
DB1
DB2
2DB1

Fig. 6

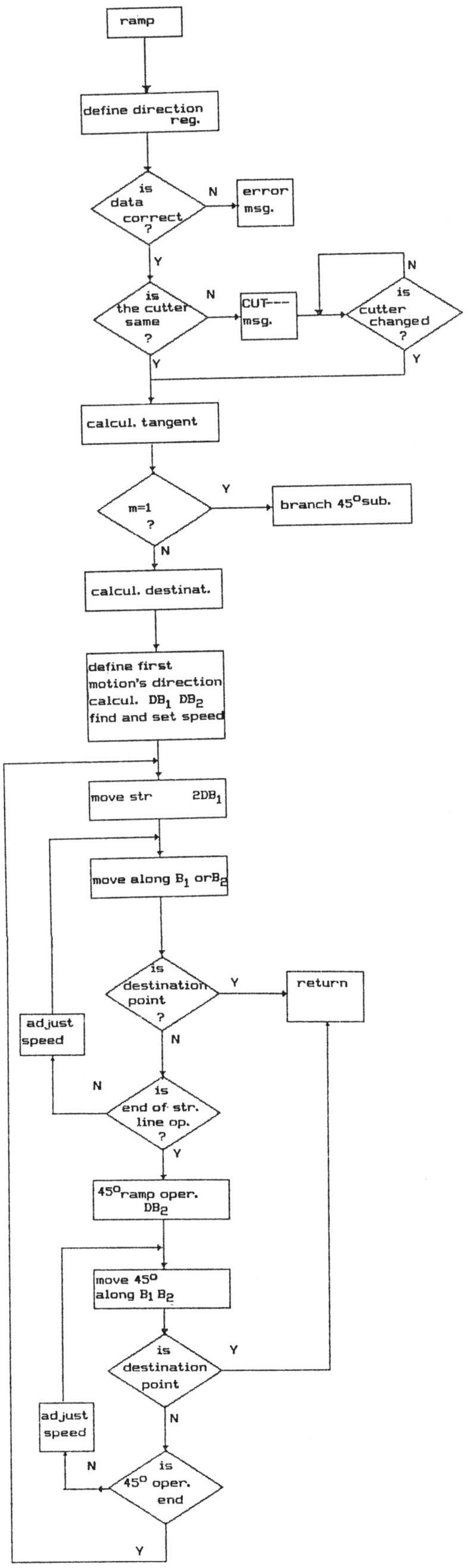
ramp
define direction reg.
is data correct ?
N
error msg.
Y
is the cutter same ?
N
CUT--- msg.
is cutter changed ?
N
Y
Y
calcul. tangent
m=1 ?
Y
branch 45° sub.
N
calcul. destinat.
define first motion's direction
calcul. DB1 DB2
find and set speed
move str 2DB1
move along B1 or B2
is destination point ?
Y
return
N
adjust speed
N
is end of str. line op. ?
Y
45° ramp oper. DB2
move 45° along B1 B2
is destination point
Y
N
adjust speed
N
is 45° oper. end
Y

Fig. 7

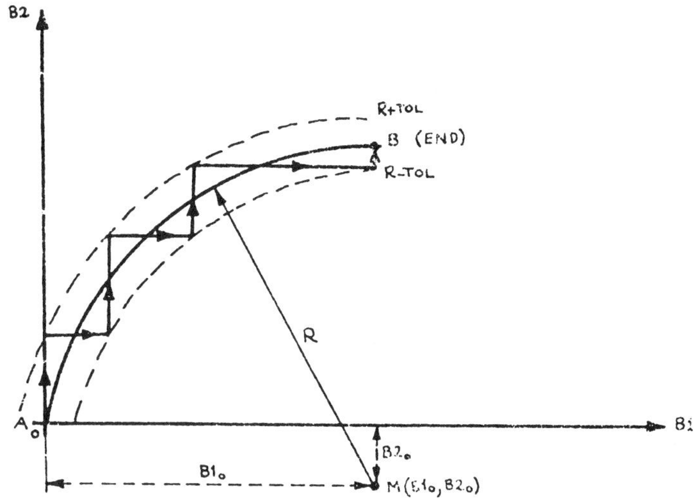
B2
R+TOL
B (END)
R-TOL
R
A
B1
B1o
B2o
M(B1o, B2o)

Fig. 8

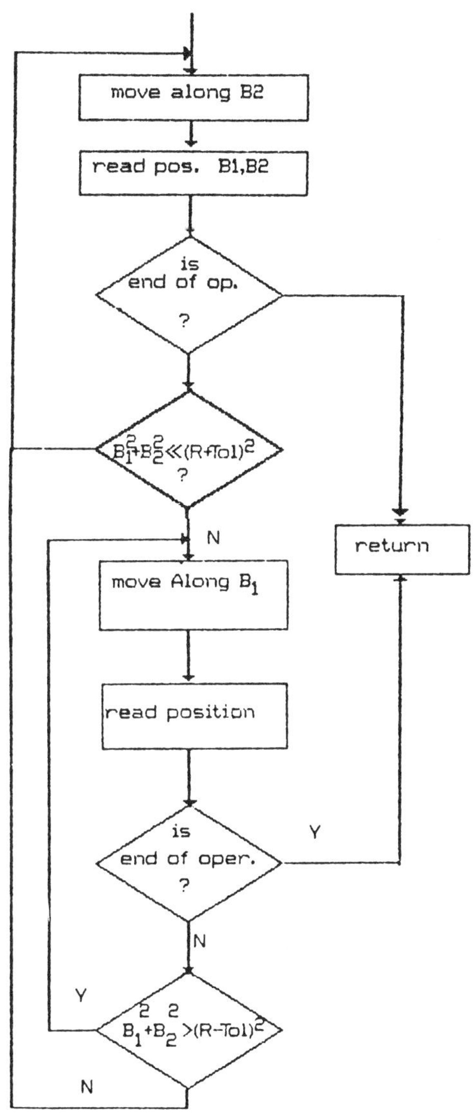
move along B2
read pos. B1,B2
is
end of op.
?
B$_1^2$+B$_2^2$≪(R+Tol)$^2$
?
N
return
move Along B$_1$
read position
is
end of oper.
?
Y
N
Y
B$_1^2$+B$_2^2$>(R-Tol)$^2$
N

Fig. 9

# AUTHOR INDEX


Adali, E. 249, 253
Akgün, B. T. 157
Ardac, M. 197
Arkut, O. 167

Bachmann, B. 137
Bailin Wang 59
Batur, C. 89
Billo, S. 163
Bock, W. J. 185
Braun, J. 89

Çalikoğlu, A. F. 167
Çelik, B. 237
Cesarini, F. 83
Cevik, K. 173
Cinar, A. 131
Csurgai, G. 215
Cuénod, M. 47, 143
Cvetko, H. 137

Duato, J. 209

Ersoy, O. 5
Ersü, E. 99, 191

Flotho, G. 117
Frittmann, L. 215

Gertler, J. 13
Good, H. 25
Gräser, A. 219
Güleç, M. 157
Guttermuth, J. 225
Guvenir, H. A. 95

Haidegger, G. 215
Harmanci, A. E. 157
Henderson, I. A. 77
Höger, W. 203
Hong Wang 63
Hulliger, P. 231

Ibrahim, A. A. 77
Itten, A. 25

Kahle, H. 123
Kaliski, M. E. 73
Kaya, A. 53
Kaynak, O. 67
Kegel, G. 191
Keyes, M. A. 53
Kohler, Ch. 231

Konar, A. F. 33
Kraif, C. 163
Kwankam, S. Y. 73
Kunz, U. 243

Latzel, W. 123

Marcuard, J.-D. 107
Martini, R. A. 131
McGhee, J. 77
Meister, A. 107
Midoğlu, H. 113
Miller, A. T. 73

Naunin, D. 173
Neddermeyer, W. 219

Oberschelp, W. 243
Olgac, N. 225

Platin, B. E. 237
Protin, L. 163

Rake, H. 197
Rathgeber, K. 179
Regli, H. 143
Reuss, H.-C. 173
Revesz, L. 215
Rouan, P. 163

Sankowski, D. 77
Sarioğlu, M. K. 157
Sarptürk, S. Z. 67
Schaefer, P. 95
Schaufelberger, W. 25
Schnell, M. 179
Schröder, D. 203
Schürmann, B. 197
Schurtenberger, M. 137, 143
Soda, G. 83

Tanyeri, G. 167
Thoma, M. 1
Tulunay, E. 113
Tunali, F. 253

Wienand, S. 99

Zappa, G. 83
Zaremba, M. B. 185
Zarzycki, R. 151
Zhao, G. 225

# SUBJECT INDEX


Adaptive control, 59, 63, 67, 83, 99, 203
Aerogenerator, 163
Agriculture, 167
Alarm systems, 143
Analog computer control, 143
Artificial intelligence, 5, 95, 99
Assembly programming, 77
Automata theory, 73
Automatic start up, 113
Automatic testing, 197
Availability requirements, 123
Axial winder, 203

Binary sequences, 77
Boilers, 151

Cartesian robot control, 179
Cement industry, 107
Centralized functional structure, 123
Centralized plant control, 167
Chemical variables control, 33
Closed loop systems, 179
Cogeneration, 53
Communications control applications, 215
Computer applications, 113, 197
Computer architecture, 73, 179, 209
Computer control, 53, 113, 151, 215, 219
Computer control energy control, 167
Computer control system, 59
Computer numerical control (CNC), 237, 253
Computer peripheral equipment, 191
Computer-aided system design, 33
Computers in control education, 25
Continuous contour control, 243
Continuous processing, 203
Control, 203
Control education, 25
Control engineering, 113
Control equipment, 143
Control system analysis, 33
Control system synthesis, 33
Controllers, 95, 143, 209, 243

Dahlin control, 89
Data acquisition, 151
dc motor, 173
Dead beat control, 89
Decentralized functional structure, 123
Decoupling, 219
Dedicated-centralized functional structure, 123
Diagnostic on chip, 179
Digital computer applications, 243
Digital control, 67, 143, 173
Direct digital control, 157, 219
Discrete systems, 67
Discrete time systems, 5
Distributed control system, 123
Distributed parameter system, 107
Dynamic property, 63

Editor, 253
Eigenvalues, 237
Energy control, 53, 117, 137

Failure isolation, 13
Failure/fault detection, 13
Fast fourier transforms, 5
Fault detection, 107
Fault-tolerant control systems, 131
Film processing, 197
Filtering, 13
Flue gas cleaning, 123
Fluidized bed combustion, 113
Fourier analysis, 5
Frequency control, 137, 143

Galvanic insulation, 163
Graphical simulation, 185

Heat exchanger control, 131
Heat systems, 113, 151
Hierarchical control, 185
Hierarchical structured control system, 107
Hierarchical systems, 117, 137, 179
Hybrid self-tuner, 83
Hydraulic turbines, 157

Identification, 77
Industrial application, 107
Industrial control, 33
Integrated plant control, 137
Interpreter, 253

Laboratory education, 25
Learning systems, 99
Lifts, 209
Load regulation, 143
Local-area control networks, 215
Lyapunor stability, 59

Machine tool software, 249
Machine tools, 231, 237, 243, 249
Man-machine systems, 137
Manufacturing processes, 215
Microcomputer application, 249
Microprocessor control, 185
Microprocessors, 95, 113, 117, 137, 143, 167, 173, 179, 209, 231
Microprogramming, 179
Model reference adaptive control, 59
Modelling, 237, 243
Motor control, 143, 173, 209
Multigenerators, 163
Multiprocessing systems, 173, 179, 215
Multiprocessor architectures, 83
Multireceiver energy unit, 163
Multitasking, 113
Multivariable control systems, 219
Multivariable systems, 33

Nonlinear control system, 99
Numerical control, 163

Observers, 203
On-line operation, 117, 143
Operator communications, 197
Optical distance sensors, 191

Optimal control, 243
Optimal control algorithm, 163
Optimization, 53, 117

Pantograph, 249
Paper cross profile control, 219
Paper industry, 219
pH control, 99
PID control, 231
Pole placement, 231
Position control, 243
Position measurement, 191
Postprocessor, 253
Power control, 143
Power converters, 163, 173
Power management, 53
Power optimization, 163
Power station control, 137
Predictive control, 99
Process control, 33, 73, 77, 117, 197
Process control with microcomputers, 131
Process diagnostics, 13
Programming languages, 253

Real-time control, 179
Robot teach-in, 191
Robotics, 185
Robots, 173, 179
Robust control, 63
Robustness, 13
Root loci, 237

Self-organising storage, 99
Sensors, 185
Sequential machines, 73
Sequential switching, 73
Servomechanisms, 231
Signal processing, 5
Sliding modes, 67
Smith dead time compensator, 89
Special purpose computers, 179
Speed control, 63, 157
Stability, 63, 89, 143, 219
State estimation, 137
State space control, 203
Statistical testing, 13
Steam plants, 151
Stepping motors, 237
System failure and recovery, 95, 209

Target adaptive control, 107
Target tracking, 185
Technology transfer, 197
Temperature control, 89
Test signals, 77
Time delay compensation, 131
Troubleshooting, 95

Variable structure systems, 67
Vision systems, 185
Voltage control, 143

Water pollution, 117
Web force, 203

ADALI & TUNALI: Microcomputer Application in Process Control.

AKASHI: Control Science and Technology for the Progress of Society, 7 Volumes

ALBERTOS & DE LA PUENTE: Components, Instruments and Techniques for Low Cost Automation and Applications

ALONSO-CONCHEIRO: Real Time Digital Control Applications

ASTROM & WITTENMARK: Adaptive Systems in Control and Signal Processing

ATHERTON: Multivariable Technological Systems

BABARY & LE LETTY: Control of Distributed Parameter Systems (1982)

BALCHEN: Automation and Data Processing in Aquaculture

BANKS & PRITCHARD: Control of Distributed Parameter Systems (1977)

BARKER & YOUNG: Identification and System Parameter Estimation (1985)

BASANEZ, FERRATE & SARIDIS: Robot Control "SYROCO '85"

BASAR & PAU: Dynamic Modelling and Control of National Economies (1983)

BAYLIS: Safety of Computer Control Systems (1983)

BEKEY & SARIDIS: Identification and System Parameter Estimation (1982)

BINDER & PERRET: Components and Instruments for Distributed Computer Control Systems

BRODNER: Skill Based Automated Manufacturing

BULL: Real Time Programming (1983)

BULL & WILLIAMS: Real Time Programming (1985)

CAMPBELL: Control Aspects of Prosthetics and Orthotics

Van CAUWENBERGHE: Instrumentation and Automation in the Paper, Rubber, Plastics and Polymerisation Industries (1980) (1983)

CHESTNUT: Contributions of Technology to International Conflict Resolution (SWIIS)

CHESTNUT, GENSER, KOPACEK & WIERZBICKI: Supplemental Ways for Improving International Stability

CHRETIEN: Automatic Control in Space (1985)

CICHOCKI & STRASZAK: Systems Analysis Applications to Complex Programs

CRONHJORT: Real Time Programming (1978)

CUENOD: Computer Aided Design of Control Systems*

DA CUNHA: Planning and Operation of Electric Energy Systems

De GIORGIO & ROVEDA: Criteria for Selecting Appropriate Technologies under Different Cultural, Technical and Social Conditions

DI PILLO: Control Applications of Nonlinear Programming and Optimization

DUBUISSON: Information and Systems

ELLIS: Control Problems and Devices in Manufacturing Technology (1980)

ELZER: Experience with the Management of Software Projects

FERRATE & PUENTE: Software for Computer Control (1982)

FLEISSNER: Systems Approach to Appropriate Technology Transfer

FLORIAN & HAASE: Software for Computer Control (1986)

GEERING & MANSOUR: Large Scale Systems; Theory and Applications (1986)

GELLIE & TAVAST: Distributed Computer Control Systems (1982)

GENSER, ETSCHMAIER, HASEGAWA & STROBEL: Control in Transportation Systems (1986)

GERTLER & KEVICZKY: A Bridge Between Control Science and Technology, 6 Volumes

GHONAIMY: Systems Approach for Development (1977)

HAASE: Real Time Programming (1980)

HAIMES & KINDLER: Water and Related Land Resource Systems

HALME: Modelling and Control of Biotechnical Processes

HARDT: Information Control Problems in Manufacturing Technology (1982)

HARRISON: Distributed Computer Control Systems (1979)

HASEGAWA: Real Time Programming (1981)*

HASEGAWA & INOUE: Urban, Regional and National Planning—Environmental Aspects

HERBST: Automatic Control in Power Generation Distribution and Protection

HRUZ & CICEL: Automatic Measurement and Control in Woodworking Industry — Lignoautomatica '86

ISERMANN: Identification and System Parameter Estimation (1979)

ISERMANN: 10th IFAC World Congress

ISERMANN & KALTENECKER: Digital Computer Applications to Process Control

JANSEN & BOULLART: Reliability of Instrumentation Systems for Safeguarding and Control

JANSSEN, PAU & STRASZAK: Dynamic Modelling and Control of National Economics (1980)

JOHANNSEN & RIJNSDORP: Analysis, Design, and Evaluation of Man-Machine Systems

JOHNSON: Modelling and Control of Biotechnological Processes

KAYA & WILLIAMS: Instrumentation and Automation in the Paper, Rubber, Plastic and Polymerization Industries (1986)

KLAMT & LAUBER: Control in Transportation Systems (1984)

KOPACEK, TROCH & DESOYER: Theory of Robots

KOTOB: Automatic Control in Petroleum, Petrochemical and Desalination Industries

LANDAU, TOMIZUKA & AUSLANDER: Adaptive Systems in Control and Signal Processing

LARSEN & HANSEN: Computer Aided Design in Control and Engineering Systems

LAUBER: Safety of Computer Control Systems (1979)

LEININGER: Computer Aided Design of Multivariable Technological Systems

LEONHARD: Control in Power Electronics and Electrical Drives (1977)

LESKIEWICZ & ZAREMBA: Pneumatic and Hydraulic Components and Instruments in Automatic Control*

MAFFEZZONI: Modelling and Control of Electric Power Plants (1984) (1986)

MAHALANABIS: Theory and Application of Digital Control

MANCINI, JOHANNSEN & MARTENSSON: Analysis, Design and Evaluation of Man–Machine Systems (1985)

MARTIN: Design of Work in Automated Manufacturing Systems

MARTOS, PAU, ZIERMANN: Modelling and Control of National Economies (1986)

McGREAVEY: Control of Distillation Columns and Chemical Reactors

MILLER: Distributed Computer Control Systems (1981)

MUNDAY: Automatic Control in Space (1979)

NAJIM & ABDEL-FATTAH: System Approach for Development (1980)

NIEMI: A Link Between Science and Applications of Automatic Control, 4 Volumes

NORRIE & TURNER: Automation for Mineral Resource Development

NOVAK: Software for Computer Control (1979)

OLLUS: Digital Image Processing in Industrial Applications—Vision Control

O'SHEA & POLIS: Automation in Mining, Mineral and Metal Processing (1980)

OSHIMA: Information Control Problems in Manufacturing Technology (1977)

PAUL: Digital Computer Applications to Process Control (1985)

PONOMARYOV: Artificial Intelligence

QUIRK: Safety of Computer Control Systems (1985) (1986)

RAMAMOORTY: Automation and Instrumentation for Power Plants

RAUCH: Applications of Nonlinear Programming to Optimization and Control*

RAUCH: Control of Distributed Parameter Systems (1986)

RAUCH: Control Applications of Nonlinear Programming

REMBOLD: Information Control Problems in Manufacturing Technology (1979)

RIJNSDORP: Case Studies in Automation related to Humanization of Work

RIJNSDORP, PLOMP & MÖLLER: Training for Tomorrow—Educational Aspects of Computerized Automation

RODD: Distributed Computer Control Systems (1983)

RODD & MULLER: Distributed Computer Control Systems (1986)

ROOS: Economics and Artificial Intelligence

SANCHEZ: Fuzzy Information, Knowledge Representation and Decision Analysis

SAWARAGI & AKASHI: Environmental Systems Planning, Design and Control

SINGH & TITLI: Control and Management of Integrated Industrial Complexes

SINHA & TELKSNYS: Stochastic Control

SKELTON & OWENS: Model Error Concepts and Compensation

SMEDEMA: Real Time Programming (1977)*

STRASZAK: Large Scale Systems: Theory and Applications (1983)

SUBRAMANYAM: Computer Applications in Large Scale Power Systems

SUSKI: Distributed Computer Control Systems (1985)

SZLANKO: Real Time Programming (1986)

TAKAMATSU: Fault Detection and Safety in Chemical Plants

TAL': Information Control Problems in Manufacturing Technology (1986)

TITLI & SINGH: Large Scale Systems: Theory and Applications (1980)

TROCH, KOPACEK & BREITENECKER: Simulation of Control Systems

UNBEHAUEN: Adaptive Control of Chemical Processes

VALADARES TAVARES & DA SILVA: Systems Analysis Applied to Water and Related Land Resources

WANG PINGYANG: Power Systems and Power Plant Control

WESTERLUND: Automation in Mining, Mineral and Metal Processing (1983)

van WOERKOM: Automatic Control in Space (1982)

YANG JIACHI: Control Science and Technology for Development

YOSHITANI: Automation in Mining, Mineral and Metal Processing (1986)

ZWICKY: Control in Power Electronics and Electrical Drives (1983)

**Out of stock—microfiche copies available. Details of prices sent on request from the IFAC Publisher.*

*IFAC Related Titles*

BROADBENT & MASUBUCHI: Multilingual Glossary of Automatic Control Technology

EYKHOFF: Trends and Progress in System Identification

ISERMANN: System Identification Tutorials (*Automatica Special Issue*)